C0-AVC-823

Reconstructing Behavior in the Primate Fossil Record

ADVANCES IN PRIMATOLOGY

Current Volumes in the Series:

ANTHROPOID ORIGINS
Edited by John G. Fleagle and Richard F. Kay

COMPARATIVE BIOLOGY AND EVOLUTIONARY RELATIONSHIPS OF TREE SHREWS
Edited by W. Patrick Luckett

EVOLUTIONARY BIOLOGY OF THE NEW WORLD MONKEYS AND CONTINENTAL DRIFT
Edited by Russell L. Ciochon and A. Brunetto Chiarelli

FUNCTION, PHYLOGENY, AND FOSSILS: Miocene Hominoid Evolution and Adaptations
Edited by David R. Begun, Carol V. Ward, and Michael D. Rose

NEW INTERPRETATIONS OF APE AND HUMAN ANCESTRY
Edited by Russell L. Ciochon and Robert S. Corroccini

NURSERY CARE OF NONHUMAN PRIMATES
Edited by Gerald C. Ruppenthal

PRIMATES AND THEIR RELATIVES IN PHYLOGENETIC PERSPECTIVE
Edited by Ross D. E. MacPhee

RECONSTRUCTING BEHAVIOR IN THE PRIMATE FOSSIL RECORD
Edited by J. Michael Plavcan, Richard F. Kay, William L. Jungers, and Carel P. van Schaik

SIZE AND SCALING IN PRIMATE BIOLOGY
Edited by William L. Jungers

Reconstructing Behavior in the Primate Fossil Record

Edited by

J. MICHAEL PLAVCAN
Department of Anthropology
University of Arkansas
Fayetteville, Arkansas

RICHARD F. KAY
Department of Biological Anthropology and Anatomy
Duke University Medical Center
Durham, North Carolina

WILLIAM L. JUNGERS
Department of Anatomical Sciences
State Unviersity of New York at Stony Brook
Stony Brook, New York

and

CAREL P. VAN SCHAIK
Department of Biological Anthropology and Anatomy
Duke University Medical Center
Durham, North Carolina

Kluwer Academic / Plenum Publishers
New York, Boston, Dordrecht, London, Moscow

Library of Congress Cataloging-in-Publication Data

Reconstructing behavior in the primate fossil record/edited by J. Michael Plavcan ... [et al.].
p. cm. — (Advances in primatology)
Includes bibliographical references and index.
ISBN 0-306-46604-X
1. Primates, Fossil. 2. Primates—Behavior. 3. Evolutionary paleobiology. I. Plavcan, J. Michael, 1962- II. Advances in primatology (Kluwer Academic / Plenum Publishers)

QE882.P7 R43 2002
569'.8—dc21

2001032668

ISBN 0-306-46604-X

233 Spring Street, New York, New York 10013

http://www.wkap.nl/

10 9 8 7 6 5 4 3 2 1

A C.I.P. record for this book is available from the Library of Congress

Printed in the United States of America

Contributors

Federico Anaya
Museo Nacional de Historia Naturale
La Paz, Bolivia

Prithijit S. Chatrath
Duke University Primate Center
Durham, North Carolina 27705

John G. Fleagle
Department of Anatomical Sciences
Health Sciences Center
State University of New York
at Stony Brook
Stony Brook, New York 11794-8081

Laurie R. Godfrey
Department of Anthropology
University of Massachusetts
Amherst, Massachusetts 01003-9278

William L. Hylander
Department of Biological Anthropology
and Anatomy
Duke University Medical Center
Durham, North Carolina 27710

Kirk R. Johnson
Department of Biological Anthropology
and Anatomy
Duke University Medical Center
Durham, North Carolina 27710

William L. Jungers
Department of Anatomical Sciences
Health Sciences Center
State University of New York
at Stony Brook
Stony Brook, New York 11794-8081

Richard F. Kay
Department of Biological Anthropology
and Anatomy
Duke University Medical Center
Durham, North Carolina 27710

Charles A. Lockwood
Institute of Human Origins
Arizona State University
Tempe, Arizona 85287-4101

Charles L. Nunn
Department of Biological Anthropology
and Anatomy
Duke University
Durham, North Carolina 27708-0383
Present address:
Section of Evolution and Ecology
University of California
Davis, California 95616

Andrew J. Petto
Division of Liberal Arts
University of the Arts
Philadelphia, Pennsylvania 19102-4994

J. Michael Plavcan
Department of Anatomy
New York College of Osteopathic Medicine
Old Westbury, New York 11568
Present address:
Department of Anthropology
University of Arkansas
Fayetteville, Arkansas 72701

Matthew J. Ravosa
Department of Cell and Molecular Biology
Northwestern University Medical School
Chicago, Illinois 60611-3008
and
Department of Zoology
Division of Mammals
Field Museum of Natural History
Chicago, Illinois

Kaye E. Reed
Institute of Human Origins and Department of Anthropology
Arizona State University
Tempe, Arizona 85281-4101
and
Bernard Price Institute of Palaeontological Research
University of the Witwatersrand
Johannesburg, South Africa

Brian G. Richmond
Department of Anthropology
George Washington University
Washington DC 20052
Present address:
Department of Anthropology
University of Illinois
Urbana, Illinois 61801

Callum F. Ross
Department of Anatomical Sciences
Health Sciences Center
State University of New York at Stony Brook
Stony Brook, New York 11794-8081

Elwyn L. Simons
Duke University Primate Center
Durham, North Carolina 27705

Michael R. Sutherland
Statistical Consulting Center
University of Massachusetts, Amherst
Amherst, Massachusetts 01003-9337

Peter Ungar
Department of Anthropology
University of Arkansas
Fayetteville, Arkansas 72701

Carel P. van Schaik
Department of Biological Anthropology and Anatomy
Duke University
Durham, North Carolina 27708-0383

Christopher J. Vinyard
Department of Biological Anthropology and Anatomy
Duke University Medical Center
Durham, North Carolina 27710

Blythe A. Williams
Department of Biological Anthropology and Anatomy
Duke University Medical Center
Durham, North Carolina 27710

Roshna E. Wunderlich
Department of Biology
James Madison University
Harrisburg, Virginia 22807

Preface

This volume brings together a series of papers that address the topic of reconstructing behavior in the primate fossil record. The literature devoted to reconstructing behavior in extinct species is overwhelming and very diverse. Sometimes, it seems as though behavioral reconstruction is done as an afterthought in the discussion section of papers, relegated to the status of informed speculation. But recent years have seen an explosion in studies of adaptation, functional anatomy, comparative sociobiology, and development. Powerful new comparative methods are now available on the internet. At the same time, we face a rapidly growing fossil record that offers more and more information on the morphology and paleoenvironments of extinct species. Consequently, inferences of behavior in extinct species have become better grounded in comparative studies of living species and are becoming increasingly rigorous.

We offer here a series of papers that review broad issues related to reconstructing various aspects of behavior from very different types of evidence. We hope that in so doing, the reader will gain a perspective on the various types of evidence that can be brought to bear on reconstructing behavior, the strengths and weaknesses of different approaches, and, perhaps, new approaches to the topic. We define behavior as broadly as we can—including life-history traits, locomotion, diet, and social behavior, giving the authors considerable freedom in choosing what, exactly, they wish to explore. Notably, we deliberately exclude issues related to later human evolution and behavior, especially with regard to archeological evidence. There are several extensive works from the archeological literature that deal well with the unique problems and opportunities available to students of the human fossil record.

We asked authors to address topics with a broad approach. We especially encouraged authors to review not only the types of evidence that can be used to reconstruct behavior, but also the limitations of the evidence. We feel that it is important to understand both what can, and cannot, be said about the behavior of extinct species. Our intent is not to cast behavioral reconstruction in a cynical light, but to emphasize the weaknesses of behavioral reconstruction as a basis for further research. We hope that by understanding both the strengths and weaknesses of behavioral reconstructions, future research will create a much more rigorous framework for understanding the adaptations and behaviors of extinct animals.

We present first a series of chapters that systematically address basic issues and types of evidence that can be used to infer behavior. These include comparative analysis and adaptation, ontogenetic evidence, paleoenvironmental and paleocommunity analysis, experimental functional analysis, and comparative socioecology. This is followed by two broad reviews of evidence for diet and social systems in primates—two of the most commonly inferred behaviors for extinct species. We end with reviews of behavioral reconstructions for extinct lemurs and *Branisella boliviana*, each of which illustrates how widely different types of evidence converge to paint a picture of the behavior and adaptations of extinct species.

We hope that students and professionals find this book interesting and useful for thinking about how to reconstruct behavior in the fossil record. We thank the authors for their thoughtful contributions and especially their patience in producing this volume. The difficulty of reviewing such a broad topic in an accessible fashion offers a real challenge, which all of the authors successfully met. This volume was initially inspired by a conference at Duke University sponsored by the Leakey Foundation, whom we thank for their generous support. John Fleagle and Ross MacPhee helped greatly in formulating the structure of the volume. Finally, we thank the many external reviewers of the papers, who provided thoughtful and insightful comments on the manuscripts, and greatly improved the quality of the volume.

J. Michael Plavcan
Richard F. Kay
William L. Jungers
Carel P. van Schaik

Old Westbury, New York
Durham, North Carolina
Stony Brook, New York
Durham, North Carolina

Contents

Reconstructing Behavior in the Primate Fossil Record

Adaptation and Behavior in the Primate Fossil Record

1

CALLUM F. ROSS, CHARLES A. LOCKWOOD, JOHN G. FLEAGLE, and WILLIAM L. JUNGERS

Introduction

Few topics in functional morphology have been discussed as extensively with so little consensus as adaptation. Controversies swirl around both theoretical and methodological issues in the study of adaptation. Persistent questions concern definitions of adaptation, criteria for recognizing adaptations, and how to integrate phylogenetic and functional data. Here we review debates about how one identifies an adaptation, discuss recent developments in the study of adaptation, and focus on how these relate to our ability to reconstruct behavior in the fossil record. In keeping with the theme of this volume we will take our examples from the vast literature on primate functional anatomy, a resource that in many respects is more extensive and detailed than that for other orders of mammals. At the same time we will draw on the literature of other animals to introduce perspectives on the study of adaptation that have not been considered extensively by primatologists, but which hold considerable promise for future research.

CALLUM F. ROSS, JOHN G. FLEAGLE, and WILLIAM L. JUNGERS • Department of Anatomical Sciences, Health Sciences Center, State University of New York at Stony Brook, Stony Brook, New York 11794-8081. CHARLES A. LOCKWOOD • Institute of Human Origins, Arizona State University, Tempe, Arizona 85287-4101.

Reconstructing Behavior in the Primate Fossil Record, edited by Plavcan *et al.* Kluwer Academic/Plenum Publishers, New York, 2002.

Defining Adaptation

The term adaptation is used to refer to (a) a thing, (b) a quality or state, and (c) a process. It seems reasonable to require that these uses apply definitions that are compatible, or at least not contradictory (Stern, 1970).

(a) *Adaptation as a thing.* Although most workers agree that adaptations are the products of evolution by natural selection (Williams, 1966; Lewontin, 1978; Gould and Vrba, 1982; Sober, 1984; Williams, 1992; but see Greene, 1986), disagreement remains concerning the role of historical factors in the definition of adaptation. These differences are represented in what some call the historical and nonhistorical conceptions of adaptation (e.g., Gould and Vrba, 1982; Amundson, 1996).

According to the *historical concept* adaptations are phenotypic attributes of organisms that have current utility, that evolved by natural selection to perform their current function, and that are maintained in their lineages by natural selection (e.g., Williams, 1966; Sober, 1984; Coddington, 1988; Baum and Larson, 1991; Harvey and Pagel, 1991). The historical concept differentiates between different kinds of "*aptations*," or traits that confer benefits to organisms. An *adaptation* is a trait that not only was fixed in a population by natural selection because of its ability to perform some specific function, but also still performs that function. If the trait subsequently acquires a new function, it becomes an *exaptation* (Gould and Vrba, 1982; see Table 1 in Larson and Losos, 1996). Adaptations improve fitness or performance. Traits that do not are *nonaptations.* Primary nonaptations do not confer an advantage and secondary nonaptations have lost any benefit that they once had (Baum and Larson, 1991). Traits that are detrimental to fitness or performance are disaptations, with primary and secondary disaptations defined as above.

Under the *nonhistorical* concept of adaptation, the historical origin of a trait is not an important part of the definition of adaptation; rather, adaptations are traits with current utility, i.e., traits that increase the adaptedness (fitness) of their possessors (Bock and von Wahlert, 1965; Kay and Cartmill, 1977; Bock, 1980; Fisher, 1985; Reeve and Sherman, 1993; Anthony and Kay, 1993). These adaptive traits are generally assumed to have arisen through natural selection, but it is not important to prove its origin by natural selection to determine whether something is or was an adaptation.

The differences between the historical and nonhistorical concepts may be summarized as follows. The historical concept is highly restrictive in the traits determined to be adaptations. It emphasizes the need to know how a trait arose, and a trait must have been produced by natural selection to be called an adaptation. The nonhistorical definition is more inclusive, saying nothing about the causal history of a trait and emphasizing the current function of a trait. The nonhistorical approach does not demand that the role of natural selection be documented and therefore has no explicit logical connection to natural selection.

We appreciate the logical rigor of the historical definition of adaptation because it is intimately tied to the theory of natural selection, and adaptation is the phenomenon that natural selection sets out to explain (Darwin, 1859). However, because it is impossible to demonstrate that selection was involved in the fixation of most traits, accepting the historical definition forces one to accept that definitive identification of adaptations is impossible in most cases (Williams, 1966: 4; Ross, 1999). This in turn implies that when studying fossil taxa, although one can study morphology and try to reconstruct behavior and function, one cannot prove or disprove a hypothesis of adaptation without making some significant assumptions about trait heritability, population structure, and past selective regimes.

(b) *Adaptation as a state of being of an organism, population, or lineage.* Organisms, populations, or lineages have relative degrees of *adaptedness* (Burian, 1983). An organism with the *propensity* to leave more offspring than another organism within the same population is better adapted; the organism that *actually* leaves more offspring is fitter (Brandon, 1978). Thus, a better adapted animal is *expected* to be fitter, although the possibility that it dies by chance means that this is not necessarily so. For example, assume for a moment (not unreasonably) that a gorilla is better adapted to a forested environment than a person. Both are standing in a clearing in the forest when lightning strikes, killing the gorilla. The fact that the gorilla is killed does not lessen the fact that it is better adapted to the forested environment. As one moves away from individuals toward populations, species and lineages of animals (e.g., gorillas and humans), chance, or rare events have no consistent effect on individual fitness, so that at large enough population sizes, or over long periods of time, individual adaptedness probably translates well into fitness. Extending this logic back in time into the fossil record, one might expect that those traits that have become fixed in a lineage and maintained in it are fair candidates for adaptations. In other words, natural selection is the most probable causal agent for trait fixation and maintenance. This expectation is implicit in the nonhistorical concept of adaptation espoused by Kay and Cartmill (1977).

In contrast, advocates of the historical perspective argue that forces other than natural selection, such as constraint, or primitive retention, genetic drift and pleiotropy, can explain the presence of traits in a lineage (e.g., Leroi *et al.*, 1994). Indeed, two of the most important questions currently confronting evolutionary morphologists are the relative importance of selection and other evolutionary forces in (1) effecting morphological change (e.g., Carroll, 1997), and (2) maintaining traits in lineages (see below).

(c) *Adaptation as a process.* Adaptation as a process must refer to the processes whereby adaptations are fixed and maintained in populations and lineages, and whereby individuals, populations, and lineages are maintained in a state of adaptedness. Therefore, by definition, the process of adaptation must be synonymized with natural selection. That other processes (such as genetic drift and pleiotropy) can conceivably produce a state of adaptedness cannot be denied, but it seems implausible that a state of adaptedness can be achieved without the influence of natural selection. Thus, although these processes should not be called adaptation *sensu stricto*, they may often be involved with the adaptation process.

Identifying Adaptations

Compared with biologists working in other areas, primatologists have generally paid little attention to rigorous definitions of adaptation and have used the term in a broad, nonhistorical sense almost synonymous with behavior (e.g., Fleagle, 1999). Likewise there is a large literature on reconstructing *behavior* in the fossil record, with the assumption that these behaviors are adaptations being only implicit (Boucot, 1990). Indeed, primatologists have been most concerned with identifying morphological features in extant primates that are correlated with and indicative of behavior. Thus the locomotor or dietary behaviors of a species or family of primates are described as adaptations without any explicit analysis of how they came into being. These morphological traits can then be used to reconstruct the behavior of fossil taxa. While most explicitly label these traits in fossils as adaptations, they use the term in its nonhistorical sense, i.e., traits that increased the adaptedness of their possessors (Kay and Cartmill, 1977; Kay and Covert, 1984).

Box 1

Components of an ideal hypothesis of adaptation according to Brandon (1990)

(1) Evidence that selection has occurred, i.e., that some types are better adapted than others in a specific environment and that this has resulted in differential reproduction.

(2) An ecological explanation for why selection occurred.

(3) Evidence that the trait is heritable.

(4) Information on population structure and gene flow.

(5) Phylogenetic information on trait polarity. Adaptations evolve because their possessors are better adapted to a particular selective environment than are possessors of other traits, thus we need to know what evolved from what.

This is in contrast with workers that insist on a historical definition of adaptation and resolutely apply it to living animals, but are generally able to say little or nothing about the study of adaptation in fossils (e.g., Coddington, 1988; Baum and Larson, 1991). In its most rigorous form, the historical approach demands the inclusion of data on the genetic structure of the populations in which a trait originated along with elimination of alternatives to natural selection, such as genetic drift and pleiotropy (Brandon, 1990; Leroi *et al.*, 1994; see Box 1). Others rest their cases primarily on phylogenetic methods of association, incorporating comparisons of performance between derived and primitive traits (Coddington, 1988; Brooks and McLennan, 1991; Larson and Losos, 1996). Still others advance criteria other than demonstrable natural selection, instead relying on some measure of design, i.e., how well a trait matches *a priori* design criteria. For example, Rudwick (1964) demands that a structure meets rigid design criteria, as embodied in a paradigmatic structure. In a less extreme advocation of the historical approach, Williams (1966, 1992; 1997) relies on "informal arguments as to whether a presumed function is served with sufficient precision, economy, efficiency, etc., to rule out pure chance as an adequate explanation" (1966: 10). In general, advocates of the historical approach are distinguished by concern for the origin of traits while those who practice a nonhistorical approach have been more concerned with identifying associations between morphological traits and behaviors.

It is easy to parody either the historical approach or the nonhistorical approach, making the former seem impossible and the latter seem hopelessly naïve (e.g., Gould and Lewontin, 1979). As one of the reviewers of this paper noted, not totally facetiously, in one extreme historical definition, an adaptation is synonymous with an apomorphy, and in an extreme nonhistorical definition, adaptation is just a synonym of function. As noted above, there are probably almost as many lists of criteria for identifying adaptations as there are papers on the subject. Moreover, there is also considerable overlap in the criteria advocated by proponents of both the historical and nonhistorical approaches. These include association between trait and function (Kay and Cartmill, 1977; Box 2), arguments of efficiency (Bock, 1980), and matching to design criteria (Hylander, 1979; Alexander, 1985, 1996), including criteria controlling structural and functional integration (Alexander, 1975; Weibel *et al.*, 1998).

Box 2

Criteria to be satisfied before a morphological trait in a fossil can be assumed to have a particular adaptive meaning (Kay and Cartmill, 1977), summarized by Kay (1984):

(1) There must be some living species which has the morphological trait. If the morphological trait of the extinct species is unique, no analogy is possible.

(2) In all extant species which possess the morphological trait, the trait must have the same adaptive role.

(3) There must be no evidence that the trait evolved in the lineage before it came to have its present adaptive role.

(4) The morphological trait in question must have some functional relationship to an adaptive role.

Rather than crafting the ultimate definition of adaptation, the goal of this paper is to review and evaluate the theoretical framework underlying attempts to reconstruct behavior in the fossil record—aspects of extinct taxa that have traditionally been called adaptations. This normally involves identifying specific morphological features such as bumps on bones or crests on teeth that are indicative of a specific behavior among extant taxa and that, when found on a fossil, can be used to reconstruct the behavior of the extinct taxon. Since all such reconstructions are in a sense probabalistic estimates of unknown and unknowable parameters, we focus on ways in which we can maximize our confidence in the association between particular morphological features and aspects of behavior. Fortunately, much of the recent debate over historical and nonhistorical definitions of adaptation bears directly on this issue. In the subsequent pages, we discuss a series of topics that are critical to all efforts to reconstruct the behavior of extinct taxa using morphological features: comparative methods; optimization and constraint; experimental data; phylogenetic information; and the use of primitive features. We will refer to these traits and associated behaviors as adaptations, in the traditional, nonhistorical sense of the term.

Association of Trait and Behavior: Comparative Methods

In a living taxon, the hypothesis that a trait is an adaptation for a specific function can be evaluated by determining whether the trait actually performs that function. When the focal taxon in a study of adaptation is a fossil, direct evidence associating the trait with a function or behavior is available only rarely, in the form of trace fossils such as footprints, or burrows, or "behavioral residues" such as dental microwear, phytoliths preserved in interstitial spaces, or preserved stomach contents (Boucot, 1990). The logical and statistical basis for inferences about trait function in fossils is the relationship between form and function in living taxa. The trait is studied in extant taxa to determine its function in these animals and its function in the fossil taxon is hypothesized to be the same. Arguably, at some level this is true of all attempts to reconstruct behavior and adaptations in fossils.

Various potential difficulties confront attempts to infer trait function in a fossil taxon (or poorly studied living form) using living analogs. These difficulties can be roughly divided into five categories: loose connection between function and structure; phylogenetic inertia; multiple functions for a single trait; problems of homology; and allometric considerations.

Structure and Function Not Tightly Linked

Any predictable relationship between morphology and behavior requires that the two be tightly linked in known forms. If there is no close link between morphology and function in extant taxa, there can be little hope for any reliable reconstruction of the behavior of fossil taxa using morphology alone. For example, in vertebrates the nervous system might act as a "wild-card," with reorganization of the motor or sensory parts of the nervous system producing radically different functions in morphologically identical structures (Lauder, 1995). Lauder even argues that this may mean that "one cannot predict function from structure in vertebrates with much certainty at all" (Lauder, 1995). As an example, Lauder showed that although all Osteoglossomorpha (a clade of teleost fish) share a structural novelty called the tongue bite, whereby the basihyal is adducted against the bottom of the skull, the taxa show differences in kinematics of basihyal movement. Lauder argues that without information on jaw kinematics, one could not infer how the jaws were actually used (i.e., function) (Lauder, 1995).

This problem can be minimized by applying a strict criterion for accepting a form–function association. For example, Kay and Cartmill (1977: 21) suggest that for a morphological trait to be a reliable indicator of function "in all extant organisms which have T [the trait], T has F [the function]." This requirement minimizes the likelihood of the equivalent of a Type II error: incorrectly accepting a false hypothesis of trait–function association. This approach is most powerful when it involves multiple instances of independent evolution of the trait–function association. The assumption of this approach is that a specific trait–function association is unlikely to have evolved numerous times independently by chance alone without the ordering hand of natural selection. For example, Ross (2002) evaluates the hypothesis that the fovea of tarsiers and anthropoids is an adaptation to visual predation (Cartmill, 1980; Ross, 1996) by collating data on the occurrence of retinal foveae among vertebrates. Foveae have evolved numerous times in vertebrates, always in association with visual predation, most often in association with diurnal visual predation, but also in the context of nocturnality (such as owls, *Sphenodon*, and some deep-sea fish). This approach has been used frequently in the study of primate adaptation (see the review by Lockwood and Fleagle, 2000) and has also been applied to test hypotheses relating the evolution of specific traits to subsequent adaptive radiations (Farrell, 1998; Hunter and Jernvall, 1995).

While one can demand a 100% association between trait and function, a less stringent approach might be to determine whether a trait–function association is statistically significant, where the null hypothesis is that the trait is distributed randomly with respect to the function. When applying statistical techniques such as these it has been pointed out that merely counting up the number of taxa showing a trait–function association might yield an inflated sample size. As Felsenstein (1985) noted, species in a comparative study of trait covariation may not be independent data points because they are within a nested hierarchy of phylogenetic relationships. Because similarity among closely related species is expected due to their shared evolutionary history, some phylogenetic correction is called for if standard statistical methods are to be employed with any confidence. Without doing so, degrees of freedom can be artificially elevated and confidence limits spuriously narrow; correlation coefficients and regression parameters may be estimated inaccurately. This problem is discussed in the next section.

Phylogenetic Inertia

The possibility that a trait and/or a function can be maintained in a clade as the passive result of a constrained process of inheritance has broader implications for the inference of function from structure in the fossil record. If living taxa can exhibit trait–function associations that are not mediated by natural selection, they will be unreliable analogs for the identification of adaptations (*sensu* the historical concept) in the fossil record. It is therefore important to determine whether the maintenance of a trait or function in a clade can be the passive result of a constrained process of inheritance, without active stabilizing or directional selection. Several phenomena can be invoked as possible causes of phylogenetic inertia or constraint, including aspects of genetic complexity such as epistasis, linkage, and pleiotropy (Bell, 1997) and developmental constraints (Maynard Smith *et al.*, 1985; Cheverud, 1988; Gould, 1989). Research into genetic variance–covariance matrices promises to provide great insight into such constraints in the future (Pigliucci and Kaplan, 2000). However, in most cases the possibility of phylogenetic inertia is considered only because of the widespread distribution of a trait within a clade, not because there is evidence of genetic constraints on variation* (Reeve and Sherman, 1993). While it might seem reasonable to imagine that some traits are maintained in lineages passively by some kind of historical constraint, some by stabilizing selection, and some by a mixture of both, an assumption that selection has played no role at all in trait maintenance assumes *no* genetic variation that selection needed to prune! This seems to us to be an unlikely, or uncommon, situation, suggesting instead that selection is arguably involved in all instances of trait maintenance, at least to some degree. Therefore, the widespread occurrence of a trait–function association can be more reasonably attributed to stabilizing selection as to some unspecified phylogenetic constraint.

One way to falsify (or at least weaken) a hypothesis of phylogenetic constraint is to invoke instances of lack of "constraint" within the clade of interest. Consider the example of the strepsirrhine tooth comb which occurs in all modern strepsirrhines except the aye-aye and some subfossil indroids (Jungers *et al.*, this volume, Chapter 10). While the functional correlates of the tooth comb are debated (see reviews by Martin, 1990; Asher, 1998), it is used in similar grooming behavior in all strepsirrhines in which it is present, and the association between grooming and possession of a toothcomb appears to

*Most other kinds of "constraint" that may lead to phylogenetic inertia, e.g., developmental constraints, are actually based in natural selection and therefore not "inertia" at all (Reeve and Sherman, 1993; Schlichting and Pigliucci, 1998). See below.

have arisen only once, there being little reason to believe that extant strepsirrhines derived their toothcomb from more than one lineage (Ross *et al.*, 1998). On one hand, the presence of this trait–function relationship among all extant strepsirrhines might be regarded as merely an inherited condition, making the number of separate instances of the toothcomb–grooming association equal to one. On the other hand, the presence of the trait–function relationship in such a large number of extant species can be considered to be the result of stabilizing selection in many independent lineages over long periods of time. The first approach ignores (or, at best, de-emphasizes) the importance of stabilizing selection, implying that the toothcomb–grooming relationship has not been subject to any variation since its fixation in the strepsirrhine stem lineage. The second approach ignores (or de-emphasizes) the possibility of phylogenetic inertia, and implies that the association of toothcomb and grooming behavior is maintained by stabilizing selection in all lineages of extant strepsirrhines. In favor of the latter approach, several strepsirrhine lineages have broken the hypothesized "constraint" of possessing the toothcomb: Indroids reduce the number of teeth in the comb to four, *Daubentonia* modifies the incisors in a completely different way, and both *Archaeolemur* and *Palaeopropithecus* independently re-evolved lower incisors that do not function as a toothcomb. On what grounds are we to assume that all other extant strepsirrhine lineages have retained a toothcomb and its association with grooming merely because of epistasis, linkage, or pleiotropy? We return to the issue of primitive retention below in the context of reconstructing behavior in the fossil record.

The possible role of selection in trait maintenance has implications for statistical methods currently being developed aimed at "correcting" for "phylogenetic inertia" (e.g., R. J. Smith, 1994; Nunn, 1995). "Independent contrasts" represent one such attempt to correct for phylogenetic influences under different evolutionary models (Felsenstein, 1985; Grafen, 1989; Harvey and Pagel, 1991; Garland *et al.*, 1992; Garland and Adolph, 1994; Garland and Ives, 2000; Martins and Hansen, 1996; Purvis and Webster, 1999). Variance components and adjusted degrees of freedom represent another (R. J. Smith, 1994; Nunn, 1995). For the most part, these valuable new methods have impacted on analyses of adaptation that focus on character covariation. We find them to be most useful when they are compared directly to analyses done without such phylogenetic corrections (e.g., Garland *et al.*, 1992). We also agree with others that this problem should be treated as an empirical issue (Gittelman *et al.*, 1996; Abouheif, 1999), i.e., phylogenetic dependence of data should be evaluated before it is "corrected."

Multiple Functions

One obvious problem is that a trait can be used for more than one thing, i.e., a trait can have more than one function. For example, there is an association between the presence of shearing crests on molar teeth and both leaf-eating (in large primates) and insect-eating (in small primates) (Kay, 1975; 1984; Kay and Hylander, 1978; Covert, 1986; Box 2). Shearing crests have two functions: in folivores to reduce cellulose in leaves and in insectivores to reduce chitin in arthropods.* If a fossil with shearing crests on its teeth were found, how could one determine whether it was a folivore or an insectivore?

In this case, the context (body size) of the trait (shearing crests on teeth) can be used to reconstruct the biological role of the trait in a particular case. Small-bodied animals (<500 g) with shearing crests are insectivorous while large-bodied animals with shearing crests (>500 g) are folivorous. Another way around this problem is to show that, rather than being functionally uniform, shearing crests may have different mechanical properties (*sensu* Bock and von Wahlert, 1965) (Lucas and Teaford, 1994; Strait, 1997). Leaves and chitin, both structural carbohydrates, have different structural properties that require different "tools" to break them apart. Chitin is brittle and contains lots of stored energy and is best broken down by short crests that initiate cracks that self-propagate to shatter the exoskeleton. Leaves, in contrast, are relatively tough and must literally be sliced into small pieces like a piece of cloth or paper. The shearing crests of faunivorous and folivorous primates are actually different structures performing different mechanical functions. Thus, although traits may often appear to have multiple functions, a close examination of the morphological context of the trait can yield clues distinguishing between these functions. In other cases this may not be possible and one might be stuck with the hypothesis that a certain trait in a fossil may be indicative of more than one behavior.

The Problem of Homology

In light of these concerns, multiple instances of independent evolution of a trait–function association currently provide the strongest basis on which trait–function associations in fossils can be hypothesized. However, this approach, labeled by Coddington (1994) the "convergence approach," presents an interesting paradox: although multiple independent instances of a trait–function association improve statistical support for a trait–function association in a fossil taxon, the more distantly related the taxa are from the fossil in question, the more likely they are to be dissimilar in ways that might confound the comparison (Wenzel and Carpenter, 1994). For example, how relevant is the function of the fovea in birds, lepidosaurs, and teleosts to the function of

*Some workers distinguish between "function" as a mechanical attribute and "biological role" as the behavioral property with which it is associated (Bock and von Wahlert, 1965). Under this scheme, the function of shearing crests would be to reduce structural carbohydrates (cellulose in leaves and chitin in arthropods) into smaller pieces to facilitate digestion and this function has two biological roles: leaf-eating and insect-eating. Here we disregard the distinction between function and biological role because it is simpler to do so and does not result in any loss of information.

the fovea in primates (Ross, 2002)? Is the phylogenetic and environmental context in which they have evolved dissimilar enough to render the comparisons irrelevant?

To minimize these problems with the convergence approach, many studies of primate comparative anatomy (e.g., Oxnard, 1975; Larson, 1998; Fleagle and Meldrum, 1988) have relied on comparisons of closely related taxa, an approach that might be termed "narrow phylogeny," by analogy to "narrow allometry" (R. J. Smith, 1980). For example, in an attempt to test hypotheses that particular musclo-skeletal features were adaptations to leaping or quadrupedal walking, Fleagle (1976, 1977) examined the musculo-skeletal anatomy and locomotor behavior of two closely related Asian leaf-monkeys. This method minimizes the likelihood that any observed differences were due to nonadaptive aspects of phylogenetic history. He was able to show that in cases where the two differed in muscular anatomy or bony morphology, the leaper always showed the hypothesized leaping morphology and the quadruped showed the hypothesized quadrupedal morphology. This approach avoids complications due to gross historical differences by comparing differences in sister-taxa (but see Garland and Adolph, 1994, for a cautionary note).

A similar way to test hypotheses of trait–function association while avoiding the thorny issues surrounding the "convergence approach" is the "homology approach" (Coddington, 1994). This method tests hypotheses of trait–function association qualitatively by optimizing the trait and the function onto a phylogeny to determine whether the novel function appears on the tree at the same time as the trait (e.g., Coddington, 1988, 1994). Disjunction between the appearance of structure and function on the internal branches of the tree falsifies a hypothesis that the trait became fixed by selection for it to perform a specific function. If the trait evolves before the function, the trait may not be related to the function at all. If the function is being performed before the trait evolves, the trait may evolve to take over the function from another trait, but the hypothesis of adaptation would be more difficult to prove.

Application of this criterion in a historical context not only requires a phylogeny that is accurate, precise, and dense, but it also requires that one can homologize behavior or function in the same way that one can homologize morphology. This is not a problem if one defines homologues as traits that were present in the last common ancestor of two taxa. However, if one requires some kind of continuity of information (Van Valen, 1982), particularly genetic information, homologizing of function or behavior can be more questionable (cf. Robson-Brown, 1999 and Greene, 1994). A particularly problematic example discussed above is the homology of muscle firing patterns. Some workers argue that one can homologize muscle firing patterns (Lauder, 1994), but Galis (1996) summarizes several general criticisms of such homologies: variability in muscle firing patterns often is obscured by experimental design; eliciting forceful activities such as powerful biting tends to result in similar muscle firing patterns; some muscles are usually not examined; heritability is seldom measured and the influence of learning on development of muscle firing patterns is little studied; and evaluation of muscle firing patterns in a phylogenetic context is often lacking. For example, it has been suggested that the four-phase jaw movement cycle of slow opening, fast opening, fast closing, and slow closing is a primitive or generalized pattern in vertebrate feeding that can be homologized across groups (e.g., Bramble and Wake, 1985). However, there is appreciable variability in jaw movement cycles across vertebrates that cannot be shoe-horned into the four-phase cycle (K. K. Smith, 1994). Moreover, the standard criterion for identifying homologies (presence in the last common ancestor) is not corroborated by optimization of these characters on a cladogram of vertebrate relationships (K. K. Smith, 1994).

In the study of primate behavior, issues of homology are often particularly difficult to address (Robson-Brown, 1999). For example, what criteria are to be used to determine whether diurnal polygynous strepsirrhines and anthropoids exhibit homologous social behaviors (Kappeler, 1999)? Are the differences indicative of two different kinds of polygyny due to convergence, or the result of divergent evolution from a common diurnal polygynous stem lineage? The question of how to homologize behavior and function across taxa is vital for attempts to study adaptation in a phylogenetic context (e.g., Di Fiore and Rendall, 1994), and further theoretical and experimental work in this area is required (Greene, 1994).

The paradox of the "homology approach," or other methods that compare sister taxa, is that the only traits that are truly the "same" as the trait in the focal taxon are those that are homologous. However, if one can use only homologous trait–function associations to infer trait–function associations in a focal taxon, statistical independence of these associations is never possible! There are some other potentially serious statistical problems with this approach (Garland and Adolph, 1994), however, it is highly intuitive and has been of great use in primatology.

The Problem of Allometry

Considerations of allometry, or the role of size, are critical for determining the relationship between morphology and function. A seductively simple but logically flawed comparative approach to the identification of adaptations can be traced to the influential opinions of Gould (1966, 1975a, 1975b). Gould (1966) initially distinguished clearly between size-required and size-dependent (i.e., correlated) relationships in interspecific scaling or "allometry," and suggested that the departure of points from the former might be characterized as "adaptations." He cautioned, however, that the explanatory value of these residuals cannot be assessed "until both the trends of normal size–shape correlation are empirically established *and the adaptive reasons for these general patterns determined*" (p. 607, emphasis added). Less than a decade later, the original distinctions were blurred and virtually any size-correlated trend among species was considered to be a viable "criterion for judgment" (Gould,

1975a) or "criterion of subtraction" (Gould, 1975b) in the diagnosis of adaptations. This so-called "null hypothesis for adaptive differences" emerged as a rationale for "size-correction" via residuals, and the possibility that information about size-related adaptation might be encoded in the allometric relationship itself was usually ignored (Biegert and Maurer, 1972; Pilbeam and Gould, 1974; Clutton-Brock and Harvey, 1979; Sweet, 1980; Pagel and Harvey, 1988; Stearns, 1992; Anthony and Kay, 1993; Martin, 1993). In other words, the observed allometry was believed to reveal little more than how organisms must change to remain "the same" in functional, mechanical, or life historical terms; the real adaptive signal was seen only in the departures from the interspecific trend.

The logical flaws behind this approach to allometry and adaptation are revealed by the general failure to appreciate that without explicit similarity criteria and *a priori* expectations for "equivalence" (e.g., Fleagle, 1985), (1) it is not clear what biological information has been purged by discarding the primary signal (*sensu* Schmidt-Nielsen, 1984), and (2) residuals with the same sign and magnitude are not automatically equivalent in any biological sense (see Smith, 1984, and Jungers *et al.*, 1995). Dramatic departures from empirical trends are certainly suggestive and worthy of further scrutiny, but allometry itself is only a description, not an explanation. In the dogged pursuit of "size-free" data via residuals, some authors acknowledge how little we currently know about genuine "size-required" modifications and freely admit that functional equivalence is being assumed without good reason. For example, Harvey and Pagel (1991) reluctantly cling to a recipe of residuals as "an absolute necessity if we are to study adaptation" (pp. 177–178), even as they confess in the same breath that "we should not place much faith in empirically derived slopes per se" (p. 178). The correlates or "effects" of size should not be automatically removed as the centerpiece of a general analytical strategy until these effects are better understood and explicitly articulated. To do so assumes that anything correlated with size is the same thing as size itself, and that is clearly not the case (Oxnard, 1978).

Faulty inferences of functional equivalence are common in primatology. For example, Pilbeam and Gould (1974; also Gould, 1975b) argued that "gracile" and "robust" australopithecines were simply scaled variants of one another because they fell on a three-point regression; they were the "same" animal expressed at different sizes despite allometry in both postcanine tooth size and brain size. We now know that their trophic adaptations are not the same (Grine, 1981) and they differ little in body size (McHenry, 1991a). Even if there were dramatic differences in body size among these early hominids, to describe them as nothing more than size variants is tantamount to arguing that gorillas are little more than functionally equivalent, scaled versions of chimpanzees. However, the strong positive allometry of postcanine tooth size in African apes (Gould, 1975b) reflects an adaptive shift in diet from frugivory to folivory, not size-required changes to remain the same (Grine, 1981). The adaptive and functional significance of numerous postcranial allometries in

African apes emphasizes the same point (Shea, 1981; Jungers and Susman, 1984).

The fallacy of assuming that interspecific scaling merely reflects necessary modifications to preserve some (usually unspecified) similarity can also be seen clearly in a consideration of long-bone cross-sectional geometry in quadrupedal primates (Jungers and Burr, 1994; Jungers *et al.*, 1998). Despite highly correlated, typically positive allometry of supporting dimensions (cortical areas, area moments, section moduli), these relationships do not reveal size-required distortions to maintain mechanical equivalence. Rather, bony strength decays as a function of body size, and additional behavioral/postural changes are required to achieve dynamic stress–strain similarity in quadrupedal mammals (Rubin and Lanyon, 1984; Biewener, 1990). There are a few clade-specific differences in these scaling patterns that are functionally explicable (e.g., terrestrial cercopithecines tend to have stronger bones at a given size than do colobines), but similar residuals do not imply comparable degrees of structural reinforcement among species differing in size. A similar caveat obtains in drawing inferences about cognitive adaptations from comparisons of relative brain size via encephalization quotients (another type of residual).

The new comparative methods discussed above (e.g., Harvey and Pagel, 1991) also have important implications for those workers wed to residuals. If there is a probable risk that uncorrected data will result in inaccurate slope values and/or inappropriately narrow confidence bands for regressions (i.e., residual scatter is inaccurately estimated), then it makes no sense whatsoever to take residuals from the uncorrected relationships and subsequently use them to generate independent contrasts in pursuit of adaptive explanations (e.g., Ross and Jones, 1999). Regardless, the adaptive signal is often in the primary allometric relationship, and it is this signal that we need to pay more attention to, rather than discarding it without explaining it first.

In sum, associations between trait and function in the focal taxon provide first indications as to the possible functions of the trait. When the focal taxon is a fossil, associations between traits and functions in extant taxa must be used to determine what the trait's function might have been. Although problems can arise when the same function can be associated with more than one biological role, these problems can be alleviated by carefully examining the context of the association, essentially subdividing the comparative sample into relevant subsamples. The possibility that structure and function can be dissociated is more difficult to circumvent, given that one can imagine that some aspect of behavior (e.g., muscle firing patterns) might have been different in the fossil. However, this criticism can be blunted by applying a strict criterion of trait–function association among numerous living taxa, especially when there is no reason to believe the association is merely the result of phylogenetic inertia. Multiple cases of independent evolution of a trait–function association in living taxa provide the strongest evidence for attribution of that same function to that trait in a fossil. The possibility that phylogenetic inertia might maintain traits and functions in lineages without natural selection is a problem that requires further study. Statistical techniques for dealing with these problems are available but should be applied with caution (see Pagel, 1999). The validity of the character optimization approach advocated by proponents of the historical concept depends on the validity of hypotheses of homology of the traits being mapped onto the phylogeny and the precision, accuracy, and density of the phylogeny in question.

Biomechanical Models: Optimization and Constraint

Phylogenetic congruence of morphological traits and function provides suggestive evidence of association, and multiple instances of trait–function association provide a logical and statistical basis for inferring trait function in living taxa that can be extended to fossils, but a hypothesis of adaptation is strengthened when the biomechanical basis of this association is understood, or at least hypothesized. For example, Kay and Cartmill (1977) require that "all the features specified in the definition of T (the trait) must have some functional relationship to F." Biomechanics can also generate hypotheses about interdependence of traits and about the functional feasibility of various transformation schemes (Ross, 1996; Galis, 1996). Most importantly in the present context is the role of biomechanics in studying trait function in the fossil record. The *locus classicus* of this approach as applied to fossils is Rudwick's paper (1964), *The inference of function from structure in fossils*. Rudwick sees such inferences as providing the only means of determining the relative contributions of natural selection and other mechanisms to the process of evolution. He was concerned with the relative importance of natural selection and Schindewolf's Typostrophic theory, but the current debate over the relative importance of microevolutionary versus macroevolutionary phenomena in evolution is also dependent on the fossil record for its resolution (Carroll, 1997). If we can identify adaptations and nonaptations in fossils, then the relative importance of natural selection and other mechanisms can be assessed. If not, it is doubtful that such questions can be answered.

Despite acknowledging the utility of comparisons for identifying possible living analogs of fossil organisms, Rudwick regards the identification of trait–function associations (what he calls "the comparative method") as a "pseudo-method." Comparisons suggest correlations between structure and function, but the interpretation of these correlations requires a mechanical analysis of the structure to determine its *mechanical fitness* for its proposed function. This requires that organisms be treated as machines: "Machines can only be described for what they are by referring to the way in which their *design* enables them to *function* for their intended *purpose*" (Rudwick, 1964: 34, italics in original). How can we determine whether a structure is designed *well enough* to be considered an adaptation?: By understanding the operational principles governing any mechanism that would fulfill the designated function. This leads to a "generalised structural specification for the function," which, when refined by a consideration of the mechanical properties of the materials involved, becomes a specification that "*describes the structure that would be capable of fulfilling the function with the maximal efficiency attainable under the limitations imposed by the nature of the materials* " (Rudwick, 1964: 36, italics in original). This specification is the *paradigm* for that function.

Alexander (1996) calls the paradigm approach "optimization theory." "Optimization is the process of minimizing costs or maximizing benefits, or obtaining the best possible compromise between the two. *Evolution by natural selection is a process of optimization*" (Alexander, 1996: 2, emphasis added). Biomechanical optimization* (like natural selection) is a process whereby the best design is continually preferred over the *available* alternatives until it predominates in the population. When the best design is fixed in the population, biomechanical optimization, produced by stabilizing selection, is a "corrective tendency" (Williams, 1992). Several good examples of biomechanical optimization analysis can be cited, such as Norberg's (1994) analysis of wing morphology in bats, and Dawkins' (1997) summary of eye evolution.

The paradigm approach articulated by Rudwick is nonhistorical, but this approach achieves its real strength when wedded with the notion of historical constraints. Indeed, contrary to Gould and Lewontin's (1979) overly simplistic characterization, the identification of constraints is implicit in the paradigm approach. Although one can sometimes identify a "best" way of solving a particular problem (a global maximum), there are usually several ways to solve a problem. The biomechanical optimization of design via natural selection can only work on the *available* alternatives; the solution that any lineage reaches will be that which was immediately accessible at the time selection acted. Therefore, biomechanical optimization analysis involves defining not only a global maximum, but local maxima as well. With this information at hand, the task is to derive hypotheses for why different lineages solve the problem in different ways, or converge on specific local maxima. In this context, hypotheses of constraint are explanations for why selection has not produced certain functionally equivalent (or better) solutions to a specific problem (Gould, 1989; Antonovics and van Tienderen, 1991; McKitrick, 1993). Therefore, a falsifiable hypothesis of constraint requires a precise definition of the alternatives and an explanation of why these alternatives are not available.

Hence, constraints are defined here as heritable attributes of a lineage that constrain the ways that natural selection can optimize, whether they were originally fixed in the lineage by natural selection or not. All constraints are phylogenetic or historical because they are intrinsic attributes of organism. The term constraint can be used in several contexts.

Constraints can be invoked to explain why an organism adopts one solution to a problem rather than another, as in the case of the dorsoventrally flattened tails of whales and the laterally compressed tails of fishes. Whereas

*We use the term biomechanical optimization to distinguish it from character optimization, used to map characters onto cladograms.

chondrichthyans, teleosts, and ichthyosaurs swim (or swam) using lateral undulations of the body, aquatic mammals swim using dorsoventral undulations. The most probable explanation for this is that the mammalian lineage reduced lateral undulations of the body in favor of dorsoventral ones in order to improve the efficiency of terrestrial locomotion (Carrier, 1987; Cowen, 1996). Consequently, when cetaceans moved into the water they utilized the optimal solution most accessible from their ancestral condition, not that used by fishes and ichthyosaurs.

If the solutions are biomechanically equivalent (i.e., in the absence of biomechanical evidence that one morphology is better than the other) these "constraints" might be more appropriately called phylogenetic bias, reserving the term "phylogenetic constraint" or "historical constraint" for a factor explaining why a lineage cannot realize an optimal condition purely because of some inherited attribute. This distinction recognizes that the phylogenetic histories of lineages always influence how they "solve problems," but that they are not always "constrained" by this. As noted above, all constraints are phylogenetic or historical in that they are encoded genomically or are epigenetic consequences of that coding. The following categories are therefore subsumed under phylogenetic constraints.

A structural or physiological constraint is a material, physiological, or biomechanical property of an organism or part of it (e.g., a tissue, a cell) that limits the alternatives available to natural selection for solving some problem posed by the environment. Uniquely biological structures such as muscle tissue and bone also have intrinsic biomechanical and material properties that limit the possible solutions to problems, usually in interaction with more general properties of the universe. For example, there are different kinds of bone, each with its own maximum strength under monotonic tension or compression, or under repetitive loads. These inherent properties of tissues derive from their constituents and their arrangements, which are often specified genetically, making them historical as well: only vertebrates have bone, making them the only organisms constrained or liberated by its material properties. Because natural selection is in part responsible for creation of bone and muscle, for example, Schlichting and Pigliucci (1998) would not classify such mechanical limitations as "constraints." However, if natural selection cannot "rewind the tape" and re-evolve "bony" materials with properties appropriate for the task at hand, the material properties surely constitute a constraint.

Developmental constraints are properties of developmental mechanisms that limit the possible outcomes of developmental processes (Maynard Smith *et al.*, 1985; Galis, 1999). For example, before gastrulation evolved in eukaryotes, organisms were restricted to simple balls of cells. This limited the number and kinds of interactions that could occur during development, thereby limiting the complexity that could arise during development (Buss, 1987). Once again, however, these constraints are also historical in that some lineage(s) evolved gastrulation, along with all the constraints and liberations that come with it.

Perhaps the most common type of limit on natural selection is functional compromise, a phenomenon falling outside of our definition of constraint because it results from natural selection satisfying two kinds of functional demand simultaneously. The need for functional compromise arises from the requirement that most structures meet multiple functional demands. The concept of functional compromise has been invoked frequently in attempts to explain, for example, the morphology of primate hands, which function both as organs of propulsion and as grasping organs (e.g., Tuttle, 1970), or the morphology of primate jaws, which function for mastication of food, food acquisition, and display (Hylander, 1979). As a result they do not seem to be "optimally designed" for either task, but nevertheless permit both activities. In this case, each of the multiple functions of a feature is a constraint on the others.

In sum, biomechanical modeling of a trait shows how its form is related to its function. When the trait under study is unique to fossils and no living analogues are available, the optimization approach can be the best means of reconstructing the possible function(s) of that trait or limiting the alternatives. By identifying other ways that this function might be served, hypotheses regarding constraints on organismal form can be generated. Although the results can be interpreted in a phylogenetic context (e.g., Crompton and Hylander, 1986, utilized by Weishampel, 1995), biomechanical optimization does not require knowledge about phylogeny. Moreover, biomechanical optimization might be better done without such knowledge. If the biomechanical optimization approach generates a hypothesis of adaptation or that specific phylogenetic constraints explain why one local optimum was reached rather than another, these hypotheses can be tested by using independently derived phylogenetic hypotheses (see Lee and Doughty, 1997, for a fuller explication of this approach).

Experimental Observation of Trait Use and Performance

Application of the comparative method can provide evidence of an association between trait and function, and biomechanics can provide a mechanical hypothesis that a trait is appropriate for a particular function. However, the strongest evidence for the identification of adaptations in living animals is detailed study of how a trait is actually used. Are the trait and function associated because the trait performs the function? The primatological literature is replete with examples of hypothesized functions for traits in fossils derived from trait–"function" associations in living animals, usually supported by some kind of biomechanical or design criterion.

Studies of trait function in extant animals, whether *in vivo* or *in vitro*, can be used to evaluate these hypotheses in several ways. They can address the question of whether the trait actually performs, or is at least involved in, the function proposed for it. Hylander *et al.*'s (1991) analysis of browridge function in catarrhine primates is illustrative. Numerous previous workers have hypothesized that the robust browridge of catarrhines, including fossil hominids, functions to reduce stresses associated with mastication or incision. Hylander *et al.*, showed that in fact bone strain magnitudes in the supraorbital region are generally very low, suggesting that the browridges could be reduced significantly in size without affecting their function during mastication and incision. Hylander and Ravosa (1992) argue that the need for a browridge arises in animals in which facial kyphosis rotates the orbits out from under the braincase (see also Shea, 1985), exposing the orbital contents to blows from above. Thus, the robust bony browridges are overdesigned for feeding behaviors because they are designed to resist less frequent forces producing higher strain rates and magnitudes than experienced during feeding. In this instance, a hypothesis about function for both fossil and extant taxa is shown to be false for extant animals and therefore without empirical support for fossils.

Interestingly, studies of function in extant animals can falsify hypotheses about function in fossil taxa without eroding their usefulness as environmental or behavioral indicators. For example, it is well documented that a tall greater tubercle on the humerus of primates is associated with terrestrial quadrupedalism, and it has been repeatedly used to reconstruct the locomotor habits of extinct species (Jolly, 1966; Fleagle and Simons, 1982; Harrison, 1989; McCrossin *et al.*, 1998). Moreover, most primatologists who discussed this feature offered the seemingly plausible functional explanation that a tall greater tubercle increases the lever arm of the supraspinatus muscle during protraction (or swing phase) of the forelimb. However, electromyographic studies by Larson and Stern (1989, 1994) have shown that supraspinatus is not active during the swing phase of locomotion in primates. Rather, supraspinatus is most active during retraction, suggesting that the tall greater tubercle aids supraspinatus in providing greater stability for terrestrial quadrupeds during support phase (propulsion). While the original hypothesis regarding the function of the enlarged greater tubercle was incorrect, this does not invalidate the well-documented association between a tall greater tubercle and terrestrial locomotion. In other words, the hypothesized function changes but the behavioral reconstruction does not.

Studies of trait function in extant animals can also determine whether the trait's function is important enough for selection to fix or maintain it in the lineage. The definition and measurement of "important enough" will vary from one context to another. One might argue that if organisms use a trait only rarely, one might expect that selection is unlikely to have favored animals with that trait for that function. However, the frequency at which a trait is used for a certain function is not always a good arbiter of that trait's most important function (Gans, 1963; Stern and Oxnard, 1973). Although Kay and colleagues (e.g., Kay and Hylander, 1978) have shown an impressive correlation between the development of shearing crests on molar teeth and the frequency of leaves in the diet of living primates, Rosenberger and Kinzey (1976) introduced the concept of "critical function" to capture the notion that a trait's form may be adapted to perform one critical, although infrequent function, vital for the organism's survival. For example, *Callicebus moloch* need well-developed shearing crests on their molars to process the leaves that constitute their sole source of protein, and a critical food item in the dry season, even though their diet is predominantly fruit (Kinzey, 1978).

Rather than frequency of use, therefore, the adaptive status of a trait in extant animals is ideally evaluated by linking trait function to reproductive success, the purview of population genetics. Arnold (1983) factored the selection acting on a trait into a performance gradient representing the effect of the trait on some aspect of performance and a fitness gradient representing the effect of performance on fitness. Under this scheme, experimental observation, or functional morphology, provides the vital link between differences in morphology on one hand and differences in performance on the other: it explains why different morphologies are associated with different performance capabilities. This is the realm of ecomorphology or ecological morphology (Dullemeijer, 1974; Alexander, 1988; Wainwright and Reilly, 1994; Losos, 1990).

Performance testing in studies of primate adaptation has only rarely been undertaken. Rather, primate functional morphologists have generally been content to relate morphological differences to reported frequencies of behaviors from naturalistic studies. This is unfortunate, because performance is a critical step in testing the validity of adaptive hypotheses in extant animals. One particularly revealing study of performance in primates is Sheine and Kay's (1977) work on shearing crests in prosimians. In order to test the hypothesis that shearing crests in insectivorous species were an adaptation to comminuting chitin into small pieces to facilitate digestion, they fed food mixed with chitin particles of a uniform size to a series of prosimian primates with different degrees of shearing crest development on their teeth. They then retrieved the chitin from the animal's stomach and looked at the particle sized after chewing and found that the species with the greater development of shearing crests had reduced the ingested chitin into smaller particles.

There is a great need for more performance testing of this nature for many of the commonly held assumptions about the adaptive nature of primate morphological features. For example, limb proportions in primates are generally argued to be adaptations for particular locomotor abilities (Jungers, 1985). However, there are only limited data relating limb lengths to performance among closely related species (Demes *et al.*, 1996; Polk, 2000) and none among individuals of the same species. Similarly, although there are good theoretical reasons (from optimization studies) to believe that longer legs should enhance leaping ability (see Demes and Gnther, 1989, for a review), there have been no studies documenting that within a species, individuals with longer legs can perform longer leaps (as has been shown in lizards, Losos, 1990), or that among similar-sized species, the ones with longer legs can perform longer leaps. Demes *et al.*, (1999) have shown that relative peak takeoff and landing forces decrease with increasing body mass in intraspecific comparisons, suggesting that animals with longer limbs may suffer lower stresses during takeoff and landing. Documentation of intraspecific performance differences would greatly strengthen arguments that specific morphological traits are adaptations resulting from selection for specific behavioral abilities and thereby assist in subsequent reconstructions of the behavior of extinct taxa.

In sum, observations on trait use tell us whether a trait actually performs a function and, in some cases, whether it performs it better than an alternative. More studies of intrapopulational or intraspecies variability in performance are needed in primates. Linking these studies to differences in fitness is desirable but probably unlikely, given the long generation times for many primates and the difficulty of gathering *in vivo* data.

Phylogenetic Information and the Study of Adaptation

As discussed in several sections above, phylogenetic information can be incorporated into the approaches listed above to add a historical and contextual dimension to the study of adaptation. However, the degree of phylogenetic resolution that is needed differs among the various phylogenetic approaches available. Instances of multiple evolution of a trait–function relationship can only be identified in a phylogenetic context, but need not require a precise phylogeny in all cases. For example, the presence in *Ateles* of features of the trunk, shoulder, and elbow related to bimanual suspensory locomotion can be invoked to corroborate the hypothesis that the same features are adaptations to suspension in extant hominoids (Larson, 1998) without knowing precisely how *Ateles* is related to other platyrrhines and how hominoids are related to other catarrhines. If there are any other platyrrhines and other catarrhines lacking these features, it is more parsimonious to conclude that they are convergent rather than homologous (Fig. 1). Similarly, one does not need a very precise phylogeny to determine that the diaphragm and endothermy of birds and mammals are convergent (Carrier, 1987): the lack of these features in lepidosaurs and suchians suffices to establish the fact of convergence. However, it is noteworthy that a fairly dense series of basal members of their two divergent lineages is necessary to determine the accuracy of the phylogeny (Gauthier *et al.*, 1988). More precise phylogenies (and, of course, accurate ones) are required for statistical techniques such as independent contrasts or generalized least-squares regression (Martins and Hansen, 1996; Purvis and Webster, 1999); less precise phylogenies in the form of taxonomies suffice for variance components analysis (R. J. Smith, 1994; Nunn, 1995). Exactly which techniques are chosen will depend on the questions at hand and the precision of the available phylogeny.

The relevance of phylogenetic information to the study of adaptation has been highlighted by recent advocates of the "homology approach" to the study of adaptation discussed above (Coddington, 1988, 1994; Baum and Larson,

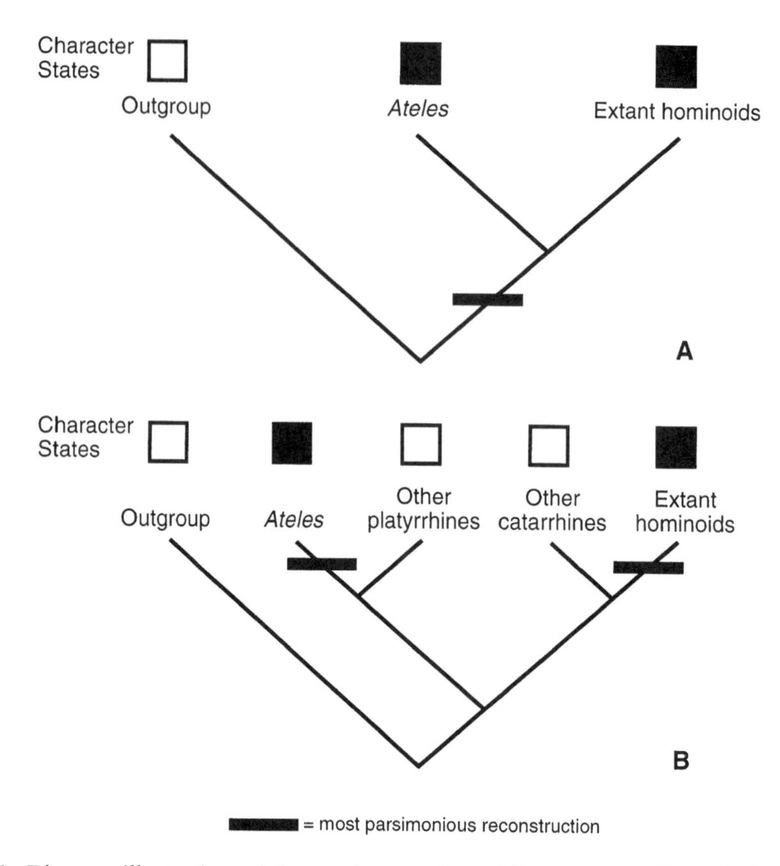

Fig. 1. Diagram illustrating minimum degree of precision necessary for a phylogenetic hypothesis in order to demonstrate that suspensory adaptations in extant hominoids and *Ateles* are convergent. Filled boxes above taxon names indicate presence of traits of interest, empty boxes indicate absence of these traits. A. Lacking information on other platyrrhines or other catarrhines, it is impossible to demonstrate convergence. B. If the outgroup lacks these traits, convergence is most parsimonious if the traits are not found in other platyrrhines or other catarrhines, regardless of relationships within these two clades.

1991; Larson and Losos, 1996). These workers suggest that not all comparisons are equal and that knowledge of phylogeny is required to identify the appropriate comparisons (Brandon, 1990). For example, Coddington defines adaptation as a derived trait (M_1) that arose at a specific position on the cladogram for the derived function (F_1) with respect to the ancestral condition (M_0) and primitive function (F_0) (Fig. 2). Functions or selective regimes are optimized onto the phylogeny along with the trait of interest setting up the comparison that will corroborate or falsify the hypothesis of adaptation (Baum and Larson, 1991; Brooks and McLennan, 1991; Larson and Losos, 1996).

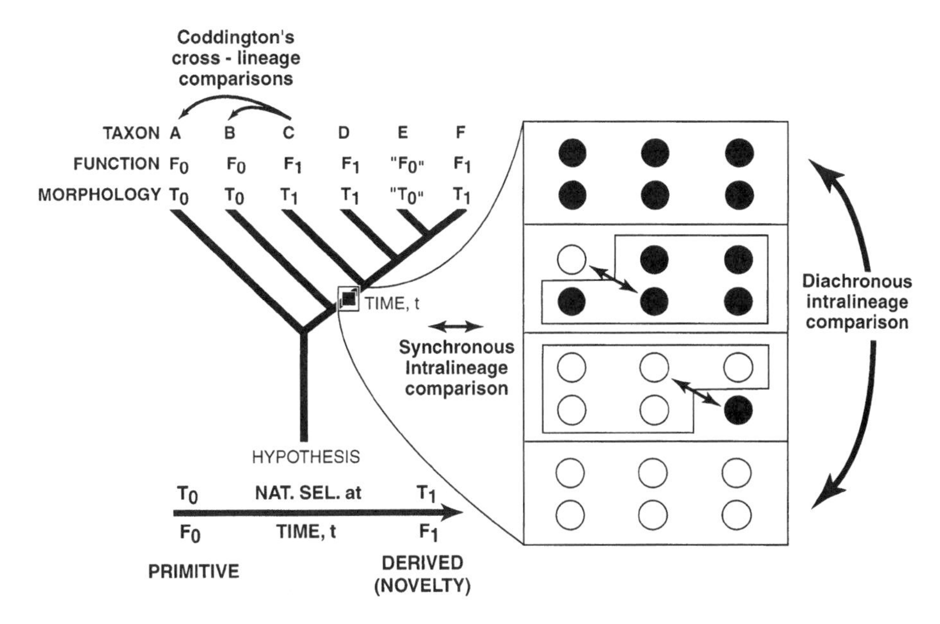

Fig. 2. Figure illustrating various kinds of comparisons discussed in the text. Adapted and expanded from Coddington (1988: Fig. 1) by addition of illustrations of possible comparions. Synchronous intralineage comparisons: comparisons between individuals that exist at the same time, in the same environment, in the same population. Diachronous intralineage comparisons: comparisons of ancestral taxon in which T_0 is still preserved, and the descendent taxon in which T_1 has already been fixed. Comparisons advocated by Coddington (1988), Baum and Larson (1991): a phylogenetically constrained cross-lineage comparison between the taxon of interest and its primitive sister taxon. See text for additional discussion.

According to Coddington it is vital to compare the "ancestral" morphology and function with the "derived" hypothesized adaptation; i.e., the appropriate hypothesis is that taxa with M_1 and F_1 are ecologically or functionally better adapted than taxa with M_0 and F_0 within a particular environment. Sometimes, information about polarity is enough to reject putative examples of adaptation) [see the rhinoceros example in Coddington (1988)]. In other cases, the polarities of trait and function are consistent with the hypothesis of adaptation, and performance or design criteria are used to compare the derived trait with the primitive antecedent. Put simply, a trait may be an adaptation if its position on the phylogeny is coincident with or subsequent to that of the function or behavior it serves.

Practical and theoretical problems confront this approach. The practical problems include the need for a well-corroborated phylogeny and, as discussed above, the difficulty of homologizing function, behavior, and "selective regimes". In addition, it is not usually possible to eliminate other possible causes of the observed relationships between trait polarity, performance, and fitness. Lauder *et al.*, (1993; Leroi *et al.*, 1994) point out that the method has no way of distinguishing between traits that arise because of selection for them and traits that arise as pleiotropic effects of selection for other traits. This problem can be minimized by fully elucidating the functional relationships between the trait, performance, and selective regime but, once again, this is not only difficult for extant taxa but next to impossible for fossils.

The theoretical problems are more fundamental and relate to the use of cladistic data to evaluate hypotheses of adaptation. Coddington (1988) and Baum and Larson (1991) argue that the appropriate comparison to make is between the primitive and derived sister taxa labeled in Fig. 1, i.e., either between C and B or C and A. Taxa A and B with T_0, F_0 must be "in some definable and measurable sense, ecologically or functionally less well adapted than C or D" with T_1, F_1. Other comparisons, such as between C and E, are inappropriate because T_0, F_0 are not homologous with T_0, F_0 in A and B. The validity of this approach is related to the degree to which the suggested comparison reconstructs the conditions under which T_1 *actually* arose (Coddington, 1994). The ideal comparison would be that represented in Fig. 1 by what we call the "synchronous intralineage comparison." Comparisons between individuals that exist at the same time, in the same environment, in the same population, control for all possible extraneous variables that might confound the analysis, in effect performing an analysis of the type recommended by Arnold (1983, see above). This approach is obviously impractical for the study of the majority of life forms, living or dead. The next best comparison would be what we call a "diachronous intralineage comparison" indicated in Fig. 1: a comparison of the ancestral taxon in which T_0 is still preserved, and the descendant taxon in which T_1 has already been fixed. Although these comparisons are to some degree removed from the actual context in which the trait arose, they can be expected to yield accurate results if nothing other than the traits in question have changed. Less direct still is the approach advocated by Coddington (1988), Baum and Larson (1991), essentially the approach used by Fleagle (1976, 1977, discussed above): a phylogenetically constrained cross-lineage comparison between the taxon of interest and its primitive sister taxon. In Fig. 1 one can see that the taxa being studied are even further removed from the context in which the trait T_1 was fixed in the lineage. The extent to which this compromises attempts to reconstruct the context of T_1's origin depends on the amount of change that has occurred along the branches leading to the terminal taxa being compared (Coddington, 1994).

The case of the basal cynodonts is illustrative: evaluating hypotheses regarding the function of the jaw joint in basal cynodonts using comparison of Reptilia and Mammalia is of dubious value considering the enormous amounts of time and morphological evolution that have occurred since the events under study (Weishampel, 1995). Exactly how much change is acceptable for the comparisons to still be valid remains to be evaluated and guidelines need to be developed. At present, we point out that the extent to which these comparisons can tell us about how a trait *actually* originated depends on how different the terminal taxa are from the ancestor–descendant pair of the diachronous intralineage comparison. In cases where the terminal taxa are not defined by any apomorphies other than the trait in question and are not separated from the lineage in question by a long period of time, this approach might indeed be valuable. The validity of this approach is inversely related to the number of apomorphies separating the terminal taxa from the node in question.

Of course, one could argue that Coddington's hypothesis is about adaptation "as an explanation of supraspecific evolutionary pattern" (1988: 3). However, if adaptation is defined according to the historical concept, it is not a supraspecific phenomenon at all and comparisons between sister taxa are not necessarily accurate surrogates for comparisons between organisms that functioned as part of the same population (Ross, 1999). And, if adaptation is not defined according to the historical approach, instead emphasizing current utility, the technique advocated by Coddington, Baum and Larson obviously is not necessary.

A more practical way to use phylogenetic information in the reconstruction of behavior in fossils is Witmer's Extant Phylogenetic Bracket approach to the reconstruction of soft tissues (1995, 1997). Witmer's approach involves three procedures: (1) the identification of causal associations between osteological and soft tissue features in extant taxa, (2) the mapping of this "causal associations" onto a phylogenetic tree using standard character optimization techniques, and (3) the identification of the osteological features in fossils. If a fossil taxon possesses the osteological correlate of a certain soft tissue feature, and both its first and second outgroups also possess both the soft tissue feature and the osteolgical correlate, then the soft tissue feature can be reconstructed in the fossil taxon with confidence (Level I inference). If only the first outgroup possesses the soft tissue feature and osteological correlate, the inference that the soft tissue was present in the fossil is less secure (level II inference). If neither of the extant outgroups of the fossil taxon possesses the soft tissue feature and osteological correlate, the soft tissue feature can still be inferred for the fossil, but with less certainty (level III inference).

The strength of Witmer's approach is in its requirement that the causal association between soft tissue and osteological structure be established in extant taxa (not necessarily the first and second outgroups of the fossil of interest) before attempting to reconstruct soft tissues in fossils. Causal associations between osteological and soft tissue features are the basis for reconstructing soft tissues in fossils; phylogenetic information is only used to assess the degree of confidence in the soft tissue reconstruction. As discussed above, character optimization techniques can be used to reconstruct any traits: behavioral, functional, or soft tissue. However, in the absence of osteological or other fossilizable correlates that can be identified in fossils, the degree of confidence in these reconstructions must be very low indeed.

In sum, recent work on adaptation emphasizes the importance of including phylogenetic data. Certainly phylogeny can provide important information on the context in which a trait evolved, but the precision necessary depends on the question at hand. Attempts to rigorously evaluate hypotheses of adaptation by mapping behavioral and "selective regime" characters onto cladograms require accurate, precise, and dense phylogenies, and that behaviors and selective regimes can be homologized. Moreover, the method is only useful when extant taxa provide good models for the lineage in which the origin of the trait is being studied. The utility of this approach remains to be demonstrated. Nevertheless, Witmer's extant phylogenetic bracket approach provides a rigorous method for reconstructing soft tissue features, an important part of reconstructing behavior in fossils.

Primitive Traits as Indicators of Behavior

Many historical approaches to adaptation focus on apomorphic traits, and tests of association that attempt to "correct for phylogeny" focus on the origins of traits, not their retention, to establish independence of data points. However, primitive traits can possess the same trait–function relationship or performance benefit as derived traits (Anthony and Kay, 1993) and therefore have as much utility for reconstructing behavior in fossil taxa as derived traits. The only situations in which a primitive correlate of behavior should be rejected are when there are reasons to believe there has been a loss or change of the behavior without loss or change of the trait. Loss of the behavior without loss of the trait produces functionless retentions, maintained in lineages by forces other than natural selection. Change of the behavior without change of the trait produces exaptations, maintained in lineages by natural selection for a different function. We discuss these two situations separately.

Functionless Retentions. The majority of traits in every taxon are primitive in the sense of being inherited from their immediate ancestor, so if behaviors are *routinely* lost while their associated traits are retained, there is essentially no basis for using morphology to reconstruct behavior in the fossil record. How is this possibility to be evaluated if extant analogues provide insufficient evidence for what happens to a particular trait when loss of the behavior occurs (e.g., many of the primitive characters associated with arboreal locomotion in the earliest hominids)?

There are at least four approaches to evaluating the possibility that primitive traits are functionless retentions, all of which involve reference to extant forms. These are signs of behavioral "residue," evidence of "frozen function" or lack thereof, compatibility or incompatibility with criteria of design and optimization, and the context of information provided by other characters. We will present examples of each of these in turn.

Behavioral "residue." As discussed earlier in the context of trait–behavior associations, it may be possible to establish whether markings or residues are left by particular behaviors and the structures used for them. Wear striae on the toothcombs of extinct strepsirrhines fall into this category (e.g., fossil lorids—Rose *et al.*, 1981; subfossil lemurs—Jungers *et al.*, this volume, Chapter 10); fur-grooming behavior leaves its unambiguous signature in the form of whip-marks and grooves on the lower anterior dentition. Attempts to reconstruct ingestive and dietary behaviors of extinct species from molar and incisor microwear analyses are less straightforward, but the principles are essentially the same (see Teaford, 1994; Ungar, this volume, Chapter 7).

Frozen function. Primitive features sometimes provide unique insights into behavior because they manifest what might be called "frozen function" (Davis, 1966). For example, curvature of both manual and pedal phalanges is probably a symplesiomorphic character of early hominids such as *Australopithecus afarensis* (McHenry, 1991b); the much straighter bones of modern humans represent a derived condition. Degree of phalangeal curvature can be linked to locomotor behavior and habitat use in extinct primates due to predictable associations between the expression of this feature and differences in behavior of living taxa (Susman *et al.*, 1984; Hamrick *et al.*, 1995; Jungers *et al.*, 1997). Through biomechanical analysis of phalangeal curvature (e.g., finite element analysis), we now also know that curved phalanges are advantageous to arborealists because they suffer lower normal and shear stresses than do straight ones during comparable antipronograde behaviors (Richmond, 1998, in press; Jungers *et al.*, this volume, Chapter 10). However, it is the analysis of the ontogeny of phalangeal curvature and positional behavior in different living species that clearly reveals curvature to be frozen function. Degree of phalangeal curvature can and does change significantly throughout the life of primates whenever gripping and climbing/suspensory behaviors also change in frequency (Richmond, 1998; Jungers *et al.*, this volume, Chapter 10). Phalanges become more curved after birth as grasping and antipronogrady increases, and they reach their maximum degrees of curvature when such behaviors also peak. They can remain curved if positional behavior changes little thereafter (e.g., in lesser apes), or they can and do straighten out later in life if such behaviors decrease in frequency (e.g., if terrestrial locomotion becomes increasingly important in older individuals). This implies that the chimpanzee-like curvature seen in australopithecines is not there as mere baggage or unverifiable pleiotropy (Lovejoy *et al.*, 1999); rather, precisely because phalangeal curvature is a "function of function," the conclusion that arboreality remained an important part of australopithecine positional repertoire is inescapable (also see Duncan *et al.*, 1994).

Design criteria. A third approach to evaluating the information content of primitive traits is to investigate situations where the presence of a trait seems to contradict known behavior, as judged by criteria of design and optimization. A good example of a trait argued to be a primitive relic based on this line of reasoning is the tarsier fovea. Tarsiers and anthropoids are the only mammals demonstrated conclusively to have a retinal fovea. Cartmill (1980) hypothesized that the haplorhine fovea evolved in a diurnal lineage as an adaptation for improved visual acuity and that it was retained in the lineage leading to extant nocturnal *Tarsius*. In support of this argument, gently sloping (concaviclivate) foveae like that of *Tarsius* are mostly found in diurnal animals and it is very unusual to find a fovea in a retina that is dominated by rods rather than cones, as in *Tarsius* (Cartmill, 1980). Tarsiers also lack a tapetum, a structure typically present in nocturnal mammals, suggesting to Cartmill that the tarsier lineage lost its tapetum in a diurnal ancestor (Cartmill, 1980). In terms of optimization and design models discussed above, the tarsier retina is not obviously designed for nocturnal vision (Ross, 2002) and not all individuals within tarsier species have a fovea (Woollard, 1925; Castenholtz, 1965), further supporting the hypothesis that it is a functionless relic. Under this model the fovea of living tarsiers is either a vestige of their diurnal ancestors or an exaptation maintained in the tarsier lineage by stabilizing selection for some new function related to nocturnality (Ross, 2002). The strength of this argument is that it explains various aspects of tarsier visual anatomy, as well as its variable presence in extant forms.

Context. The last situation in which primitive traits may be considered unimportant for reconstructing behavior is when their message is inconsistent with the message of other traits, i.e., the total morphological pattern. A recent example of this approach is Richmond and Strait (2000), who concluded that some characters of the early hominid radius, such as a distally projecting dorsal ridge, are indicative of a knuckle-walking heritage but not the behavior itself. The taxa on which these features are observed, *A. anamensis* and *A. afarensis*, lack other, more definitive, signs of knuckle-walking. The latter include pronounced dorsal metacarpal ridges, which, if present, can be argued to be "frozen function" in much the same way as curved phalanges were above (Susman, 1979). Moreover, there is "clearly derived morphological evidence for bipedalism" in these early hominids, which further suggested to Richmond and Strait (2000: 383) that knuckle-walking was an unlikely behavior.

Although different in emphasis, this approach is not unlike that of authors who more consistently dismiss early hominid primitive traits in behavioral reconstructions, these primitive traits typically being evidence for some arboreal behavior (e.g., Latimer and Lovejoy, 1989). These authors see a dichotomy between bipedalism and arboreal locomotion that allows for few intermediate stages, and so evidence from derived traits—those that indicate bipedalism as the predominant mode of locomotion—is favored, to the exclusion of other evidence. This approach clearly depends on the context provided by the "total morphological pattern." In this way, it is unlike the approaches discussed above, which focus on the intrinsic properties of the trait, such as its ontogeny.

These examples illustrate the importance of making explicit the criteria on which primitive traits are judged, whether they are concluded to be evidence for behavior or simply retentions from an ancestor. If a trait has been demonstrated through other means to be an adaptation for a function, we would consider it to be evidence of behavior regardless of its status as primitive or derived, unless it is shown otherwise. Lack of discernible function and variable presence in living animals are perhaps the best criteria for identifying a feature as a potentially functionless relic (e.g., Wenzel and Carpenter, 1994; Ross, 2002). Ontogenetic responses to behavioral shifts, or behavioral residue left by use of the trait, are the best evidence to affirm that a primitive trait correctly reflects behavior. The context of other traits is useful, but this is clearly an iterative process, as the interpretation of each trait ultimately depends on all others.

Most importantly, we emphasize that for any taxon, most of the traits under consideration are primitive and, at present, we have little evidence of the extent to which characters are retained in lineages without selection for a function. In the absence of such evidence, approaches that dismiss the evidence of primitive traits *a priori* should be applied with caution. This includes the approaches discussed above, such as "independent contrasts" and other methods aimed at "phylogenetic correction" of data. If natural selection maintains primitive trait–function associations in multiple lineages after divergence from a common ancestor, the extant endpoints of these lineages could be argued to constitute biologically independent examples of the trait–function association, even though the association was present in their last common ancestor. Under this line of reasoning, the many millions of years over which a trait–function association is maintained are as important as (or more so than?) the shorter period of time over which selection acted to fix the association in the lineage in the first place. Pervasive selection of this type clearly invalidates the null model of "Brownian motion" (Felsenstein, 1985), and analytical alternatives need to be considered (Pagel, 1999). Whether the endpoints of these independent lineages are independent in a statistical sense is a separate issue.

Exaptations. Primitive traits will be unreliable indicators of behavior if they can acquire new functions after their origin; in other words, that they are *exaptations* rather than adaptations (*sensu* Gould and Vrba, 1982; Larson and Losos, 1996). For exaptations to frustrate those interested in behavioral reconstruction, constancy of structure must be maintained through a change in function. In other words, trait T should not change into a derived trait (an adaptation) that performs the function better. Coddington (1988: 19) phrased this requirement in a stronger way: "such a feature [the exaptation] would have to exhibit from the beginning the kind of efficient design adaptations usually do, so that subsequent change does not take place.*

The first step in demonstrating exaptation is to show that a trait is used for multiple functions (see above under "Association between Trait and

* Constancy of structure was not strictly required in the original concept of exaptation (Goule and Vrba, 1982) but if the trait changes with the change in function it can be identified as a new trait and would not be an exaptation.

Function: Comparative Methods"). An excellent example of this phenomenon in primates is the use of molar shearing crests broadly defined. As discussed earlier in the context of trait functions, both folivorous and insectivorous primates have extensive shearing crests (Kay, 1975; 1984; Kay and Hylander, 1978; Covert, 1986). In folivores they are used to reduce cellulose in leaves, and in insectivores they are used to reduce chitin in arthropods but in both cases the *function* of the shearing crests can be argued to be the same: to reduce structural carbohydrates (such as cellulose or chitin) into smaller pieces to facilitate digestion.

If we imagine a situation where a descendant is more folivorous and less insectivorous than its ancestor,* the trait "lots of shearing" would remain unchanged and the shearing crests might be considered to be exaptations for folivory. At the same time, the trait only informs us that the animal had a diet that required the breakdown of structural carbohydrates. It does not tell us, by itself, whether the animal ate insects or leaves. On the other hand, it is possible to focus on the biological role of the trait, which is to facilitate eating leaves by breaking down cellulose or to facilitate eating insects by breaking down chitin. In this case the trait is an adaptation for the biological role it serves in the ancestor (in which the trait originated), and an exaptation for the new biological role it serves in the descendant. It does not tell us what its role was in either animal, i.e., whether the descendant or ancestor ate insects or leaves. The information content of these two situations is the same, even though in one case we have called the trait an adaptation for a single function, and in the other we recognized it as a potential exaptation for one of two biological roles. To call it one or the other thing is semantic with respect to the question of reconstructing behavior.

In actual fact, however, there is some evidence that shearing crests used for folivory are different from those used in insectivory (Lucas and Teaford, 1994; Strait, 1997). Chitin is brittle, containing large amounts of stored energy, and is best broken down by short shearing crests that initiate self-propagating cracks that shatter the exoskeleton. Leaves, in contrast, are relatively tough and must literally be sliced into small pieces like a piece of cloth or paper. Thus, the shearing crests of insectivorous and folivorous primates are in many cases recognizably different traits. Thus, it is imprecise to say that shearing crests are not modified through a transition to a different function, and therefore the trait cannot be viewed as an exaptation.

More common examples and hypotheses of exaptation can be discussed from a similar perspective: are there good examples of exaptations wherein the structure remains unchanged and the function changes, such that attempts to reconstruct behavior are confounded? Gould and Vrba (1982) saw bird feathers as an exaptation for flight, because the first expression of feathers

*Constraints of body size on diet suggest that a transition from an insectivorous diet to a predominantly folivorous one requires an intermediate stage where fruit is the primary resource and tooth shape changes accordingly (e.g., Kay, 1984). Thus, shearing crests are unlikely to show a progression from use in consumption of insects to use in consumption of leaves.

appeared to serve the function of insulation or possibly predation. But they also recognize changes in feather structure, and particularly skeletal features, that occurred with the origin of flight. Only in a very general sense are feathers exaptations for flight. Specific types of feathers may be adaptations for flight, along with specific wing designs.

The concept "exaptation" has intuitive value in pointing out that traits serving novel, derived functions are often variations on primitive traits. However, if the trait serving the novel function is in any way different from the primitive condition, it is actually a new feature and an adaptation. Moreover, in the latter case the trait is informative with regard to behavior, while in the case of true exaptation (constancy of structure despite a change in function) it is not. The challenge is to investigate the empirical evidence for exaptations, or lack thereof. This would go some way toward resolving whether "primitive" traits are likely to be informative with regard to behavior.

In sum, one of the most important questions in evolutionary morphology is the degree to which primitive traits are maintained in lineages by forces other than natural selection when they lose or change their function. How common are functionless retentions and exaptations? If they are both very common, then phylogenetic information will be vital for the study of adaptation, function, and behavior in fossils. If they are not, phylogenetic information might be less important than currently believed.

The Best Formula

What then are the most appropriate criteria for identifying adaptations in extant or fossil taxa? There is a tendency in the recent literature (particularly the literature promoting a historical approach) to advocate one method as better than all others. In contrast, we believe that the study of adaptation requires application of all methods that can yield insight into the evolution of the trait in question, while keeping their various strengths and weaknesses in mind. Consilience between diverse approaches provides compelling support for a hypothesis of adaptation: lack of consilience suggests the need for more work (Lee and Doughty, 1997). Here we summarize the uses, advantages, and disadvantages of the various criteria outlined above.

In studying extant taxa, observations on what a trait is used for are essential for determining whether it performs the function hypothesized for it. A biomechanical explanation of how a trait performs its function is fundamental to the determination of whether or not the trait is an adaptation because it elucidates the causal basis underlying the trait–function association. Without such a study there is always the possibility that a trait–function association is spurious, i.e., a correlation without causation. Biomechanical optimization analysis, i.e., an argument from design, also generates hypotheses regarding the selective advantage of the trait and why the trait may not meet a universal optimum. Such optimization should be the first step in the development of hypotheses of historical constraint. Identification of multiple instances in which this trait–function association has evolved convergently increases confidence in the hypothesis of adaptation. Placing the trait and its function in a phylogenetic context enables their association to be tested in another way: did they evolve at the same time or was their appearance disjunct? If all prior tests corroborate the hypothesis that the trait–function association is an adaptation, the disjunct appearance of trait and function implies that the trait is either an exaptation for the function, or it replaced another trait that was previously performing the function. The latter case represents an adaptation if the trait that was replaced performed the function less effectively than the new trait. The validity of this historical approach clearly depends on the validity of hypotheses of homology of the traits being mapped onto the phylogeny and the precision, accuracy, and density of the phylogeny in question. Despite the potentially important temporal and contextual information that phylogeny can add to the study of adaptation, hypotheses of adaptation need not be restricted to the first appearance of apomorphies. The presence of trait–function associations in lineages over substantial periods of time and through many cladogenetic events strongly suggests that they are under strong stabilizing selection, and behavioral inferences are relatively secure.

When the focal taxon is a fossil, associations between traits and functions in extant taxa must be used to determine what the trait's function might have been. Although problems can arise when the same function can be associated with more than one biological role, these problems can often be alleviated by carefully examining the morphological, behavioral, and functional details of the situation more closely. Lauder's criticism of structure–function associations is more difficult to circumvent, given that one can imagine that some aspect of behavior (e.g., muscle firing patterns) might have been different in the fossil. However, its power is effectively negated by applying a strict criterion of trait–function association among numerous living taxa, especially when there is no reason to believe the association is merely the result of phylogenetic inertia. Because of the latter concern, multiple cases of independent evolution of a trait–function association in living taxa provide the strongest evidence for attribution of the same function to that trait in a fossil. Once function has been established for the fossil taxon, further evaluation of a hypothesis of adaptation relies on experimental observations of extant taxa and biomechanical arguments. We also should acknowledge that some details of function (such as muscle firing patterns) can never be verified for fossils.

How Can Fossil Taxa Be Used to Study Adaptation?

Although we have emphasized how the study of adaptation in living taxa forms the basis for our interpretations of fossils, it is also worthwhile to point out that a full understanding of adaptation can involve the use of fossil taxa.

Most researchers recognize that fossil taxa are useful in phylogenetic studies because they often represent intermediate forms that are not captured in modern ranges of variation and combine traits in a way that may be unexpected based on the complexes of traits in modern species. In the same manner, some fossil taxa occupy behavioral regimes not seen among modern taxa or have novel trait–function relationships. Theoretically, this information could increase the explanatory power of hypotheses of adaptation. However, there are practical problems, the first of which is circularity.

The usual process of inference with fossil taxa is to infer behavior from morphology. The opportunity to perform a study of adaptation that relies on the observation of functions, selective regimes, or performance will usually not be available for extinct species. It would be circular to infer behavior from morphological characters and then interpret the characters as adaptations based on the inferred behavior.

However, there are some situations, such as dental microwear or fossilized footprints (see Boucot, 1990), where the behavior of the fossil taxon is not necessarily inferred based on morphological structure. In these cases, the external evidence of behavior can be combined with the evidence of anatomy in the fossil taxon to test hypotheses of adaptation in a normal manner. Discussion of the fossil organism may be little different from that involving an extant organism, as long as there is reason for confidence in the evidence for behavior.

In addition, geological or paleoecological information can be used as external evidence for the ecological context in which a trait evolved (e.g., Kay *et al.*, this volume, Chapter 9). This can facilitate arguments of adaptation, although knowledge of behavior will usually be fairly vague. A prominent example of such inference is the evolution of human bipedalism. Early hominids were predominantly bipedal animals, although there is disagreement concerning the degree to which arboreal locomotion played a supplementary role. Whatever the answer to the latter question, arguments about the adaptive component of bipedalism have centered on the first animals to adopt this behavior. Many alternative explanations have been proffered, some of which are locomotor efficiency for long-distance travel, greater height to see predators (or prey), thermoregulatory efficiency in hot, arid climates, and the benefit of freeing the hands for activities other than locomotion (reviewed in Fleagle, 1999: pp. 225–228). Of these, the first three explanations were conceived with a savanna or open woodland environment in mind. In other words, the savanna environment offered a broad specification of selective regime in which a number of alternative designs might be successful. At the same time, the viability of these hypotheses seemed to affirm the likelihood that the first hominids were savanna-dwelling animals.

However, it is now becoming widely accepted that environments occupied by the earliest hominids were not savannas (see Potts, 1998; Reed, this volume, Chapter 6 for reviews). The study of hominid paleoenvironments is currently a burgeoning field, and the degree of cover is seen to vary among hominid-bearing sites. Adaptive explanations for the origin of bipedalism should now be tested against the criteria demanded by optimal behaviors in wooded environments. This example highlights both the importance and difficulty of adaptive explanations that rely on fossil forms. These cases will always be iterative in nature, as inferences of anatomy, behavior, and environment are refined independently and used to judge one another.

Summary and Conclusions

Recent developments in the study of adaptation have had mixed success. Problems such as tautology that plagued definitions of adaptation in the early 1970s have been resolved: historical definitions that link adaptation to natural selection are true to evolutionary theory. Unfortunately they are virtually impossible to apply, particularly to the fossil record. Thus, while the historical definition discussed above is most accurate in theory, it is almost useless in practice. The most productive approach to date has been the nonhistorical approach. Less concerned with trait *origination*, this approach has arguably more realistic goals: the reconstruction of behavior for its own sake. It aims to determine how traits actually function in living taxa through studies of trait–function association, experimental observation of performance, optimization studies, and comparative methods. These data are then used to evaluate hypotheses regarding trait function in fossil taxa: inferences about fossils are based on rules and principles identified from biomechanical analyses of living animals. The problem with this approach is that it is heavily dependent on the study of living animals for derivation of its rules, with the attendant danger that fossil taxa were actually unlike living animals. This danger can be minimized by appropriate modeling and optimization studies and by not invoking principles that are too far removed from the physical and chemical laws that are likely to have been uniform through time. Uniformitarianism must be applied cautiously (Conway Morris, 1995).

The role of phylogeny reconstruction in the study of adaptation is much discussed. Coddington's approach and that of Baum and Larson must be applied very critically if the aim is to investigate events of which there is no direct record. Attempts to use these cross-lineage comparisons as surrogates for synchronous or diachronous intralineage comparisons are fraught with peril. However, cross-lineage comparisons are still powerful tools for identifying rules and principles of animal function and trait–function association. These rules and principles can then be used either to reconstruct behavior in fossil taxa that are close in time to the actual events under study, or to define the context in which these events might have taken place. The most important role of phylogenetic pattern in the study of adaptation is in its essential component in modern applications of the comparative method. Although much still needs to be done to delineate appropriate statistical techniques, accurate, precise, and dense phylogenies, with appropriate estimates of possible ranges of error, have the potential to provide a solid foundation for much of what many primatologists do in the future.

One of the most crucial areas for future research is the relative importance of phylogenetic inertia and stabilizing selection in trait maintenance. It has implications for issues surrounding degrees of freedom in interspecific analyses, the applicability of data on living taxa to the study of fossils, and the degree to which adaptations can be defined as current utility. If phylogenetic inertia is a relatively unimportant phenomenon, then the distinctions between the historical and nonhistorical concepts of adaptation are mostly semantic. If, however, stabilizing selection is not important in the maintenance of traits in lineages over time, then we can be pessimistic about our chances of identifying adaptations in the fossil record.

ACKNOWLEDGMENTS

We thank Matt Carrano, Larry Witmer, and Mike Plavcan for useful comments and John Hunter for a detailed, insightful review of the manuscript. Frietson Galis provided stimulating discussion on several topics addressed here. Luci Betti-Nash prepared the figures.

References

Abouheif, E. 1999. A method for testing the assumption of phylogenetic independence in comparative data. *Evol. Ecol. Res.* **1:**895–909.

Alexander, R. McN. 1975. Evolution of integrated design. *Am. Zool.* **15:**419–425.

Alexander, R. McN. 1985. The ideal and the feasible: physical constraints on evolution. *Biol. J. Linn. Soc.* **26:**345–358.

Alexander, R. McN. 1988. The scope and aims of functional morphology. *Neth. J. Zool.* **38:**3–22.

Alexander, R. McN. 1996. *Optima for Animals.* Princeton University Press, Princeton, N.J.

Amundson, R. 1996. Historical development of the concept of adaptation. In: M. R. Rose and G. V. Lauder (eds.), *Adaptation.* Academic Press, San Diego.

Anthony, M. R. L., and Kay, R. F. 1993. Tooth form and diet in ateline and alouattine primates: Reflections on the comparative method. *Am. J. Sci.* **293A:**356–B382.

Antonovics, J., and van Tienderen, P. H. 1991 Ontoecogenophyloconstraints? The chaos of constraint terminology. *Trends Ecol. Evol.* **6:**166–168.

Arnold, S. 1983. Morphology, performance and fitness, *Am. Zool* **23:**347–361.

Asher, R. J. 1998. Morphological diversity of anatomical strepsirrhinism and the evolution of the lemuriform toothcomb. *Am. J. Phys. Anthropol.* **105:**355–367

Baum, D. A., and Larson A. 1991. Adaptation reviewed: a phylogenetic methodology for studying character macroevolution. *Syst. Zool.* **40:**1–18.

Bell, G. 1997. *Selection: The Mechanism of Evolution.* Chapman and Hall, New York.

Biegert, J., and Maurer, R. 1972. Rumpfskelettlange, Allometrien und Korperproportionen bei catarrhinen Primaten. *Folia Primatol.* **17:**142–156.

Biewener, A. 1990. Biomechanics of mammalian terrestrial locomotion. *Science* **250:**1097–1103.
Bock, W. J. 1980. The definition and recognition of biological adaptation. *Am. Zool.* **20:**217–227
Bock W. J., and Von Wahlert, G. 1965. Adaptation and the form-function complex. *Evolution* **19:**269–299.
Boucot, A. J. 1990. *Evolutionary Paleobiology of Behavior and Coevolution.* Elsevier, Amsterdam.
Bramble, D. and Wake, D. B. 1985. Feeding mechanisms in lower vertebrates. In: M. Hildebrand, D. M. Bramble, D. B. Wake, and K. F. Liem (eds.), *Functional Vertebrate Morphology*, pp. 230–261. Belknap Press, Cambridge.
Brandon, R. N. 1978. Adaptation and evolutionary theory. *Stud. Hist. Philos. Sci.* **9:**181–206.
Brandon, R. 1990. *Adaptation and Environment.* Princeton University Press, Princeton.
Brooks, D. R., and McLennan, D. A. 1991. *Phylogeny, Ecology, and Behavior: A Research Program in Comparative Biology.* University of Chicago Press, Chicago.
Burian, R. 1983. Adaptation. In: M. Grene (ed.), *Dimensions of Darwinism.* pp. 287–314. Cambridge University Press, Cambridge, Mass.
Buss, L. 1987. *The Evolution of Individuality.* Princeton University Press, Princeton.
Carrier, D. R. 1987. The evolution of locomotor stamina in tetrapods: circumventing a mechanical constraint. *Paleobiology* **13:**326–341.
Carroll, R. L. 1997. *Patterns and Processes of Vertebrate Evolution.* Cambridge University Press, Cambridge.
Cartmill, M. 1980. Morphology, function and evolution of the anthropoid postorbital septum. In: R. L. Ciochon and A. B. Chiarelli (eds.), *Evolutionary Biology of the New World Monkeys and Continental Drift.* pp. 243–274, Plenum Press, New York.
Castenholtz, E. 1965. Über die Struktur der Netzhautmitte bei Primaten. *Z. Zellforsch.* **65:**646–661
Cheverud, J. 1988. A comparison of genetic and phenotypic correlations. *Evolution* **42:**958–968.
Clutton-Brock, T. H., and Harvey, P. H. 1979. Comparison and adaptation. *Proc. R. Soc. London Ser.B* **205:**547–565.
Coddington, J. A. 1988. Cladistic test of adaptational hypotheses. *Cladistics* **4:**3–22.
Coddington, J. A. 1994. The roles of homology and convergence in studies of adaptation. In P. Eggleton, and R. I. Vane-Wright (eds), *Phylogenetics and Ecology.* pp. 53–78. Academic Press for the Linnaen Society of London, London.
Conway Morris, S. 1995. Ecology in deep time. *TREE* **10:**290–294.
Covert, H. H. 1986. Biology of Early Cenozoic primates. In: D. R. Swindler and J. Erwin (eds.), *Comparative Primate Biology, Volume 1: Systematics, Evolution and Anatomy*, pp. 335-359, Alan R. Liss, New York.
Cowen, R. 1996. Locomotion and respiration in aquatic air-breathing vertebrates. In: D. Jablonski, D. H. Erwin, and J. H. Lipps (eds.), *Evolutionary Paleobiology.* University of Chicago Press, Chicago.
Crompton, A. W., and Hylander, W. L. 1986. Changes in mandibular function following the acquisition of the dentary-squamosal jaw articulation. In: N. Hotton III, P. D. MacLean, J. J. Roth, and E. C. Roth (eds.). *The Ecology and Biology of Mammal-like Reptiles*, pp. 263-282. Smithsonian Institution Press, Washington, D.C.
Darwin, C. 1859. *The Origin of Species by Means of Natural Selection, or the Preservation of Favoured Races in the Struggle for Life.* John Murray, London. (Reprint of First edition, Harmondsworth: Penguin Books, 1981.)
Davis, D. D. 1966. Non-functional anatomy. *Folia biotheoretrica (Series B)* **6:**5–8.
Dawkins, R. 1997. *Climbing Mount Improbable.* W. W. Norton and Company, New York.
Demes, B., and Günther, M. M. 1989. Biomechanics of allometric scaling in primate locomotional morphology. *Folia Primatol.* **52:**58–69.
Demes, B., Jungers, W. L., Fleagle, J. G., Wunderlich, R. E., Richmond, B. G., and Lemellin, P. 1996. Body size and leaping kinematics in Malagasy vertical clingers and leapers. *J. Hum. Evol.* **31:**367–388.
Demes, B., Fleagle, J. G., and Jungers, W. L. 1999. Takeoff and landing forces of leaping strepsirhine primates. *J Hum. Evol.* **37:**279–292.
Di Fiore, A., and Rendall, D. 1994. Evolution of social organization: A reappraisal by using phylogenetic methods. *Proc. Natl. Acad. Sci. USA* **91:**9941–9945.

Dullemeijer, P. 1974. *Concepts and Approaches in Animal Morphology.* Van Gorcum, The Netherlands.
Duncan, A. S., Kappelman, J, and Shapiro, L. J. 1994. Metatarsophalangeal joint function and positional behavior in *Australopithecus afarensis. Am. J. Phys. Anthropol.* **93:**67–81.
Farrell, B. D. 1998. "Inordinate fondness" explained: Why are there so many beetles? *Science* **281:**555–559.
Felsenstein, J. 1985. Phylogenies and the comparative method. *Am. Nat.* **125:**1–15.
Fisher, D. C. 1985. Evolutionary morphology: beyond the analogous, the anecdotal, and the ad hoc. *Paleobio* **11:**120–138.
Fleagle, J. G. 1976. Locomotor behavior and skeletal anatomy of sympatric malaysian leaf-monkeys (*Presbytis obscura* and *Presbytis melalophos*). *Yearbk. Phys. Anthropol.* **20:**440–453
Fleagle, J. G. 1977. Locomotor behavior and muscular anatomy of sympatric malaysian leaf-monkeys (*Presbytis obscura* and *Presbytis melalophos*). *Am. J. Phys. Anthropol.* **46:**297–308.
Fleagle, J. G. 1985. Size and adaptation in primates. In: W. L. Jungers (ed.) *Size and Scaling in Primate Biology*. pp. 1–19, Plenum Press: New York.
Fleagle, J. G. 1999. *Primate Adaptation and Evolution.* Academic Press, New York.
Fleagle, J. G., and Meldrum, D. J. 1988. Locomotor behavior and skeletal morphology of two sympatric pitheciine monkeys, *Pithecia pithecia* and *Chiropotes satanas. Am. J. Primatol.* **16:**227–249.
Fleagle, J. G., and Simons, E. L. 1982. The humerus of *Aegyptopithecus zeuxis:* A primitive anthropoid. *Am. J. Phys. Anthropol.* **59:**175–193.
Galis, F. 1996. The application of functional morphology to evolutionary studies. *TREE* **11:**124–129.
Galis, F. 1999. Why do almost all mammals have seven cervical vertebrae? Developmental constraints, Hox genes, and cancer. *J. Exp. Zool. (Mol. Dev. Evol.)* **285:**19–26.
Gans, C. 1963. Functional analysis within a single adaptive radiation. *Proc. XIV Congr. Zool.* **3:**278–282.
Garland, T. Jr., and Adolph, S. C. 1994. Why not to do two-species comparative studies: limitations on inferring adaptation. *Physiol. Zool.* **67:**797–828.
Garland, T. and Ives, A. R. 2000. Using the past to predict the present: confidence intervals for regression equations and phylogenetic comparative methods. *Am. Nat.* **155:**346–364.
Garland, T., Harvey, P. H., and Ives, A. R. 1992. Procedures for the analysis of comparative data using phylogenetically independent contrasts. *Syst. Biol.* **41:**18–32.
Gauthier, J., Kluge, A. G., and Rose, T. 1988. Amniote phylogeny and the importance of fossils. *Cladistics* **4:**105–209.
Gittleman, J. L., Anderson, C. G., Kot, M., and Luh, H-K. 1996. Phylogenetic lability and rates of evolution: a comparison of behavioral, morphological and life history traits. In: E. P. Martins, (ed.), *Phylogenies and the Comparative Method in Animal Behavior.* pp. 166–205. Oxford University Press, Oxford.
Gould, S. J. 1966. Allometry and size in ontogeny and phylogeny. *Biol. Rev.* **41:**587–640.
Gould, S. J. 1975a. Allometry in Primates, with emphasis on scaling and the evolution of the brain. In: F. Szalay, (ed.) *Approaches to Primate Paleobiology.* pp. 244–292. Karger, Basel.
Gould, S. J. 1975b. On the scaling of tooth size in mammals. *Am. Zool.* **15:**351–362.
Gould, S. J. 1989. A developmental constraint in *Cerion*, with comments on the definition and interpretation of constraint in evolution. *Evolution* **43:**516–539.
Gould, S. J., and Lewontin, R. C. 1979. The Spandrels of San Marco and the Panglossion Paradigm: A critique of the adaptationist programme. *Proc. R. Soc. London Ser. B* **205:**582–598.
Gould, S. J., and Vrba, E. 1982. Exaptation: a missing term in the science of form. *Paleobiology* **8:**4–15.
Grafen, A. 1989. The phylogenetic regression. *Phil. Trans. R. Soc. London.* **326:**119–156.
Greene, H. W. 1986. Diet and arboreality in the emerald monitor, *Varanus prasinus*, with comments on the study of adaptation. *Fieldiana Zool. New Series* **31:**1–12.
Greene, H. W. 1994. Homology and behavioral repertoires. In: B. K. Hall (ed.), *Homology. The Hierarchical Basis of Comparative Biology*, pp. 369–391. Academic Press, San Diego.
Grine, F. E. 1981. Trophic differences between "gracile" and "robust" australopithecines: a scanning electron microscope analysis of occlusal events. *S. Afr. J. Sci.* **77:**203–230.

Hamrick, M. W., Meldrum, D. J, and Simons E. L. 1995. Anthropoid phalanges from the Oligocene of Egypt. *J. Hum. Evol.* **28:**121–145.

Harrison, T. 1989. New postcranial remains of *Victoriapithecus* from the middle Miocene of Kenya. *J. Hum. Evol.* **18:**3–54

Harvey, P. H. and Pagel, M. D. 1991. *The Comparative Method in Evolutionary Biology.* Oxford University Press, Oxford.

Hunter, J. P. and Jernvall, J. The hypocone as a key innovation in mammalian evolution. *Proc. Natl. Acad. Sci.* **92:**10718–10722.

Hylander, W. L. 1979. The functional significance of primate mandibular form. *J. Morphol.* **160:**223–240.

Hylander, W. L., and Ravosa, M. J. 1992. An analsyis of the supraorbital region of primates: A morphometric and experimental approach. In: P. Smith and E. Tchernov (eds.), *Structure and Function of Teeth.* pp. 223–255. Freund Ltd., Tel Aviv.

Hylander, W. L., Picq, P. G., and Johnson, K. R. 1991. Masticatory-stress hypotheses and the supraorbital region of Primates. *Am. J. Phys. Anthropol.* **86:**1–36.

Jolly, C. J. 1966. The evolution of the baboons. In: H. Vagborg (ed.), *The Baboon in Medical Research, Vol. II.* pp. 23–50. University of Texas Press, Austin.

Jungers, W. L. 1985. Body size and scaling of limb proportions in primates. In W. L. Jungers (ed.), *Size and Scaling in Primate Biology,* pp 345–381. Plenum Press, New York.

Jungers, W. L., and Burr, D. B. 1994. Body size, long bone geometry and locomotion in quadrupedal monkeys. *Z. Morphol. Anthropol.* **80:**89–97.

Jungers, W. L., and Susman, R. L. 1984. Body size and skeletal allometry in African apes. In: R. L. Susman, (ed.), *The Pygmy Chimpanzee,* pp 131–177. Plenum Press, New York.

Jungers, W. L., Burr, D. B., and Cole, M. S. 1998. Body size and scaling of long bone geometry, bone strength, and positional behavior in cercopithecoid primates. In: E. Strasser, J. Fleagle, A. Rosenberger, and H. McHenry (eds.), *Primate Locomotion.* pp. 309–330. Plenum Press, New York.

Jungers, W. L., Falsetti, A. B., and Wall, C. E. 1995. Shape, relative size and size-adjustments in morphometrics. *Yearb. Phys. Anthropol.* **38:**137–161.

Jungers, W. L., Godfrey, L. R., Simons, E. L., and Chatrath, P. S. 1997. Palangeal curvature and positional behavior in extinct sloth lemurs (Primates, Palaeopropithecidae). *Proc. Natl. Acad. Sci. USA* **94:**11998–12001.

Kappeler, P. 1999. Lemur social structure and convergence in primate socioecology. In: P. C. Lee (ed.), *Comparative Primate Socioecology,* pp. 273–299. Cambridge University Press, Cambridge.

Kay, R. F. 1975. The functional adaptations of primate molar teeth. *Am. J. Phys. Anthropol.* **43:**195–216.

Kay, R. F. 1984. On the use of anatomical features to infer foraging behavior in extinct primates. In: P. S. Rodman and J. G. Cant (eds.), *Adaptations for Foraging in Nonhuman Primates,* pp. 21–53. Columbia University Press, New York.

Kay, R. F., and Cartmill, M. 1977. Cranial morphology and adaptations of *Palaechthon nacimienti* and other Paromomyidae (Plesiadapoidea, ?Primates), with description of a new genus and species. *J. Hum. Evol.* **6:**19–53.

Kay, R. F., and Covert, H. H. 1984. The use of anatomical features to infer foraging behavior in extinct primates. In: D. J. Chivers, B. A. Wood, and A. Bilsborough (eds.), *Food Acquisition and Processing in Primates,* pp. 467–508. Plenum Press, New York.

Kay, R. F., and Hylander, W. L. 1978. The dental structure of mammalian folivores with special reference to primates and phalangeroidea (Marsupiala). In: G. Montgomery (ed.), *The Ecology of Arboreal Folivores,* pp 173–191. Smithsonian Institution Press, Washington, D.C.

Kinzey, W. G. 1978. Feeding behavior and molar features in two species of titi monkey. In: D. J. Chivers and J. Herbert (eds.), *Recent Advances in Primatology, Vol. 1: Behaviour.* pp. 373–385. Academic Press, London.

Larson, A., and Losos, J. B. 1996. Phylogenetic systematics of adaptation. In: M. R. Rose and G. V. Lauder (eds.), *Adaptation,* pp. 187–220. Academic Press, San Diego.

Larson, S. G. 1998. Parallel evolution in the hominoid trunk and forelimb. *Evol. Anthropol.* **6(3):**87–99.

Larson, S. G., and Stern, J. T. Jr. 1989. Role of supraspinatus in the quadrupedal locomotion of vervets (*Cercopithecus aethiops*): implications for interpretation of humeral morphology. *Am. J. Phys. Anthropol.* **79:**369–377.

Larson, S. G., and Stern, J. T. Jr. 1994. Further evidence for the role of supraspinatus in quadrupedal quadrupedal monkeys. *Am. J. Phys. Anthropol.* **87:**359–363.

Latimer, B., and Lovejoy, C. O. 1989. The calcaneus of *Australopithecus afarensis* and its implications for the evolution of bipedality. *Am. J. Phys. Anthropol.* **78:**369–386.

Lauder, G. V. 1994. Homology, form, and function. In: B. K. Hall (ed.), *Homology. The Hierarchical Basis of Comparative Biology*, pp. 151–196. Academic Press, San Diego.

Lauder, G. V. 1995. On the inference of function from structure. In: J. J. Thomason (ed.), *Functional Morphology in Vertebrate Paleontology.* pp. 1–18. Cambridge University Press, Cambridge.

Lauder, G. V., Leroi, A. M., and Rose, M. R. 1993. Adaptations and history. *Tree* **8:**294–297.

Lee, M. S. Y., and Doughty, P. 1997. The relationship between evolutionary theory and phylogenetic analysis. *Biol. Rev.* **72:**471–495.

Leroi, A. M., Rose, M. R., and Lauder, G. V. 1994. What does the comparative method reveal about adaptation? *Am. Nat.* **143:**381–402.

Lewontin, R. C. 1978. Adaptation. *Scientific American* **239:**156–169.

Lockwood, C. A., and Fleagle, J. G. 2000. The recognition and evaluation of homoplasy in primate and human evolution. *Yearb. Phys. Anthropol.* **42:**189–232.

Losos, J. B. 1990. Ecomorphology, performance capability, and scaling of Eest Indian *Anolis* lizards: an evolutionary analysis. *Ecol. Mono.* **60:**369–388.

Lovejoy, C. O., Cohn, M. J., and White, T. D. 1999. Morphological analysis of the mammalian postcranium: A developmental perspective. *Proc. Natl. Acad. Sci. USA* **96:**13247–13252.

Lucas, P. W., and Teaford, M. F. 1994. Functional morphology of colobine teeth. In: A. G. Davies and J. F. Oates (eds.) *Colobine Monkeys: Their Ecology, Behaviour and Evolution*, pp. 173–203. Cambridge University Press. New York.

Martin, R. D. 1990. *Primate Origins and Evolution: A Phylogenetic Reconstruction.* Princeton University Press, Princeton, NJ.

Martin, R. D. 1993. Allometric aspects of skull morphology in *Theropithecus.* In: N. Jablonski, (ed.), *Theropithecus: The Rise and Fall of a Primate Genus.* pp. 273–298. Cambridge University Press, Cambridge.

Martins, E. O., and Hansen, T. F. 1996. The statistical analysis of interspecific data: a review and evaluation of phylogenetic comparative methods. In: E. P. Martins, (ed.), *Phylogenies and the Comparative Method in Animal Behavior.* pp. 22–75. Oxford University Press, Oxford.

Maynard Smith, J., Burian, R., Kauffman, S., Alberch, P., Campbell, J., Goodwin, B., Lande, R., Raup, D., and Wolpert, L. 1985. Developmental constraints and evolution. *Q. Rev. Biol.* **60:**265–287.

McCrossin, M. L., Benefit, B. R., Gitau, S. N., Palmer, A. K., and Blue, K. T. 1998. Fossil evidence of terrestriality among Old World higher primates. In: E. Strasser, J. Fleagle, A. Rosenberger, and H. McHenry, (eds.) *Primate Locomotion: Recent Advances*, pp. 353–396. Plenum Press, New York.

McHenry, H. M. 1991a. The petite bodies of the "robust" australopithecines. *Am. J. Phys. Anthropol.* **86:**445–454.

McHenry, H. M. 1991b. First steps? Analysis of the postcranium of early hominids. In: Y. Coppens, and B. Senut, (eds.), *Origine(s) de la Bipedie Chez les Hominids*, pp. 133–141. Editions du CNRS, Paris.

McKitrick, M. C. 1993. Phylogenetic constraint in evolutionary theory: Has it any explanatory power? *Annu. Rev. Ecol. Syst.* **24:**307–330.

Norberg, R. A. 1994. Wing design, flight performance, and habitat use in bats. In: P. C. Wainwright and S. M. Reilly (eds.), *Ecological Morphology*, pp. 205–239. University of Chicago Press, Chicago.

Nunn, C. L. 1995. A simulation test of Smith's "degrees of freedom" correction for comparative studies. *Am. J. Phys. Anthropol.* **98:**355–367.

Oxnard, C. E. 1975. *Uniqueness and Diversity in Human Evolution.* University of Chicago Press, Chicago.

Oxnard, C. E. 1978. One biologist's view of morphometrics. *Annu. Rev. Ecol. Syst.* **9:**219–241.

Pagel, M. D. 1999. Inferring the historical patterns of biological evolution. *Nature* **401:**877–884.

Pagel, M. D, and Harvey, P. H. 1988. Recent developments in the analysis of comparative data. *Q. Rev. Biol.* **63:**413–440.

Pigliucci, M., and Kaplan, J. 2000. The fall and rise of Dr Pangloss: Adaptationism and the Spandrels paper 20 years later. *TREE* **15:**66–69.

Pilbeam, D., and Gould, S. J. 1974. Size and scaling in the evolution of man. *Science* **186:**892–901.

Polk, J. 2000. The kinematics of cursoriality: how patas monkeys differ from other primate quadrupeds. *Am. J. Phys. Anthropol.* Suppl. **30:**252.

Potts, R. 1998. Environmental hypotheses of hominin evolution. *Yearb. Phys. Anthropol.* **41:**93–136.

Purvis, A., and Webster, A. J. 1999. Phylogenetically independent comparisons and primate phylogeny. In: P. C. Lee (ed.), *Comparative Primate Socioecology.* pp. 44–68. Cambridge University Press, New York.

Reeve, H. K., and Sherman P. W. 1993. Adaptation and the goals of evolutionary research. *Q. Rev. Biol.* **68:**1–32.

Richmond, B. G. 1998. Ontogeny and biomechanics of phalangeal form in primates. Doctoral Dissertation, SUNY at Stony Brook, NY.

Richmond, B. G. In press. Finite element methods in paleoanthropology: the case of phalangeal curvature. *J. Human. Evol.*

Richmond, B. G., and Strait, D. S. 2000. Evidence that humans evolved from a knuckle-walking ancestor *Nature* **404:**382–385.

Robson-Brown, K. 1999. Cladistics as a tool in comparative analysis. In: P. C. Lee (ed.), *Comparative Primate Socioecology*, pp. 23–43. Cambridge University Press, Cambridge.

Rose, K. D., Walker, A. C., and Jacobs, L. L. 1981. Function of the mandibular tooth comb in living and extinct mammals. *Nature* **289:**583–585.

Rosenberger, A. L., and Kinzey, W. G. 1976. Functional patterns of molar occlusion in platyrrhine primates. *Am. J. Phys. Anthropol.* **45:**281–298

Ross, C., and Jones, K. E. 1999. Socioecology and the evolution of primate reproductive rates. In: P. C. Lee (ed.) *Comparative Primate Socioecology*, pp. 73–110. Cambridge University Press, Cambridge.

Ross, C. F. 1996. An adaptive explanation for the origin of the Anthropoidea (Primates). *Am. J. Primatol.* **40:**205–230.

Ross, C. F. 1999. How to carry out functional morphology? *Evolutionary Anthropology* **7:**217–222.

Ross, C. F. 2002. The tarsier fovea: functionless vestige or nocturnal adaptation? In: C. F. Ross and R. F. Kay (eds.), *Anthropoid Origins: New Visions.* Kluwer Academic/Plenum, New York.

Ross, C. F., Williams, B. A., and Kay, R. F. 1998. Phylogenetic analysis of anthropoid relationships. *J. Hum. Evol.* **35:**221–306.

Rubin, C. T., and Lanyon, L. E. 1984. Dynamic strain similarity in vertebrates: an alternative to allometric limb bone scaling. *J. Theor. Biol.* **107:**321–327.

Rudwick, M. J. S. 1964. The inference of function from structure in fossils. *Br. J. Philos. Sci.* **15:**27–40.

Schlichting, C. D., and Pigliucci, M. 1998. *Phenotypic Evolution. A Reaction Norm Perspective.* Sinauer Associates, Sunderland, MA.

Schmidt-Nielsen, K. 1984. *Scaling: Why is Animal Size So Important?* Cambridge University Press, Cambridge.

Shea, B. T. 1981. Relative growth of the limbs and trunk in the African apes. *Am. J. Phys. Anthropol.* **56:**179–201.

Shea, B. 1985. On aspects of skull form in African apes and orang-utans, with implications for hominoid evolution. *Am. J. Phys Anthropol.* **68:**329–342.

Sheine, W. S., and Kay, R. F. 1977. Analysis of chewed food particle-size and its relationship to molar structure in primates *Cheirogaleus medius* and *Galagosenegalensis* and insectivoran *Tupaia glis. Am. J. Phys. Anthropol.* **47:**15–20.

Smith, K. K. 1994. Are neuromotor systems conserved in evolution? *Brain Behav. Evol.* **43:**293–305.

Smith, R. J. 1980. Rethinking allometry. *J. Theor. Biol.* **87:**97–111.
Smith, R. J. 1984. Determination of relative size: the "criterion of subtraction" problem in allometry. *J. Theor. Biol.* **108:**131–142.
Smith, R. J. 1994. Degrees of freedom in interspecific allometry: an adjustment for the effects of phylogenetic constraint. *Am. J. Phys. Anthropol.* **93:**95–107.
Sober, E. 1984. *The Nature of Selection.* MIT Press, Cambridge, MA.
Stearns, S. C. 1992. *The Evolution of Life Histories.* Oxford University Press, Oxford.
Stern J. T. Jr. 1970. The meaning of "adaptation" and its relation to relation to the phenomenon of natural selection. *Evol. Biol.* **4:**39–66.
Stern, J. T. Jr., and Oxnard, C. E. 1973. Primate locomotion: some links with evolution and morphology. *Primatologia* **4:**1–93.
Strait, S. G. 1997. Tooth use and the physical properties of food. *Evol. Anthropol.* **5**(6):199–211.
Susman, R. L. 1979. The comparative and functional morphology of hominoid fingers. *Am. J. Phys. Anthropol.* **50:**215–236.
Susman, R. L., Stern, J. T. Jr. and Jungers, W. L. 1984. Arboreality and bipedality in the Hadar hominids. *Folia Primatol.* **43:**113–156.
Sweet, S. S. 1980. Allometric inferences in morphology. *Am. Zool.* **20:**643–652.
Teaford, M. 1994. Dental microwear and dental function. *Evol. Anthropol* **3:**17–30.
Tuttle, R. H. 1970. Postural, propulsive, and prehensile capabilities in the cheridia of chimpanzees and other great apes. In: *The Chimpanzee*, Vol. 2, pp. 167–253. Karger, Basel, New York.
Van Valen, L. 1982. Homology and causes. *J. Morphol.* **173:**305–312.
Wainright, P. C., and Reilly, S. M. 1994. *Ecological Morphology: Integrative Organismal Biology.* Univesity of Chicago Press, Chicago.
Weibel, E. R., Tayler, C. R., and Bolis, L. 1998. *Principles of Animal Design. The Optimization, Symmorphosis Debate.* Cambridge University Press, Cambridge.
Weishampel, D. B. 1995. Fossils, function and phylogeny. In: J. J. Thomason (ed.), *Functional Morphology in Vertebrate Paleontology.* pp. 34–54. Cambridge University Press, Cambridge.
Wenzel, J. W., and Carpenter, J. M. 1994. Comparing methods: adaptive traits and tests of adaptation. In:. P. Eggleton and R. Vane-Wright (eds.) *Phylogenetics and Ecology*, pp. 79–101. Academic Press, New York.
Williams, G. C. 1966. *Adaptation and Natural Selection.* Princeton University Press, Princeton.
Williams, G. C. 1992. *Natural Selection: Domains, Levels, and Challenges.* Oxford University Press, New York.
Williams, G. C. 1997. *The Pony Fish's Glow: and Other Clues to Plan and Purpose in Nature.* Basic Books, New York.
Witmer, L. M. 1995. The extant phylogenetic bracket and the importance of recontructing soft tissues in fossils. In: J. J. Thomason (ed.), *Functional Morphology in Vertebrate Paleontology*, pp 19–33. Cambridge University Press, Cambridge.
Witmer, L. M. 1997. The evolution of the antorbital cavity of archosaurs: A study in soft-tissue reconstruction in the fossil record with an analysis of the function of pneumaticity. *J. Vert. Paleont.* **17,** Supplement to number **1:**1–73.
Woollard, H. H. 1925. The anatomy of *Tarsius spectrum. Proc. Zool. Soc. London* **1925:**1071–1184.

Functional Morphology and *In Vivo* Bone Strain Patterns in the Craniofacial Region of Primates: Beware of Biomechanical Stories about Fossil Bones

2

WILLIAM L. HYLANDER and KIRK R. JOHNSON

Introduction

It is frequently assumed, either implicitly or explicitly, that the concentration, distribution, and geometry of bone mass within the craniofacial region is largely or exclusively determined by or associated with routine and habitual forces associated with mastication, incision, or isometric biting (cf. Seipel, 1948; Weinman and Sicher, 1955; Scott, 1967; Wolpoff, 1980, 1996; Russell, 1985; Rak, 1986; Preuschoft *et al.*, 1986; Demes, 1987; and many others). For example, much of the work dealing with the functional significance of craniofacial form in Neandertals is based on this commonly made but frequently

WILLIAM L. HYLANDER and KIRK R. JOHNSON • Department of Biological Anthropology and Anatomy, Duke University Medical Center, Durham, North Carolina 27710.

Reconstructing Behavior in the Primate Fossil Record, edited by Plavcan *et al.* Kluwer Academic/Plenum Publishers, New York, 2002.

erroneous assumption (e.g., Wolpoff, 1980, 1996; Rak, 1986; Demes, 1987). The intent of this paper is to demonstrate that reconstructing the masticatory behavior* and biomechanics of primates from the fossilized remains of the craniofacial skeleton is extremely problematic, particularly when done in the complete absence of a detailed understanding of the biomechanical environment of the craniofacial region of living primates. We intend to demonstrate that the routine acceptance of the above assumption is unjustified, and for that reason a considerable amount of what has been said about the functional significance of various fossil skulls is likely incorrect.

Most paleontologists would agree that the ability to reconstruct primate masticatory behavior and biomechanics from the fossilized remains of the craniofacial skeleton is dependent on an excellent understanding of bone biology and the functional anatomy of primates. For many years, the comparative anatomical approach has figured importantly in our understanding of the functional anatomy of the craniofacial region. Within the last 25 years, however, many investigators have taken other approaches. Analyses of jaw movement and the electromyography of the jaw muscles during mastication have yielded particularly useful information about craniofacial biomechanics (e.g., Kay and Hiiemae, 1974; Luschei and Goodwin, 1974; Hylander and Johnson, 1985, 1994; Hylander *et al.*, 2000). Furthermore, an understanding of how masticatory loads are countered in the mammalian face is an area in which there has been both intense interest and steadily increasing progress (e.g., Hylander, 1977a, 1979a; Hylander *et al.*, 1987a; Hylander *et al.*, 1991a; Hylander and Johnson, 1997a; Hylander *et al.*, 1998; Herring and Mucci, 1991; Oyen and Tsay, 1991; Ross and Hylander, 1996). As many workers over the years have suggested that the functional significance of various bony features of the face are related to masticatory stress dissipation, this intense interest in stress analysis is not surprising (e.g., Sicher, 1950; Tappen, 1953; Biegert, 1963; Simons, 1972; Hylander, 1972, 1977b; Cartmill, 1974, 1980; Wolpoff, 1980; Greaves, 1985, 1988; Russell, 1985; Preuschoft *et al.*, 1986; Russell and Thomason, 1993; and many others).

There are two approaches to stress analysis: theoretical and experimental. Both have been used for problems related to craniofacial biomechanics with varying degrees of success. As one might expect, the theoretical approach was initially employed most frequently, and it is responsible for many of our working hypotheses (cf. Hylander, 1977b; Greaves, 1985; Russell and Thomason, 1993; Daegling, 1993). However, in more recent years, in order to both test and refine these hypotheses, there has been an increasing emphasis on experimental stress analysis.

There are two general approaches to experimental stress analysis: full-field methods and limited-field methods. Full-field methods provide data for

* Masticatory behavior is used here in its broadest sense in that it refers to all behaviors associated with the masticatory apparatus during mastication, incision, and isometric biting.

a relatively large area of a structure. In general, full-field methods are particularly useful for exploratory studies, i.e., studies designed to locate highly stressed or strained areas. Techniques such as photoelasticity and brittle coatings are examples of full-field methods. Although these sorts of techniques may be useful for attacking many biological problems, they cannot be used *in vivo* on fully alert subjects. Therefore, full-field methods require that an isolated bone or model be subjected to simulations that mimic *in vivo* loads; this is a major disadvantage as the nature of these loads (magnitude, direction, timing, and point of application) is generally poorly understood and frequently controversial.

Finite-element analysis deserves special mention here as it is a powerful technique that is gaining in popularity among biologists, and in the near future will likely play a major role in biomechanical analyses of the craniofacial region (cf. Korioth *et al.*, 1992; Korioth and Hannam, 1994; Chen and Chen, 1998). Currently, however, even the best of the finite-element analyses have been restricted to analyzing static isometric biting, rather than analyzing continuous dynamic loading patterns associated with mastication and incision. This is partly because finite-element analysis requires an accurate simulation of muscle, bite, and condylar reaction forces, and the nature and timing of these loads for most species are essentially unknown. Moreover, in finite-element analysis there is the additional problem of accurately modeling complex biological shapes as well as the difficulties associated with modeling tissues whose mechanical properties are poorly understood, e.g., periodontal ligament, trabecular bone, and the articular disk of the temporomandibular joint.

In contrast to full-field methods, limited-field methods, as the name implies, provide data for a very limited area of a structure. The use of electrical-resistance foil strain gauges is an example of a limited-field method as it provides information only about surface strains along the area immediately below the attached strain gauge. Although this technique is unable to provide a complete characterization of strain along an entire structure, and therefore is not as useful as full-field techniques for locating highly stressed and strained areas, it is very useful for testing hypotheses regarding loading regimes (Hylander, 1979a). Furthermore, there are two additional major advantages of the resistance strain gauge technique. First, it can be used to record rapid, continuously changing strains due to continuously altered states of stress, and second, this technique can be used *in vivo* in fully alert subjects (Lanyon and Smith, 1970). Thus, this technique allows one to do a biomechanical analysis during dynamic as well as static conditions, e.g., throughout an entire chewing sequence or simply during an isometric biting episode (Hylander *et al.*, 1987a; Hylander *et al.*, 1991a). The fact that it can be used *in vivo* means that it is generally unnecessary to make prior assumptions about the direction, magnitude, point of application, and timing of muscle and reaction forces, and it does not require that stress-bearing structures such as the periodontium be modeled properly.

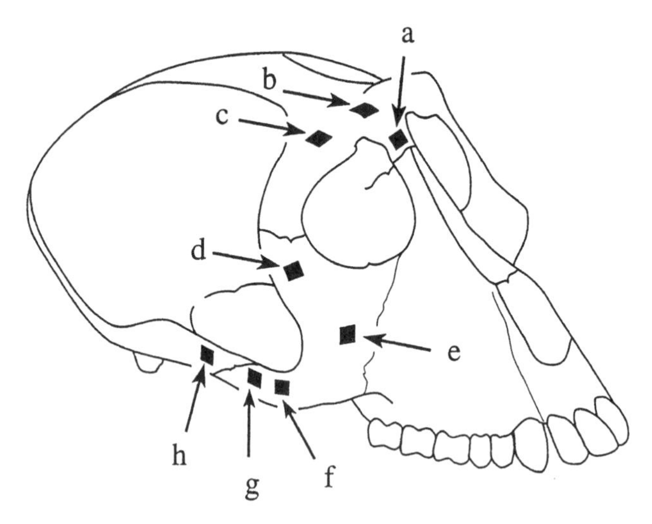

Fig. 1. The labeled arrows point to the location of the following strain-gauge sites: (a) the rostral interorbital region; (b) the dorsal interorbital region; (c) the dorsal orbital region; (d) the postorbital bar region; (e) the anterior root of the zygoma; (f, g, and h) the anterior, middle, and posterior portions of the zygomatic arch, respectively (from Hylander and Johnson, 1997b).

In recent years the resistance strain gauge technique has been used to test several hypotheses as to how the browridge region of nonhuman primates is loaded by masticatory forces (Hylander *et al.*, 1987b, 1991a; Hylander and Johnson, 1997a). The primary intent of that work was to test the hypothesis that enlarged browridges in primates are a structural adaptation to counter powerful masticatory loads. In order to do this, *in vivo* bone strain patterns were recorded from 3-element rosette strain gauges bonded to the browridge (supraorbital region), postorbital bar, anterior root of the zygoma (infraorbital region), and the zygomatic arch of macaques and baboons (Fig. 1).

One of the most interesting things to emerge from this body of work is the finding that during incision and mastication there are steep strain gradients present throughout the face of macaques and baboons. That is, during chewing and biting, some areas of the face experience relatively large strains while others experience relatively small (and apparently insignificant) strains. This finding has important implications relative to the well-accepted hypothesis that facial bones are optimized for countering masticatory loads, and that functional strains have a significant influence on the growth, development, and maintenance of the facial skeleton. The purpose of this paper is to discuss these implications. Prior to doing so, however, we will first provide a brief review of the nature of strain patterns throughout different regions of the primate face. Following this review, we will discuss the significance of these data for understanding the functional anatomy of craniofacial form, which in turn has a direct and obvious influence on the reconstruction of masticatory behavior and biomechanics of extinct primates when based on the fossilized remains of the craniofacial skeleton.

In Vivo Bone Strain Patterns

The experimental sets from which the comparative strain data are drawn consist of two types: *multipple-site strains* and *single-site strains.* The first type, multiple-site strains, refers to those experiments where two or three 3-element rosette strain gauges were used to measure *simultaneous* strains along cortical bone surfaces. These experimental sets allow direct comparisons of strain levels between two or more facial regions during mastication and incision, regardless of what and how powerfully the subject masticates or incises. In contrast, the second type, single-site strains, refers to data sets derived primarily from experiments which employed only a single 3-element rosette strain gauge bonded to cortical bone.

The use of the single-site strains for comparative purposes have a major disadvantage because the actual strain comparisons are based on data recorded during different chewing and biting episodes. Thus, there are obvious problems with interpreting a comparison of, for example, dorsal-interorbital strains recorded during mastication of apple-skin, with zygomatic-arch strains recorded during the mastication of a much harder (or softer) food object. It is for this reason that we limit the comparisons of single-site strains to data recorded during apple-skin mastication. Limiting the comparisons in this way enables us to control in part how forcefully the subject chewed since the food chewed has nearly identical mechanical properties. Nevertheless, as the strain comparisons are based on different episodes of chewing and biting, these data are presumably not as useful as the multiple-site strain data.

For purposes of comparison, we have no compelling reason (at least at this time) to emphasize the exclusive use of either the maximum principal strain (principal tension) or the minimum principal strain (principal compression) in our analysis. Therefore, in the following discussion of regional strain differences in the face, we use maximum shear strain as it is derived equally from both principal tension and principal compression strain values. Furthermore, since bone is relatively weak when exposed to shearing loads (Nordin and Frankel, 1989), perhaps for our purposes shear strain magnitudes are in fact the most relevant comparative measure of bone strain.

Strain (ε), a dimensionless unit, equals the change in length of an object (ΔL) divided by the original length of the object (L), i.e., $\varepsilon = \Delta L/L$. The unit of strain is microstrain ($\mu\varepsilon$), which equals 1×10^{-6} inches/inch or mm/mm, etc.

By convention, tensile strain is a positive value and compressive strain is a negative value. The maximum principal strain (ε_1) is the largest tensile value (principal tension). The minimum principal strain (ε_2) is the largest compressive strain value (principal compression). Maximum shear strain (γ_{max}) equals ε_1 minus ε_2.

Regional Comparisons of In Vivo Bone Strain

Overall, our data indicate that relative to the mandible, zygomatic arch, and anterior root of the zygoma, the browridge region experiences relatively small peak functional strains during mastication and incision. Moreover, within the browridge itself, there are varying amounts of strain. For example, a comparison of multiple-site strains demonstrates that there is significantly more strain along the dorsal interorbital region than along the dorsal orbital region during both mastication and incision (Hylander *et al.*, 1991a) (Fig. 2). On average, there is about 2.5 times more γ_{max} along the dorsal interorbital region during mastication, and about 4.5 times more γ_{max} during incision. In contrast to these two regions, strains recorded from the rostral interorbital region are even smaller. Although not based on simultaneous strains, the data indicate that compared to the rostral interorbital region, the dorsal interorbital region experiences about 3 times more γ_{max} during mastication and about 7 times more γ_{max} during incision (Hylander *et al.*, 1991a).

Although the dorsal interorbital region is clearly strained much more than both the dorsal orbital and rostral interorbital regions, compared to the zygomatic arch and anterior root of the zygoma, the dorsal interorbital region is strained much less (Hylander *et al.*, 1991a). The multiple-site strain data indicate that on average γ_{max} along the middle portion of the working-side zygomatic arch exceeds strain values from the dorsal interorbital region by about 4 times during mastication, and by about 3 times during incision. The multiple-site strain data also indicate that γ_{max} along the working-side anterior root of the zygoma exceeds values along the dorsal interorbital region by nearly 3 times during mastication. Finally, multiple-site strain data also indicate that γ_{max} along the working-side anterior root of the zygoma exceeds strain from the dorsal orbital region on average by nearly 6 times during mastication (Fig. 3) (Hylander *et al.*, 1991a).

A comparison of single-site strains indicates that compared to the middle portion of the zygomatic arch and to the lateral aspect of the mandibular corpus, the dorsal interorbital region is strained relatively little. The data indicate that there is, on average, over 4 times more strain along the zygomatic arch than along the dorsal interorbital region during mastication. This result is very similar to differences detected for *multiple-site* strains between these two regions (grand mean = 4.1). The single-site data also indicate that there is much more strain along the lateral corpus of the mandible than along the dorsal interorbital region, with about 6 times more strain along the mandible

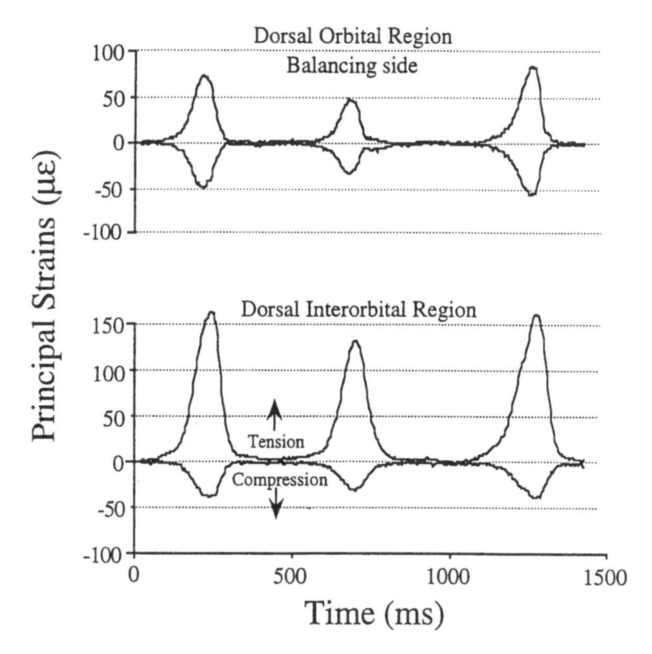

Fig. 2. Plot of the principal strains recorded simultaneously from the dorsal interorbital and dorsal orbital regions of an adult macaque during mastication. The zero (0) level of strain is indicated. The principal strains are in microstrain units ($\mu\varepsilon$), and time is in milliseconds (ms). The maximum principal strains are positive (tensile) and the minimum principal strains are negative (compressive). Therefore, principal tension is plotted above and principal compression is plotted below the zero level of strain. The maximum shear strains can be determined by simply adding together the absolute values of the two principal strains ($\gamma_{max} = \varepsilon_1 - \varepsilon_2$). The dorsal orbital strains are from the left side and the subject is chewing apple with skin on the right side. Overall, strains from the interorbital region are much larger than those from the orbital region (from Hylander *et al.*, 1991a).

during mastication, and about 7 times more strain during incision (Hylander *et al.*, 1991a).

A preliminary analysis of bone strain in the postorbital bar indicates that strain levels within this region may be intermediate between those from the dorsal interorbital region and zygomatic arch. The nature of these strain differences can be seen in Fig. 4. More recently it has been demonstrated there is also a steep anteroposterior strain gradient along the zygomatic arch. As indicated in Fig. 5, multiple-site strains along the working-side zygomatic arch during mastication are much larger along the anterior portion of the arch than along its posterior portion (Hylander and Johnson, 1997a). On average there is about 3 times more strain anteriorly than posteriorly, and strains along the middle are more or less intermediate in magnitude.

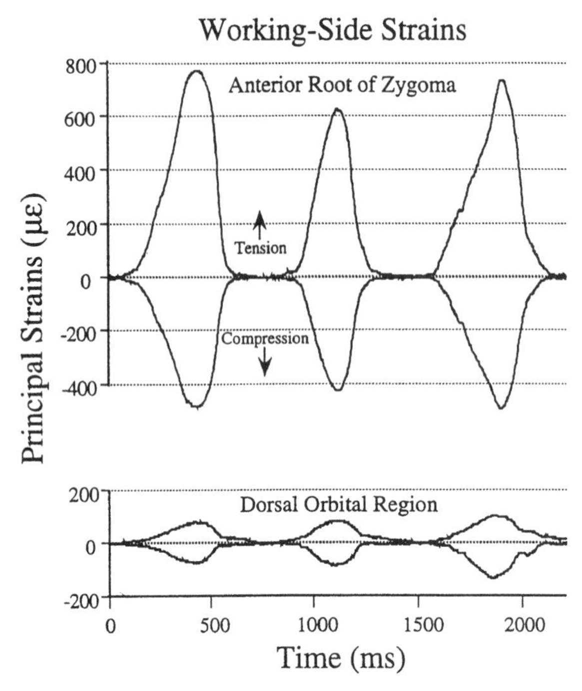

Fig. 3. Plot of the principal strains recorded simultaneously from the dorsal orbital region and anterior root of the zygoma of an adult macaque during mastication. The strains are from the left side and the subject is chewing a hard prune on the left side. Note that the zygoma strains are much larger than the dorsal orbital strains. See Figure Legend 2 for additional information (from Hylander and Johnson, 1997b).

Peak functional strains also vary somewhat throughout the macaque mandible during mastication, and the most lingual aspect of the mandibular symphysis is thought to be one of the most highly strained areas in the macaque face (cf. Hylander, 1979b, c; 1984). In contrast, recent preliminary work also indicates the occurrence of rather uniform simultaneous strains along the alveolar process and basal portion of the macaque mandibular corpus in the region of the first mandibular molar (Fig. 6). Figure 7 indicates these near-uniform levels of strain along the *working-side* mandibular corpus during rhythmic biting of a small wooden rod. Interestingly, these data are quite unlike the steep strain gradients predicted from finite-element analyses of the mandible (Knoell, 1977). Figure 8 indicates less uniform levels of functional strain for these areas along the *balancing-side* corpus during rhythmic biting. Figure 8 also clearly demonstrates that the alveolar process is loaded even along the nonbiting side of the mandible (cf. Hylander, 1979b). Finally, Fig. 9 indicates simultaneous levels of strain along the lateral aspect of the mandibular corpus and the labial aspect of the mandibular symphysis. When comparing these two regions, on average, there is much more strain

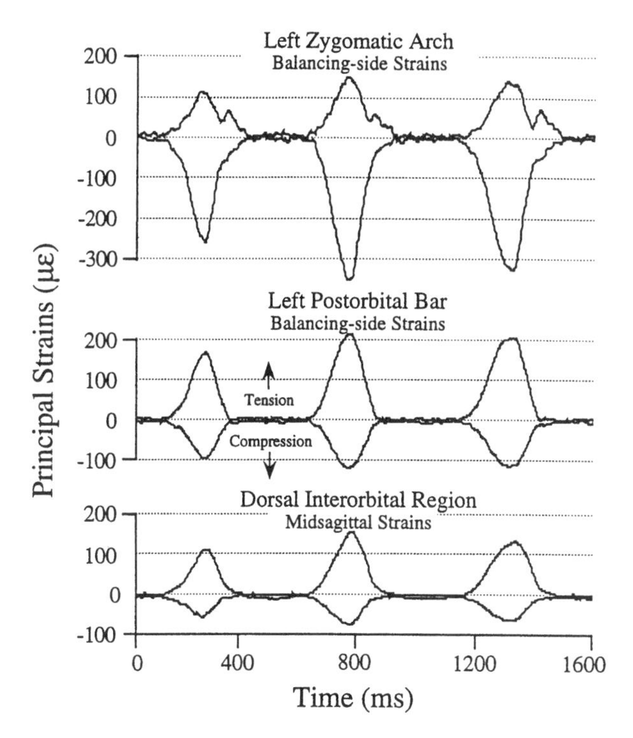

Fig. 4. Plot of the principal strains recorded simultaneously from the zygomatic arch, postorbital, and dorsal interorbital regions of an adult macaque during mastication. The postorbital bar and zygomatic-arch strains are from the left side and the subject is chewing dried apricot on the right side. Note that the postorbital bar strains are larger than the dorsal-interorbital strains but smaller than the zygomatic-arch strains. During chewing on the opposite side (the left side), the zygomatic-arch strains tend to be larger while the overall magnitude of strains from the dorsal interorbital and postorbital regions tend to be more or less unchanged (from Hylander and Johnson, 1997b). See Figure Legend 2 for additional information.

along the lateral mandibular corpus during mastication. In contrast, there is substantial evidence to suggest that there is considerably more strain along the lingual aspect than along the labial aspect of the symphysis (Hylander, 1984, 1985).

In summary, the comparison of *in vivo* bone strain patterns indicates the presence of widely varying amounts of strain within the facial bones of primates during mastication and incision. The high-strain areas are located along the more anterior and lateral aspect of the zygomatic arch, the posterior and lateral aspect of the mandibular corpus, the lingual aspect of the mandibular symphysis, and the facial aspect of the anterior root of the zygoma. In contrast, the dorsal orbital, dorsal interorbital, and rostral interorbital surfaces in the browridge region and the posterior and lateral part of the zygomatic

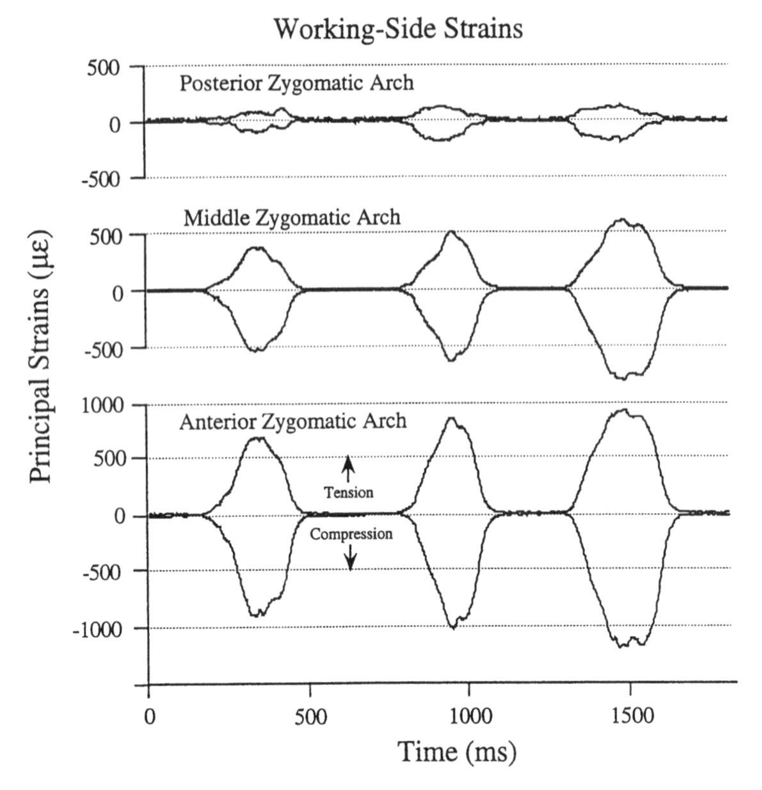

Fig. 5. Plot of the principal strains recorded simultaneously from the anterior, middle, and posterior portions of the zygomatic arch of an adult macaque during mastication. The zygomatic-arch strains are from the right side and the subject is chewing a prune seed on the same side. Note that the strains along the anterior part of the zygomatic arch are much larger than those from the posterior part of the arch (Hylander and Johnson, 1997a). In this subject the strains from the middle portion of the zygomatic arch tend to be more or less intermediate (from Hylander and Johnson, 1997b). See Figure Legend 2 for additional information.

arch are strained much less. Strains along the facial aspect of the postorbital bar may be more or less intermediate between these relatively high- and low-strain areas, although insufficient data are available for this region. There is also considerable variation in bone strain magnitudes throughout the low-strained browridge region. Of the three areas analyzed, the largest browridge strains are located in the dorsal interorbital region while the smallest strains are located in the rostral interorbital region. Finally, a comparison of the highest and lowest strained areas indicate that peak functional strains along the mandibular corpus and anterior portion of the zygomatic arch are about 15–20 times larger than peak strains along the rostral interorbital region during mastication and incision.

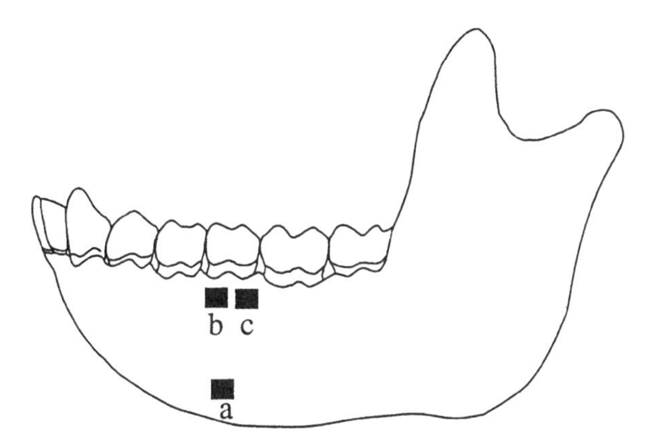

Fig. 6. The labeled squares indicate the location of the (a) basal corpus, (b) the mesial alveolar, and (c) distal alveolar, strain-gauge sites (from Hylander and Johnson, 1997b).

Discussion

Strain Gradients and the Optimal Design of Facial Bones

It was originally expected that *in vivo* bone strain magnitudes would be more or less similar throughout different regions of the facial skeleton (Hylander *et al.*, 1987b, 1991a). This expectation was based largely on the well-known fact that loading conditions play an important role in the maintenance of bone mass (Liskova and Hert, 1971; Morey and Baylink, 1978; Bouvier and Hylander, 1981; Lanyon and Rubin, 1985; Kimmel, 1993; and many others). This ability of bone to respond to cyclical mechanical loading, which is often called "functional adaptation" (e.g., Roux, 1881; Murray, 1936; Goodship *et al.*, 1979; Lanyon and Rubin, 1985; Martin and Burr, 1989), is thought to have the effect of maintaining bone strain magnitudes within a particular range of values (Fig. 10), e.g., the "optimal strain environment" of Rubin (1984) and Lanyon and Rubin (1985) (also cf. Bassett, 1968; Pauwels, 1980; Lanyon *et al.*, 1982; Lanyon, 1991). Thus, this concept of how bone responds to routine cyclical loads leads one to believe that there must be a more or less uniform distribution of functional strain throughout the bony face (Fig. 10).

In addition, the concept of "functional adaptation" and "the optimal strain environment" also leads one to believe that bones are optimized for countering routine cyclical loads, i.e., that they presumably exhibit minimum material and maximum strength for load-bearing purposes. Not surprisingly, many workers have explicitly hypothesized that the mammalian appendicular skeleton is indeed optimized for load-bearing purposes (e.g., Roux, 1881; Koch, 1917; Weinman and Sicher, 1955; Kummer, 1972; Lanyon, 1973; Pauwels, 1980;

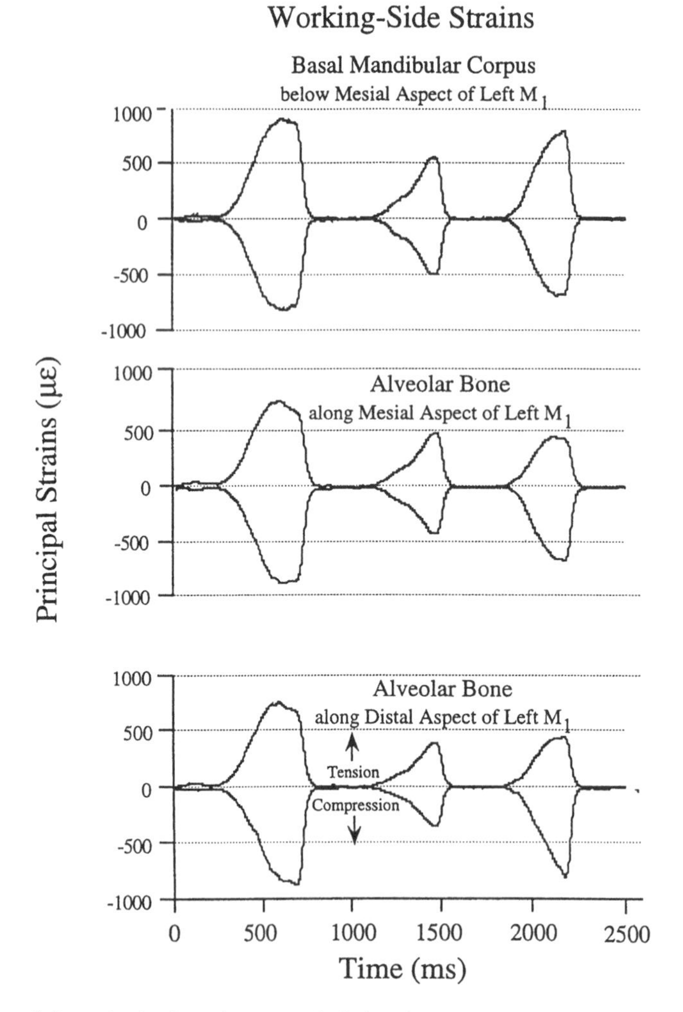

Fig. 7. Plot of the principal strains recorded simultaneously from the basal corpus, and mesial and distal alveolar strain-gauge sites of an adult macaque during rhythmic biting on a wooden rod. The mandibular strains are from the left side and the subject is biting on the same side. Therefore, these are working-side strains. Note that the peak strains from all three sites are near-uniform to one another (from Hylander and Johnson, 1997b). See Figure Legend 2 for additional information.

Rubin, 1984; Frost, 1986; Kimmel, 1993; Rubin *et al.*, 1994). Similarly, it is commonly thought (either implicitly or explicitly), as expressed by DuBrul (1988), that the facial skeleton consists of "an optimal force-resisting framework for masticatory stress" (page 55) (also cf. Seipel, 1948; Weinman and Sicher, 1955; Scott, 1967; Hylander, 1979a, 1985; Greaves, 1985; Russell,

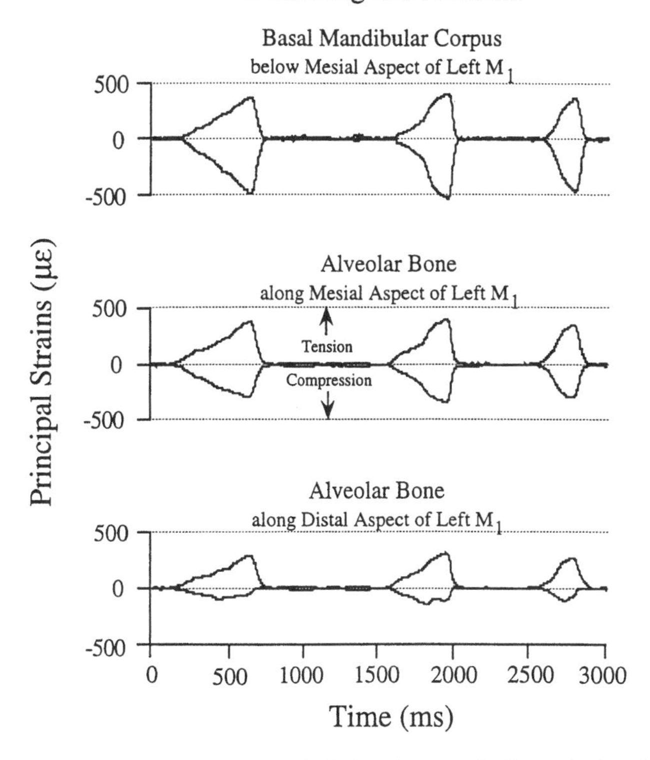

Fig. 8. Plot of the principal strains recorded simultaneously from the basal corpus, and mesial and distal alveolar strain-gauge sites of an adult macaque during rhythmic biting on a wooden rod. The mandibular strains are from the left side and the subject is biting on the right side. Therefore, these are balancing-side strains. Note that the peak strains from alveolar sites are somewhat less than those from the basal corpus. This is particularly evident along the distal alveolar site (from Hylander and Johnson, 1997b). See Figure Legend 2 for additional information.

1985; Preuschoft *et al.*, 1986; Demes, 1987; Roberts *et al.*, 1992; Biknevicius and Ruff, 1992; and many others).

At this point it may be useful to consider the definition of the "optimal strain environment." Although the optimal strain environment was not precisely defined by Rubin and Lanyon, it is clear that these authors consider peak functional principal strains to be an important component of this environment (Rubin, 1984; Rubin and Lanyon, 1984; Lanyon and Rubin, 1985). For example, while discussing their "dynamic strain similarity model," Lanyon and Rubin have repeatedly pointed out that peak functional strains within long bones of vertebrates tend to lie within the 2000 to 3000 $\mu\varepsilon$ level. This means that for vertebrate long bones natural selection

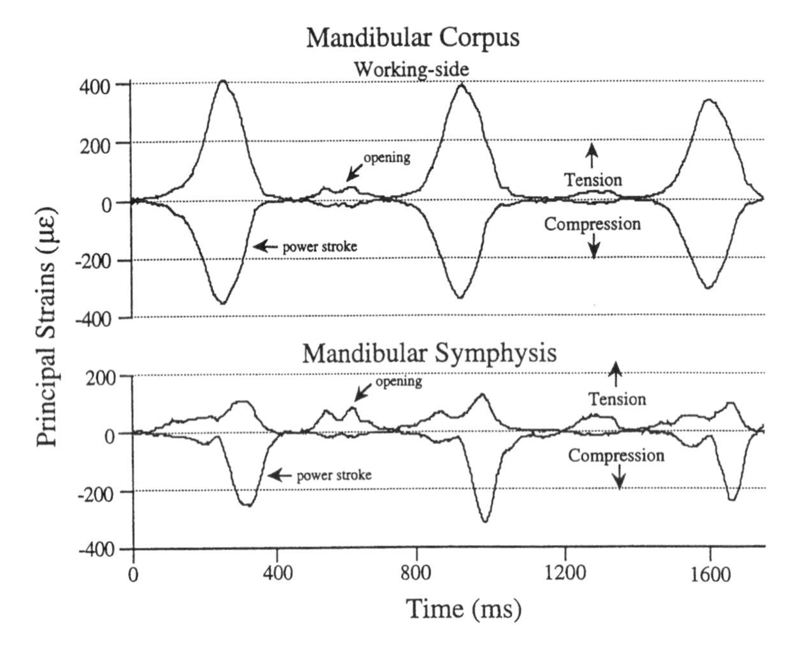

Fig. 9. Plot of the principal strains recorded simultaneously from the lateral aspect of the working-side mandibular corpus and labial aspect of the mandibular symphysis during mastication. The subject is chewing a piece of monkey chow. Unlike bone strain in the cranium, mandibular bone strain is markedly biphasic during mastication. Note that peak power stroke strains along the corpus are larger than those along the symphysis while peak opening strains along the symphysis tend to be larger than those along the corpus. Overall, peak functional strains tend to be larger along the mandibular corpus (from Hylander and Johnson, 1994).

"has produced a margin of safety between functional and yield strains of between two and three times" (Rubin and Lanyon, 1984, page 323). On the other hand, it is also clear that these authors (and many others as well) do not believe that peak functional strains are necessarily the only or even the most important of the various physical stimuli that have osteoregulatory capabilities (cf. Lanyon and Rubin, 1985; Martin and Burr, 1989; Rubin *et al.*, 1994). As pointed out by Rubin *et al.*, (1994), strain magnitude, the fabric tensor, strain frequency, strain rate, strain gradients, electrokinetics, piezoelectricity, strain history, strain energy density, etc. have all been thought to be important osteoregulatory stimuli. Thus, in its narrowest sense, we interpret the concept of the optimal strain environment as simply meaning that peak principal strains within bones will be about 2000–3000$\mu\varepsilon$ during strenuous but routine behaviors, and certainly will not regularly exceed these limits. If they did exceed these limits, there would be a greatly increased risk of bone failure following routine functional behaviors. It

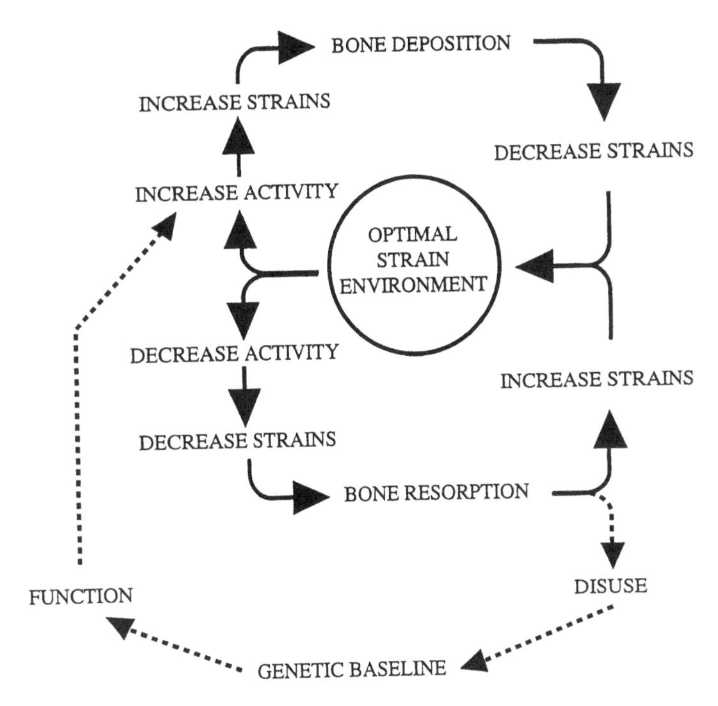

Fig. 10. The effects of an increase or decrease in functional strains on bone mass so as to achieve an optimal strain environment. Increased activity that causes functional strains to rise above the optimal strain environment stimulates bone deposition. This increases skeletal mass, which in turn decreases functional strains so that the altered strains now fall within the limits of the optimal strain environment. The reduction in bone mass following decreased activity also alters functional strains so that they also fall within the limits of the optimal strain environment. This figure is redrawn from Lanyon and Rubin (1985).

is for this reason that our study focuses on peak strain magnitudes, and not because we believe that they are necessarily an osteoregulatory stimulus of overriding importance.

One test of the optimal design hypothesis involves the characterization of *in vivo* bone strain throughout the primate facial skeleton. As optimized load-bearing structures by definition exhibit maximum strength with a minimum amount of material, this hypothesis predicts that during forceful chewing and biting there should be relatively high (cf. Hylander *et al.*, 1991a) and near-uniform amounts of strain throughout the facial skeleton. If levels of strain in certain areas of the facial skeleton during these behaviors are relatively low while in other areas they are relatively high, this indicates that the low-strain areas are not optimized for countering masticatory stress simply because the bone mass in the low-strain areas could be reduced without endangering its structural competence.

The *in vivo* bone strain data recorded in our earlier work demonstrate that peak functional strains differ greatly from one region of the bony face to the next (cf. Hylander *et al.*, 1991a). The fact that the overall magnitude of peak functional strains is highly variable from one region of the skull to the next indicates that it is unlikely that all facial bones are especially designed to minimize bone tissue and maximize strength for countering masticatory loads (Hylander and Johnson, 1992). As we have argued elsewhere (Hylander *et al.*, 1991a), the apparent concentration of bone mass within certain regions of the face need not necessarily bear any relationship to the presence of unusually large routine masticatory loads. Thus, rather than being optimized for countering masticatory stress, the overall morphology and concentration of bone mass in these areas most likely has been selected or influenced mainly by factors unrelated to the dissipation or countering of chewing and biting forces (cf. Hylander *et al.*, 1991a). For example, some portions of the primate skull may be designed so as to counter or dissipate relatively infrequent unintentional traumatic loads (Tappen, 1978; Hylander *et al.*, 1991a). Moreover, perhaps the functional significance of certain portions of the skull are altogether unrelated to countering mechanical forces, and instead are due to spatial factors (Moss and Young, 1960; Ravosa, 1989), i.e., are due to the position of the face relative to the braincase. Alternatively, perhaps some skeletal features are simply secondary sex characters (Lanyon and Rubin, 1985). Whatever the case, these results in turn suggest that the bone distribution in the low-strained areas is not significantly influenced by "functional adaptation."

Bone Strain Levels and Morphology

The process of "functional adaptation", as indicated in Fig. 10, predicts that increases (or decreases) in bone mass adjust levels of increased (or decreased) functional strains so that eventually these strains will equilibrate within a particular uniform range of values. If these strains were to equilibrate within a narrow range of values, there should be a strong positive correlation between forces (axial or shear) or moments (bending or twisting) and the amount (and geometry) of bone mass. The bone strain analysis, however, demonstrates the absence of even a remotely uniform "optimal strain environment," indicating that this optimization process is not operating throughout the entire craniofacial region. Thus, it appears unlikely that there will be a strong positive correlation between routine daily forces or moments and bone mass in the craniofacial region. This in turn indicates the importance of having access to *in vivo* data because it is unlikely that one can reliably estimate biomechanical conditions associated with normal routine masticatory loading patterns from only an examination of gross facial bone morphology. Therefore, without such data it becomes unlikely that one can accurately reconstruct or predict important aspects of the masticatory behavior or biomechanics of primates from their craniofacial skeletons.

Based on known levels of functional strains, what sorts of predictions, if any, can be made regarding the distribution of bone mass within the craniofacial region? As already noted, the posterior portion of the zygomatic arch of macaques experiences much less strain than does the anterior portion. Although the external dimensions of the macaque zygomatic arch do differ somewhat between anterior and posterior portions, what about the distribution of cortical and trabecular bone within these areas? If forces and moments associated with the masseter muscle were more or less evenly distributed along the entire zygomatic arch, one would predict that the low strains from the posterior portion of the arch are due to its being much stiffer than the anterior portion, and if so, this must be mainly because of much larger external dimensions and/or a much denser concentration of bone mass posteriorly. These forces and moments, however, are probably not evenly distributed along the entire zygomatic arch. Instead, they are concentrated along the anterior portion because the zygomatic arch is best modeled as an irregular asymmetric beam with fixed ends and loaded by masseter muscle force that is concentrated along the more anterior portion of the arch (Hylander and Johnson, 1997a). Therefore, perhaps there is no difference in the overall internal architecture of the anterior and posterior portions of the arch, and the differences in functional strains are due simply to the more anteriorly-positioned muscle force and moments associated with the masseter muscle.

The cross-sectional anatomy of the adult macaque zygomatic arch does not support either of these hypothetical expectations. As demonstrated in Fig. 11, while the anterior portion of the arch is near-solid cortical bone, the posterior portion is made up of trabecular bone surrounded by a relatively thin layer of cortical bone. Furthermore, the cortical bone is particularly thin along the lateral aspect of the zygomatic arch, which is the location from where the strains were recorded. Thus, although functional strains are much smaller along the posterior portion of the arch, there is also much less bone mass in this area. If the density of bone mass was comparable to what is found anteriorly, there would be even less strain posteriorly.

Do these results indicate that low-strain areas in the face will always be characterized by trabecular bone surrounded by a thin layer of cortical bone? Following a consideration of strain levels and morphology of the rostral interorbital region, the answer to this question is negative. Although functional strain levels along the rostral interorbital region are lower than in any other area yet examined by us (Hylander *et al.*, 1991a), this area is made up of a very thick dense layer of cortical bone (Fig. 12). Furthermore, contrary to Lanyon (1991), these observations make it quite unlikely that bone mass is strictly determined by peak functional strains. Instead, as noted by many workers, hereditary factors have a major influence on bone mass (e.g., Slemenda *et al.*, 1991; Morrison *et al.*, 1994).

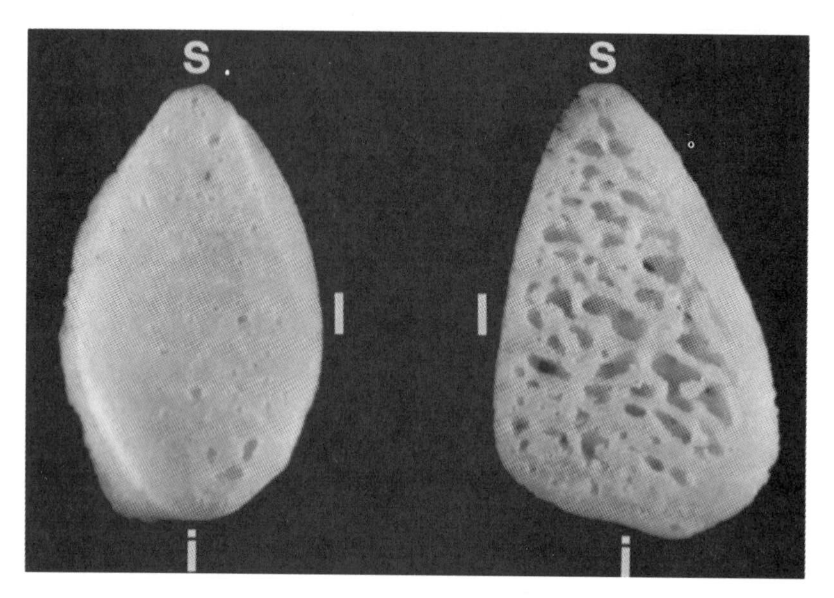

Fig. 11. Photographs of cross sections of the left zygomatic arch of a large adult male *Macaca fascicularis* (WLH private collection); l, s, and i indicate the lateral, superior (dorsal), and inferior (ventral) surfaces of the arch, respectively. The section on the left is from the anterior portion of the zygomatic arch and the section on the right is from the posterior portion. The anterior section is almost all compact bone while the posterior section has a considerable amount of trabecular bone (from Hylander and Johnson, 1997a).

What about the architectural correlates of high-strain areas? Levels of functional strains are relatively high in the mandibular corpus and symphysis, anterior root of the zygoma, and anterior portion of the zygomatic arch, and these areas also exhibit relatively thick areas of densely compacted cortical bone (Fig. 2 in Hylander, 1979a, and Fig. 11). Nevertheless, until functional strains have been recorded in the upper jaw, it may be premature to conclude that high-strain areas in the primate craniofacial region are always associated with a thick layer of dense cortical bone. Thus, although high levels of strain may be associated with densely compacted cortical bone, the presence of low levels of strain clearly do not allow one to predict anything about the internal architecture of the underlying bone. Why is this so?

As previously noted, during chewing and biting certain areas of the face experience relatively high functional strains. Moreover, these same areas are made up primarily of dense cortical bone. If bone mass in these areas were to be uniformly reduced by say 30–40% (as in the reduction in bone mass from the anterior to posterior portion of the zygomatic arch), functional strain levels during powerful chewing and biting would increase and perhaps reach dangerously high levels. Thus, one can convincingly argue that the morphologies

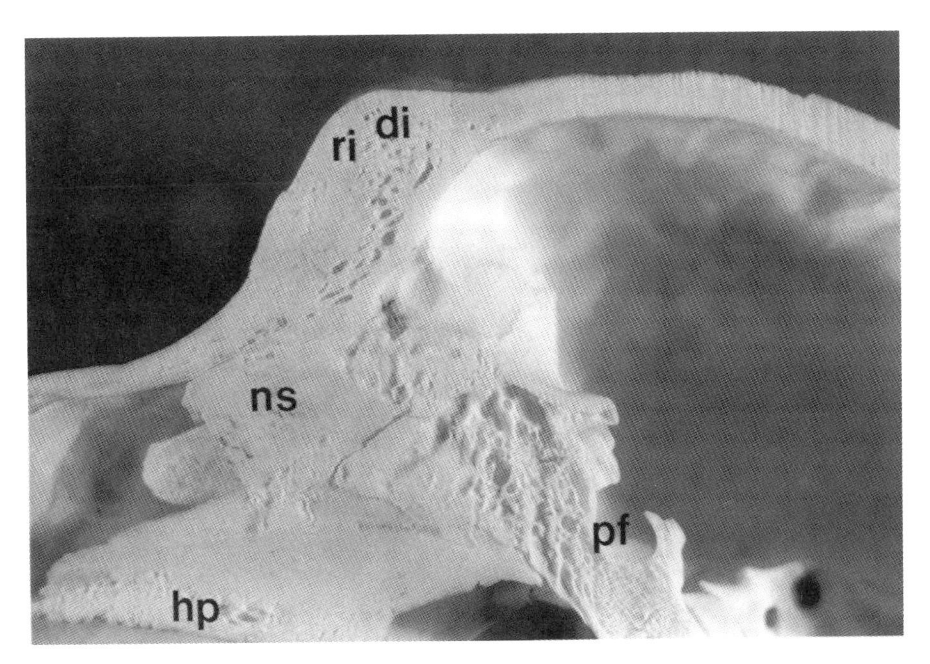

Fig. 12. Photograph of a midsagittal section of the skull of a large adult male *Macaca fascicularis* (WLH private collection). Abbreviations: rostral interorbital region (ri); dorsal interorbital region (di); pituitary fossa (pf); nasal septum (ns), and hard palate (hp). See text for discussion (from Hylander and Johnson, 1997a).

of the relatively high-strain areas are importantly related to countering masticatory loads (Hylander *et al.*, 1991a), and if allowance is made for an appropriate safety factor (cf. Alexander, 1981; Currey, 1984; Rubin, 1984; Lanyon and Rubin, 1985; Biewener, 1993), then these areas may function in part as near-optimized structures. This of course does not mean that the entire bone or region is optimized for load bearing, e.g., significant portions of the macaque mandibular corpus are not exposed to relatively large bending and twisting moments during chewing and biting (Daegling, 1993).

But what about the low-strain areas such as the posterior portion of the zygomatic arch, browridge, and the rostral interorbital region? Why has there not been selection to reduce the bone mass in these regions? Although functional strains during chewing and biting might double or triple in magnitude following such a reduction, these regions would still be strong enough to counter masticatory loads, and they would then more closely approximate the presumably desirable condition of maximum strength with minimum material.

Perhaps one reason why bone mass has not been further reduced is simply because facial bones must *also* be structurally adapted to counter infrequent (rare?) nonmasticatory traumatic loads. For example, although a significant reduction in bone mass of the posterior portion of the zygomatic arch would have little bearing on its continued structural competence for countering routine masticatory loads, this reduction would surely weaken this region to blows delivered to the orbit and zygomatic arch, and of course for most mammals a broken zygomaxillary complex would seriously diminish an individual's overall reproductive fitness because of its decreased ability to nourish itself due to a painful and malfunctioning masticatory apparatus. A similar argument can be made for the browridge and interorbital region, although structural failure in these instances does not result in a malfunctioning masticatory apparatus, but instead may result in damage to the contents of the orbit or neurocranium (cf. Hylander *et al.*, 1991a).

Maximum Levels of Functional Strains

An interesting finding relates to peak levels of functional strain. As much as $2100\mu\varepsilon$ in *tension* was recorded from the anterior root of the zygoma during unilateral molar biting. Comparable levels of *compressive* strain during molar biting have been recorded in the mandible of galagos and macaques (i.e., $-2100\mu\varepsilon$ and $-1500\mu\varepsilon$ respectively) (Hylander, 1979c). As the strain gages along the mandible were not positioned along the presumed maximally stressed areas, these values are minimum estimates of the maximum strains. In addition, tensile strains in the range of 2000 to $2500\mu\varepsilon$ apparently occur along the lingual aspect of the macaque symphysis during vigorous mastication (Hylander, 1984). Furthermore, compressive strain magnitudes as high as $-2100\mu\varepsilon$ have been recorded from the anterior half of the zygomatic arch during powerful incision (Hylander and Johnson, 1997a). The above data are interesting as Rubin (1984) and Lanyon and Rubin (1985) have noted that compressive bone strain magnitudes in the range of -2000 to $-3000\mu\varepsilon$ appear to be near the upper limit of functional strains that occur in vertebrates during normal routine behaviors (i.e., swimming, running, flying, and biting). This strain limit is thought to be importantly related to factors of safety in the design of skeletal tissues.

Safety Factors and Structural Failure of the Facial Skeleton

If safety factors to minimize injuries are built into craniofacial structures, what is the nature of these factors? Unlike various postcranial bones that occasionally fail due to fatigue fractures during vigorous or increased levels of locomotor activity (cf. Nunamaker *et al.*, 1990), with the possible exception of pathological fractures associated with tumors or cysts, we are unaware of any instances where facial bones have experienced structural failure (fatigue failure) during periods of chewing or biting. This suggests that not all bones have the same safety factor for countering routine cyclical loads (cf. Alexander, 1981; Rubin, 1984; Lanyon and Rubin, 1985; Biewener, 1993). This is hardly surprising as the consequences of fractured bones must vary considerably. For example, for highly active mammals such as primates, a fractured rib is ordinarily less of a problem than is a broken femur or mandible. In fact, healed rib fractures are a common occurrence among wild-shot great apes, while healed femoral fractures are less commonly found and healed mandibular fractures are absent or very rare (Lovell, 1990). An important reason for this distribution of healed fractures, of course, is that primates with mandibular and femoral fractures are less apt to survive long enough for the fracture to heal, and therefore are less apt to be included in these study samples. If they are less apt to survive, then it is also likely that natural selection has resulted in the safety factors for the femur and mandible to be significantly larger than those for ribs, and therefore a fractured mandible or femur is also less apt to occur.

Lanyon and Rubin (1985) have suggested that safety factors of some bones are influenced by the process of "functional adaptation." Moreover and not surprisingly, traumatic blows to the face do not invariably result in facial bone fractures, so obviously there are safety factors for facial bones to these sorts of loads. If an organism has yet to experience these loads to the face, however, then obviously the process of "functional adaptation" cannot possibly adjust bone mass so as to provide an appropriate safety factor (for these particular loads). This is simply because "functional adaptation" cannot respond to loads it has not yet experienced (cf. Lanyon and Rubin, 1985). Of course an increase in bone mass in response to an increase in frequency and magnitude of masticatory force will result in a more robust and stronger masticatory apparatus (Bouvier and Hylander, 1981), which in turn will be better able to counter nonmasticatory traumatic loads. This, however, must be an ineffective way to increase the safety factor for such traumatic loads because the bone mass will likely not be optimally distributed if it is responding solely to masticatory loads. More importantly, modifying safety factors (to nonmasticatory loads) in this manner means that the modification must be linked to modifying chewing and biting behaviors. That is, safety factors for accidental loads must be increased by habitually chewing and biting more forcefully. Instead of altering safety factors only by "functional adaptation," it seems more likely that the distribution of bone mass throughout most regions of the cranium must be influenced significantly by natural selection on the genetic "blueprint" (Lanyon and Rubin, 1985).

What about the actual size of safety factors in skeletal tissues? It has been suggested by Biewener, Lanyon, Rubin, and their colleagues (Rubin and Lanyon, 1984; Lanyon and Rubin, 1985; Biewener, 1993; and Rubin *et al.*, 1994) that as maximum functional principal strains in the vertebrate skeleton are in the range of 2000 to 3000$\mu\varepsilon$ and because cortical bone fails (yields) at about 6800$\mu\varepsilon$, safety factors for vertebrate bone are for the most part in the range of 2 to 4, e.g., bone strain at failure (6800$\mu\varepsilon$)/observed peak strain for the macaque anterior root of the zygoma ($2100\mu\varepsilon$) = 3.2. It would appear, however, that since cortical bone experiences fatigue failure at levels well below $6800\mu\varepsilon$ (e.g., at about $3000\mu\varepsilon$ following one million loading cycles) (Carter *et al.*, 1977; Carter and Spengler, 1978; Carter, 1984), and since for some mammals behaviors such as mastication (or running, flying, etc.) involve tens of thousands of loading cycles each day (Stobbs and Cowper, 1972), perhaps these safety factor estimates are often simply too high because bone-strain levels associated with fatigue failure were not used in the calculations? On the other hand, perhaps secondary Haversian remodeling helps prevent structural failure of bones by the selective elimination and replacement of damaged bone tissue (Hylander, 1979c). If so, then presumably some combination of functional strain magnitude, total number of loading cycles, and number of loading cycles per day (and perhaps also strain rate) should be used in combination with data on the fracture mechanics of bone for determining the size of safety factors. Finally and very importantly, there are certain areas of the face where the peak functional principal strains are mechanically trivial, i.e., they do not exceed $200\mu\varepsilon$ and therefore these bony areas have *fatigue* safety factors for routine cyclical loads in excess of 15.

Thus, it would appear that safety factors for vertebrate bones are not restricted to a narrow range of 2 to 4 since the magnitude of functional strains varies considerably and the consequences and types of bone failures must vary appreciably, depending on the species and the involved bone. One could argue that safety factors in the range of 2 to 4 are confined to load-bearing bones (Lanyon and Rubin, 1985), but then this raises the issue of how one defines a load-bearing bone. This is particularly problematic when dealing with the skull. For example, although the browridge region is strained during chewing and biting, and therefore is load-bearing during these behaviors, its bony architecture is not optimally designed to counter these loads as its safety factor (for masticatory loads) is in excess of 20. Furthermore, in those situations involving a single traumatic load, for some bones or areas (such as the rostral interorbital region) it may not matter that it is loaded well beyond the yield point because what is crucial is that the bone tissue remains structurally intact long enough for the yielded area to heal (model and remodel). In these instances, perhaps safety factor estimates should be based on the fact that bone experiences ultimate failure at $16{,}000\mu\varepsilon$, as opposed to monotonic yield failure of $6800\mu\varepsilon$, or fatigue failure (following one million loading cycles) at about $3000\mu\varepsilon$?

In summary, although it is well recognized that vertebrate bone must have design features that include safety factors related to loads experienced during behaviors such as swimming, flying, running, and chewing (cf. Alexander, 1981; Currey, 1984), there appears to be less appreciation for safety factors related to nonroutine infrequent traumatic loads (although cf. Lanyon and Rubin, 1985). This may be partly because the concept of safety factors for skeletal structures has been developed primarily by workers interested in the structural failure of long bones (Alexander, 1981), and the failure of long bones occasionally occurs during loading regimes that may not differ greatly from the loads associated with normal locomotion.

In contrast to long bones, traumatic loads to the skull are probably *always* quite different from routine functional loads associated with masticatory forces, and therefore if appropriate and necessary safety factors for traumatic loads are built into the craniofacial skeleton, these are probably done so by factors other than the process of "functional adaptation," simply because the process of "functional adaptation" is unable to anticipate loads it has yet to experience. Moreover, this process is also unable to optimize the morphology of the bones so as to minimize the occurrence of these types of fractures if it links modifying safety factors for nonmasticatory traumatic loads solely to exposure to masticatory loads.

Presumably the process of "functional adaptation" also does not entirely account for safety factors for many portions of the postcranial skeleton because, for example, long bone fractures among arboreal primates frequently occur during accidental falls from trees, and presumably these fractures result from loading regimes that do not closely approximate loads encountered during routine ordinary locomotor behaviors. Thus, surely safety factors have been selected on the basis of more than simply the loads encountered during routine locomotor behaviors (Alexander, 1981). Thus, it is likely that a particular bone or bony region in fact has multiple safety factors, depending on whether one considers normal routine cyclical loads, or various single or multiple traumatic loads that are quite unlike the routine cyclical loads of daily life.

Finally, if indeed safety factors for both masticatory loads and accidental or traumatic loads are built into the craniofacial complex, then it becomes readily apparent why there are large differences in strain magnitudes throughout the primate face during powerful chewing and biting. That is, although the low-strain areas are in effect over-engineered for countering masticatory loads, perhaps these areas are near-optimized for countering traumatic nonmasticatory loads. In contrast, the high-strain areas are apparently near-optimized for countering masticatory loads, and if so, these areas must range from being near-optimized to being well over-engineered for countering nonmasticatory loads. The latter may be partly related to why traumatic fractures of some facial areas are rarely encountered.

Facial Bone Morphology and Nonmechanical Factors

The functional significance of some features of the facial skeleton may be related primarily to nonmechanical factors. For example, browridge morphology is thought by many workers to be influenced mainly by spatial and/or allometric factors, i.e., the position of the orbits relative to the neurocranium (Moss and Young, 1960; Ravosa, 1989). Nevertheless, the overall architecture of the browridge and surrounding bone must also be structurally adapted so

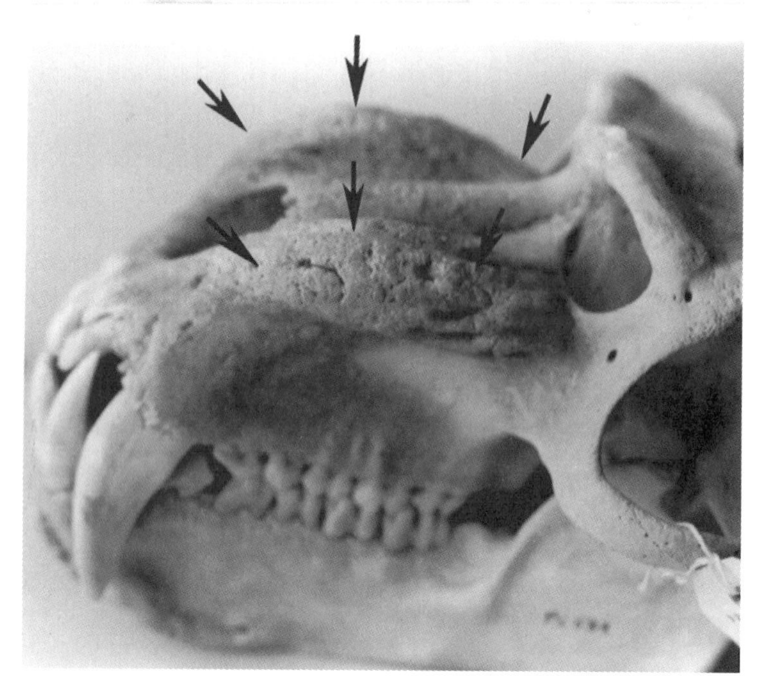

Fig. 13. Photograph of the skull of an adult male mandrill (*Mandrillus sphinx*) (Field Museum of Natural History, 121292 FMNH). As indicated by the black arrows, note the large bulbous concentration of highly vascular bone along the dorsal and lateral aspect of the snout. This concentration of bone is not related to housing the root of the canine, nor is it related to any other biomechanical function (see text for discussion) (from Hylander and Johnson, 1997b).

as to avoid failure due to infrequent nonmasticatory traumatic loads (Hylander *et al.*, 1991a; Hylander and Ravosa, 1992). Therefore, browridge morphology is in fact probably influenced by both nonmechanical and mechanical factors, although the mechanical factors have nothing to do with the mechanics of chewing and biting.

On the other hand, a good example of a strictly nonmechanical function for bone is seen in the face of adult male mandrills (Fig. 13). Note in this figure the large bulbous concentration of highly vascularized bone along the lateral and dorsal aspects of the mandrill snout. This concentration of bone, which is absent in the adult female mandrill, lies immediately below the very colorful skin of the adult male. Its function is arguably to be related to attracting female mates or displacing other males by enhancing the male's overall impressive and dazzling appearance. These bony growths do not owe their existence to mechanical factors, and are not likely to be influenced by "functional adaptation."

Strain Gradients and Bone Remodeling

The presence of the steep strain gradient within the face of macaques prompted a search for certain microanatomical correlates of *in vivo* strain (Hylander *et al.*, 1991b; Bouvier and Hylander, 1994, 1996a, b). Analyzing the relationship between bone strain magnitudes and the distribution of secondary osteons or Haversian systems is particularly promising, because many believe that previous strain patterns have a major influence on this aspect of bone remodeling (e.g., Enlow, 1962, 1963; Hert *et al.*, 1972; Lanyon and Baggott, 1976; Bouvier and Hylander, 1981; Lanyon and Rubin, 1985; Martin and Burr, 1989, and many others). Thus, perhaps the microanatomy of fossil and recent bone can provide clues of primate feeding behaviors. Our data indicate that in young growing macaques the number of newly-formed Haversian systems is positively correlated with shear strain magnitudes (Bouvier and Hylander, 1994, 1996a, b). Thus, these data indicate the potential of using fossil bone microstructure for reconstructing *in vivo* strain gradients, which in turn offers the potential for reconstructing certain aspects of the biomechanical environment associated with previous feeding behaviors (Bouvier and Hylander, 1996a).

Conclusions

Many workers have hypothesized that the mammalian facial skeleton is optimized for countering or dissipating masticatory stress. One test of this optimal design hypothesis involves the characterization of *in vivo* bone strain throughout the facial skeleton of primates. As optimized load-bearing structures by definition exhibit maximum strength with a minimum amount of material, this hypothesis predicts that during chewing and biting there should be relatively high and near-uniform amounts of strain throughout the facial skeleton.

An analysis of *in vivo* bone strain throughout the face of macaques indicates the clear absence of a high and near-uniform strain environment. These data indicate that as levels of functional strains during chewing and biting are highly variable from one region of the face to the next, it appears unlikely that all facial bones are especially designed so as to minimize bone tissue and maximize strength for countering masticatory loads. Thus, the functional significance of the morphology of certain facial bones need not necessarily bear any important or special relationship to routine and habitual cyclical mechanical loads associated with chewing or biting. Furthermore, the presence of these steep strain gradients within the facial skeleton suggests that the amount of bone mass in the low-strain areas may be largely determined by factors unrelated to processes frequently referred to as "functional adaptation." Nevertheless, it is quite clear that functional strains have an important influence on the development and maintenance of certain facial bones, such as the mandible (Bouvier and Hylander, 1981; and many others). On the other hand, a significant but undefined portion of the anatomy of the facial skeleton of primates is due to more than simply the countering and dissipating of masticatory force. Finally, these data indicate that reconstructing aspects of the feeding behavior of extinct primates from fossilized facial bone remains is much more difficult than originally anticipated simply because not all facial bones are optimized to counter routine masticatory loads encountered during feeding.

Acknowledgments

This study was supported by the Department of Biological Anthropology and Anatomy, Duke University. It was also supported by research grants from NIH (MERIT Award DE04531) and NSF (SBR-9420764) to WLH. We would like to thank Dr. Christine Wall for her insightful comments on an earlier draft of this manuscript.

References

Alexander, R. M. 1981. Factors of safety in the structure of animals. *Sci. Prog.* **67:**109–130.

Bassett, C. A. L. 1968. Biological significance of piezoelectricity. *Calcif. Tissue Res.* **1:**252–272.

Biegert, J. 1963. The evaluation of characteristics of the skull, hands, and feet for primate taxonomy. In: SL Washburn (ed.), *Classification and Human Evolution*, pp. 116–145. Aldine Press, Chicago.

Biewener, A. A. 1993. Safety factors in bone strength. *Calcif. Tissue Int.* **53** (Suppl 1):S68–S74.

Biknevicius, A. R., and Ruff, C. B. 1992. The structure of the mandibular corpus and its relationship to feeding behaviours in extant carnivorans. *J. Zool. London.* **228:**479–507.

Bouvier, M., and Hylander, W. L. 1981. Effect of bone strain on cortical bone structure in macaques (Macaca mulatta). *J. Morphol.* **167:**1–12.

Bouvier, M., and Hylander, W. L. 1994. Bone remodeling responses to strain gradients in the zygomatic arch of macaques. *J. Dent. Res.* **73:**195.

Bouvier, M., and Hylander, W. L. 1996a. The function of secondary osteonal bone: mechanical or metabolic? *Arch Oral Biol.* **41:**941–950.

Bouvier, M., and Hylander, W. L. 1996b. Strain gradients, age, and levels of modeling and remodeling in the facial bones of *Macaca fascicularis*. In: Z. Davidovitch and L. A. Norton (eds.), *The Biological Mechanisms of Tooth Movement and Craniofacial Adaptation.* pp. 407–412. Harvard Society for the Advancement of Orthodontics, Boston.

Carter, D. R. 1984. Mechanical loading histories and cortical bone remodeling. *Calcif. Tissue Int.* **36:**S19–S24.

Carter, D. R., and Spengler, D. M. 1978. Mechanical properties and composition of cortical bone. *Clin. Orthop.* **135:**192–217.

Carter, D. R., Spengler, D. M., and Frankel, V. H. 1977. Bone fatigue in uniaxial loading at physiologic strain rates. *IRCS Med. Sci.* **5:**592.

Cartmill, M. 1974. *Daubentonia, Dactylopsila*, woodpeckers and klinorhynchy. In: R. D. Martin, G. A. Doyle, and A. C. Walker (eds.), *Prosimian Biology*. pp. 655–670. Gerald Duckworth and Co. Ltd., London.

Cartmill, M. 1980. Morphology, function, and evolution of the anthropoid postorbital septum. In: R. L. Ciochon and A. B. Chiarelli (eds.), *Evolutionary Biology of New World Monkeys and Continental Drift*. pp. 243–274 Plenum Press, New York.

Chen, X. and Chen, H. 1998. The influence of alveolar structures on the torsional strain field in a gorilla corporeal cross-section. *J. Hum. Evol.* **35:**611–633.

Currey, J. 1984. *The Mechanical Adaptations of Bones.* Princeton University Press, Princeton, NJ.

Daegling, D. J. 1993. The relationship of *in vivo* bone strain to mandibular corpus morphology in *Macaca fascicularis*. *J. Hum. Evol.* **25:**247–269.

Demes, B. 1987. Another look at an old face: biomechanics of the neandertal facial skeleton reconsidered. *J. Hum. Evol.* **16:**297–303.

DuBrul, E. L. 1988. *Sicher and DuBrul's Oral Anatomy.* Ishiyaku EuroAmerica, St Louis.

Enlow, D. H. 1962. Functions of the haversian system. *Am. J. Anat.* **110:**269–282.

Enlow, D. H. 1963. *Principles of Bone Remodeling. An Account of Post-Natal Growth and Remodeling Processes in Long Bones and the Mandible.* Charles C. Thomas, Springfield, Illinois.

Frost, H. M. 1986. *Intermediary Organization of the Skeleton.* CRC Press, Boca Raton.

Goodship, A. E., Lanyon, L. E., and McFie, H. 1979. Functional adaptation of bone to increased stress. An experimental study. *J. Bone Jt. Surg.* **61A:**539–546.

Greaves, W. S. 1985. The mammalian postorbital bar as a torsion-resisting helical strut. *J. Zool. London* **207:**125–136.

Greaves, W.S. 1988. A functional consequence of an ossified mandibular symphysis. *Am. J. Phys. Anthropol.* **77:**53–56.

Herring, S. W. and Mucci, R. J. 1991. *In vivo* strain in cranial sutures: The zygomatic arch. *J. Morphol.* **207:**225–239.

Hert, J., Pribylova, E., and Liskova, M. 1972. Reaction of bone to mechanical stimuli. Part 3. Microstructure of compact bone of rabbit tibia after intermittent loading. *Acta Anat.* **82:**218–230.

Hylander, W. L. 1972. *The Adaptive Significance of Eskimo Craniofacial Morphology.* Ph.D. Thesis, University of Chicago.

Hylander, W. L. 1977a. *In vivo* bone strain in the mandible of *Galago crassicaudatus*. *Am. J. Phys. Anthropol.* **46:**309–326.

Hylander, W. L. 1977b. The adaptive significance of eskimo craniofacial morphology. In: A. A. Dahlberg and T. M. Graber (eds.), *Orofacial Growth and Development.* pp 129–170. Mouton, The Hague.

Hylander, W. L. 1979a. Mandibular function in *Galago crassicaudatus* and *Macaca fascicularis*: An *in vivo* approach to stress analysis of the mandible. *J. Morphol.* **159:**253–296.

Hylander, W. L. 1979b. An experimental analysis of temporomandibular joint reaction force in macaques. *Am. J. Phys. Anthropol.* **51:**433–456.

Hylander, W. L. 1979c. The functional significance of primate mandibular form. *J. Morphol.* **160:**223–240.

Hylander, W. L. 1984. Stress and strain in the mandibular symphysis of primates: A test of competing hypotheses. *Am. J. Phys. Anthropol.* **64:**1–46.

Hylander, W. L. 1985. Mandibular function and biomechanical stress and scaling. *Am. Zool.* **25:**315–330.

Hylander, W. L., and Johnson, K. R. 1985. Temporalis and masseter muscle function during incision in macaques and humans. *Int. J. Primatol.*, **6:**289–322.

Hylander, W. L., and Johnson, K. R. 1992. Strain gradients in the craniofacial region of primates. In: Z. Davidovitch (ed.), *The Biological Mechanisms of Tooth Movement and Craniofacial Adaptation*, pp. 559–569. Ohio State University College of Dentistry, Columbus, Ohio.

Hylander, W.L., and Johnson, K. R. 1994. Jaw muscle function and wishboning of the mandible during mastication in macaques and baboons. *Am. J. Phys. Anthropol.* **94:**523–547.

Hylander, W. L. and Johnson, K. R. 1997a. *In vivo* bone strain patterns in the zygomatic arch of macaques and the significance of these patterns for functional interpretations of cranial form. *Am. J. Phys. Anthropol.*, **102:**203–232.

Hylander, W. L., and Johnson, K. R. 1997b. *In vivo* bone strain patterns in the craniofacial region of primates. In: C. McNeill, A. Hannam, and D. Hatcher (eds.), *Occlusion: Science and Practice.* pp. 165–178. Quintessence Publishing Co., Chicago.

Hylander, W. L., and Ravosa, M. J. 1992. An analysis of the supraorbital region of primates: A morphometric and experimental approach. In: P. Smith and E. Tchernov (eds.), *Structure, Function and Evolution of Teeth.* pp. 223–255. Freund, London.

Hylander, W. L., Johnson, K. R., and Crompton, A. W. 1987a. Loading patterns and jaw movements during mastication in *Macaca fascicularis*: A bone-strain, electromyographic, and cineradiographic analysis. *Am. J. Phys. Anthropol.* **72:**287–314.

Hylander, W. L., Picq, P. G., and Johnson, K. R. 1987b. A preliminary stress analysis of the circumorbital region in *Macaca fascicularis. Am. J. Phys. Anthropol.* **72:**214.

Hylander, W. L., Picq, P. G., and, Johnson, K. R. 1991a. Masticatory-stress hypotheses and the supraorbital region of primates. *Am. J. Phys. Anthropol.* **86:**1–36.

Hylander, W. L., Rubin, C. T., Bain, S. D., and Johnson, K. R. 1991b. Correlation between haversian remodeling and strain magnitude in the baboon face. *J. Dent. Res.* **70:**360.

Hylander, W. L., Ravosa, M. J., Ross, C., and Johnson, K. R. 1998. Mandibular corpus strain in Primates: Further evidence for a functional link between symphyseal fusion and jaw-adductor muscle force. *Am. J. Phys. Anthropol.* **107:**257–271.

Hylander, W. L., Ravosa, M. J., Ross, C. F., Wall, C. E., and K. R. Johnson. 2000. Symphyseal fusion and jaw-adductor muscle force: An EMG study. *Am. J. Phys. Anthropol.* **112:** 469–492.

Kay, R. F., and Hiiemae, K. M. 1974. Jaw movement and tooth use in recent and fossil primates. *Am. J. Phys. Anthropol.* **40:**227–256.

Kimmel, D. B. 1993. A paradigm for skeletal strength homeostasis. *J. Bone Mineral Res.* **8** (Suppl. 2):S515–S522.

Knoell, A. C. 1977. A mathematical model of an *in vitro* human mandible. *J. Biomech.* **10:**159–166.

Koch, J. C. 1917. The laws of bone architecture. *J. Anat.* **21:**177–298.

Korioth, T. W. P., and Hannam, A. G. 1994. Deformation of the human mandible during simulated tooth clenching. *J. Dent. Res.* **73:**56–66.

Korioth, T. W. P., Romilly, D. P., and Hannam, A. G. 1992. Three-dimensional finite element stress analysis of the dentate human mandible. *Am. J. Phys. Anthropol.* **88:**69–96.

Kummer, B. K. F. 1972. Biomechanics of bone: Mechanical properties, functional structure, functional adaptation. In: Y. C. Fung, N. Perrone, and M. Anliker (eds), *Biomechanics: Its Foundations and Objectives.* pp. 237–271. Prentice Hall Inc., Englewood Cliffs.

Lanyon, L. E. 1973. Analysis of surface bone strain in the calcaneus of sheep during normal locomotion. *J. Biomech.* **6:**41–49.

Lanyon, L. E. 1991. Biomechanical properties of bone and response of bone to mechanical stimuli: Functional strain as a controlling influence on bone modeling and remodeling behavior. In: B. K. Hall (ed.), *Bone Matrix and Bone Specific Products*, pp. 79–108. CRC Press, Ann Arbor, Michigan.

Lanyon, L. E., and Baggott, D. G. 1976. Mechanical function as an influence on the structure and form of bone. *J. Bone Jt. Surg.* **58B:**436–443.

Lanyon, L. E., and Rubin, C. T. 1985. Functional adaptation in skeletal structures. In: M. Hildebrand, D. M. Bramble, K. F. Liem, and D. B. Wake (eds.), *Functional Vertebrate Morphology.* pp. 1–25. Harvard University Press, Cambridge.

Lanyon, L. E., and Smith, R. N. 1970. Bone strain in the tibia during normal quadrupedal locomotion. *Acta Orthop. Scand.* **41:**238–248.

Lanyon, L. E., Goodship, A. E., Pye, C. J., and MacFie, J. H. 1982. Mechanically adaptive bone remodelling. *J. Biomech.* **15:**141–154.

Liskova, M., and Hert, J. 1971. Reaction of bone to mechanical stimuli. Part 2. Periosteal and endosteal reaction of tibial diaphysis in rabbit to intermittent loading. *Folia Morphol. (Prague)* **19:**301–317.

Lovell, N. C. 1990. *Patterns of Injury and Illness in Great Apes.* Smithsonian Institution Press, Washington, DC.

Luschei, E. S., and Goodwin, G. M. 1974. Patterns of mandibular movement and jaw muscle activity during mastication in the monkey. *J. Neurophysiol.* **37:**954–966.

Martin, R. B., and Burr, D. B. 1989. *Structure, Function, and Adaptation of Compact Bone.* Raven Press, New York.

Morey, E. R., and Baylink, D. J. 1978. Inhibition of bone formation during space flight. *Science* **201:**1134–1141.

Morrison, N. A., Qi, J. C., Tokita, A., Kelly, P. J., Crofts, L., Nguyen, T. V., Sambrook, P. N., and Eisman, J. A. 1994. Prediction of bone density from vitamin D receptor alleles. *Nature* **367:**284–287.

Moss, M. L., and Young, R. W. 1960. A functional approach to craniology. *Am. J. Phys. Anthropol.* **18:**281–292.

Murray, P. D. F. 1936. Bones. *A Study in the Development and Structure of the Vertebrate Skeleton.* Cambridge University, Cambridge.

Nordin, M., and Frankel, V. H. 1989. Biomechanics of bone. In: M. Nordin and V. H. Frankel (eds.), *Basic Biomechanics of the Musculoskeletal System.* pp. 3–29. Lea & Febiger, Philadelphia.

Nunamaker, D. M., Butterweck, D. M., and Provost, M. T. 1990. Fatigue fractures in thoroughbred racehorses: Relationships with age, peak bone strain, and training. *J. Orthop. Res.* **8:**604–611.

Oyen, O. J., and Tsay, T. P. 1991. A biomechanical analysis of craniofacial form and bite force. *Am. J. Orthod. Dentofac. Orthop.* **99:**298–309.

Pauwels, F. 1980. *Biomechanics of the Locomotor Apparatus.* Springer-Verlag, Berlin, New York.

Preuschoft, H., Demes, B., Meyer, M., and Bärr, H. F. 1986. The biomechanical principles realised in the upper jaw of long-snouted primates. In: J. G. Else and P. C. Lee (eds.), *Primate Evolution.* pp. 249-264. Cambridge University Press, Cambridge.

Rak, Y. 1986. The neanderthal: A new look at an old face. *J. Hum. Evol.* **15:**151–164.

Ravosa, M. J. 1989. Browridge development in anthropoid primates. *Am. J. Phys. Anthropol.* **78:**287–288.

Roberts, W. E., Garetto, L. P., and Katona, T. R. 1992. Principles of orthodontic biomechanics: Metabolic and mechanical control mechanisms. In: D. S. Carlson and S. A. Goldstein (eds.), *Bone Biodynamics in Orthodontic and Orthopedic Treatment,* pp. 189–255. University of Michigan, Ann Arbor, Michigan.

Ross, C., and Hylander, W. L. 1996. *In vivo* and *in vitro* bone strain in owl monkey circumorbital region and the function of the postorbital septum. *Am. J. Phys. Anthropol.* **101:**183–216.

Roux, W. 1881. *Der zuchtende Kampf der Teile, oder die 'Teilauslese' im Organismus. (Theorie der 'funktionellen Anpassung').* Wilhelm Engelmann, Leipzig.

Rubin, C. T. 1984. Skeletal strain and the functional significance of bone architecture. *Calcif. Tissue Int.* **36:**S11–S18.

Rubin, C. T., and Lanyon, L. E. 1984. Dynamic strain similarity in vertebrates: an alternative to allometric limb bone scaling. *J. Theor. Biol.* **107:**321–327.

Rubin, C. T., Gross, T. S., Donahue, H. J., Guilak, F., and McLeod, K. J. 1994. Physical and environmental influences on bone formation. In: C. Brighton, G. Friedlander, and J. Lane (eds.), *Bone Formation and Repair.* pp. 61–78. American Academy of Orthopaedic Surgeons.

Russel, A. P., and Thomason, H. J. 1993. Mechanical analysis of the mammalian head skeleton. In: J. Hanken and B. K. Hall (eds.), *The Skull,* Vol. 3: *Functional and Evolutionary Mechanisms,* pp. 345–383. University of Chicago, Chicago.

Russell, M. D. 1985. The supraorbital torus: "A most remarkable peculiarity." *Curr. Anthropol.* **26:**337–360.

Scott, J. H. 1967. *Dento-facial Development and Growth.* Pergamon, New York.

Seipel, C. M. 1948. Trajectories of the jaws. *Acta Odontol.* Scand. **8:**81–191.

Sicher, H. 1950. *Oral Anatomy.* C.V. Mosby, St. Louis.

Simons, E. L. 1972. *Primate Evolution: An Introduction to Man's Place in Nature.* Macmillan, New York.

Slemeda, C. W., Christian, J. C., Williams, C. J., Norton, J.A., and Johnston, C. C. Jr. 1991. Genetic determinants of bone mass in adult women: A reevaluation of the Twin Model and the potential importance of gene interaction on heritability estimates. *J. Bone Mineral Res.* **6:**561–567.

Stobbs, T. H., and Cowper, L. J. 1972. Automatic measurement of the jaw movements of dairy cows during grazing and rumination. *Trop. Grassl.* **6:**107–112.

Tappen, N. C. A. 1953. A functional analysis of the facial skeleton with split-line technique. *Am. J. Phys. Anthropol.* **11:**503–532.

Tappen, N. C. 1978. The vermiculate surface pattern of brow ridges in Neandertal and modern crania. *Am. J. Phys. Anthropol.* **49:**1–10.

Weinman, J. P., and Sicher, H. 1955. *Bone and Bones.* C. V. Mosby Co., St. Louis.

Wolpoff, M. H. 1980. *Paleoanthropology.* Alfred A. Knopf, New York.

Wolpoff, M. H. 1996. *Human Evolution.* The McGraw-Hill Companies, Inc., College Custom Series, New York.

On the Interface between Ontogeny and Function

3

MATTHEW J. RAVOSA and
CHRISTOPHER J. VINYARD

Introduction

Owing to the unique importance of fossils in depicting the evolutionary history of a clade, it is not surprising that there is continued interest in the behavior and morphology of extinct organisms (e.g., Alexander, 1989; Rayner and Wootton, 1991; Thomason, 1995). From a phylogenetic perspective, fossil taxa can present us with an amalgam of character states, some of which are intermediate between living clades and no longer expressed in such extant sister taxa (e.g., basal anthropoids—Simons, 1989, 1992; Beard *et al.*, 1996; Jaeger *et al.*, 1999). Extinct species may also range beyond the size limits or morphospace of modern relatives, thus potentially affording us a window on the evolution of novel form:function complexes. As part of an emerging body of work on the role of growth and development in evolution, it has become increasingly evident that ontogenetic data can bring another dimension to the study of fossils. There are several benefits of an ontogenetic approach to the morphology and behavior of extinct taxa, most important of which is the

---

MATTHEW J. RAVOSA • Department of Cell and Molecular Biology, Northwestern University Medical School, Chicago, Illinois 60611-3008 and Department of Zoology, Division of Mammals, Field Museum of Natural History, Chicago, Illinois. CHRISTOPHER J. VINYARD • Department of Biological Anthropology and Anatomy, Duke University Medical Center, Durham, North Carolina 27710.

Reconstructing Behavior in the Primate Fossil Record, edited by Plavcan *et al.* Kluwer Academic/Plenum Publishers, New York, 2002.

ability to contrast changes in form:function relationships during *ontogeny* across *phylogeny*.

For instance, evolutionary arguments about the structural and functional transformations from one character state to another, especially as they pertain to the influence of overall body size on the proportions of a skeletal feature, are questions often better elucidated within an ontogenetic framework. This conclusion highlights two putative facts often disregarded by paleontologists and neontologists. First, juvenile form is functional and can be evolutionarily adapted to an environment via natural selection. Second, information on morphological change throughout ontogeny is superior to data based solely on adults as it ultimately provides a more comprehensive understanding of the evolution of a form:function complex. This is, in part, because hypotheses about size and scaling are inherently concerned with the development of phenotypic and genetic covariance patterns and how such intrinsic networks are expressed across taxa—what limits or directs the occupation of morphospace. Moreover, as selection can target early growth stages, ontogenetic comparisons among sister taxa can furnish tests of character homology.

An inherent difficulty with assessing adaptation and behavior in fossil taxa is that one cannot investigate the *biological role* of a feature—what an organism actually does with a given morphology during its lifetime (Bock and von Walhert, 1965). In extinct species, function and behavior must be inferred from living analogs with similar morphologies (Kay and Cartmill, 1977). In the absence of data on the biological role, we advocate an explicitly broad comparative ontogenetic framework for investigating the pattern of character evolution within and among sister taxa, one that better discerns processes influencing functional versus adaptive sources of morphological variation. As this volume focuses on behavioral reconstruction in fossils, we review three interrelated avenues of inquiry in which an ontogenetic perspective can better inform interpretations of the form and function of living and extinct taxa—ontogenetic scaling, biomechanical scaling, and heterochrony (Fig. 1). While all these approaches can be applied to the paleontological record, perhaps the best explanation for why comparative ontogenetic studies of extinct and extant taxa are exceedingly rare is that they require a considerable amount of data.

Ontogeny as a Criterion of Subtraction

Ontogenetic allometry is used as a criterion of subtraction to understand the patterning of morphological variability among a group of close relatives that vary in size, shape, and/or function (Fig. 1). This approach typically entails the comparison of bivariate and/or multivariate growth trajectories among sister taxa. In an ontogenetic analysis, a given morphology is interpreted differently depending on whether the taxa of interest show concordant (ontogenetic scaling) or divergent (biomechanical scaling) relative growth patterns (*sensu* Gould, 1966, 1975a). The comparison of growth trajectories for closely related taxa differing in size offers a criterion of subtraction for assessing the *likelihood* of adaptive explanations for allometric modifications of a structure or character complex across the interspecific sample. If relative growth trajectories are *ontogenetically scaled* (Gould, 1975a), then variation in adult skeletal form among sister taxa likely represents pleiotropic changes in shape linked to selection for body-size differentiation, or perhaps genetic drift (Gould, 1966, 1977; Jungers and Fleagle, 1980; Cheverud, 1982; Lauder, 1982; Shea, 1985, 1995; Ravosa, 1992; Ravosa *et al.*, 1993; Klingenberg, 1998). Among ontogenetically scaled relatives, the novel proportions of certain structures not thought to be under selection likely reflect correlated responses tied to pleiotropy with the feature experiencing selection and linkage disequilibrium. As such novel shapes do not necessarily indicate the presence of other selective forces, they are less likely to be viewed as independent characters (Lande, 1979; Cheverud, 1982, 1984; Atchley, 1987; Shea, 1988; Atchley and Hall, 1991; Price and Langen, 1992).

Though some would have us believe that a functional perspective to the study of ontogeny is frequently overlooked (Godfrey *et al.*, 1998), it has long been widely recognized that allometric variation among ontogenetically scaled taxa is of potential functional importance (Gould, 1966, 1971; Dodson, 1975a,b; Shea, 1986; Ravosa *et al.*, 1993; Ravosa, 1998; Vinyard and Ravosa, 1998). Thus, in following Gould's characterization of ontogenetic and biomechanical scaling, we want to emphasize that this dichotomy does not represent a strict division between biomechanics and ontogeny, rather it is a beneficial heuristic device for partitioning scaling patterns. However, without an *a priori* biomechanical, experimental, or ecological benchmark, it is often difficult to evaluate the functional significance of a relative growth pattern. The phylogenetic underpinnings of a growth trajectory may be more evident (Shea, 1995). That is, such a framework makes an explicit distinction between if a specific shape results from size-related functional differentiation along a common trajectory or a departure from ontogenetic scaling among sister taxa. In the latter example of *biomechanical scaling*, this apomorphic pattern or *adaptation* may reflect selection to uncouple ancestral genetic covariance patterns, which themselves may have resulted from past selection* (Gould, 1975a). This critical distinction between *function* and *adaptation* is fundamental to the use of ontogeny as a criterion of subtraction and was largely misunderstood in a recent discussion of relative growth (Godfrey *et al.*, 1998).

When incomplete ontogenetic series or only adult data are available for a fossil taxon, data on the development of a living close relative can be employed

*In this case, we are referring to a change in morphology linked to selection for a change (uncoupling) of underlying relative growth patterns in a single taxon, an *evolutionary adaptation*. This differs from the physiological and morphological responses of various biological systems during the ontogeny of an organism (perhaps due to changes in behavior), a *functional adaptation* (cf. Lanyon and Rubin, 1985).

to study the functional, adaptive, and phylogenetic significance of skeletal form. If dimensions of homologous structures are available, it is unnecessary to calculate body-mass estimates as they often vary widely (Roth, 1990; Ravosa, 1992, 1996d). Moreover, while information on body size may be important for assessing general, or in some instances specific, aspects of the biology of an organism, measures of a functional and/or structural unit at a lower hierarchical level are equally valid foci of research (*contra* Smith, 1993). If the growth trajectory for one taxon intersects the scatter for the adult fossils or the fossil data fall within the 95% prediction interval for a regression line, then one can

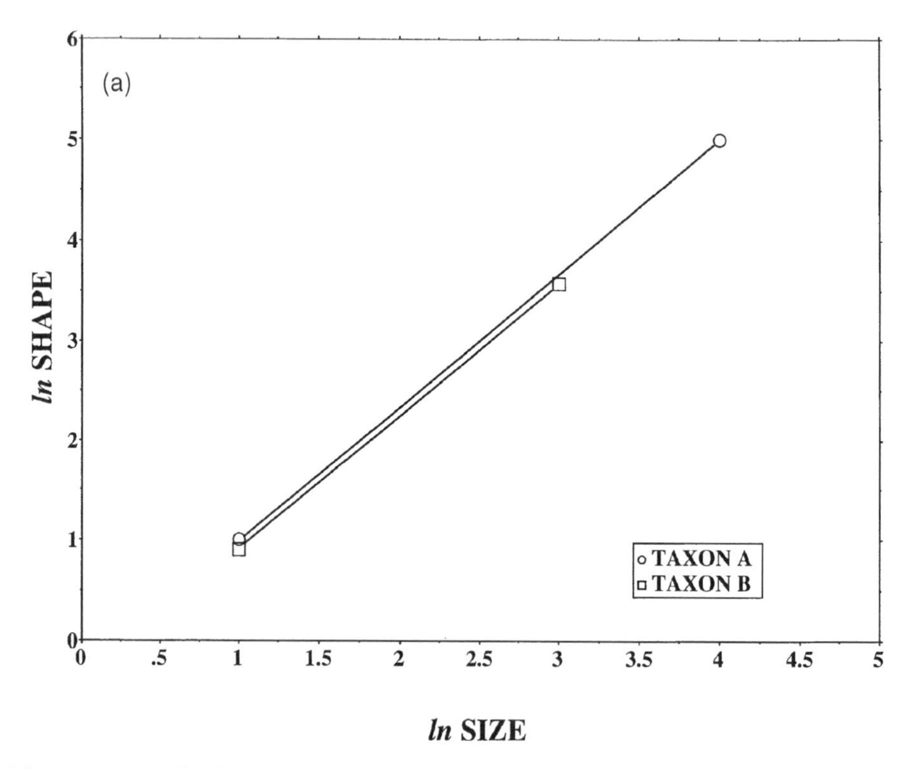

Fig. 1. Ontogenies for a measure of interest and a size measure in two sister taxa. In one example, the trajectories are coincidental and the taxa are ontogenetically scaled due likely to selection for body-side differentiation (1a). If there is allometry in this system, then the line may describe functional differentiation as well. In the other cases growth patterns are dissociated, due to an intercept difference (1b) and/or a slope difference (1c). These latter two examples of biomechanical scaling appear to result from the presence of selection for a novel shape : size relationship and, due to their apomorphic nature, can be considered to be adaptations. [Note that the slight intercept transposition between taxa in Fig. 1a is an heuristic device to facilitate identification of both concordant trajectories.]

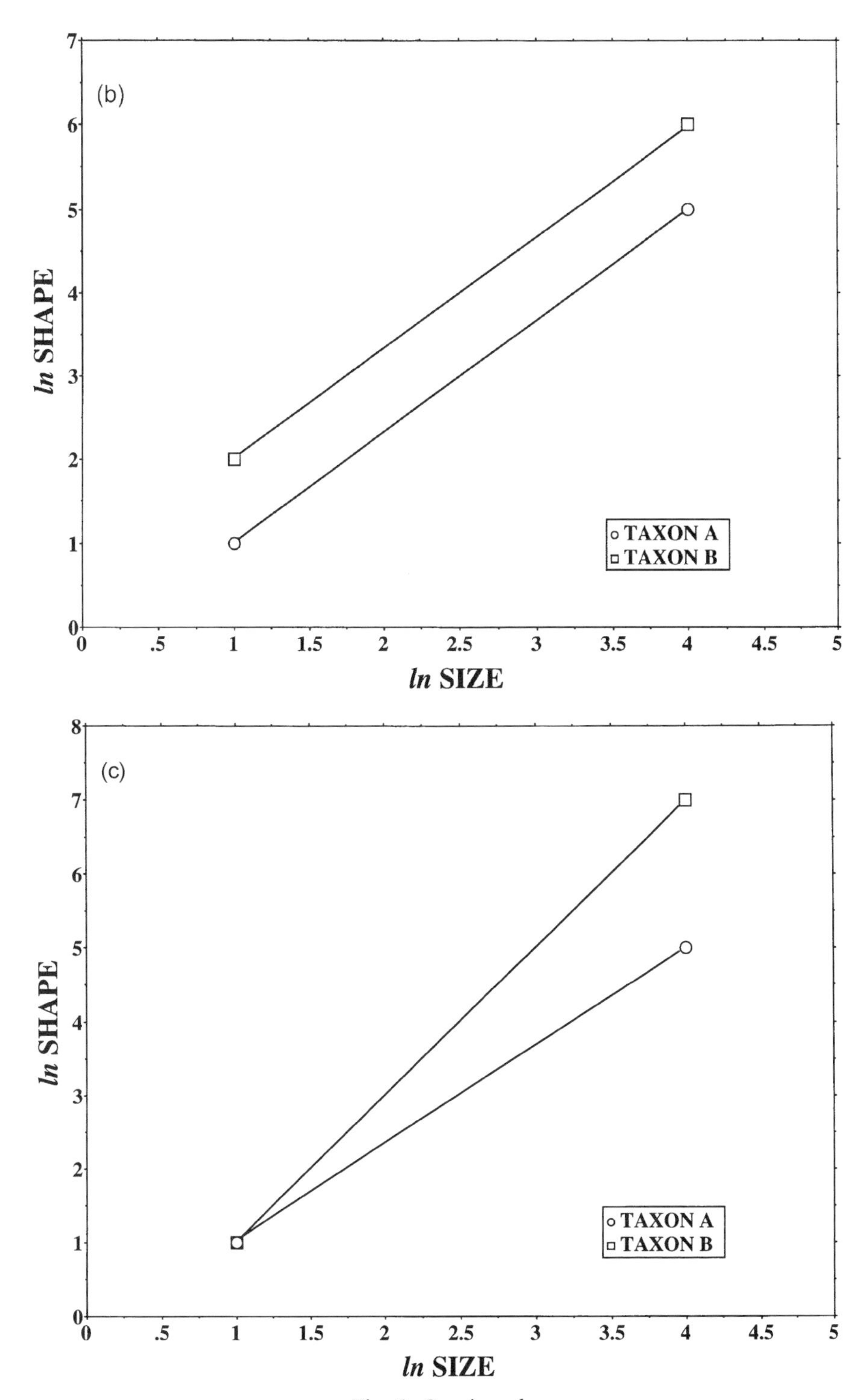

Fig. 1. Continued.

infer that the morphology of these taxa results from the differential extension and/or truncation of shared relative growth patterns—ontogenetic scaling. Obviously, such comparisons are rendered more difficult if the characters of interest exhibit complex ontogenetic patterns requiring multivariate methods, e.g., complex allometry, Gompertz curves.

Below, we review the most prevalent type of ontogenetic research to date—cases in which ontogeny is employed as a criterion of subtraction. In assessing the ecological, phylogenetic, developmental, and adaptive bases of variation in form, we focus on the specific biological implications of a pattern of ontogenetic scaling.

Ecogeographic Size Variation and Ontogenetic Scaling

Among sister taxa that differ in adult body size, variation in postcranial and cranial proportions is often the consequence of ontogenetic scaling (Fig. 1). In such instances, it is likely that the interspecific patterning of adult form results from body-size differentiation via natural selection (or genetic drift). In the former case, one can infer that ecological factors played a significant role in phyletic size change across a clade. Furthermore, this size-related variation may have occurred via evolutionary modifications in the timing and/or rate of development—heterochrony (e.g., Gould, 1966, 1975a, 1977; Shea, 1985, 1988; Wake and Roth, 1989; Ravosa *et al.*, 1993; McNamara, 1995; Klingenberg, 1998). Such examples exist in primates.

Work on the evolution of body-size differences in *Propithecus* indicate that all three sifaka species display an ecogeographic size pattern and that cranial and postcranial measures are ontogenetically scaled (Fig. 2) (Ravosa, 1992; Ravosa *et al.*, 1993). In the larger-bodied *P. diadema*, a successful ecological strategy of coping with the lower-quality forage of the eastern rainforests of Madagascar has been through an evolutionary increase in adult body size, which reduces relative metabolic and dietary demands. Allometric differentiation in sifakas develops via heterochronic changes in growth rate, with the larger *P. diadema* growing faster* due to reduced resource seasonality and thus less nutritional and energetic constraints on overall growth rate (Ravosa *et al.*, 1993).

In other extant and extinct lemurs, cranial-size differentiation via ontogenetic scaling is demonstrated for 1-kg *Hapalemur griseus* and 2.5-kg *H. simus* as well as, to a lesser extent, for *Varecia* and its larger-bodied extinct sister taxon, *Pachylemur* (Ravosa, 1992). Subfossil lemurs also exhibit ecogeographic size variation (Albrecht *et al.*, 1990), however it is unclear the extent to which

*As used in this heterochronic context, a "faster" rate of postnatal development refers to *growth-in-time* not to different coefficients—regression slopes or PCA loadings—of *relative or allometric growth*.

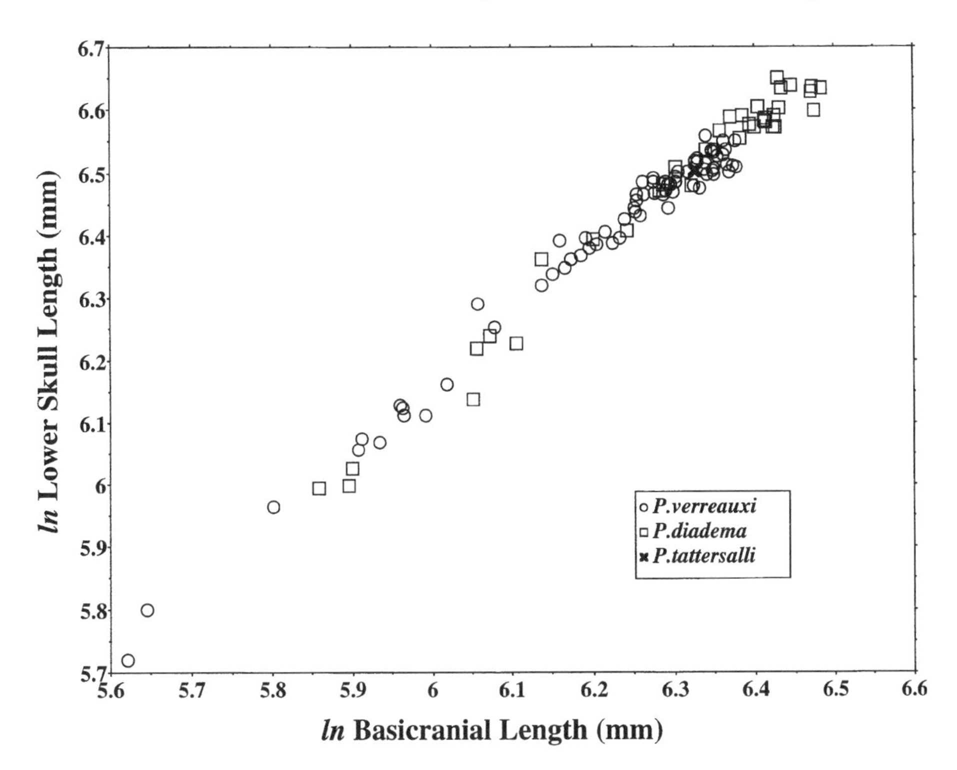

Fig. 2. A plot of *ln* lower skull length (mm) versus *ln* basicranial length (mm). Ontogenetic series for *Propithecus verreauxi* and *P. diadema* are coincidental, such that variation in skull form is due to the differential extension of shared allometric patterns. The only available adult cranial specimen of *P. tattersalli* plots with adult *P. verreauxi* and thus all three taxa appear to be ontogenetically scaled. This pattern of body-size differentiation typifies the vast majority of sifaka cranial and postcranial scaling comparisons (adapted from Ravosa, 1992).

morphological differences in indriids and their larger extinct relatives, palaeopropithecids and archaeolemurids, are due to ontogenetic scaling. Preliminary comparisons of postnatal growth series for *Indri*, *Propithecus*, and *Avahi* with mostly static adult data for two larger-bodied palaeopropithecid genera—*Babakotia* and *Mesopropithecus*—indicate that certain indrioid cranial proportions are ontogenetically scaled (Fig. 3) (Ravosa, unpubl.). Therefore, the evolution of geographic size variation in (some clades of) Malagasy primates appears to result from the ecological modification of common underlying relative growth patterns. Given the spectacular range of body-size variation encompassed by the eight large-bodied subfossil genera, lemurs are in need of a comprehensive ontogenetic and ecological analysis of morphological variation.

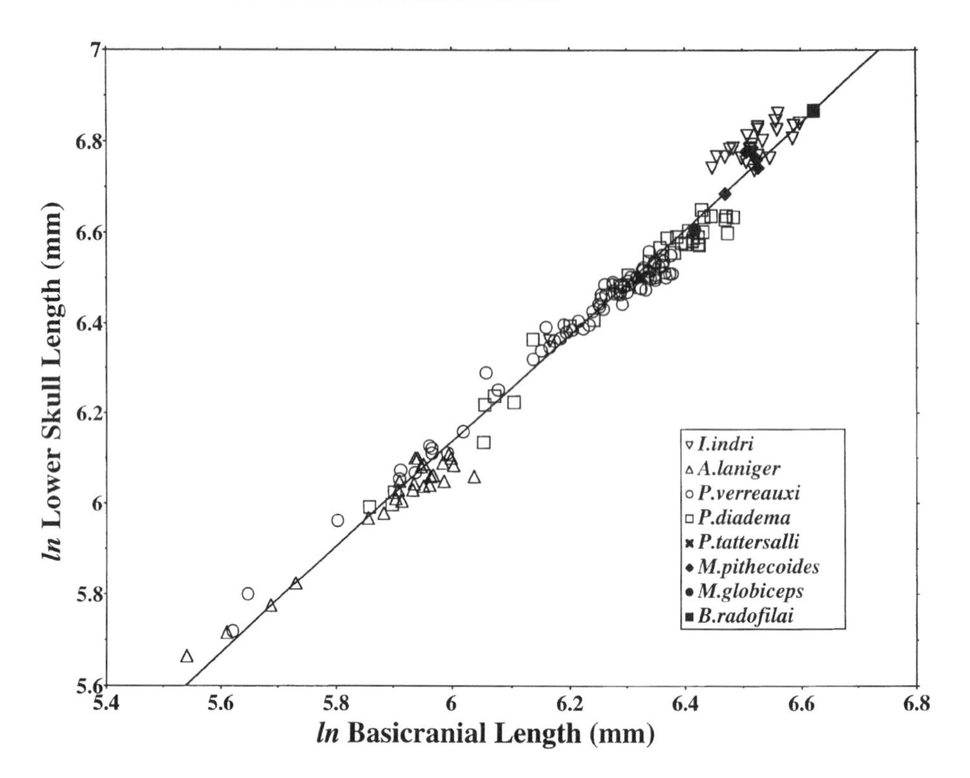

Fig. 3. A plot of *ln* lower skull length (mm) versus *ln* basicranial length (mm) in which ontogenetic series for living indriids—*Avahi laniger, Propithecus verreauxi, P. tattersalli, P. diadema, Indri indri*—intersect the mostly adult data for three larger-bodied extinct palaeopropithecids—*Babakotia radofilai, Mesopropithecus pithecoides, M. globiceps.* This indicates that some component of the variation in indrioid craniofacial proportions is ontogenetically scaled (Ravosa, unpubl.).

Phylogenetic Patterning of Form and Evolutionary Adaptation

When ecological factors drive the evolution of closely related size variants along a single ontogenetic trajectory, one can assume that the trajectory is inherited from a common ancestor and indicative of phylogenetic affinities, i.e., shared covariance patterns (Freedman, 1962; Gould, 1966; Jolly, 1972; Dodson, 1975a,b; Shea, 1988, 1995; Ravosa and Profant, 2000). It follows that a more direct case for functional *and* adaptive specialization can be made when a relative growth pattern deviates from that of other sister taxa as this is evidence that selection has acted to uncouple a specific ontogenetic covariance pattern (Gould, 1966, 1975a; Lauder, 1982; Jungers and Fleagle, 1980; Shea, 1985, 1995). The inclusion of fossils in an ontogenetic analysis facilitates a better knowledge of the antiquity and polarity of the underlying developmental basis of morphological variation across a clade. Only with an independent determination of the ancestral trajectory, however, can such comparisons offer a biologically interpretable baseline for identifying adaptation and the polarity of character state change. If it is unknown which growth pattern is primitive, all that can be inferred is the homology or lack of homology of a pre- and/or post-natal process leading to a given adult morphology.

The presence of ontogenetic scaling does not necessarily imply the absence of functional differentiation among sister taxa. Based on their body size (Kay, 1975), smaller-bodied slow lorises (*Nycticebus pygmaeus*, *N. coucang menagensis*) should be more insectivorous while the largest form (*N. c. bengalensis*) should have a tougher, more herbivorous diet (Ravosa, 1998). The positive allometry of measures of masticatory force resistance—symphysis and corpus height—across *Nycticebus* ontogenies may support this claim, as this indicates that larger slow lorises develop relatively deeper jaws than smaller forms via ontogenetic scaling, a pattern perhaps linked to a tougher diet requiring greater forces and repetitive loading during mastication. Due to the positive scaling of mandibular depth in *Nycticebus*, functional differentiation perhaps has occurred *without* an uncoupling of shared growth trajectories, such that closely related size variants appear to be adequately designed for diets with different mechanical properties (Ravosa, 1998). Clearly, data on the ontogeny of feeding behavior in all *Nycticebus* taxa would help resolve the functional bases of within- and among-taxa mandibular scaling patterns. This underscores why it is important to consider the *functional* significance of allometric patterns among ontogenetically scaled sister taxa as well as the *adaptive* significance of the lack of allometric concordance in certain structures (e.g., Gould, 1966, 1975a; Dodson, 1975a,b; Jungers and Fleagle, 1980; Shea, 1981, 1985, 1995; Ravosa, 1991d, 1992, 1998; Cole, 1992; Ravosa *et al.*, 1993; Ravosa and Ross, 1994; Biknevicius and Leigh, 1997; Inouye and Shea, 1997; Vinyard and Ravosa, 1998).

It is thus important to identify skeletal elements and functional systems where morphological variation among sister taxa does not occur via ontogenetic scaling (Fig. 1). For instance, close relatives differing in body size often have dissimilar diets, with larger forms ingesting tougher, lower-quality foods which result in elevated peak loads and greater repetitive loading of the mandible (Hylander, 1979b, 1985, 1988; Scapino, 1981; Beecher, 1983; Ravosa, 1991a, 1996a,b, 1998, 1999, 2000). Therefore, features of the masticatory apparatus linked to food processing and stress resistance may not be ontogenetically scaled (Ravosa, 1991d, 1992; Cole, 1992). In bamboo lemurs, recent and subfossil *Hapalemur simus* exhibit relatively deeper and thicker symphyses and corpora than smaller-bodied *H. griseus* (Ravosa, 1992). The more robust jaw proportions of *H. simus* are adaptations to resist greater masticatory stresses due to the processing of the fibrous pith of the giant bamboo, while *H. griseus* ingests the delicate shoots and leaves of smaller bamboo plants (cf. Tan, 1999). Extinct *Pachylemur* possess more robust mandibles than their more diminutive sister taxon—*Varecia*—which also appear

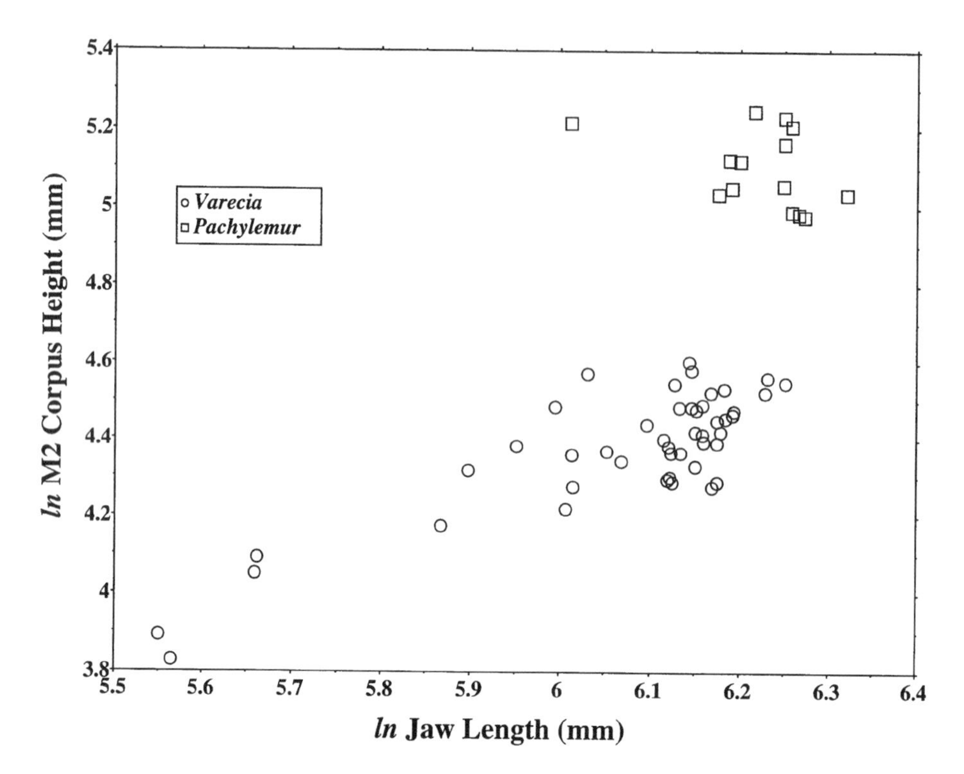

Fig. 4. A plot of *ln* M_2 corpus height (mm) versus *ln* basicranial length (mm) in *Varecia* and *Pachylemur*. As opposed to most cranial analyses, the ontogenetic scaling trajectories for *Varecia* and *Pachylemur* do not appear to correspond. Instead, the data scatter for *Pachylemur* tend to lie above the ontogenetic trajectory for the ruffed lemur, which suggests an underlying dietary and thus adaptive basis of these morphological differences (adapted from Ravosa, 1992).

functionally related to the tougher diet inferred for the fossil genus (Fig. 4) (Ravosa, 1992).

A similar use of ontogeny as a criterion of subtraction has been applied to analyses of dental:dietary adaptations in African apes (Shea, 1983a) and Malagasy lemurs (Ravosa, 1992). In both cases, due to the processing of a tougher/harder diet, larger-bodied species are observed to exhibit relatively larger postcanine teeth than expected based on the extrapolation of trajectories for smaller-bodied sister taxa.

Preliminary work on galagos ranging in body size by two orders of magnitude suggest a pervasive pattern of ontogenetic scaling in skull length (Fig. 5) (Ravosa, unpubl.). Cranial growth patterns for two of 11 taxa are discordant, however. At a common size, *Galagoides demidovii* possesses a relatively longer skull than other galagos, which apparently reflects the elongate nasal aperture of this species. In addition, *Euoticus elegantulus* exhibits

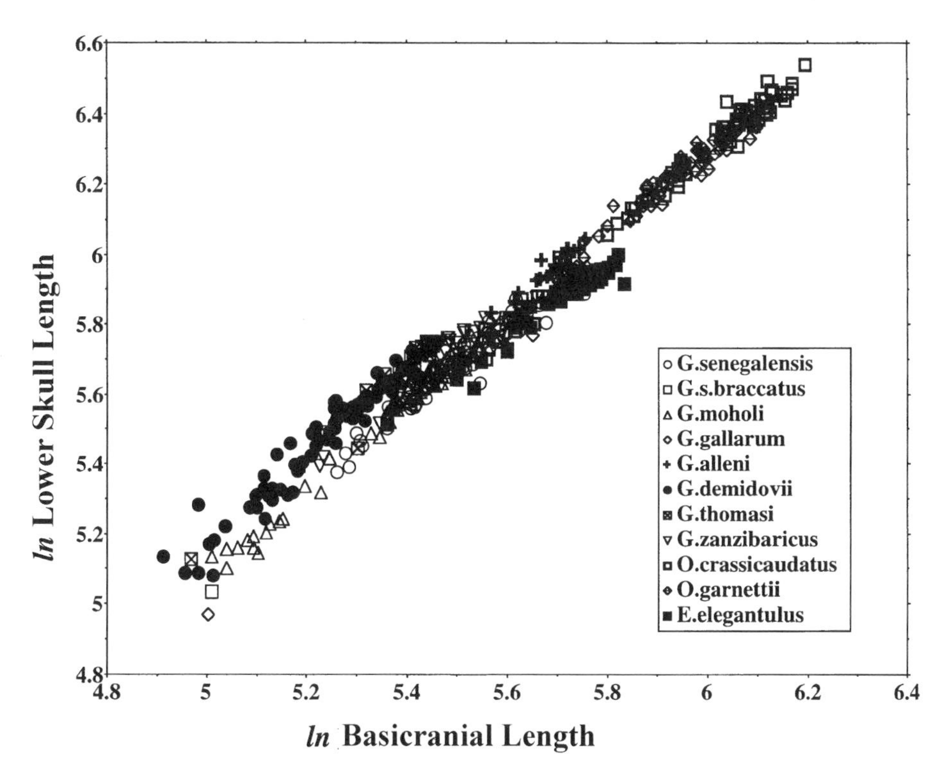

Fig. 5. A plot of *ln* lower skull length (mm) versus *ln* basicranial length (mm) in galagids. As the growth series and adult data are largely coincidental, this indicates that some interspecific variation in galago skull morphology is due to ontogenetic scaling. Unfortunately, the ecological bases of such variation are unknown. Those cases where trajectories are discordant identify potential examples of dietary adaptations—*Euoticus elegantulus*—and nasal specializations—*Galagoides demidovii*. Both changes occur via transpositions with a retention of primitive shape:size covariation patterns (Ravosa, unpubl.).

a relatively shorter skull (due to a shorter face) which may serve to increase the mechanical advantage of the jaw adductors during tree gouging, a behavior unique among galagids (Williams and Wall, 1999). While both species are transposed relative to a hypothetical ancestral ontogeny, they nonetheless appear to retain similar coefficients of relative growth (Fig. 5). The inclusion of similar data for early fossil galagids would offer both a functional and phylogenetic perspective on the evolution of skull form across this diverse clade.

In a comparative study of postcranial ontogeny and function in the stockier *Cebus apella* versus the more gracile *C. albifrons*, Jungers and Fleagle (1980) employ similar methods to detail salient differences in limb scaling as it relates to locomotor and positional behaviors. In particular, they find that at a common body size, *Cebus albifrons* possesses relatively longer limbs associated with more cursorial behaviors.

To test the claim that scapular form and function in "Lucy"—*Australopithecus afarensis*—is as expected for an African ape, but not for a human, of its size (Susman *et al.*, 1984), Inouye and Shea (1997) examined the ontogeny of glenoid orientation in African hominoids. Inouye and Shea observed that humans exhibit allometry of scapular form during growth, such that smaller and/or younger humans possess a more cranial orientation of the glenoid; Lucy's bar-glenoid angle lies on this glenoid ontogenetic trajectory. Susman and colleagues used a regression line described only by adult human data and instead found that Lucy's bar-glenoid angle is significantly below the adult human trajectory; thus, Lucy's morphology and locomotor behavior were interpreted to be distinct from humans. Interestingly, no ontogenetic changes in scapular form are noted for the African apes *and* Lucy falls within the prediction interval for the ape growth pattern (Inouye and Shea, 1997). Thus, in a hominoid of Lucy's small size, ontogenetic trajectories for chimps, gorillas, *and* humans overlap, with scapular proportions diverging only between adult *Pan* and *Gorilla* on one hand and adult *Homo* on the other. Inouye and Shea argue cogently that as African apes and humans have different locomotor behaviors early in ontogeny, but the bar-glenoid angle is similar during such stages, *by itself* this character is unreliable as a functional indicator of more ape-like arboreal behaviors (cf. Kay and Cartmill, 1977 regarding criteria for inferring function in fossils based on living analogs). Their study also points to the theoretical and methodological pitfalls of using only intraspecific adult data as a criterion of subtraction for allometric extrapolation and functional interpretation (cf. Gould, 1966; Shea, 1981; Cheverud, 1982).

Apart from highlighting patterns of functional covariations, ontogenetic data can offer unique insight into underlying structural linkages during development. To test the claim that the deeper face and thicker hard palate of *Paranthropus* are adaptations to resist elevated masticatory stresses linked to broader, flaring zygoma and larger, more vertical adductor forces (Rak, 1983), McCollum (1994) analyzed the postnatal ontogeny of these and other masticatory and structural parameters in the African apes. Accounting for ontogenetic variation in skull size, bizygomatic breadth and facial height were found to be uncorrelated in *Gorilla* and *Pan*, a finding contrary to Rak's predictions. Employing additional data, McCollum (1994) instead argued that the development of a thicker hard palate in *Paranthropus* facilitates a greater superior displacement of the upper face relative to the cranial vault. This structural, functional, and allometric framework has also been applied to the multifactorial nature of primate browridge formation (Ravosa, 1988, 1991b–d).

An ontogenetic criterion of subtraction can also inform questions concerning intraspecific sexual differences in craniodental form and function. Work on sexual dimorphism indicates that relative growth patterns are largely concordant in males and females, such that dramatic intersexual differences in adult cranial proportions are simple allometric correlates of larger adult male sizes (Gould, 1975a; Cochard, 1985; Shea, 1985). For instance, marked sexual differences in macaque browridge size and shape are due to ontogenetic scaling coupled with the fact that anthropoid browridge dimensions scale with strong positive allometry in ontogenetic and interspecific comparisons (Ravosa, 1991b–d). On the other hand, anthropoid male canines are relatively larger than in females of similar size (Cochard, 1985, 1987; Shea, 1985, 1995; Ravosa, 1991d; Ravosa and Ross, 1994) due to the differential importance of canines in male agonistic displays (Plavcan and van Schaik, 1992). Similar analyses should be extended to more fossil taxa so as to assess the numerous causes of intraspecific differentiation in skeletal form, function, and ontogeny.

In fact, the degree of morphological differentiation among size variants is directly linked to the coefficient of relative growth (cf. Dodson, 1975a,b). In other words, higher levels of adult morphological variation (e.g., CVs) in certain skeletal elements may simply reflect the amount of ontogenetic allometry inherent to a given measure or system. For those structures growing with stronger positive allometry, selection for an unit increase in size will result in a greater degree of relative shape variation between two ontogenetically scaled morphs. Thus, in formulating adaptive explanations for morphological differences in sexually dimorphic taxa or in testing if two fossil morphs represent one or two taxa, it is critically important to characterize allometric trajectories for the characters of interest as well as determine if morphological variation among the sexes or species is ontogenetically scaled.

Brain and Dental Scaling

Allometric analyses constitute a common methodology employed to assess the functional or ecological significance of relative brain size as well as the dietary correlates of dental proportions. In both avenues of study, there are several morphological consequences of evolutionary differentiation via ontogenetic scaling (whether driven by ecogeographic or sexual factors). If sister taxa indicate a consistent pattern of ontogenetic scaling of cranial dimensions, brain size may demonstrate a dissociation of shared patterns of relative growth. This is because selection for body-size change often targets primarily postnatal systemic growth. Therefore, differences in brain size, an organ that develops mostly during prenatal growth stages and is thus subject to a different suite of genetic and epigenetic controls, should be minimal as compared to differences in skeletal elements that enlarge postnatally (Gould, 1975a; Lande, 1979; Shea, 1983b, 1988; Riska and Atchley, 1985; Shea *et al.*, 1987; Ravosa, 1991d, 1992; Ravosa and Ross, 1994). While it remains to be demonstrated that following selection on body size (*and* regardless of the direction of size change), "brain and body size relations will eventually return to the functional line" (Deaner and Nunn, 1999, p. 687), the above ontogenetic factors serve to underscore why phyletic dwarfs and phyletic giants are, respectively, more and less encephalized. In the absence of comparative evidence on development, one should still exercise caution in interpreting the adaptive nature of relative brain size (and characters correlated therewith) across a size series.

Patterns of dental allometry also mirror those for neural size across and within ontogenetically scaled taxa, such that variation in postcanine tooth size among morphs is less than that for other facial parameters. In ontogenetically scaled size series, regardless of whether selection acts to increase or decrease adult body size, postcanine tooth size should scale negatively. Thus, larger forms should have relatively smaller teeth than more diminutive sister taxa, an argument that explains the relatively larger cheek teeth of human pygmies (Shea and Gomez, 1988). This apparently occurs because genetic and epigenetic controls on dental morphogenesis operate more independently of systemic effects on skeletal growth (Gould, 1975b; Cochard, 1985, 1987; Shea, 1988; Shea and Gomez, 1988; Shea *et al.*, 1990; Ravosa, 1991d, 1992; Ravosa and Ross, 1994). In the absence of paleoecological evidence on diet, this highlights the need to distinguish the effects of such developmental processes from size-related functional differentiation across species.

Summary

The examples discussed above touch upon the myriad ways in which ontogeny is used as a criterion of subtraction. As is common of allometric research using ontogenetic and especially interspecific data, more attention is focused on the scatter or dissociation about the line than on the biomechanical, functional, or behavioral significance of the trajectory itself. An understanding of the scaling coefficient is equally important, particularly if fossil taxa, for instance, lie along an extension of a common growth pattern and fall well beyond the size range of other extant sister taxa. However, poor sampling or low levels of taxonomic diversity can greatly influence the slope of an interspecific regression line, in turn significantly affecting the interpretation of deviations or residuals from such a scaling trajectory.

In comparisons where ontogenetic trajectories for a structure are discordant among sister taxa, it is important to distinguish if such interspecific variation is due to y-intercept transpositions and/or different slope coefficients (Cock, 1966; Gould, 1966, 1971; Dodson, 1975a; Shea, 1981; Cole, 1992; Ravosa *et al.*, 1993; Vinyard and Ravosa, 1998). The former pattern may reflect a process of evolutionary adaptation where the maintenance of a shared covariance pattern—common slope—is coupled with selection for a similar shape across taxa of different sizes. The latter may represent a case where the proportions of a structure are similar among neonates of closely related taxa and allometric trajectories differ during later postnatal stages due to the cumulative effects of increasingly divergent interspecific differences in behaviors and loading regimes. More work is needed to evaluate the frequency of these patterns vis-a-vis the processes of functional and evolutionary adaptation.

In addition, the invasion of previously unoccupied size ranges has the potential for the evolution of novel morphologies and functions, especially if underlying relative growth patterns are strongly allometric and considerable transformation in shape occurs for each unit increase in the size of a structure (Gould, 1966; Dodson, 1975a,b; Shea, 1988). On the other hand, the presence of low levels of allometry in an ancestral growth pattern may characterize those cases where an uncoupling of such trajectories occurs more frequently so as to attain a particular shape (cf. Gould, 1966, 1971, 1975a; Dodson, 1975a,b).

Lacking an experimental or theoretical prediction regarding the meaning of the coefficient of a particular scaling trajectory, the use of ontogeny as a criterion of subtraction is perhaps the best alternative means of interpreting the functional and adaptive significance of skeletal form in fossil sister taxa. Thus, while ratio-based methods of size adjustment (Jungers *et al.*, 1995) have as yet been applied to comparative studies of development, we caution against the use of such shape measures as they would conflate functional variation along a scaling trajectory with adaptive variation about such an ontogeny. Depending on whether sister taxa are ontogenetically scaled, one can also infer the polarity of a morphological transformation and thus gain unique insights into evolutionary processes. In using an ontogenetic criterion of subtraction to infer the adaptive importance of a specific trajectory, it is vital to compare development among sister taxa. Without such a phylogenetic control, one is constrained to offer relatively uninformative statements regarding interspecific variation in ontogenetic patterns of morphological (e.g., Richtsmeier *et al.*, 1993) and functional (e.g., Godfrey *et al.*, 1998) change.

Biomechanical Scaling and Functional Equivalence

As ontogeny often entails both changes in size and shape, it is important to detail the biomechanical factors affecting the postnatal expression and modification of biological form. One purpose of a study of growth and function is to understand the nature and role of functional design criteria and behaviors in the morphogenesis of a skeletal element or structural system across a range of body sizes (Fig. 1). In this regard, *functional equivalence* is the maintenance of similar uses or actions of a feature or character complex across a morphospace (cf. Biewener and Bertram, 1993a; Emerson *et al.*, 1994; Vinyard and Ravosa, 1998). As allometric changes in a morphological feature may entail concordant size-required shifts in a functionally linked parameter, selection for equivalent functions during growth or across taxa may necessitate the uncoupling of shared relative growth patterns such that the interspecific pattern of shape change among sister taxa reflects biomechanical, rather than ontogenetic, scaling (e.g., Gould, 1975a; Jungers and Fleagle, 1980; Shea, 1985). A distinction among size-correlated, size-required, and size-invariant influences is requisite to a more informed consideration of *if* and *why* functional equivalence is maintained over a range of sizes and ages (Fleagle, 1985). The elucidation of biomechanical rules governing how morphospace is filled (Alexander, 1989; Biewener, 1991), especially in the case of phyletic dwarfism or gigantism, is critical to a knowledge of why a scaling trajectory changes as it does and why taxa do or do not deviate therefrom. While ontogenetic analyses may be applied only rarely to fossil taxa, an understanding of such patterns has implications for the analysis of skeletal function and scaling in fossil primates, as it makes explicit statements about biomechanical, behavioral, and structural constraints that may apply to members of a clade. In other words, if you know what a trajectory describes, one can interpret the data relative to this experimental and/or theoretical expectation. Clearly, a detailed rendering of the biomechanical and/or ecological underpinnings of an allometric pattern is one of the more difficult tasks facing morphologists and behaviorists.

In this section, we review theoretical, experimental, and comparative work on perhaps the best example of functional equivalence and biomechanical scaling in the cranium—the allometry of the papionin mandibular symphysis as it relates to a "wishboning" loading regime during unilateral or one-sided mastication. The second and third cases represent equally rare examples of functional equivalence in the postcranium—femoral ontogeny and locomotor behavior in iguanodontian dinosaurs and limb scaling and the biomechanics of climbing in arboreal primates.

Wishboning during Mastication and Mandibular Allometry

Bone-strain analyses indicate that the most important functional determinant of symphyseal form in papionins is wishboning stress during unilateral molar biting and chewing (Fig. 6) (Hylander, 1984, 1985, 1988; Hylander *et al.*, 1987, 1998, 2000; Hylander and Johnson, 1994). Wishboning is the simultaneous pulling apart or lateral transverse bending of the mandibular corpora at the end of tooth–food–tooth contact. This loading regime results in high stress concentrations and strains* along the symphysis, especially at its inner or lingual surface. Wishboning is best resisted by increasing the labiolingual thickness of the symphysis (Hylander, 1984, 1985, 1988; Ravosa, 1991a, 1996a,b, 2000; Daegling, 1993; Vinyard and Ravosa, 1998).

A key determinant of symphyseal stress concentrations during wishboning is symphyseal curvature. If the symphysis is more curved in a transverse plane, the neutral axis—that point at which internal stresses and strains are zero—in lateral bending moves progressively closer to the lingual border of the symphysis. This results in increasingly elevated stress concentrations along this

*Strain is a dimensionless unit that measures the change in length of an object relative to its original length. By convention, tensile strain is a positive value and compressive strain is negative. Stress is the amount of force per unit area. Stress is directly proportional to strain (via the elastic modulus) when the tissue is isotropic and loaded in a normal elastic physiological range. A loading regime is described by tensile and/or compressive strain directions at a homologous region during a given behavior. The relative importance of several loading regimes at a site can be assessed by comparing peak-strain levels.

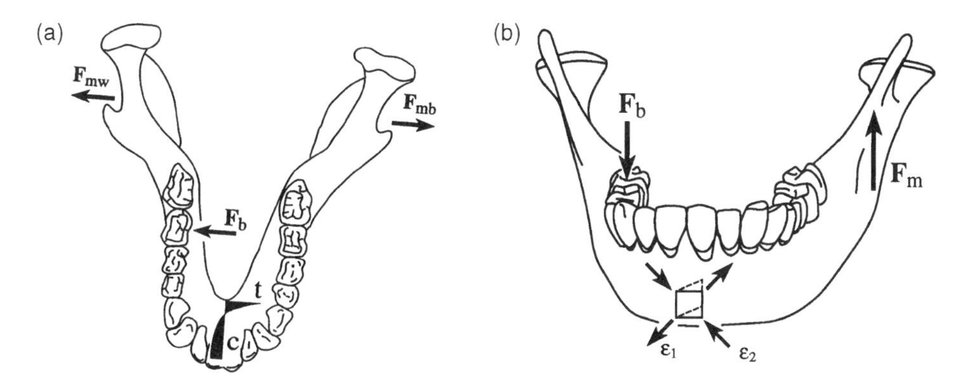

Fig. 6. Superior (6a) and frontal (6b) views of an adult anthropoid mandible with a fused symphysis. There are two primary symphyseal loading regimes experienced during unilateral mastication. Wishboning (6a) is due to a laterally directed component of jaw-adductor muscle force on the balancing side (F_{mb}) and a laterally directed component of bite force (F_b) and jaw-muscle force on the working side (F_{mw}). As wishboning results in high tensile stresses and high stress concentrations at the lingual surface of a fused symphysis, these stresses are best countered by increasing the labiolingual thickness of the symphysis and/or orienting the long axis of the symphysis more horizontally. The relative size of the force (F) arrows does not indicate the amount of forces on either side of the jaw. Dorsoventral shear (6b) is proportional to the amount of the vertical component of jaw-adductor muscle force (F_m) transmitted across the symphysis from the balancing to the working side of the mandible (F_b = vertical component of molar bite force). Greater relative recruitment of balancing-side jaw-muscle force causes elevated levels of dorsoventral shear. This stress is best resisted by increasing the number of ligaments spanning a partially fused symphysis and by increasing the amount of cortical bone along a symphyseal cross-section in a fully fused joint. Arrows ε_1 and ε_2 indicate tensile and compressive strain directions (adapted from Ravosa and Hylander, 1994).

inner surface due to a higher strain gradient (Fig. 6). Theoretical and interspecific analyses suggest that allometric changes between mandibular length and arch width affect the degree of symphyseal curvature and thus the distribution of wishboning stress at the symphysis (Hylander, 1984, 1985, 1988). It is now evident that mammalian jaw length typically increases with positive allometry relative to jaw arch width, such that the mandible is more elongate and narrow and the symphysis is more curved in larger taxa (Hylander, 1984, 1985, 1988; Ravosa, 1991a,d, 1992, 1996a–c, 1998, 1999, 2000; Daegling, 1993; Ravosa and Ross, 1994; Vinyard and Ravosa, 1998; Hogue and Ravosa, 2001). To offset greater wishboning stress concentrations due to size-related increases in symphyseal curvature, symphysis measures thus scale positively.

Hylander's (1985) biomechanical model regarding symphyseal scaling and wishboning details how individual elements of a functional complex vary with size so as to maintain functional equivalence of the system. As symphyseal stress concentrations increase with size due to the scaling of symphyseal curvature, positive allometry of symphyseal dimensions occurs to counter such relatively higher stresses and thus maintain similar levels of resistance to wishboning over a range of sizes. While this remains to be tested experimentally in the skull, theoretical and *in vivo* analyses of postcranial function and scaling in vertebrates as diverse as birds and horses suggest that, independent of variation in body size and locomotor behavior, load-bearing limb elements are designed to ensure similar strain levels and hence constant safety factors* (Biewener, 1982, 1991, 1993; Rubin and Lanyon, 1984; Lanyon and Rubin, 1985; Biewener *et al.*, 1986; Keller and Spengler, 1989; Selker and Carter, 1989; Biewener and Bertram, 1993b). In this *optimal strain environment*, routine physiological loads of a sufficiently high level and specific direction are required to maintain a given amount of tissue in a skeletal section. Thus, the genetic and epigenetic processes regulating the distribution of cortical bone offer a capacity for functional adaptation to a limited range of dynamic strains at an homologous site (Lanyon and Rubin, 1985).

Using the formula for stress along the concave aspect of a curved beam (cf. Hylander, 1985), we tested Hylander's hypothesis of functional equivalence regarding wishboning and symphyseal scaling in an ontogenetic context (Vinyard and Ravosa, 1998). To ensure that all juvenile papionins chosen for study had fully developed levels of neuromuscular coordination and thus employed "adult-like" masticatory behaviors, the postnatal onset of adult loading regimes was inferred for individuals with the first permanent molars erupted† (Vinyard and Ravosa, 1998).

*A *safety factor* is a measure of the relative strength of an element to a loading regime. It is the ratio of the bone-strain value at yield, structural, or ultimate failure divided by the observed peak-strain level during a given behavior. The safety factor to yield at midshaft for cortical bone in vertebrate limbs averages 2.8 at the trot–gallop transition (Lanyon and Rubin, 1985; Biewener, 1993).

†Even though they may be the same size, closely related juveniles and adults may not function alike. If similar functional and morphological patterns are to be compared one should choose only those stages in which juveniles possess adult levels of neuromuscular coordination, and thus adult behaviors and loading regimes. Otherwise, an allometric "pattern" may track different behaviors and functions at different sizes/ages (i.e., it is size-correlated) rather than size-required functional equivalence over a size/age series. Data on feeding behavior indicate that juvenile mammals develop adult-like adductor recruitment patterns soon after weaning *and* less variation exists among older individuals (Herring, 1985; Weijs *et al.*, 1989; Herring *et al.*, 1991; Iinuma *et al.*, 1991; Westneat and Hall, 1992; Huang *et al.*, 1994). This strongly suggests that adult loading regimes and corresponding morphological responses to adult-like masticatory stresses are likewise present early in ontogeny (Rigler and Mlinsek, 1968; Trevisan and Scapino, 1976a,b; Hirschfeld *et al.*, 1977; Ravosa, 1991a, 1992, 1996a, 1999; Cole, 1992; Ravosa and Simons, 1994; Biknevicius and Leigh, 1997; Vinyard and Ravosa, 1998). As it is correlated with weaning (Smith, 1991), primate M1 eruption appears to delineate adult-like behaviors. The early development of this form:function link is not unexpected as juvenile primates often ingest foods similar to adults (Watts, 1985). To gain a better functional understanding of this postnatal transition, it would be important to assess if juveniles and adults process foods of similar mechanical properties (immature versus leaves) and relative bolus size as well as the daily proportions of foods of different properties.

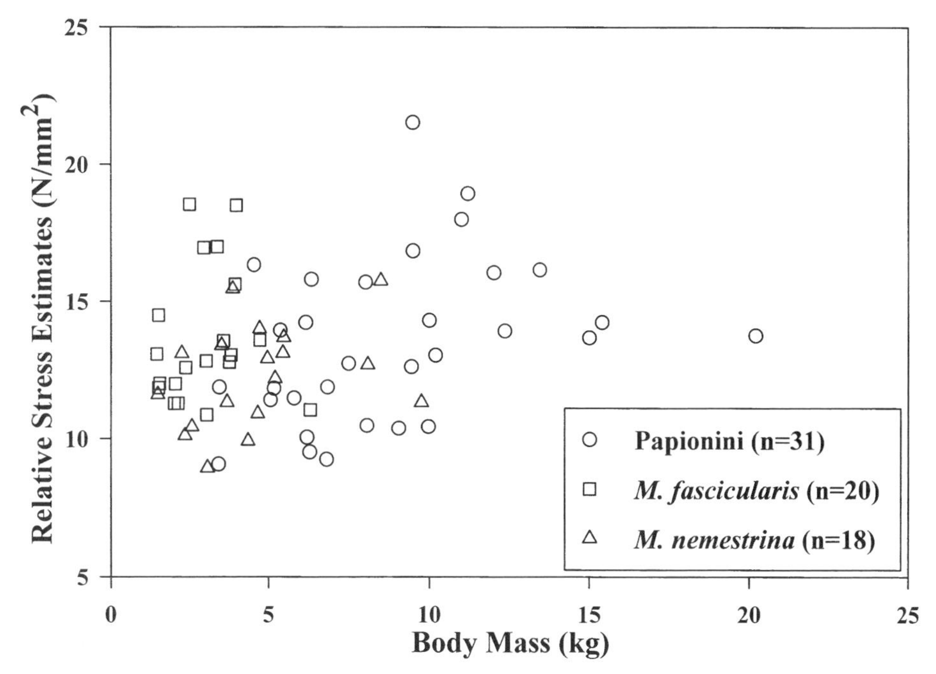

Fig. 7. Plot of relative symphyseal stress estimates for two macaque ontogenetic series and for 20 closely related papionin species. Bivariate analyses indicate that relative stress estimates do not vary with body mass during macaque post-weaning development. Stress values for the growth series also do not differ from each other or with interspecific estimates. Allometric changes in symphyseal form occur primarily to maintain similar levels of resistance to wishboning stress over a spectrum of sizes and ages (adapted from Vinyard and Ravosa, 1998).

In two macaque species, relative symphyseal stress estimates do not change significantly during growth (Fig. 7). Moreover, symphyseal stress values for both ontogenetic series do not differ from each other or from symphyseal stress estimates for 20 papionin sister taxa (Vinyard and Ravosa, 1998). Thus, symphyseal stress magnitudes due to wishboning appear to be similar during macaque post-weaning ontogeny *and* across an interspecific papionin sample. As first noted for an adult size series (Hylander, 1985), allometric changes in symphyseal form occur largely to maintain functional equivalence vis--vis size-related increases in symphyseal curvature and wishboning stress concentrations (and safety factors related thereto).

Although much of the variation in papionin cranial proportions results from ontogenetic scaling (Freedman, 1962; Profant, 1995; Ravosa and Profant, 2000), allometric changes in symphyseal curvature *do not* occur due to the differential extension or truncation of macaque mandibular dimensions (Vinyard and Ravosa, 1998). Rather, the symphysis in larger-bodied adult monkeys is

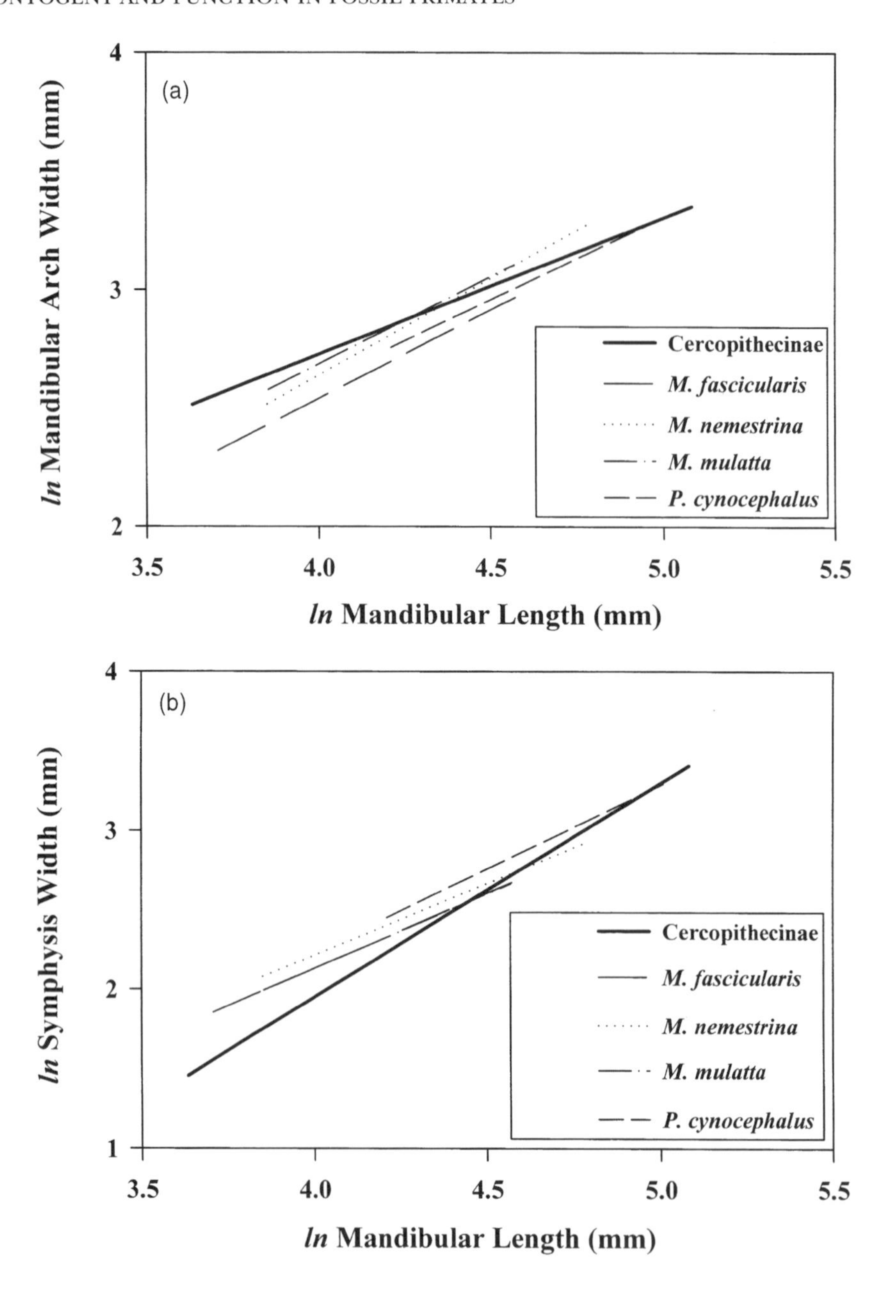

Fig. 8. Plots of *ln* M_1 arch width (mm) (8a) and *ln* symphysis width (mm) (8b) versus *ln* mandibular length (mm) for ontogenetic series of *Macaca fascicularis*, *M. nemestrina*, *M. mulatta* and *Papio cynocephalus anubis* and an interspecific series of 35 Cercopithecinae. Across cercopithecines, allometric changes in curvature and stress concentrations do not result from ontogenetic scaling. Larger-bodied taxa have more curved symphyses when controlling for jaw size, with interspecific variation due mainly

more curved than would be predicted due to ontogenetic increases in skull size, with interspecific differences due to the successive transposition of growth trajectories for increasingly larger taxa below those for smaller sister taxa (Fig. 8a). Thus, an interspecific line connecting adult species means is lower in slope than each ontogenetic line. In papionins, this relative increase in curvature demonstrates that, holding other factors constant, symphyseal stress concentrations are elevated as size increases interspecifically, i.e., across papionin ontogenies. Due to the transpositions for symphyseal curvature, symphyseal dimensions are likewise transposed such that the symphysis is more robust than would be expected due to ontogenetic scaling (Fig. 8b). As there is no reason to predict that the positive allometry of symphyseal curvature and the concomitant production of relatively greater stress concentrations are linked to selection for resistance to symphyseal wishboning, it is likely that the negative allometry of mandibular arch width reflects other adaptive or nonadaptive factors influencing papionin cranial evolution (Vinyard and Ravosa, 1998). The fact that papionin symphyseal dimensions are not ontogenetically scaled perhaps provides one of the best examples of adaptation in the primate cranium. That is, the interspecific scaling pattern for the symphysis is a manifestation of a common ontogenetic process of adaptation among sister taxa of different terminal adult sizes (Vinyard and Ravosa, 1998). Similar data for the large-bodied fossil baboons would offer a further test of this model. Interestingly, this represents another case whereby allometric coefficients for close relatives are similar but y-intercepts differ (Cock, 1966; Gould, 1966, 1971; Dodson, 1975a; Shea, 1981; Ravosa *et al.*, 1993).

Locomotor Ontogeny and Femoral Biomechanics

To infer the ontogeny of bipedal locomotor behavior in *Dryosaurus lettowvorbecki*, an iguanodontian dinosaur, Heinrich *et al.* (1993) employed a biomechanical analysis of postnatal changes in the relative amount and distribution of cortical bone from femoral cross-sections. Although they expected primarily to observe regular allometric changes in cortical bone thickness vis-a-vis ontogenetic increases in body mass, the pattern across three age/size classes was not linear. Versus smaller forms, medium-sized morphs

---

to successive transpositions of growth trajectories (8a). Thus, the interspecific line connecting the adult means is lower in slope than individual growth trajectories. Given the transpositions for symphyseal curvature, symphyseal dimensions are likewise not concordant, such that the symphysis is more robust than would be predicted due to ontogenetic scaling (8b). In sum, the interspecific symphyseal scaling pattern may be largely a manifestation of a shared ontogenetic process of functional adaptation expressed in sister taxa of differing adult sizes (adapted from Vinyard and Ravosa, 1998).

had a greater amount of cortical bone relative to total subperiosteal area and a larger index of bending strength indicative of a single, predominant bending regime. Due to increases in body size throughout *Dryosaurus* growth, Heinrich *et al.* expected changes evident between small and medium morphs to continue into later stages. However, in spite of an estimated 300% increase in body mass over medium-sized morphs, larger forms had the same percent cortical bone and the index of bending strength was lower (like the smaller morphs). Such changes in femoral form imply the presence of marked age-/size-related variation in femoral loading patterns. Throughout ontogeny, the axis of maximum bending strength remained constantly oriented anteroposteriorly (AP), indicative of either bipedal or quadrupedal behaviors (Heinrich *et al.*, 1993).

Therefore, Heinrich *et al.* considered the relative growth of dinosaur body proportions vis-à-vis locomotor behavior and function. Neonates of two *Dryosaurus* sister taxa possessed relatively larger crania and smaller, shorter tails (Horner and Weishampel, 1988), such that the body's center of gravity was located more cranially and farther from the hip joint in smaller, younger *Dryosaurus*. Given negative scaling of head size, positive allometry of tail size, and positive allometry of hindlimb to forelimb size, the body's center of gravity should shift caudally in larger, older juveniles. As such, there should have been a size-related decrease in AP bending of the femur and the index of bending strength should have decreased correspondingly. Based on the index of bending strength, however, AP bending stresses were most pronounced at medium sizes and this is coupled with sharp increases in relative cortical area.

Heinrich *et al.* (1993) reconcile such seemingly disparate results by arguing that the ontogeny of body proportions imposes a constraint on the development of bipedality at small sizes and that morphological changes between small and medium juveniles indicate a shift from quadrupedal to bipedal locomotion. In comparing the two younger morphs, differences in percent cortical bone are thought to be due to a shift in the number of load-bearing supports, from four to two in the medium forms, and this resulted in differences in cortical bone distribution due to a center of gravity located more cranially to the hip joint and a correlated increase in AP bending stress. Variation in cortical bone distribution between medium and large morphs is linked to a center of gravity placed closer to the hip in older forms (which thus experience lower AP bending stresses). Younger, smaller *Dryosaurus* were obligate quadrupeds because they could not counterbalance a more craniad center of gravity. Due to a caudad postnatal shift in the center of gravity, and presumably greater neuromotor coordination, medium-sized juvenile *Dryosaurus* could adopt adult-like locomotor behaviors. Based on avian growth models, Heinrich *et al.* (1993) infer this functional constraint lasted the first several months of life. It would be interesting to have similar data from humeral cross-sections so as to ascertain the extent and timing of changes in forelimb loading patterns. Nonetheless, such a study offers us a glimpse of what we might find were we to possess similar evidence for early hominids, a group with bipedal locomotor behaviors and certain similar constraints.

Vertical Climbing and Intermembral Allometry

In a consideration of the biomechanics of vertical climbing in arboreal taxa lacking claws, Cartmill (1974, 1985) developed a model to explain the ratio of forelimb-to-hindlimb length as a means of maintaining adequate levels of pedal and manual friction. In order to examine the functional significance of limb proportions in *Megaladapis*, a subfossil lemur well beyond the body-size range of extant lemurs, Jungers (1978) applied an allometric framework to test predictions about the biomechanics of climbing and size-required changes in the intermembral index. As the center of gravity of a large-bodied arboreal primate is positioned absolutely farther away from the substrate due to its absolutely longer limbs, such taxa should exhibit relatively longer forelimbs than smaller close relatives (Jungers, 1978, 1984, 1985). This positive allometry of the intermembral index is necessary to maintain similar climbing and clinging abilities on vertical supports over a wide size spectrum.

Cartmill and Junger's model regarding the allometry of limb proportions in arboreal primates is supported in more controlled phylogenetic and ontogenetic comparisons. In an analysis of relative growth in African ape limbs, Shea (1981) observed that larger-bodied adult *Gorilla gorilla* possess the highest intermembral indices, followed next by the smaller *Pan troglodytes* and then in turn by the even-smaller *P. paniscus*. In Malagasy lemurs, larger-bodied *Propithecus diadema* evince a higher intermembral index than smaller adult *P. verreauxi* (Ravosa *et al.*, 1993). A similarly intriguing postcranial scaling pattern occurs among *P. verreauxi* subspecies, with larger adult *P. v. coquereli* exhibiting higher intermembral indices than smaller adult *P. v. verreauxi* (Ravosa *et al.*, 1993). It would be interesting to extend the ontogenetic analysis of indriid intermembral scaling to the larger, fossil sloth lemurs.

Relative growth patterns also have important implications for understanding the evolutionary development of the intermembral index within a clade. Allometric changes in the intermembral index *across* taxa occur mainly via negative allometry of hindlimb length versus body size (e.g., Shea, 1981, 1984; Jungers, 1984, 1985). It might be assumed that a larger intermembral index results postnatally from strong negative allometry of hindlimb length relative to forelimb length. However, most scaling coefficients for sifaka arm and leg dimensions indicate slight negative allometry to slight positive allometry (Ravosa *et al.*, 1993). The combination of slight allometry of appendicular growth and positive allometry of intermembral indices across taxa suggests that sifaka limb proportions are determined prenatally in large part, rather than being attained via higher postnatal allometry of the forelimb and/or diminished relative growth of the hindlimb. This conclusion is supported by the fact that intermembral indices for diademed and golden-crowned sifakas appear to be present at birth and maintained with little postnatal change (Ravosa *et al.*, 1993), a scaling pattern demonstrated for the limb skeleton in a wide variety of anthropoid clades (Shea, 1981, 1984; Jungers, 1984, 1985; Jungers and Hartman, 1988; Falsetti and Cole, 1992; Jungers and Cole, 1992).

As with the previous two examples, this observation has implications for understanding the application of ontogeny as a criterion of subtraction. Due to less postnatal variation in the intermembral index and given that larger species require larger indices so as to maintain comparable climbing abilities, close relatives show discordant forelimb versus hindlimb allometries (although they are ontogenetically scaled within a limb element or in various cranial regions). Therefore, while some structures in an integrated system may be ontogenetically scaled, in order to maintain functional equivalence across sizes, other components must change adaptively and thus become uncoupled from ancestral relative growth patterns. Such sister taxa often exhibit similar scaling coefficients but different y-intercepts (Cock, 1966; Gould, 1966, 1971; Dodson, 1975a; Vinyard and Ravosa, 1998).

Summary

These three examples illustrate how biomechanical scaling analyses help explain ontogenetic trajectories and highlight how the study of relative growth can help unravel the processes and behaviors underlying interspecific variation in form. Although it is fundamentally important to understand the functional significance of a scaling trajectory (Gould, 1966; Dodson, 1975a,b; Fleagle, 1985; Carrier, 1996), the study of such factors remains intractable in fossils as this requires data on the biological role of an organism (Bock and von Walhert, 1965). As form and behavior are not always perfectly correlated, one potential problem inherent to fossil analyses is that function/behavior is inferred from morphology (Lauder, 1995). Therefore, to more closely detail functional and adaptive inputs into allometric variation among extinct sister taxa, it is incumbent upon us to utilize information on living analogs (Kay and Cartmill, 1977) as well as incorporate available paleoecological evidence.

In the example of papionin symphyseal scaling, experimental information on adult jaw-loading patterns provides a reasonable proxy for the biological role, facilitating a more direct consideration of functional equivalence. However, *in vivo* ontogenetic data on bone-strain patterns and jaw-adductor activity are needed to fully examine this hypothesis. As limb performance has yet to be tested in the field or laboratory, the model of intermembral scaling is based on theoretical and empirical evidence, thus necessitating further research to demonstrate that limb allometry occurs to maintain functional equivalence and adaptation. In this regard, it would be critical to investigate relevant morphological changes vis-a-vis the ontogeny of locomotor behavior at corresponding developmental stages. In fact, such a question served as the major aim of Doran's (1992) field-based tests of Shea's (1981, 1986) allometric studies of the growth and form of the African ape postcranium. Lastly, in the case of dinosaur ontogeny and locomotion, there are relevant *in vivo* strain data on the development of postcranial loading patterns in bipedal birds (Biewener *et al.*, 1986).

In integrating a biomechanical scaling analysis with information on related behaviors, one can better assess if functional parameters are correlated with and/or are required to change with size. Such a perspective is especially important if an extinct species lies outside the size range of its living sister taxa, such as a phyletic giant, or the morphological complex of interest is underlain by considerable and/or complex allometry. In this situation, if scaling analyses indicate that function may vary with body size and if relevant experimental or field data inferring biological role do not exist for a structure, then it is more difficult to determine the evolutionary implications of a newly invaded size range.

Heterochrony

Heterochronic investigations address changes in the timing, rate, and/or sequence of developmental events which underlie global and/or local patterns of morphological differentiation among ancestral and descendent sister taxa (Gould, 1977; Shea, 1988; McNamara, 1995; Klingenberg, 1998). Selection during ontogeny can have novel adaptive or functional implications regarding evolution across clades and the study of such heterochronic transformations can facilitate insight into processes of evolutionary change. Although this need not be the case, heterochronic research is often combined with analyses of relative growth in skeletal elements.

One of the difficulties with assessing heterochrony in fossil taxa is that such analyses require a chronological or temporal framework. While some researchers have used size as proxy for age, this can be problematic as the relationship between size and age can differ between sister taxa (e.g., chimps versus gorillas, or within taxa, such as human sexual dimorphism—Shea, 1988; see also Klingenberg, 1998). Ontogenetic data can be grouped by dental eruption sequence, although caution must be exercised as "dental age" could be uncoupled from chronological age (cf. Ravosa and Ross, 1994), much like when body size is employed as an age surrogate. Identifying individuals by dental-age classes is helpful, however such practices may obscure more subtle details of heterochronic change. In fact, though some assume it is fundamental to have data on "age" to study heterochrony, all one requires is comparative information on the developmental event of interest (Gould, 1977).

Below, we discuss experimental, theoretical, and ontogenetic analyses of the functional and heterochronic significance of primate symphyseal fusion. As such, we aim to show how data on symphyseal ontogeny enrich our understanding of adaptive transformations and phylogenetic affinities among living and fossil anthropoids.

Developmental Timing, Phylogeny, and the Evolution of Morphological Novelties

The phylogenetic and adaptive significance of symphyseal fusion has figured prominently in numerous discussions of anthropoid origins. Crown anthropoids differ from several Eocene basal anthropoids and extant strepsirhines in evincing the derived character state of a fully fused or ossified mandibular symphysis (Fig. 9). Middle Eocene Eosimiidae (*Eosimias*, *Bahinia*) possessed the primitive primate and mammalian condition of an unfused symphysis, while Late Eocene Qatraniinae (*Arsinoea*) and Oligopithecidae (*Catopithecus*) had only partially fused symphyses (Simons, 1989, 1992; Ravosa and Hylander, 1994; Beard *et al.*, 1996; Ravosa, 1996a, 1999; Simons and Rasmussen, 1996; Jaeger *et al.*, 1999). There are also salient differences in the ontogenetic timing of symphyseal fusion which distinguish the Parapithecinae (*Simonsius*, *Apidium*) from catarrhine and platyrrhine anthropoids.

Before discussing symphyseal fusion and heterochrony, we briefly review experimental and morphological investigations regarding the functional significance of symphyseal fusion in primate and nonprimate mammals. Such studies indicate that two jaw-loading regimes and underlying jaw-adductor muscle recruitment patterns during unilateral mastication are the primary determinants of anthropoid symphyseal form and fusion—wishboning (Fig. 6a) and dorsoventral shear (Fig. 6b) (Beecher, 1977, 1979, 1983; Hylander, 1979a,b, 1984, 1985, 1988; Hylander *et al.*, 1987, 1998, 2000; Ravosa, 1991a, 1996a–c, 1999, 2000; Daegling, 1993; Hylander and Johnson, 1994; Ravosa and Hylander, 1994)*. Variation in symphyseal fusion is continuous and directly proportional to the amount of stress along the symphyseal joint during mastication *and* such variation can be parceled into a series of discrete character states each with unique functional underpinnings.

Dorsoventral shear of the symphyseal joint is due to the downward movement of the working-side corpus and the oppositely directed upward movement of the balancing-side corpus. Histological analyses of the mammalian symphysis indicate that the calcified or ossified ligaments in a partially fused joint are those best oriented only to resist dorsoventral shear and that

*Although symphyseal stress during incisal biting is still considered to be important in the evolution of anthropoid symphyseal fusion (Fleagle, 1999), *in vivo* and comparative analyses do not support this view.

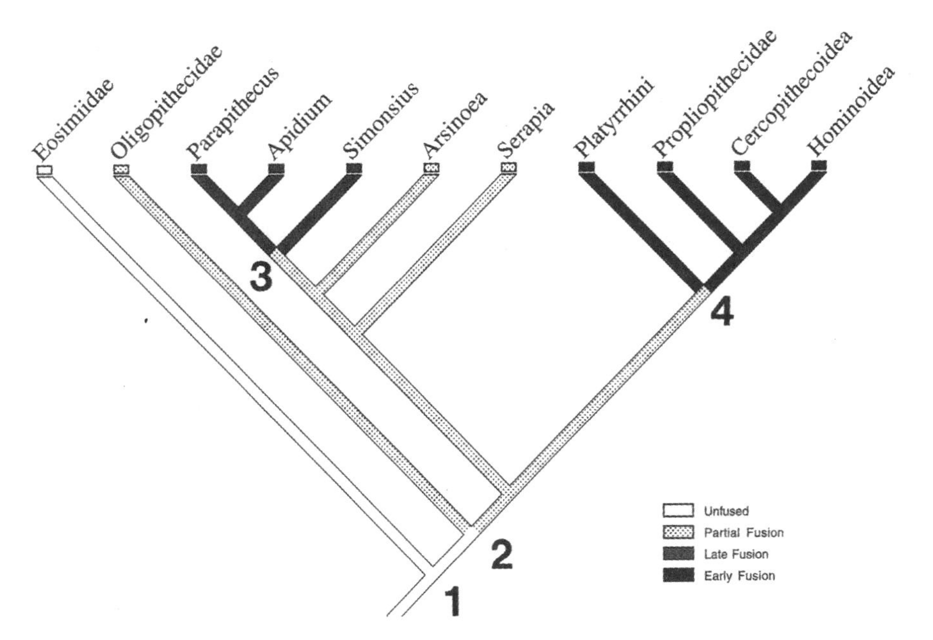

Fig. 9. Evolutionary relationships among major anthropoid clades (cf. Beard *et al.*, 1996; Kay *et al.*, 1997). Associated symphyseal fusion character states for each group are at each node (#1–4). Middle Eocene Eosimiidae evinced the primitive anthropoid condition of an unfused symphysis (#1), while qatraniine parapithecids and oligopithecids of the Late Eocene had partial fusion (#2). The development of a late-fusing symphysis in Early Oligocene parapithecine parapithecids (#3) is independent of the evolution of the early-fusing pattern of platyrrhines and catarrhines (#4). Propliopithecidae possessed an early-fusing symphysis and, as the evolution of this clade occurred after the split between the two anthropoid infraorders, it is parsimonious to infer that all family members also had the derived jaw-adductor activity and jaw-loading patterns characteristic of living catarrhines and platyrrhines (Ravosa, 1999). Note that a new putative eosimiid—*Bahinia*—had an unfused symphysis like *Eosimias* (Jaeger *et al.*, 1999). (Character states for *Parapithecus* and *Serapia* are inferred based on the condition for respective sister taxa.)

there are absolutely greater numbers of these ligaments in such a symphysis (Trevisan and Scapino, 1976a,b; Beecher, 1977, 1979; Hirschfeld *et al.*, 1977; Scapino, 1981). Primates with partial fusion often exhibit discontinuous rugosities which variably span the symphysis and articulate with the opposite dentary, a morphology well-designed solely to counter significantly greater levels of dorsoventral shear (Rigler and Mlinsek, 1968; Scapino, 1981; Beecher, 1983; Ravosa, 1991a, 1996a, 1999; Ravosa and Hylander, 1994; Ravosa and Simons, 1994).

Of the two main symphyseal loading regimes during mastication, wishboning results in higher strain magnitudes and higher stress concentrations than dorsoventral shear (Hylander, 1984; Hylander *et al.*, 1987, 1998). Comparative, theoretical, and *in vivo* evidence also shows that only a fully fused symphysis is sufficiently strong to resist significant wishboning (Rigler and Mlinsek, 1968; Scapino, 1981; Beecher, 1983; Hylander, 1984, 1985; Hylander *et al.*, 1987, 1998, 2000; Ravosa, 1991a, 1996a–c, 1999, 2000; Daegling, 1993; Hylander and Johnson, 1994; Ravosa and Hylander, 1994; Ravosa and Simons, 1994; Vinyard and Ravosa, 1998; Hogue and Ravosa, 2001).*

Recent analyses of symphyseal growth and form have yielded insights into our knowledge of primate symphyseal function (Ravosa and Simons, 1994; Ravosa, 1996a, 1999; Vinyard and Ravosa, 1998). Of specific relevance is information on the postnatal timing of symphyseal ossification relative to the onset of adult jaw-muscle activity patterns and jaw-loading regimes (see discussed above). Such information is critical for interpreting the functional and behavioral implications of heterochronic changes in symphyseal fusion in a clade and are thus basic to an informed characterization of evolutionary transformations of functional-adaptive complexes in the skull of living and fossil primate clades.

The postnatal onset of adult masticatory behaviors and functions (weaning) can be inferred as present in individuals with the first permanent molars erupted (Iinuma *et al.*, 1991; Smith, 1991; Vinyard and Ravosa, 1998). In this regard, the majority of mammals with complete symphyseal fusion by adulthood exhibit late postnatal development of this derived character state (Szalay and Delson, 1979; Beecher, 1983; Gingerich and Sahni, 1984; Ravosa and Hylander, 1994; Ravosa and Simons, 1994; Simons *et al.*, 1995; Ravosa, 1996a, 1999). From a functional standpoint, "late-fusing" mammals experience a prolonged period following weaning in which adult-like masticatory behaviors are routinely employed while the symphysis is partially fused and, in turn, well designed only for countering significant dorsoventral shear. If wishboning were present in a late-fusing mammal,† as the juvenile symphysis of this taxon would have lacked the bony morphology necessary to resist significant wishboning, such forces must have been low in level and less important as a

* Lieberman and Crompton (2000) argue that symphyseal fusion occurs to stiffen the symphyseal joint so as to facilitate the transfer of transversely directed balancing-side jaw-muscle force during mastication. There is substantial evidence from recent *in vivo* (Hylander *et al.*, 1998, 2000) and comparative (Ravosa and Hylander, 1994; Ravosa *et al.*, 2000; Hogue and Ravosa, 2001) investigations demonstrating that fusion instead serves to strengthen the symphysis against symphyseal stress during chewing and biting.

† Due to the rarity of juvenile parapithecine specimens, all we know is that subadult *Simonsius* with P_3, P_4, and M_3 unerupted had fused symphyses (Kay and Simons, 1983; Ravosa, 1999) and subadult *Apidium* also evinced symphyseal fusion (Simons, 1974). Given that the early-fusing pattern is quite rare in nonprimates and, especially, in primates (Ravosa, unpubl.), it is more conservative to infer that parapithecines developed the late-fusing condition found in *Notharctus* (Beecher, 1983), *Leptadapis*, *Adapis* (Ravosa, 1996a), *Sivaladapis* (Gingerich and Sahni, 1984), and *Aframonius* (Simons *et al.*, 1995). This contrasts with the derived early-fusing pattern of catarrhines and platyrrhines (Ravosa, 1999).

functional determinant of symphyseal morphology* (Ravosa and Hylander, 1994; Ravosa and Simons, 1994; Ravosa, 1996a, 1999).

Interestingly, if the ancestral timing of symphyseal ossification is shifted earlier in ontogeny in descendent species to a stage preceding the onset of weaning, this would facilitate the evolution of a jaw-adductor activity pattern resulting in significant wishboning stress during mastication. Considered in an heterochronic context, the change from an unfused or partially fused state to a more derived fully fused condition is the recapitulation of complete symphyseal fusion via acceleration (cf. Gould, 1977). From a functional perspective, this evolutionary transformation is necessary because (1) juvenile mammals develop adult-like jaw-muscle activity and symphyseal loading patterns soon after weaning, (2) subadult and adult motor behaviors do not differ appreciably from those characterizing juveniles, and (3) significant wishboning is best countered by complete synostosis. An "early-fusing" pattern where complete fusion is shifted and established prior to the eruption of the first molar uniquely typifies a representative sample of several dozen platyrrhines and catarrhines, and this differs from the late-fusing condition found in the majority of nonanthropoid mammals with ossified symphyses (Fig. 9) (Ravosa, unpubl.).

Likewise, the evolution of a jaw-adductor activity pattern resulting in marked dorsoventral shear can occur only in species with the development of partial fusion shifted to a postnatal period preceding the onset of stereotypical adult masticatory patterns. The recapitulation of partial symphyseal fusion via acceleration is likely to be common in descendent taxa that ingest greater amounts of obdurate food items (Ravosa and Hylander, 1994; Ravosa, 1996a, 1999). Because relatively greater balancing-side jaw-muscle activity occurs during the mastication of a tougher diet (Hylander *et al.*, 1992, 1998, 2000), this results in elevated dorsoventral shear of the symphysis (Hylander, 1979a,b, 1988; Bouvier, 1986; Cole, 1992; Daegling, 1993; Ravosa, 1996b, 2000). As larger taxa tend to have tougher and/or harder diets than smaller relatives (Kay, 1975) and this results in an allometric pattern of dorsoventral shear, symphyseal fusion often increases with body size across a clade (Beecher, 1977, 1979, 1983; Scapino, 1981; Ravosa, 1991a, 1996a, 1999; Ravosa and Hylander, 1994).

Given that jaw-adductor activity and mandibular loading regimes are similar in catarrhine and platyrrhine anthropoids (Ravosa, 1996c; Hylander *et al.*, 1998, 2000) and that both clades uniformly exhibit early ontogenetic fusion

*Given that the lingual surface of the anthropoid symphysis experiences high strains and high stress concentrations during mastication (Hylander, 1984, 1985), this region should ossify earlier in ontogeny if a taxon also encountered significant wishboning. In fact, the labial-to-lingual gradient of synostosis in juvenile adapids (*Adapis*, *Leptadapis*—Ravosa, 1996a) and subfossil lemurs (*Archaeolemur*, *Hadropithecus*—Ravosa and Simons, 1994) is opposite to such a wishboning prediction. As this ontogenetic pattern of fusion is inconsequential for resisting dorsoventral shear, one can infer that ossification in these and other late-fusing mammals is due mainly to significant dorsoventral shear.

of the symphysis, it is parsimonious to assume that these functional patterns are inherited from a common ancestor and thus homologous (Ravosa, 1999; Ravosa *et al.*, 2000). Employing the *extant phylogenetic bracket* for reconstructing soft-tissue form, function, and behavior (Witmer, 1995), one can further infer that all anthropoid clades postdating the platyrrhine:catarrhine split also possessed the derived wishboning deep-masseter activity pattern and jaw-loading regime. This seems justified for propliopithecids as the symphyseal joint of juvenile *Propliopithecus chirobates* with the first molar erupting exhibited the early-fusing condition of living anthropoids. As catarrhines and platyrrhines also experience significant dorsoventral shear, it is likely that the last common ancestor of this anthropoid group itself developed from an ancestor with a late-fusing symphysis. The evolution of this pattern apparently occurred independent of that for Early Oligocene parapithecines (Ravosa, 1999).

Such ontogenetic information has systematic implications. If oligopithecids and parapithecids are both outgroups to a crown anthropoid clade (Kay *et al.*, 1997), partial fusion in Late Eocene qatraniines and oligopithecids is a character state intermediate functionally, developmentally, and phylogenetically between the plesiomorphic unfused symphysis of Middle Eocene eosimiids and the apomorphic early fusion of a platyrrhine:catarrhine last common ancestor (Fig. 9) (Ravosa, 1999). This interpretation is at odds with recent claims that oligopithecids and catarrhines are sister taxa due to the shared presence of only two premolars (Simons and Rasmussen, 1996; Fleagle, 1999). In this alternative phylogenetic scenario, symphyseal fusion in platyrrhines and catarrhines represents a case of functional convergence *and* oligopithecids (with partially fused symphyses) are sister taxa to a catarrhine ancestor with a fused symphysis. This conclusion assumes that the "anthropoid" pattern of fusion is a simple change in a single character state similar to the loss, for example, of a second premolar. To the contrary, symphyseal fusion in catarrhines and platyrrhines *actually* consists of two integrated suites of derived neuromotor and morphological patterns developed in two stages, the first related to dorsoventral shear and the second to wishboning. It is therefore more parsimonious to interpret the sole character linking oligopithecids and catarrhines as homoplasic.

If oligopithecids were derived from larger-bodied basal catarrhines (Simons and Rasmussen, 1996; Fleagle, 1999), this morphological transformation differs in one significant way from that depicted in the similar evolution of smaller body size and greater insectivory in callitrichids. That is, this scenario depicts oligopithecids as evolving partially fused symphyses from a larger ancestor with a greater degree of ossification, while callitrichids retain the early fusing pattern and masticatory loading regimes inherited from a larger ancestor (Ravosa, 1999).

In fact, due to the greater rarity of the early-fusing pattern and no apparent evidence of reversal in the degree of ossification in a variety of primate (Fig. 9) and nonprimate mammals (Ravosa, 1999, unpubl.), any phylogenetic hypothesis placing oligopithecids and parapithecids as outgroups to a platyrrhine:catarrhine clade (e.g., Kay *et al.*, 1997) is in agreement with the heterochronic, experimental, and theoretical information on anthropoid symphyseal fusion. The phylogenetic importance of symphyseal fusion may not be trivial. Apart from characterizing larger-bodied primates with tougher diets—bigger adapids, subfossil lemurs, parapithecines, crown anthropoids—those taxa exhibiting the greatest degree of ossification are typically more recent and represent the terminal members of a lineage (Ravosa, 1999). Thus, while symphyseal fusion has occurred independently at least 10 times during primate evolution, the apparent lack of reversal in character states within a given clade provides little support for scenarios proposing that any anthropoid *or* adapid with a fused symphysis can be ancestral to an anthropoid with a lesser degree of fusion (*contra* Simons and Rasmussen, 1996; Fleagle, 1999).

Summary

A consideration of the ontogenetic and functional significance of primate symphyseal fusion points to the important role of heterochrony in the development of the early-fusing pattern of extant anthropoids. In this situation, a shift in the timing of an ontogenetic event of functional importance appears to be linked to the evolution of a novel form:function complex—the derived wishboning jaw-muscle activity and jaw-loading pattern. Postnatal data on the onset of primate symphyseal fusion represent a case in which the correlation among functional/behavioral, not chronological, events are important for interpreting the heterochronic bases of evolutionary change. In this case, a consideration of heterochrony in fossil anthropoids has been greatly facilitated by the *in vivo* study of masticatory behavior, and in particular symphyseal form and function, in extant mammals. As opposed to more commonly cited instances of heterochrony in primates (cf. Shea, 1988), early symphyseal fusion constitutes an example of local, not global, heterochrony.

Because heterochronic analyses ultimately focus on various morphological transformations between ancestral and descendent sister taxa, information on the ontogenetic timing of symphyseal fusion in fossil taxa has important implications for understanding the antiquity of evolutionary changes. Determination of the clade in which a novel character state and form:function relationship first evolved facilitates the identification of the proximate cause(s) of adaptive change(s). In crown anthropoids, early postnatal onset of ossification and associated masticatory patterns represent an adaptation occurring at a *juvenile* growth stage. Insights into anthropoid symphyseal fusion offered by a phylogenetic and heterochronic approach make it possible to infer masticatory function and behavior in fossil catarrhines and platyrrhines as well as interpret homology and the polarity of morphological changes in the skull of basal anthropoids. For instance, while homology of symphyseal fusion would have been assessed between parapithecines *and* catarrhine and platyrrhine anthropoids based solely on adult joint morphology, a characterization of the ontogenetic process of symphyseal fusion instead highlights a parapithecine pattern which is not homologous with that of crown anthropoids.

Conclusions

The application of experimental, theoretical, and morphological analyses to living taxa provides a logical first step in figuring out how a biological system functions in fossil species. This is because a much wider variety of sources of information—soft-tissue anatomy, feeding or locomotor behavior—is available for larger, more complete samples of extant (sister) taxa. Although this remains at the core of Kay and Cartmill's (1977) criteria for assessing the functional and behavioral significance of a character or character complex in fossil taxa, this is not always feasible. For instance, the form of an extinct species may differ significantly from that of living taxa, thus diminishing the likelihood that appropriate contemporary analogs exist. Likewise, to the extent possible, it is important to test if the same rules governing the design of adult form can be applied to (fossil) juveniles.

Growth trajectories can serve as a biologically meaningful baseline for characterizing the functional, adaptive, and phylogenetic bases of morphological variation within and among sister taxa. Before undertaking an explicitly ontogenetic study of fossils, there are several methodological issues that require consideration. Some of these apply to comparative neontological analyses as well. First, are there living sister taxa with which to make ontogenetic comparisons? Second, ontogenetic studies require reasonably complete samples and, as the data are typically metric and the analyses can be multivariate, the morphology of the fossil specimens must be relatively intact and possess diagnostic features. Thus, fossil crania may be more amenable to ontogenetic analysis than postcranial specimens. Lastly, to construct a meaningful developmental perspective with which to investigate the adaptive basis of a specific shape, multiple-taxon comparisons necessitate a phylogenetic context. Such a framework can be uniquely enriched via a consideration of fossil evidence.

It is not surprising that we conclude by arguing for an increased emphasis on the use of ontogeny in the investigation of biological form in extant and extinct organisms. To be sure, interspecific analyses of adult data are not invalid or irrelevant, just that without an *in vivo* or theoretical framework, such scaling comparisons can infer function but not necessarily how differences in adult shape arose. A knowledge of development offers unique insight into the evolutionary patterning of biological form, character homology, and functional linkages among skeletal elements (e.g., Lauder, 1982, 1995; Sattler, 1994; Wake and Roth, 1989). Moreover, it facilitates a consideration of the extent to which adult interspecific patterns are epiphenomena of ontogenetic processes as well as the functional consequences of size and size-related changes in shape. While such data are critical to a more informed perspective on inferring function and behavior in extinct taxa, the nature of research on relative growth and heterochrony realistically precludes a consideration of certain kinds of fossil evidence. Thus, from a practical standpoint, interspecific analyses of adults will often be necessary due to the frequent absence of ontogenetic series for many species, especially those of great antiquity and rarity.

ACKNOWLEDGMENTS

We would like to thank the editors for graciously affording us the opportunity to expound on our ideas regarding ontogeny and function. B. Shea, W. Hylander, M. Dagosto, S. Herring, R. Kay, B. Payseur, M. Plavcan, J. Fleagle, W. Jungers, C. Ross, L. Profant, V. Noble, A. Hogue, A. Yoder, C. van Schaik, and several reviewers offered help, comments, and/or information. Personal research discussed in this chapter was supported by the NIH, NSF, Leakey Foundation, American Philosophical Society, Northwestern University, Boise Fund, and Field Museum of Natural History. This is Duke University Primate Center publication #672.

References

Albrecht, G. H., Jenkins, P. D., and Godfrey, L. R. 1990. Ecogeographic size variation among the living and subfossil prosimians of Madagascar. *Am. J. Primatol.* **22:**1–50.

Alexander, R. M. 1989. *Dynamics of Dinosaurs and other Extinct Giants.* Columbia University Press, New York.

Atchley, W. R. 1987. Developmental quantitative genetics and the evolution of ontogenies. *Evol.* **41:**316–330.

Atchley, W. R., and Hall, B. K. 1991. A model for development and evolution of complex morphological structures. *Biol. Rev.* **66:**101–157.

Beard, K. C., Tong, Y., Dawson, M. R., Wang, J., and Huang, J. 1996. Earliest complete dentition of an anthropoid primate from the late Middle Eocene of Shanxi Province, China. *Science* **272:**82–85.

Beecher, R. M. 1977. Function and fusion at the mandibular symphysis. *Am. J. Phys. Anthropol.* **47:**325–336.

Beecher, R. M. 1979. Functional significance of the mandibular symphysis. *J. Morphol.* **159:**117–130.

Beecher, R. M. 1983. Evolution of the mandibular symphysis in Notharctinae (Adapidae, Primates). *Int. J. Primatol.* **4:**99–112.

Biewener, A. A. 1982. Bone strength in small mammals and bipedal birds: Do safety factors change with body size? *J. Exp. Biol.* **98:**289–301.

Biewener, A. A. 1991. Musculoskeletal design in relation to body size. *J. Biomech.* **24:**19–29.

Biewener, A. A. 1993. Safety factors in bone strength. *Calc. Tissue Int.* **53:**568–574.

Biewener, A. A., and Bertram, J. E. A. 1993a. Mechanical loading and bone growth *in vivo.* In: B. K. Hall (ed.), *Bone. Volume 7: Bone Growth*, pp. 1–36. CRC Press, Boca Raton.

Biewener, A. A., and Bertram, J. E. A. 1993b. Skeletal strain patterns in relation to exercise training during growth. *J. Exp. Biol.* **185:**51–69.

Biewener, A. A., Swartz, S. M., and Bertram, J. E. A. 1986. Bone modeling during growth: Dynamic strain equilibrium in the chick tibiotarsus. *Calc. Tissue Int.* **39:**390–395.

Biknevicius, A. R., and Leigh, S. R. 1997. Patterns of growth of the mandibular corpus in spotted hyenas (*Crocuta crocuta*) and cougars (*Puma concolor*). *Zool. J. Linn. Soc.* **120:**139–161.

Bock, W. J., and von Walhert, G. 1965. Adaptation and the form-function complex. *Evolution* **19:**269–299.

Bouvier, M. 1986. Biomechanical scaling of mandibular dimensions in New World monkeys. *Int. J. Primatol.* **7:**551–567.

Carrier, D. R. 1996. Ontogenetic limits on locomotor performance. *Physiol. Zool.* **69:**467–488.

Cartmill, M. 1974. Pads and claws in arboreal locomotion. In: F. A. Jenkins (ed.), *Primate Locomotion*, pp. 45–83. Academic Press, New York.

Cartmill, M. 1985. Climbing. In: M. Hildebrand, D. M. Bramble, K. F. Liem, and D. B. Wake (eds.), *Functional Vertebrate Morphology*, pp. 73–88. Harvard University Press, Cambridge.

Cheverud, J. M. 1982. Relationships among ontogenetic, static, and evolutionary allometry. *Am. J. Phys. Anthropol.* **59:**139–149.

Cheverud, J. M. 1984. Quantitative genetics and developmental constraints on evolution by selection. *J. Theor. Biol.* **110:**155–171.

Cochard, L. R. 1985. Ontogenetic allometry of the skull and dentition of the rhesus monkey (*Macaca mulatta*). In: W. L. Jungers (ed.), *Size and Scaling in Primate Biology*, pp. 231–256. Plenum Press, New York.

Cochard, L. R. 1987. Postcanine tooth size in female primates. *Am. J. Phys. Anthropol.* **74:**47–54.

Cock, A. G. 1966. Genetical aspects of metrical growth and form in animals. *Q. Rev. Biol.* **41:**131–190.

Cole, T. M. 1992. Postnatal heterochrony of the masticatory apparatus in *Cebus apella* and *Cebus albifrons*. *J. Hum. Evol.* **23:**253–282.

Daegling, D. J. 1993. Functional morphology of the human chin. *Evol. Anthropol.* **1:**170–177.

Deaner, R. O., and Nunn, C. L. 1999. How quickly do brains catch up with bodies? A comparative method for detecting evolutionary lag. *J. Zool., London* **266:**687–694.

Dodson, P. 1975a. Relative growth in two sympatric species of *Sceloporus*. *Am. Midl. Nat.* **94:**421–450.

Dodson, P. 1975b. Functional and ecological significance of relative growth in *Alligator*. *J. Zool., London* **175:**315–355.

Doran, D. M. 1992. The ontogeny of chimpanzee and pygmy chimpanzee locomotor behavior: A case study of morphological paedomorphism and its behavioral correlates. *J. Hum. Evol.* **23:**139–157.

Emerson, S. B., Greene, H. W., and Charnov, E. L. 1994. Allometric aspects of predator-prey interactions. In: P. C. Wainwright and S. M. Reilly (eds.), *Ecological Morphology*, pp. 123–139. University of Chicago Press, Chicago.

Falsetti, A. B., and Cole, T. M. 1992. Relative growth of the postcranial skeleton in callitrichids. *J. Hum. Evol.* **23:**79–92.

Fleagle, J. G. 1985. Size and adaptation in primates. In: W. L. Jungers (ed.), *Size and Scaling in Primate Biology*, pp. 1–19. Plenum Press, New York.

Fleagle, J. G. 1999. *Primate Adaptation and Evolution.* Second Edition Academic Press, New York.

Freedman, L. 1962. Growth of muzzle length relative to calvaria length in *Papio*. *Growth* **162:**117–128.

Gingerich, P. D., and Sahni, A. 1984. Dentition of *Sivaladapis nagrii* (Adapidae) from the late Miocene of India. *Int. J. Primatol.* **5:**63–79.

Godfrey, L. R., King, S. J., and Sutherland, M. R. 1998. Heterochronic approaches to the study of locomotion. In: E. Strasser, J. Fleagle, A. Rosenberger, and H. McHenry (eds.), *Primate Locomotion. Recent Advances*, pp. 277–307, Plenum Press, New York.

Gould, S. J. 1966. Allometry and size in ontogeny and phylogeny. *Biol. Rev.* **41:**587–640.

Gould, S. J. 1971. Geometric similarity in allometric growth: A contribution to the problem of scaling in the evolution of size. *Am. Nat.* **105:**113–136.

Gould, S. J. 1975a. Allometry in primates, with emphasis on scaling and the evolution of the brain. In: F. S. Szalay (ed.), *Approaches to Primate Paleobiology. Contributions to Primatology, Volume 5*, pp. 244–292. S. Karger, Basel.

Gould, S. J. 1975b. On the scaling of tooth size in mammals. *Am. Zool.* **15:**351–362.

Gould, S. J. 1977. *Ontogeny and Phylogeny*. Belknap Press, Cambridge.

Heinrich, R. E., Ruff, C. B., and Weishampel, D. B. 1993. Femoral ontogeny and locomotor biomechanics of *Dryosaurus lettowvorbecki* (Dinosauria, Iguanodontia). *Zool. J. Linn. Soc.* **108:**179–196.

Herring, S. W. 1985. The ontogeny of mammalian mastication. *Am. Zool.* **25:**339–349.

Herring, S. W., Anapol, F. C., and Wineski, L. E. 1991. Motor-unit territories in the masseter muscle of infant pigs. *Arch. Oral Biol.* **36:**867–873.

Hirschfeld, Z., Michaeli, Y., and Weinreb, M. M. 1977. Symphysis menti of the rabbit: Anatomy, histology, and postnatal development. *J. Dent. Res.* **56:**850–857.

Hogue, A. S., and Ravosa, M. J. 2001. Transverse masticatory movements, occlusal orientation and symphyseal fusion in selenodont artiodactyls. *J. Morphol.* **249** (in press).

Horner, J. R., and Weishampel, D. B. 1988. A comparative embryological study of two ornithischian dinosaurs. *Nature* **332:**256–257.

Huang, X., Zhang, G., and Herring, S. W. 1994. Age changes in mastication in the pig. *Comp. Biochem. Physiol.* **107A:**647–654.

Hylander, W. L. 1979a. Mandibular function in *Galago crassicaudatus* and *Macaca fascicularis*: An *in vivo* approach to stress analysis of the mandible. *J. Morphol.* **159:**253–296.

Hylander, W. L. 1979b. The functional significance of primate mandibular form. *J. Morphol.* **160:**223–240.

Hylander, W. L. 1984. Stress and strain in the mandibular symphysis of primates: A test of competing hypotheses. *Am. J. Phys. Anthropol.* **64:**1–46.

Hylander, W. L. 1985. Mandibular function and biomechanical stress and scaling. *Am. Zool.* **25:**315–330.

Hylander, W. L. 1988. Implications of *in vivo* experiments for interpreting the functional significance of "robust" australopithecine jaws. In: F. E. Grine (ed.), *Evolutionary History of the "Robust" Australopithecines*, pp. 55–83. Aldine de Gruyter, New York.

Hylander, W. L., and Johnson, K. R. 1994. Jaw muscle function and wishboning of the mandible during mastication in macaques and baboons. *Am. J. Phys. Anthropol.* **94:**523–547.

Hylander, W. L., Johnson, K. R., and Crompton, A. W. 1987. Loading patterns and jaw movements during mastication in *Macaca fascicularis*: A bone-strain, electromyographic and cineradiographic analysis. *Am. J. Phys. Anthropol.* **72:**287–314.

Hylander, W. L., Johnson, K. R., and Crompton, A. W. 1992. Muscle force recruitment and biomechanical modeling: An analysis of masseter muscle function during mastication in *Macaca fascicularis*. *Am. J. Phys. Anthropol.* **88:**365–387.

Hylander, W. L., Ravosa, M. J., Ross, C. F., and Johnson, K. R. 1998. Mandibular corpus strain in primates: Further evidence for a functional link between symphyseal fusion and jaw-adductor muscle force. *Am. J. Phys. Anthropol.* **107:**257–271.

Hylander, W. L., Ravosa, M. J., Ross, C. F., Wall, C. E., and Johnson, K. R. 2000. Symphyseal fusion and jaw-adductor muscle force: An EMG study. *Am. J. Phys. Anthropol.* **112:**469–492.

Iinuma, M., Yoshida, S., and Funakoshi, M. 1991. Development of masticatory muscles and oral behavior from suckling to chewing in dogs. *Comp. Biochem. Physiol.* **100A:**789–794.

Inouye, S. E., and Shea, B. T. 1997. What's your angle? Size-correction and bar-glenoid orientation in "Lucy" (A.L. 288-1). *Int. J. Primatol.* **18:**629–650.

Jaeger, J. J., Thein, T., Benammi, M., Chaimanee, Y., Soe, A. N., Lwin, T., Tun, T., Wai, S., and Ducrocq, S. 1999. A new primate from the Middle Eocene of Myanmar and the Asian early origin of anthropoids. *Science* **286:**528–530.

Jolly, C. J. 1972. The classification and natural history of *Theropithecus* (*Simopithecus*) (Andrews, 1916), baboons of the African Plio-Pleistocene. *Bull. Br. Mus. Nat. Hist.* **22:**1–123.

Jungers, W. L. 1978. The functional significance of skeletal allometry in *Megaladapis* in comparison to living prosimians. *Am. J. Phys. Anthropol.* **19:**303–314.

Jungers, W. L. 1984. Aspects of size and scaling in primate biology with special reference to the locomotor skeleton. *Yearb. Phys. Anthropol.* **27:**73–97.

Jungers, W. L. 1985. Body size and scaling of limb proportions in primates. In: W. L. Jungers (ed.), *Size and Scaling in Primate Biology*, pp. 345–382. Plenum Press, New York.

Jungers, W. L., and Cole, M. S. 1992. Relative growth and shape of the locomotor skeleton in lesser apes. *J. Hum. Evol.* **23:**93–105.

Jungers, W. L., and Fleagle, J. G. 1980. Postnatal growth allometry of the extremities in *Cebus albifrons* and *Cebus apella*: A longitudinal and comparative study. *Am. J. Phys. Anthropol.* **53:**471–478.

Jungers, W. L., and Hartman, S. E. 1988. Relative growth of the locomotor skeleton in orang-utans and other large-bodied hominoids. In: J. H. Schwartz (ed.), *Orangutan Biology*, pp. 347–359. Oxford University Press, Oxford.

Jungers, W. L., Falsetti, A. B., and Wall, C. E. (1995) Shape, relative size, and size-adjustments in morphometrics. *Yearb. Phys. Anthropol.* **38:**137–161.

Kay, R. F. 1975. The functional adaptations of primate molar teeth. *Am. J. Phys. Anthropol.* **43:**195–215.

Kay, R. F., and Cartmill, M. 1977. Cranial morphology and adaptations of *Palaechthon nacimienti* and other Paromomyidae (Plesiadapoidea, ?Primates), with a description of a new genus and species. *J. Hum. Evol.* **6:**19–35.

Kay, R. F., and Simons, E. L. 1983. Dental formulae and dental eruption patterns in Parapithecidae (Primates, Anthropoidea). *Am. J. Phys. Anthropol.* **62:**363–375.

Kay, R. F., Ross, C., and Williams, B. A. 1997. Anthropoid origins. *Science* **275:**797–804.

Keller, T. S., and Spengler, D. M. 1989. Regulation of bone stress and strain in the immature and mature rat femur. *J. Biomech.* **22:**1115–1127.

Klingenberg, C. P. 1998. Heterochrony and allometry: The analysis of evolutionary change in ontogeny. *Biol. Rev.* **73:**79–123.

Lande, R. 1979. Quantitative genetic analysis of multivariate evolution, applied to brain:body size allometry. *Evol.* **33:**402–416.

Lanyon, L. E., and Rubin, C. T. 1985. Functional adaptation in skeletal structures. In: M. Hildebrand, D. M. Bramble, K. F. Liem, and D. B. Wake (eds.), *Functional Vertebrate Morphology*, pp. 1–25. Harvard University Press, Cambridge.

Lauder, G. V. 1982. Historical biology and the problem of design. *J. Theor. Biol.* **97:**57–67.

Lauder, G. V. 1995. On the inference of function from structure. In: J. J. Thomason (ed.), *Functional Morphology in Vertebrate Paleontology*, pp. 1–18. Cambridge University Press, Cambridge.

Lieberman, D. E., and Crompton, A. W. 2000. Why fuse the mandibular symphysis? A comparative analysis. *Am. J. Phys. Anthropol.* **112:**517–540.

McCollum, M. A. 1994. Mechanical and spatial determinants of *Paranthropus* facial form. *Am. J. Phys. Anthropol.* **93:**259–273.

McNamara, K. J. (ed). 1995. *Evolutionary Change and Heterochrony*. Wiley, London.

Plavcan, J. M., and van Schaik, C. P. 1992. Intrasexual competition and canine dimorphism in anthropoid primates. *Am. J. Phys. Anthropol.* **87:**461–477.

Price, T., and Langen, T. 1992. Evolution of correlated characters. *TREE* **7:**307–310.

Profant, L. 1995. Historical allometric inputs to interspecific patterns of craniofacial diversity in the cercopithecine tribe Papionini. *Am. J. Phys. Anthropol.* **20:**175.

Rak, Y. 1983. *The Australopithecine Face*. Academic Press, New York.

Ravosa, M. J. 1988. Browridge development in Cercopithecidae: A test of two models. *Am. J. Phys. Anthropol.* **76:**535–555.

Ravosa, M. J. 1991a. Structural allometry of the mandibular corpus and symphysis in prosimian primates. *J. Hum. Evol.* **20:**3–20.

Ravosa, M. J. 1991b. Ontogenetic perspective on mechanical and nonmechanical models of primate circumorbital morphology. *Am. J. Phys. Anthropol.* **85:**95–112.

Ravosa, M. J. 1991c. Interspecific perspective on mechanical and nonmechanical models of primate circumorbital morphology. *Am. J. Phys. Anthropol.* **86:**363–396.

Ravosa, M. J. 1991d. The ontogeny of cranial sexual dimorphism in two Old World monkeys: *Macaca fascicularis* (Cercopithecinae) and *Nasalis larvatus* (Colobinae). *Int. J. Primatol.* **12:**403–426.

Ravosa, M. J. 1992. Allometry and heterochrony in extant and extinct Malagasy primates. *J. Hum. Evol.* **23:**197–217.

Ravosa, M. J. 1996a. Mandibular form and function in North American and European Adapidae and Omomyidae. *J. Morphol.* **229:**171–190.

Ravosa, M. J. 1996b. Jaw morphology and function in living and fossil Old World monkeys. *Int. J. Primatol.* **17:**909–932.

Ravosa, M. J. 1996c. Experimental analysis of masticatory function in capuchin monkeys. *Am. J. Phys. Anthropol. Suppl.* **22:**194.

Ravosa, M. J. 1996d. Jaw scaling and biomechanics in fossil taxa. *J. Hum. Evol.* **30:**159–160.

Ravosa, M. J. 1998. Cranial allometry and geographic variation in slow lorises (*Nycticebus*). *Am. J. Primatol.* **45:**225–243.

Ravosa, M. J. 1999. Anthropoid origins and the modern symphysis. *Folia Primatol.* **70:**65–78.

Ravosa, M. J. 2000. Size and scaling in the mandible of living and extinct apes. *Folia Primatol.* **71:**305–322.

Ravosa, M. J., and Hylander, W. L. 1994. Function and fusion of the mandibular symphysis in primates: Stiffness or strength? In: J. G. Fleagle, and R. F. Kay (eds.), *Anthropoid Origins*, pp. 447–468. Plenum Press, New York.

Ravosa, M. J., and Profant, L. P. 2000. Evolutionary morphology of the skull in Old World monkeys. In: P. F. Whitehead, and C. J. Jolly (eds.), *Old World Monkeys*, pp. 237–268. Cambridge University Press, Cambridge.

Ravosa, M. J., and Ross, C. F. 1994. Craniodental allometry and heterochrony in two howler monkeys: *Alouatta seniculus* and *A. palliata*. *Am. J. Primatol.* **33:**277–299.

Ravosa, M. J., and Simons, E. L. 1994. Mandibular growth and function in *Archaeolemur*. *Am. J. Phys. Anthropol.* **95:**63–76.

Ravosa, M. J., Meyers, D. M., and Glander, K. E. 1993. Relative growth of the limbs and trunk in sifakas: Heterochronic, ecological, and functional considerations. *Am. J. Phys. Anthropol.* **92:**499–520.

Ravosa, M. J., Vinyard, C. J., Gagnon, M., and Islam, S. A. 2000. Evolution of anthropoid jaw loading and kinematic patterns. *Am. J. Phys. Anthropol.* **112:**493–516.

Rayner, J. M. V., and Wootton, R. J. (eds.) 1991. *Biomechanics in Evolution.* Cambridge University Press, Cambridge.

Richtsmeier, J. T., Corner, B. D., Grausz, H. M., Cheverud, J. M., and Danahey, S. E. 1993. The role of postnatal growth pattern in the production of facial morphology. *Syst. Biol.* **42:**307–330.

Rigler, L., and Mlinsek, B. 1968. Die Symphyse der Mandibula beim Rinde. Ein Beitrag zur Kenntnis ihrer Struktur und Funktion. *Anat. Anz.* **122:**293–314.

Riska, B., and Atchley, W. R. 1985. Genetics of growth predicts patterns of brain-size evolution. *Science* **229:**668–671.

Roth, V. L. 1990. Insular dwarf elephants: A case study in body mass estimation and ecological inference. In: J. Damuth, and B. J. MacFadden (eds.), *Body Size in Mammalian Paleobiology: Estimation and Biological Implications*, pp. 151–179. Cambridge University Press, Cambridge.

Rubin, C. T., and Lanyon, L. E. 1984. Dynamic strain similarity in vertebrates: An alternative to allometric limb bone scaling. *J. Theor. Biol.* **107:**321–327.

Sattler, R. 1994. Homology, homeosis, and process morphology in plants. In: B. K. Hall (ed.), *Homology: The Hierarchical Basis of Comparative Biology*, pp. 423–475. Academic Press, New York.

Scapino, R. P. 1981. Morphological investigation into functions of the jaw symphysis in carnivorans. *J. Morphol.* **167:**339–375.

Selker, F., and Carter, D. R. 1989. Scaling of long bone fracture strength with animal mass. *J. Biomech.* **22:**1175–1183.

Shea, B. T. 1981. Relative growth of the limbs and trunk in the African apes. *Am. J. Phys. Anthropol.* **56:**179–201.

Shea, B. T. 1983a. Size and diet in the evolution of African ape craniodental form. *Folia Primatol.* **40:**32–68.

Shea, B. T. 1983b. Phyletic size change and brain/body allometry: A consideration based on the African pongids and other primates. *Int. J. Primatol.* **4:**33–62.

Shea, B. T. 1984. An allometric perspective on the morphological and evolutionary relationships between pygmy (*Pan paniscus*) and common (*Pan troglodytes*) chimpanzees. In: R. L. Susman (ed.), *The Pygmy Chimpanzee. Evolutionary Biology and Behavior*, pp. 89–130. Plenum Press, New York.

Shea, B. T. 1985. Ontogenetic allometry and scaling: A discussion based on the growth and form of the skull in African apes. In: W. L. Jungers (ed.), *Size and Scaling in Primate Biology*, pp. 175–205. Plenum Press, New York.

Shea, B. T. 1986. Scapular form and locomotion in chimpanzee evolution. *Am. J. Phys. Anthropol.* **70:**475–488.

Shea, B. T. 1988. Heterochrony in primates. In: M. L. McKinney (ed.), *Heterochrony in Evolution*, pp. 237–266. Plenum Press, New York.

Shea, B. T. 1995. Ontogenetic scaling and size correction in the comparative study of primate adaptations. *Anthropologie*. **33:**1–16.

Shea, B. T., and Gomez, A. M. 1988. Tooth scaling and evolutionary dwarfism: An investigation of allometry in human pygmies. *Am. J. Phys. Anthropol.* **77:**117–132.

Shea, B. T., Hammer, R. E., and Brinster, R. L. 1987. Growth allometry of the organs in giant transgenic mice. *Endocrinology*. **121:**1924–1930.

Shea, B. T., Hammer, R. E., Brinster, R. L., and Ravosa, M. J. 1990. Relative growth of the skull and postcranium in giant transgenic mice. *Genet. Res., Cambr.* **56:**21–34.

Simons, E. L. 1974. *Parapithecus grangeri* (Parapithecidae, Old World Higher Primates): New species from the Oligocene of Egypt and the initial differentiation of Cercopithecoidea. *Postilla* **166:**1–12.

Simons, E. L. 1989. Description of two genera and species of Late Eocene Anthropoidea from Egypt. *Proc. Natl. Acad. Sci. USA* **86:**9956–9960.

Simons, E. L. 1992. Diversity in the early Tertiary anthropoidean radiation in Africa. *Proc. Natl. Acad. Sci., USA* **89:**10743–10747.

Simons, E. L., and Rasmussen, D. T. 1996. Skull of *Catopithecus browni*, an early Tertiary catarrhine. *Am. J. Phys. Anthropol.* **100:**261–292.

Simons, E. L., Rasmussen, D. T., and Gingerich, P. D. 1995. New cercamoniine adapid from Fayum, Egypt. *J. Hum. Evol.* **29:**577–589.

Smith, B. H. 1991. Age of weaning approximates age of emergence of the first permanent molar in nonhuman primates. *Am. J. Phys. Anthropol. Suppl.* **12:**163–164.

Smith, R. J. 1993. Categories of allometry: Body size versus biomechanics. *J. Hum. Evol.* **24:**173–182.

Susman, R. L., Stern, J. T., and Jungers, W. L. 1984. Arboreality and bipedality in the Hadar hominids. *Folia Primatol.* **43:**113–156.

Szalay, F. S., and Delson, E. 1979. *Evolutionary History of the Primates*. Academic Press, New York.

Tan, C. L. 1999. Group composition, home range size, and diet of three sympatric bamboo lemur species (Genus *Hapalemur*) in Ranomafana National Park, Madagascar. *Int. J. Primatol.* **20:**547–566.

Thomason, J. J. (ed.). 1995. *Functional Morphology in Vertebrate Paleontology*. Cambridge University Press, Cambridge.

Trevisan, R. A., and Scapino, R. P. 1976a. Secondary cartilages in growth and development of the symphysis menti in the hamster. *Acta Anat.* **94:**40–58.

Trevisan, R. A., and Scapino, R. P. 1976b. The symphyseal cartilage and growth of the symphysis menti in the hamster. *Acta Anat.* **96:**335–355.

Vinyard, C. J., and Ravosa, M. J. 1998. Ontogeny, function, and scaling of the mandibular symphysis in papionin primates. *J. Morphol.* **235:**157–175.

Wake, D. B., and Roth, G. (eds.). 1989. *Complex Organismal Functions: Integration and Evolution in Vertebrates*. S. Bernhard, Dahlem.

Watts, D. P. 1985. Observations on the ontogeny of feeding behavior in mountain gorillas (*Gorilla gorilla beringei*). *Am. J. Primatol.* **8:**1–10.

Weijs, W. A., Brugman, P., and Grimbergen, C. A. 1989. Jaw movements and muscle activity during mastication in growing rabbits. *Anat. Rec.* **224:**407–416.

Westneat, M. W., and Hall, W. G. 1992. Ontogeny of feeding motor patterns in infant rats: An electromyographic analysis of suckling and chewing. *Behav. Neurol.* **106:**539–554.

Williams, S. H., and Wall, C. E. 1999. Morphological correlates of gummivory in the skull of prosimian primates. *Am. J. Phys. Anthropol. Suppl.* **28:**278.

Witmer, L. M. 1995. The extant phylogenetic bracket and the importance of reconstructing soft tissues in fossils. In: J. J. Thomason (ed.), *Functional Morphology in Vertebrate Paleontology*, pp. 19–33. Cambridge University Press, Cambridge.

Dental Ontogeny and Life-History Strategies: The Case of the Giant Extinct Indroids of Madagascar

4

LAURIE R. GODFREY, ANDREW J. PETTO, and MICHAEL R. SUTHERLAND

Introduction

While investigating ecogeographic size variation in the extant and extinct lemurs of Madagascar (Albrecht *et al.*, 1990; Godfrey *et al.*, 1990), one of us (LRG) noted the occurrence in the collections of the Académie Malgache of several unusually small demimandibles of *Mesopropithecus*, an extinct "sloth lemur" (or palaeopropithecid) from southwest Madagascar. The specimens appeared, at first glance, to belong to adults; all of the permanent teeth were fully erupted. But the jaws were little more than two-thirds the size of those of adult *Mesopropithecus globiceps* from the same localities. They did not appear to belong to a new and smaller species of *Mesopropithecus*, because the teeth were identical in size and morphology to those of *M. globiceps*.

An examination of jaws of other palaeopropithecids, including the chimpanzee-sized *Palaeopropithecus maximus* and the gorilla-sized *Archaeoindris fontoynontii*, revealed the same phenomenon. Despite considerable variation in the

---

LAURIE R. GODFREY • Department of Anthropology, University of Massachusetts, Amherst, Massachusetts 01003-9278. ANDREW J. PETTO • Division of Liberal Arts, University of the Arts, Philadelphia, Pennsylvania 19102-4994. MICHAEL R. SUTHERLAND • Statistical Consulting Center, University of Massachusetts, Amherst, Massachusetts 01003-9337.

Reconstructing Behavior in the Primate Fossil Record, edited by Plavcan *et al.* Kluwer Academic/Plenum Publishers, New York, 2002.

size of the jaws of individuals belonging to single species at single subfossil localities, not one specimen bore either maxillary or mandibular milk teeth. Only one retained an alveolus for an (accidentally missing) deciduous tooth. In contrast, most nonpalaeopropithecid species (including extremely rare ones such as *Hadropithecus stenognathus*) had known milk dentitions. This prompted Godfrey to launch an investigation of the dental developmental schedules of living and extinct lemurs (Godfrey, 1993; Seachrist, 1996; Godfrey *et al.*, 1997, 1998; Samonds *et al.*, 1999; Godfrey *et al.*, 2001). The goal was to discover the adaptive foundation for what seemed a most un-primate-like adaptation—run-away dental development in some of the largest-bodied primates ever to have lived.

The fossil record is our most direct window into the biological diversity of the past. Ontogenies and life histories of extinct species are important components of that diversity. Even when natural selection specifically targets adult phenotypes, it is the genetic blueprints for the development of those phenotypes that evolve. Life histories also evolve; they are, in effect, the manifestations of ontogenies played out within population contexts. Evolutionary changes in ontogenetic trajectories can affect the relative abundance in any population of individuals at different life-cycle stages; these in turn impact population growth. Thus, the population dynamics of species (the demographic and population genetic consequences of developmental variation) can be said to evolve. If paleontologists can extract ontogenetic pathways from the fossil record, then we are not constrained to see life histories and population dynamics solely in terms of extant organisms. It is often the novel, the unanticipated, that best informs theory.

This chapter explores strategies for inferring the behavioral ontogenies, life histories, and population dynamics of extinct species. We examine two realms of developmental research—evolutionary developmental biology and life-history studies—and show how each might help us to better understand the interface between skeletons and behavior, and, through that interface, the bizarre world of the giant extinct indroids of Madagascar. Jungers *et al.* (this volume, Chapter 10) discuss other studies of subfossil lemurs, and their implications for behavioral reconstruction.

Development and the Reconstruction of the Behavior and Life Histories of Extinct Primates: Approaches and Methodologies

Growth, Development, and Behavior

Developmental biologists study the generation of form. Evolutionary developmental biologists study the evolution of species differences in developmental trajectories—i.e., how developmental differences arise. This field has deep historic roots. Ontogeny was the centerpiece of evolutionary biology at the turn of the 20th century and during the early 1900s; the most prominent evolutionary biologists of that era (e.g., Huxley, 1932) argued that differences in adult form must be analyzed and understood in terms of how they are generated. With the advent of neo-Darwinism, ontogeny lost its centrality in evolutionary theory to genetics and population biology. The birth of molecular and developmental genetics effectively brought ontogeny back to the forefront, and fostered a new "Developmental Synthesis" at the intersection of many fields (morphology, ecology, systematics, developmental genetics, and population genetics); see Alberch, 1985; Gilbert *et al.*, 1996; Gilbert, 1997.

A fundamental task of evolutionary developmental biology is the discovery of "*dissociable*" or "*modular*" parts (Gould, 1977; McKinney and McNamara, 1991). Modular parts are suites of traits that are *functionally*- and/or *developmentally-integrated*. Functional and developmental integration are types of *morphological integration* (phenotypic interdependency) operating at the individual level (Cheverud, 1996). Developmental integration occurs when trait linkages are epigenetically preserved. Developmental modularity can be elucidated through the study of cellular interactions and tissue induction sequences in early development (essentially, developmental cascades, or causal sequences) or through the study of the relative development (in different taxa and at standardized points throughout ontogeny) of suites of traits. Functional integration occurs when traits contribute to the same functional unit (e.g., occluding teeth). Functional and developmental integration can result in *evolutionary integration*. This occurs when selection cannot operate on some traits without affecting others. Thus, in effect, *different* traits within single modular units function as single traits from the perspective of evolution, but selection operates independently on traits belonging to different modular units.

Olson and Miller (1958) hypothesized that the degree to which traits show morphological integration (or phenotypic correlation) in adults depends on their functional and/or developmental integration. In other words, traits that *appear* (on the basis of their strong adult correlations) to have co-evolved, are indeed linked by common function and/or development. A large body of research (Cheverud, 1982, 1995, 1996; Atchley, 1984; Atchley *et al.*, 1985; Zelditch, 1996; Churchill 1996) has tended to support the Olson/Miller hypothesis. Phenotypic correlations can be fairly well predicted by trait function or by the dynamics of trait development; functionally- and developmentally-related traits do tend to evolve as coordinated modular units. Adult form arises through a mosaic of developmentally-integrated modules following at least partially independent clocks (Gould, 1977; Tanner, 1990; Morbeck, 1997). Patterns of developmental integration can themselves evolve through the "*decoupling*" of previously-linked traits or through the evolution of new trait associations (Lauder, 1981; Zelditch, 1988, 1996; Fink and Zelditch, 1995). *Trait dissociation* occurs when *ancestral* developmental linkages are broken in the descendant. Particular combinations of developmentally-integrated complexes comprise the historical constraints of alternative life-history strategies (Thomas and Reif 1993).

Normally, developmental modularity is diagnosed through the comparative analysis of ontogenetic trajectories. But because development can affect trait correlations among adults, clade-specific differences in adult trait correlations can provide indirect evidence of developmental modularity. In addition, variation in the relative development of groups of traits sometimes provides indirect clues to the *absolute* pace of growth. For example, species with *relatively* early completion of dental eruption (cf. overall somatic growth and maturation) tend to exhibit early cessation of growth (or, at least brain growth) on an *absolute* scale (see Smith, 1992). Variation in the absolute pace of growth can in turn affect ecology and behavior. Leigh and Terranova (1998) suggest that skeletal dimorphism may be constrained by rates of postnatal growth. Similarly, ecogeographic patterns of body size variation may be so affected (see below).

A huge literature exists on mating behavior, agonism, and skeletal dimorphism (see Plavcan this volume, Chapter 8). Low levels of canine or size dimorphism (or indeed monomorphism) do not necessarily imply low agonism in intrasexual encounters (Plavcan *et al.* 1995; Kappeler, 1996b). However, assuming selection operates independently on the weapon teeth of males and females (Greenfield, 1996; Plavcan, 1998), the intensity of agonism in males and females may be accessible through an analysis of variation in the size of weapon teeth relative to some size surrogate (Plavcan *et al.* 1995).* Sex role differentiation in agonistic encounters can be assessed through a comparison of absolute or relative canine size in males and in females.

Life-History Strategies

In many ways, the connection between development and life history is obvious. The events and milestones that we mark when studying life histories are precisely the same as those that we use to explore developmental phenomena. These include gestation lengths, maternal–infant mass ratios, postpartum growth rates, age at weaning, age at first reproduction, reproductive span, and interbirth intervals. Variation for any of the systems of the body in the duration or rate of growth, or age at maturation, may have profound effects on a species' life history. Individual variation in the scheduling of life-cycle events may allow some individuals to avoid predators or other mortality risks better than others, particularly at specific stages of their development. Furthermore, even minor differences in the age at sexual maturity can have profound effects on the

*Among lemurs, male and female canines may covary as a by-product of their common genetic control (Plavcan, 1998). For our purposes, however, the important observation is that, among lemurs, male and female conspecifics do tend to have comparable levels of agonism. Whatever the underlying genetic control, relative canine size reflects agonism in both sexes.

life-time reproductive output of individuals, particularly when considered within the context of species-typical litter sizes, interbirth intervals, and infant survival profiles (Cole, 1954). Populations and species show variation within typical ranges for all life-history parameters.

Life-history *strategies* evolve as relationships among life-cycle *stages* change. A species' life-history strategy depends on how the species constructs and utilizes the relationships among life-cycle stages (however the latter are defined—i.e., whether by somatic growth, dental emergence and wear or reproductive maturation).* It is the outcome of a set of dynamic relationships within and among stages. The number of individuals in any life-cycle stage depends on their survival rates from previous stages, as well as competition among individuals within that stage and the length of time necessary to complete it.

Among different stages, there are several important interactions. Each stage contributes positively to the others by reproduction (adults add new infants to the population), recruitment (adults help infants to survive and be recruited into the juvenile stage) or by survival. There can also be competition among stages, especially under resource scarcity. Individuals in two stages might compete directly for the same resources, or adults may forego infant care when their own survival is threatened by the demands of rearing young. Adults may impact infant abundance negatively through infanticide. Caswell (1989) advocated the use of life-history diagrams to model and visualize the types and consequences of such stage interactions (see also Benton and Grant, 1999).

Life-cycle stage interactions differ among species, and distributions of individuals by stage (or age) reflect those differences. In other words, demographic variation in the numbers of individuals and durations of time spent in any particular life-cycle stage is not random across species. Rather, it reflects the ways in which the interactions among and within stages produce species-"typical" demographic units. This demographic variation also reflects the ways in which species respond (in terms of population growth rate and population stability) to perturbations in the relative abundances of individuals in each life-cycle stage.

In essence, a life history is the direct outcome of the interaction between developmental variables (growth rate, age at skeletal maturation, etc.) and demographic variables (survival, reproduction, population growth, life-cycle stage census counts, etc.). The distribution of individuals in a population by life-cycle stage is the outcome of the developmental program played out within the context of the population's ecological present and evolutionary past. An observed distribution of individuals in different life-cycle stages reflects how,

* Life *history* refers to the events that we observe in a particular situation or the outcome of a particular strategy; and *life cycle* refers to events that are common to living organisms—conception, gestation, birth, infancy, weaning, juvenescence, adulthood, reproduction, death. Life-cycle *stages* are ideally assessed for the two sexes separately; thus, for example, subadult males may be distinguished from subadult females. Precise definitions of life-cycle stages are left up to the investigator.

at a certain point in time and under a given set of environmental conditions, characteristic reproductive and survival schedules of individuals at each stage affect and are affected by those of individuals in other life-cycle stages (see, for example, Cole 1954).

Lefkovitch (1965) showed that one could study life-cycle stages using the matrix approach that Leslie (1945, 1948) had designed for age-based data. Lefkovitch thus laid the theoretical groundwork for the application of these mathematical tools to life-history dynamics (see the Appendix for methodological details). Caswell (1989) illustrated several methods for deriving estimates of *stable population distributions* and *reproductive values* of life-cycle stages from *standing distributions* of single populations. Standing population distributions are the *observed* abundances of individuals in each life-cycle stage. Stable population distributions are the "ideal" proportional distributions of individuals by stage given the reproductive and survival characteristics of the species in question. In other words they are the population distributions that generate and maintain the species-typical population structure or characteristics (e.g., the observed values for population growth rate, stage-specific mortality rates, stage-specific fertility rates, and generation time). The reproductive value of a stage represents that stage's remaining reproductive contribution to the next generation. *Population growth rates* equal one when population size remains constant; they are less than one when the population is declining, and greater than one when the population is growing.

One advantage of Lefkovitch's technique (over model life tables) is that it allows researchers to evaluate how any single stage's survival rates and reproductive value affect the survival and reproduction of any other stage (Caswell, 1978). Canonical (idealized) representations of the stage-specific survival rates and reproductive values may be used to determine what are called the *sensitivities* of the population growth rate to life-cycle stage interactions. Population growth-rate sensitivities reveal how much the population growth rate changes as a function of perturbations in the products of the reproductive values and abundances of the various life-cycle stages. If a population is subdivided into four life-cycle stages (say, infants, juveniles, adult males, and adult females) sixteen population growth-rate sensitivities can be calculated (a 4 by 4 matrix). The elements of the sensitivity matrix are the partial derivatives of the population growth rate with respect to the interactions (or products) between the reproductive values of any stage and the abundance of any other. The sensitivity matrix includes the interactions within life-cycle stages as well; the diagonal cells of the matrix contain the sensitivities of the population growth rate to interactions between the reproductive value of any stage and the abundance of individuals in that same stage. Note that sensitivities can take on any positive value and that the sensitivity matrix is asymmetric. This is because the interaction of, say, adult female reproductive value with infant survival (or abundance) is not the same as that between adult female abundance and infant reproductive value. The advantage of sensitivity analysis is that the resultant matrix allows us to determine *which interaction* among or within life-cycle stages contributes most to changes in the population growth rate. However, sensitivity matrices may be difficult to compare across populations or species, because reproductive value and survival probabilities are measured on different scales (see the Appendix). One solution to this problem is to combine and rescale sensitivities, thus expressing the relationships among life-cycle stages as *population growth-rate elasticities*.

Population growth-rate elasticities (van Groenendael *et al.*, 1988; Benton and Grant, 1999) estimate the proportional contribution to the population growth rate of sets of sensitivities involving particular pairs of life-cycle stages. Unlike sensitivity matrices, elasticity matrices are symmetric and all elements sum to one. For our hypothetical population divided into four life-cycle stages, four population growth-rate elasticities can be calculated for each life-cycle stage (on the basis of all interactions of individuals in that and the other three life-cycle stages). *Cumulative elasticities* (the column sums) measure the extent to which perturbations in the population growth rate are likely to be dependent on perturbations in the interactions involving individuals in particular life-cycle stages. Thus, for example, the sum of the effects on population growth of all interactions involving adult females is the cumulative elasticity for the adult female stage, and the sum of the cumulative elasticities for all life-cycle stages is 1.00. Once cumulative elasticities have been obtained for every life-cycle stage, it is possible to generate (on the basis of the effects of interactions within and among stages) a summary of the proportional effects on population growth of changes in stage-specific vital rates (i.e., the survival rate and reproductive value of individuals in that stage). The advantage of elasticity analysis is that it allows a direct comparison among populations (or species) of the degree to which different life-cycle stages might produce changes in the population growth rate. In effect, population growth rate elasticities capture an important aspect of life-history strategies.

Life-history strategies reflect species' differences in the length of infant dependency, the age at sexual maturity, litter size, interbirth intervals, female–infant ratios, etc. These differences affect species-typical ontogenetic trajectories, as well as the way in which each species actualizes its population growth rate. Classic "r-strategists" are said to exhibit opportunistically fluctuating population growth rates (with boom-and-bust cycles) while classic "K-strategists" are said to exhibit steady rates of population growth closely tracking environmental quality (Pianka, 1970, 1971). While primate species are sometimes characterized as K-strategists with prolonged periods of infant and juvenile development, there is a tremendous amount of variation in primate life-history strategies and in their demographic and ontogenetic manifestations. Elasticity analysis allows investigators to explore the complex manner in which reproductive value and survival of individuals in different life-cycle stages *actually interact*, and thus to avoid artificially simplistic dichotomization of species into short-lived r-strategists that rely on reproductive resilience and long-lived K-strategists that depend on survival.

Application to Skeletal Materials in the Fossil Record

Ideally, aspects of the behavioral ontogeny of extinct species can be inferred from ontogenetic series of skeletons and from patterns of adult variation. Possible sources of information include: (1) skeletal evidence of variation in the relative timing of the development of sets of traits; (2) clade-specific patterns of adult trait correlation; and (3) skeletal evidence of variation in the absolute pace of development.

Modularity of structural complexes can be inferred from skeletons of extinct species using methodologies parallel to those employed in the study of extant taxa. The limiting factors are the availability of ontogenetic series (particularly very young individuals) and of *associated* skeletal elements. Absolute age is not easily accessible for fossil materials, although techniques for assessing absolute age have been developed and applied with some success to fossil hominids (see Bromage and Dean, 1985; Dean, 1987; Smith, 1992). Nevertheless, as with extant organisms, different developmental markers can be benchmarked against one another and degrees of skeletal and cranial development, at any selected developmental stage, compared. The fossil record can thus supply direct anatomical evidence of trait-complex modularity. Such evidence will appear as shifts, at given developmental stages, in the relative degrees to which suites of traits (apparent members of independent modular units) are developed.

Paleontologists can also examine the population consequences of variation in ontogenetic trajectories and the correlation structure of traits in adults. Tools such as Finite Mixture Analysis (Pearson 1894; Robertson and Fryer, 1969; Titterington *et al.*, 1985) are useful in exploring sexual dimorphism when males and females cannot be distinguished on the basis of morphological differences. While Finite Mixture Analysis may not work as well as other techniques in estimating dimorphism when dimorphism is strong, it provides good estimates of the maximum theoretically-possible bimodality embedded in unimodal distributions (see Flury *et al.*, 1992; Godfrey *et al.*, 1993; Plavcan, 1994; Josephson *et al.*, 1996; Acuña and Muñoz, 1998; Rehg and Leigh, 1999).

Under ideal circumstances, elasticity analysis can be applied to paleontological populations. Several conditions must hold. First, samples must contain geologically associated or contemporaneous individuals. Second, death assemblages must be reasonably representative of the biocoenosis—the living assemblage. Third, skeletal or dental parameters must be appropriate markers of the boundaries of life-cycle stages.

The process of modeling life-history strategies of extinct species requires that individuals in a fossil assemblage be viewed as members of a population. If individuals in fossil assemblages can be shown to be synchronous (for example, members of a group that was buried in a single catastrophic event or in a series of related events), then the biological age (or life-cycle stage) and sex distribution of individuals in that assemblage can be used to estimate a number of other characteristics. These in turn allow us to model life-history strategies for that population.

The standing distributions of individuals in a death assemblage may not reflect population standing distributions. For example, large numbers of juveniles may reflect high juvenile mortality rather than high juvenile abundance. Biases may be introduced due to the way in which the death assemblages formed. Additional biases may be introduced because of the differential postmortem preservation of individuals at different life-cycle stages. For a variety of taphonomic reasons (for example, the vulnerability of bones of small individuals to pre-depositional destruction by carnivores and scavengers and to post-depositional acidic dissolution due to their high surface-to-volume ratios), there is a size bias against small individuals (including infants) at fossil sites (see Shipman, 1981; Lyman, 1994, for a review). Such biases can be corrected before Caswell's method is applied.

Extinct species' life-cycle "stages" can be constructed in a variety of ways. One can follow Sauer and Slade (1985), for example, in using body size as a "demographic" variable. Specific developmental structures can be used to define stage boundaries (see Caswell, 1989; Caswell and Twombly, 1989)—the eruption of particular teeth, or the closure of particular sutures, for example. If adult skeletons are sexually dimorphic, one can separate adult males from females, while combining sexes in the infant and juvenile categories. What is important is that the markers used to divide samples into stages be ecologically meaningful. That this is not always easy or straightforward has already been discussed. Suppose, for example, that we were to take the completion of dental eruption as a marker of reproductive maturation (and thus adulthood) in sifakas and macaques. Sifakas complete their dental eruption when they are about a year old—long before breeding first occurs (i.e., ages 3–6). Macaque females first breed at around age 4 years; yet dental eruption is not complete until several years later. Thus, the significance of full dental eruption as a skeletal marker of life-cycle stage differs for sifakas and macaques.

Lefkovitch's method also requires time-successive samples of single populations. This condition is virtually never satisfied for skeletal assemblages; fossil vertebrate assemblages generally do not provide the stratigraphic control to allow comparisons of samples drawn from the same population at consecutive points in time. Lefkovitch's method can be altered, however, to overcome this obstacle. Several adjustments are possible. The simplest is to use the available skeletal assemblage to represent both ends of the time series. This assumes that the population is stable—i.e., that there is zero population growth and there is no significant change in the population's age-sex distribution.

A second solution is to use the values of life-cycle parameters obtained for extant relatives, along with the single sample obtained for the extinct species, to construct plausible time-successive series. The derivation of reproductive values for each life-cycle stage typically requires an estimate of the reproductive output of individuals in each stage (see Charlesworth, 1980). Because reproductive outputs of individuals in each life-cycle stage cannot be obtained unless single populations can be analyzed at successive points in time, they must somehow be estimated whenever time-successive samples are unavailable. The idea here is to select at least one (preferably closely related) extant species that exhibits a similar survival profile (or standing distribution of individuals at each life-cycle stage). Alternatively, one might select an extant species with similar patterns of development, morphological characteristics, ecological adaptations, or some combination of the above. In essence, extant species are used as analogs. The life-history dynamics of the chosen analog or analogs can be used not merely to derive estimates for parameters that cannot be measured directly for the extinct species, but to test the assumption of demographic stability for the extinct species.

To calculate cumulative elasticities for skeletal populations, we must reconstruct stage-specific population distributions. This entails: (1) defining (using skeletal parameters) the life-cycle "stages" that we want to use; (2) assigning specimens to "populations" (controlling, if possible, for any temporal, spatial, and taphonomic factors that might affect extinct species assemblages). Then, using the skeletally-defined life-cycle "stages," we can assess the standing distribution of individuals at each stage. By assuming that the extinct population was stable or that it shared population growth characteristics with specified extant relatives, one has all the information needed to calculate stage-specific population growth-rate elasticities. Stage-specific growth-rate elasticities obtained from such analyses can be compared to those obtained for extant taxa using actual census data.

Development and the Reconstruction of Behavior in Extinct Primates: An Example

The subfossil lemurs of Madagascar offer an excellent test of the utility for behavioral inference of the methods described above. Because they date only to the late Pleistocene and Holocene specimens are not fossilized. Many hundreds of subfossil lemur bones have been collected at dozens of sites, and some species are represented by excellent ontogenetic series.

The Palaeopropithecidae and Archaeolemuridae comprise two of the five families of Malagasy lemurs containing extinct species (Table I). They are believed to be related to the extant Indridae and are sometimes included, as subfamilies, within the family Indridae; alternatively they are considered members, with the Indridae, of the superfamily Indroidea. Archaeolemurids are similar to indrids in aspects of their facial and basicranial (but not dental) anatomy, while palaeopropithecids are more similar to indrids in their dental (but not necessarily basicranial or facial) morphology (Tattersall, 1982). Neither family is very like indrids postcranially (see Jungers *et al.*, this volume, Chapter 10). Indeed, palaeopropithecids are called "sloth lemurs" because of their remarkable postcranial convergences with arboreal sloths (Jungers, 1980;

Table I. Taxonomy of the Extinct Lemurs of Madagascar

Palaeopropithecidae
Archaeoindris fontoynontii
Palaeopropithecus maximus
Palaeopropithecus ingens
Palaeopropithecus sp. nov.
Babakotia radofilai
Mesopropithecus pithecoides
Mesopropithecus globiceps
Mesopropithecus dolichobrachion
Archaeolemuridae
Archaeolemur edwardsi
Archaeolemur majori
Hadropithecus stenognathus
Megaladapidae
Megaladapis (Peloriadapis) edwardsi
Megaladapis (Megaladapis) grandidieri
Megaladapis (Megaladapis) madagascariensis
Daubentoniidae
Daubentonia robusta
Lemuridae
Pachylemur insignis
Pachylemur jullyi

Jungers *et al.*, 1991, 1997; Simons *et al.*, 1992; Jungers *et al.*, this volume, Chapter 10). No archaeolemurid or palaeopropithecid is particularly anthropoid-like, yet each is larger than the largest indrid—often considerably so. To the extent that body size is an important determinant of behavior, anthropoids should make better extant models than indrids. However, details of the morphology and development of trait complexes suggest more appropriate extant models within the much smaller-bodied Indridae or other strepsirrhines.

Dental Ontogeny and Behavior of Extinct Indroids

Like adults of the family Indridae, adult palaeopropithecids have only four teeth in their "tooth combs" and two (rather than three) pairs of maxillary and mandibular premolars. The developmental basis for the loss of a pair of permanent premolars and (apparently) the permanent mandibular canine can be found in histological studies of odontogenesis in extant indrids.

Luckett (1984) compared fetal dental development in the indrid, *Propithecus*, and the lemurid, *Lemur*. The mandibular deciduous dental formula is identical in the two, but one deciduous maxillary premolar (probably dp^2) that is present in *Lemur* is missing in *Propithecus*. At 25 mm HL, fetal sifakas display relative retardation of the development of the third and the fifth deciduous teeth (dc_1 and dp_3) in the lower jaw. While all of the other deciduous teeth (including the anteriormost premolar, dp_2 and the posteriormost premolar, dp_4) are well calcified, these teeth still lack dentin and enamel. The two developmentally "retarded" teeth are situated close to oral epithelium while all other deciduous teeth extend deeper into the jaw. The dev/eloping roots of di_2, dp_2, and dp_4 are in close apposition, leaving little space for the development of dc_1 or dp_3. Relative retardation of the development of these two teeth in indrids results in their failure to develop successional laminae and in the consequent agenesis of their successional teeth. In contrast, in fetal specimens of *Lemur*, there is no relative retardation of dc_1 or dp_3; all of the deciduous teeth develop normally and are evenly spaced. There is no loss of a pair of adult premolars, nor of the lateral tooth of the toothcomb.

The milk dentition of the Indridae is also unique among living lemurs in that all of the deciduous teeth, including the molar-like maxillary and mandibular last deciduous premolars, are very small in comparison to the size of the first permanent molars. Sifakas are born with their milk dentition virtually fully erupted and with the crowns of the first and second permanent molars in an advanced state of calcification (Fig. 1). The size of the deciduous teeth is constrained by the small size of the jaws of neonates and the large size of the permanent molar crowns. The early and rapid growth of the crowns of the molars contributes substantially to the crowding of the developing deciduous teeth into the very anterior portion of the neonatal jaw. The result is that all of the deciduous teeth are small and the relatively "retarded" third and fifth deciduous teeth in the mandible are extremely small. Molar megadonty is achieved at the expense of the deciduous dentition, which is allocated insufficient space in the fetal jaws to attain large size (Table II).

Dental developmental precocity in sifakas results in the emergence of all postcanine permanent teeth by age one year—at a time when less than two-thirds and perhaps only half, of adult body mass has been attained, and a full two years (or more) prior to first breeding. It also results in an altered relationship between the developing dentition and the rest of the cranium. To fit all of the permanent teeth into still-small jaws, premolar eruption is staggered, with the distal moieties of more mesial mandibular teeth overlapping the mesial portions of more distal teeth. As the face and mandible continue to grow, the permanent premolars spread out (gradually assuming normal adult alignment), the jaw deepens, and the cranium widens. Figure 2 shows the skull of an eight or nine month old *Propithecus v. verreauxi* from Ampotaka. All of its permanent teeth except C^1 have erupted.

The indrid pattern, only more extreme, is manifested in the extinct sloth lemurs. Figure 3 shows the skulls of a young *Palaeopropithecus ingens* from Ankazoabo Cave (in Southwest Madagascar) and a fully-grown *Palaeopropithecus* sp. from Anjohibe Cave (in the Northwest). The dental development of the

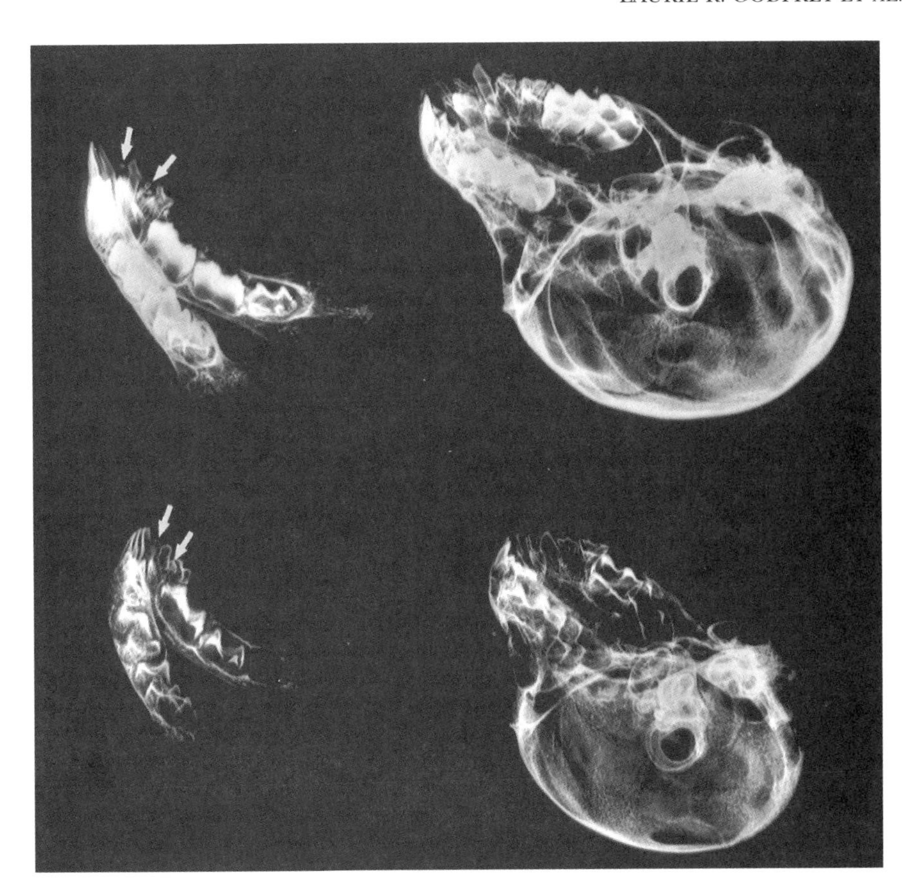

Fig. 1. Radiographs of skulls and mandibles of infant sifakas. Top: BMNH 30.3.15.1 (*Propithecus verreauxi verreauxi* from Ampoza near Ankazoabo; about 2 months old). Bottom: BMNH 67.1365 (*Propithecus verreauxi*, captivity; neonate). White arrows point to the vestigial dc_1 and dp_3. Note the advanced state of calcification of the first and second molars in the neonate.

young *Palaeopropithecus* is identical to that of the eight-month-old sifaka shown in Fig. 2. Standing (1908, pp. 86–88) measured, described, and figured a skull of yet another palaeopropithecid, *Mesopropithecus pithecoides* (from Ampasambazimba in Central Madagascar), at a similar stage of dental development. The latter has its M^3 erupting, its permanent premolars erupted, and is "just cutting" (p. 88) its upper canine. Figure 4 shows the percent completion of bizygomatic growth in a wide cross-section of primate species at a comparable dental developmental stage (our "Dental Stage 9," representing full crown eruption of ca. 95% of the maxillary and mandibular primary and replacement teeth). The indrids and especially palaeopropithecids stand out as least craniofacially mature.

Table II. Maxillary and Mandibular dp4/M1 Length Ratios (in %) for Selected Primates

Family	Genus and Species	dp^4/M^1 length ratio	dp_4/M_1 length ratio
Indridae	*Propithecus verreauxi*	54.2	41.2
	Propithecus diadema	53.8	45.5
	Avahi laniger	54.9	42.9
	Indri indri	52.8	45.2
Archaeolemuridae	*Hadropithecus stenognathus*	92.6	92.5
	Archaeolemur edwardsi	95.8	102.6
	Archaeolemur majori	95.3	108.2
Megaladapidae	*Megaladapis edwardsi*	74.2	71.2
	Megaladapis madagascariensis	74.7	64.7
	Megaladapis grandidieri	—	66.3
Lepilemuridae	*Lepilemur ruficaudatus*	78.4	66.3
Lemuridae	*Lemur catta*	89.3	96.00
	Eulemur fulvus	100.6	100.6
	Eulemur mongoz	97.4	99.2
	Varecia variegata	102.0	105.3
Cercopithecidae	*Macaca fascicularis*	86.5	88.7
	Macaca nemestrina	82.5	84.6
	Papio anubis	75.5	83.9
	Papio cynocephalus	81.1	82.4
	Kasi vetulus	90.3	89.5
	Trachypithecus cristata	90.3	89.5
	Trachypithecus obscura	94.6	94.7
	Nasalis larvatus	78.8	80.8
	Colobus guereza	87.3	88.4
	Piliocolobus badius	83.0	84.8

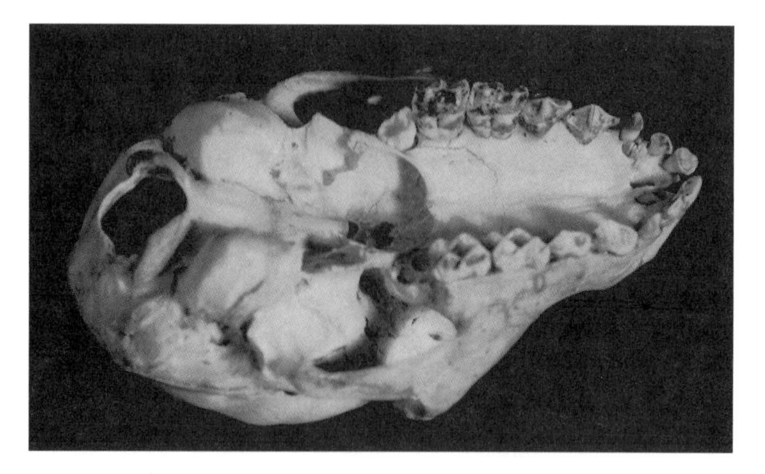

Fig. 2. Skull of *Propithecus verreauxi verreauxi* (AMNH 100551) from Ampotaka, approximately 8–9 months old. M^3 is erupting and the deciduous upper canines are unreplaced. The mandible of this individual has its full adult dentition.

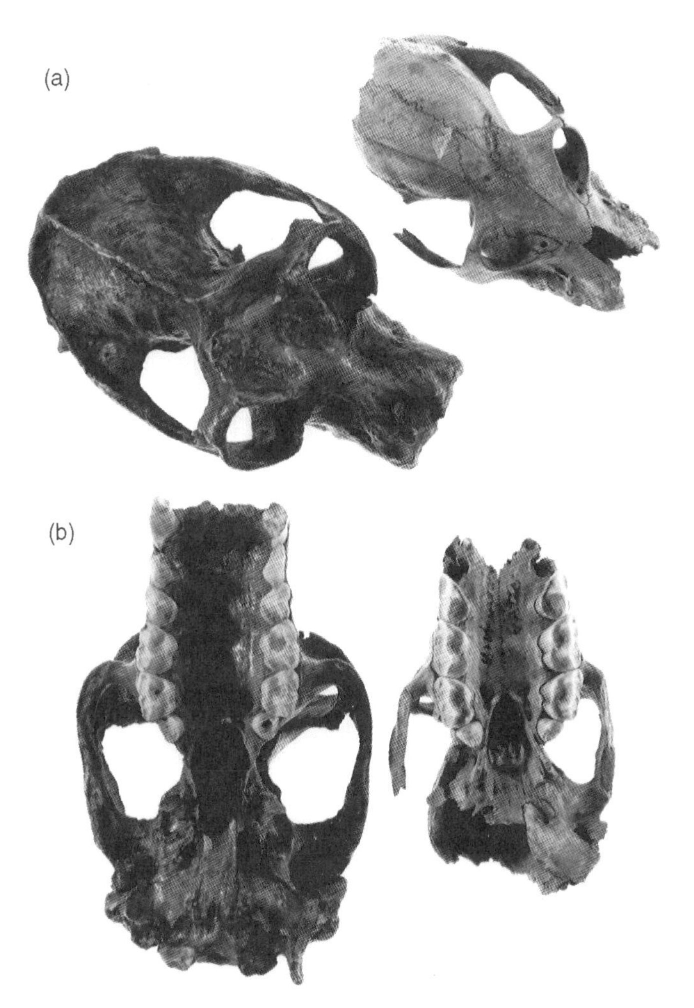

Fig. 3. a. Dorsal views of an older (left) and younger (right) *Palaeopropithecus*. The skull of the full adult (UA-UM 8136) was recently discovered at Anjohibe in Northwest Madagascar. It belongs to a species that is currently being described. The juvenile is a *P. ingens* (UA-AM PPH1) from Ankazoabo Cave in Southwest Madagascar. b. Occlusal views of the same two specimens. The deciduous canines were unreplaced.

There are other skeletal correlates of dental developmental acceleration in the indrids and the palaeopropithecids. Table III documents the CV's for mandibular corpus height of samples of primates *with full eruption of the mandibular permanent dentition*. We controlled for sex in sexually dimorphic species, and for geographic variation whenever possible. The CV (15%) for mandibular corpus height in our sample of mandibles of *Palaeopropithecus maximus* from Ampasambazimba is high in comparison to those of other

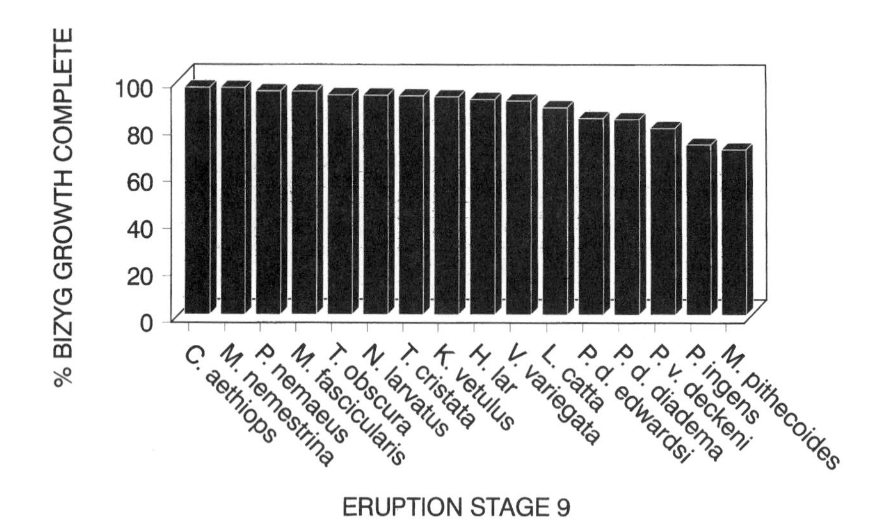

Fig. 4. Mean percent completion of bizygomatic growth, at "Eruption Stage 9" (when approximately 95% of all teeth, primary and secondary, have erupted), for a variety of primates, including two lemurids, three indrids, and two palaeopropithecids. To derive mean % completion of bizygomatic growth, the mean bizygomatic breadth of Stage 9 skulls was divided by the mean for full adults from the same sites or geographic regions. For sexually dimorphic species, only females were considered. Genera are (left to right): *Chlorocebus*, *Macaca*, *Pygathrix*, *Macaca*, *Trachypithecus*, *Nasalis*, *Trachypithecus*, *Kasi*, *Hylobates*, *Varecia*, *Lemur*, *Propithecus* (three subspecies), *Palaeopropithecus*, and *Mesopropithecus*. The latter two are palaeopropithecids.

species, and to the same sample's CV's for molar crown dimensions (e.g. 5.8% for the length of the first mandibular molar and 3.1% for the length of the mandibular molar series). Mandibular corpus height ranges from 27–53.9 mm in this sample.

Full adult *Palaeopropithecus maximus* normally bear a diastema between the mandibular premolars P_2 and P_4 (Tattersall 1982). Young dental adults from Ampasambazimba exhibit diastemata of varying lengths (0 to 9.5 mm in our sample of 15 mandibles for which diastema length could be measured). This variation is largely ontogenetic; individuals lacking or bearing small diastemata also have incompletely fused mandibular symphyses, shallow mandibular corpi, and little tooth wear. The sample CV for diastema length is 61.5%. Diastema length is related to mandibular corpus height but not to molar series length (Fig. 5).

The gorilla-sized *Archaeoindris fontoynontii* shows the same developmental pattern. Only two mandibles belonging to this extremely rare species are known. Their molar series lengths are virtually identical—61.0 mm for the type specimen (uncatalogued) and 60.1 mm for the younger individual (UA-AM 6237) with less worn teeth. However, their jaw depths are 52.1 and

Table III. Coefficients of Variation for Mandibular Corpus Depth among Individuals with Fully Erupted Mandibular Dentitions

Family	Genus and species	Sample	Sample size	Coefficient of variation
Palaeopropithecidae	*Palaeopropithecus maximus*	Pooled sexes, Ampasambazimba, Central Madagascar	19	15.0%
	Mesopropithecus globiceps	Pooled sexes, Southwest Madagascar	6	12.7%
	Mesopropithecus pithecoides	Pooled sexes, Ampasambazimba, Central Madagascar	6	11.8%
Indridae	*Propithecus verreauxi*	Pooled sexes, pooled sites in Western Madagascar	23	10.5%
Archaeolemuridae	*Archaeolemur edwardsi*	Pooled sexes, pooled sites in Central and Northern Madagascar	37	8.3%
	Archaeolemur majori	Pooled sexes, pooled sites in Southwest Madagascar	16	6.4%
Cercopithecidae	*Macaca fascicularis*	Females only, Kinabatangan, Borneo	10	5.7%
	Trachypithecus obscura	Females only, pooled sites, Southeast Asia	24	6.9%
	Colobus guereza	Females only, pooled sites, Tanzania	10	6.0%

46.7 mm, respectively, and the separation between mandibular premolars is almost 10 mm greater in the older individual.

As in extant indrids, the smaller-bodied palaeopropithecid species (i.e. *Babakotia radofilai*, the three species of *Mesopropithecus*, and the smallest-bodied species of *Palaeopropithecus* from the Northwest) fail to develop a diastema between P_2 and P_4. Yet, even *Avahi occidentalis* (the smallest-bodied at ca. 0.75–0.80 kg of extant indrids; see Smith and Jungers, 1997) exhibits strongly staggered eruption of the adult mandibular premolars. A short mandibular premolar-series diastema sometimes develops in *Palaeopropithecus ingens* (which is only slightly smaller than *P. maximus*) although some clear adults (with worn teeth and deep mandibular corpi) lack this feature. A mandibular diastema develops only in the largest-bodied members of the palaeopropithecid-indrid clade, and even then only relatively late in ontogeny.

Dental developmental precocity does not characterize all of the giant extinct lemurs. Neither the Megaladapidae nor the Archaeolemuridae follow the indrid pattern of dental development. The Archaeolemuridae match the living Lemuridae in bearing large posterior deciduous premolars (and the Megaladapidae match the living Lepilemuridae in bearing somewhat reduced

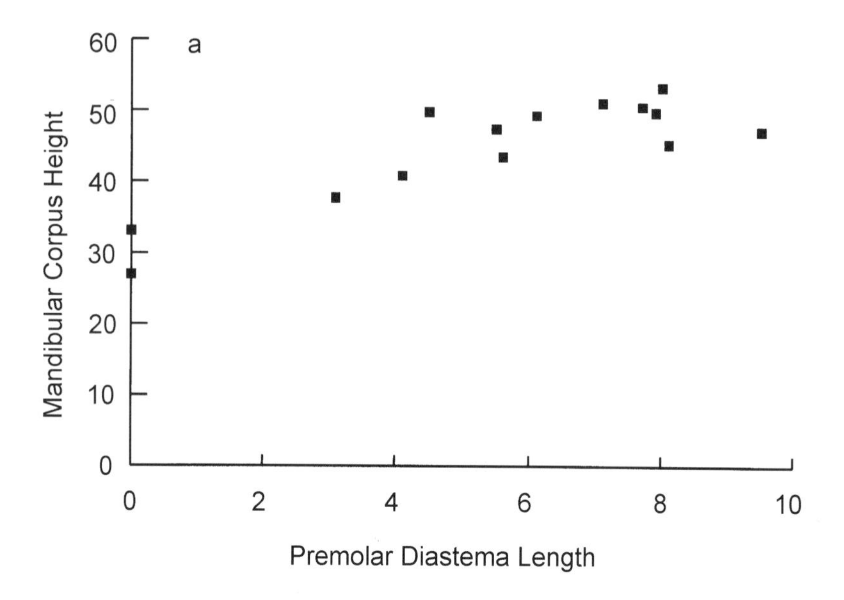

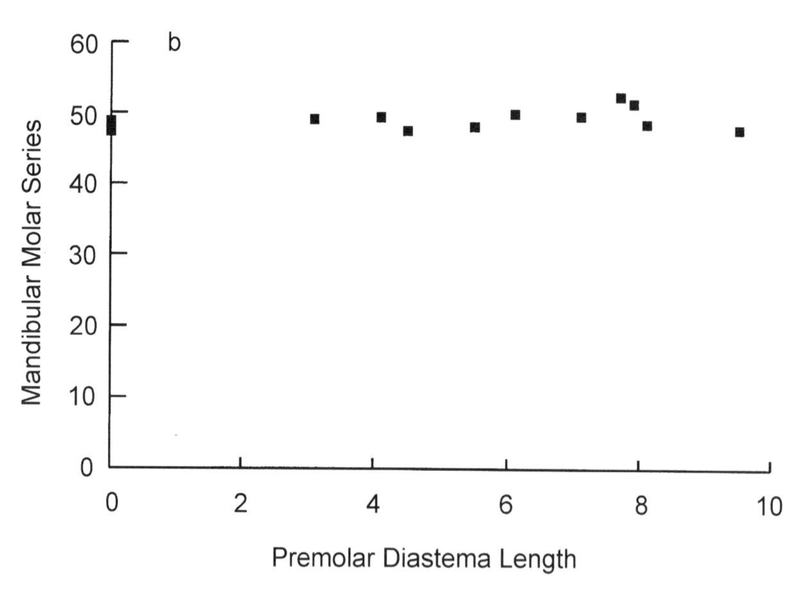

Fig. 5. a. Mandibular corpus height vs. diastema length in *Palaeopropithecus maximus* (from Ampasambazimba) with full adult mandibular dentitions. Pearson's r = 0.862 (n = 14, P < .01). b. Mandibular molar series length vs. diastema length in *Palaeopropithecus maximus* (from Ampasambazimba) bearing full adult mandibular dentitions. Pearson's r = 0.367 (n = 17, NS).

dp4s; Table II). In the case of the Archaeolemuridae the posteriormost deciduous premolars are almost as large as the first molars in occlusal area; they are often longer in their mesiodistal dimension. The craniofacial growth trajectories of *Archaeolemur* spp. also resemble those of extant lemurids and not indrids. In *A. edwardsi*, for example, bizygomatic breadth attains almost three-quarters of its adult value before the eruption of M1. This exceeds that of *Palaeopropithecus ingens* or *Mesopropithecus pithecoides* in *advanced* stages of dental eruption. The coefficients of variation for mandibular corpus height in samples of *Archaeolemur* "dental adults" from single sites are considerably lower than those of palaeopropithecids or extant indrids (Table III).

Two corollaries of accelerated dental development in the extant Indridae and in the Palaeopropithecidae are: (1) a sharp break in the morphological gradient of the permanent cheek teeth; and (2) first molar megadonty. The premolar and molar series are very different in morphology. The break in the morphological gradient emerges from the difference in space allocated to the tooth buds developing from the primary and secondary dental laminae.

Table IV documents the size (relative to both skull and palate length) of the first maxillary molar in a sample of primate species. Across the Order Primates, relative molar length varies independently of body size.* The first maxillary and mandibular molars are especially large in the Palaeopropithecidae and Indridae. This clearly results, in these species, from the acceleration of the development (from initial tooth bud formation to eruption) of the entire dental series, coupled with competition between the the tooth germs budding from the primary and secondary laminae for limited space in the fetal jaw. Given the small size of the jaws of young indrids and palaeopropithecids, the molar crowns can grow rapidly to adult size *only at the expense of the deciduous dentition*, which is, effectively, developmentally accelerated *and* eclipsed. The simple morphology of the replacement teeth apparently results from the early loss of developmental competence of the premolar progenitor cells.

The correlation structure of the adult dentition provides an indirect test of the developmental dissociation of the milk teeth (and their replacements) and the permanent molars. For example, Table V shows a partial correlation matrix for four mandibular cheek teeth of *Propithecus*. Partial correlations are conditional correlations; partial correlation matrices indicate the strengths of the relationships between any two variables when the effects of all other variables considered are controlled. Partial correlation matrices can thus provide some indication of patterns of morphological integration (or phenotypic interdependency). In this case, the two premolars and the two molars display the highest partial correlations and the premolar–molar correlations are lower. The premolars and molars are weakly-dissociated modules.

*We found isometry for the regressions of first maxillary molar length on palate length and first maxillary molar length on skull length, for a pooled-sex sample of 534 adult individuals belonging to 31 primate species ranging in size from *Galagoides demidoff* to *Pongo pygmaeus*.

Table IV. Molar Megadonty

Family	Genus and species	M^1 length as a percentage of skull length	M^1 length as a percentage of palate length
Palaeopropithecidae	*Palaeopropithecus maximus*	8.9	19.4
	Palaeopropithecus ingens	10.2	20.9
	Palaeopropithecus sp. nov.	9.8	23.2
	Mesopropithecus pithecoides	8.7	22.3
	Mesopropithecus globiceps	8.0	21.4
	Mesopropithecus dolichobrachion	8.2	—
	Babakotia radofilai	7.8	19.1
	Archaeoindris fontoynontii	8.3	18.8
Indridae	*Propithecus verreauxi*	8.2	20.8
	Propithecus diadema	8.0	21.5
	Propithecus tattersalli	8.6	22.7
	Avahi laniger	7.9	22.7
	Indri indri	7.4	20.2
Lepilemuridae	*Lepilemur ruficaudatus*	7.0	19.3
Archaeolemuridae	*Hadropithecus stenognathus*	7.7	20.8
	Archaeolemur edwardsi	5.7	15.9
	Archaeolemur majori	6.0	16.5
Lemuridae	*Lemur catta*	5.8	14.1
	Eulemur fulvus	6.6	15.4
	Eulemur mongoz	5.9	14.6
	Varecia variegata	6.6	14.5
Cercopithecidae	*Macaca fascicularis*	5.7	14.1
	Macaca nemestrina	5.4	12.3
	Papio anubis	5.6	12.4
	Papio cynocephalus	5.7	12.6
	Cercopithecus aethiops	5.5	15.3
	Kasi vetulus	6.4	17.1
	Trachypithecus cristatus	5.9	17.2
	Trachypithecus obscura	5.9	16.9
	Nasalis larvatus	5.8	17.0
	Colobus guereza	6.3	15.0
	Piliocolobus badius	5.9	15.6

Multiple regression analysis effectively provides the same analytic message, as the regression coefficients are nothing more than scaled partial correlations. We used multiple regression analysis to ascertain the modularity patterns for the teeth at the premolar/molar junction of indrids and other taxa (Palaeopropithecidae, Archaeolemuridae, and Cercopithecidae; Table VI). Separate multiple regressions of the first permanent molar and last permanent premolar on the adjacent mesial and distal teeth were calculated. Increased developmental independence of the premolar and molar series should result in reduced correlations *across* the premolar/molar boundary. Indeed, we found that, in the Indridae and Palaeopropithecidae, the lengths of the distalmost

Table V. Partial Correlation Matrix for the Mandibular Cheek Teeth of *Propithecus*

	P_2 md	P_4 md	M_1 md	M_2 md
P_2 md	—	0.38	0.20	0.12
P_4 md	0.38	—	0.24	0.25
M_1 md	0.20	0.24	—	0.58
M_2 md	0.12	0.25	0.58	—

permanent premolars are best predicted by the lengths of the neighboring (more mesial) premolars, while the lengths of the first molars are best predicted by the lengths of the neighboring (more distal) molars. However, this modularity pattern is not manifested in the Archaeolemuridae or the Cercopithecidae, for which maxillary and mandibular P4 and M1 lengths are consistently better predicted by the lengths of the next more distal tooth in the dental arcade.

Ecological correlates of differences in the timing of dental eruption in extant lemurs were first explored by Eaglen (1985). Because most lemur species grow in cohorts (due to reproductive synchrony), selection can fine-tune the timing of the eruption of particular teeth to coincide with particular phenological events. Weaning itself is timed to correspond to the season of greatest availability of young leaves and other foods (Meyers and Wright, 1993; Wright, 1999). Selection operates on dental development independently of the development of the rest of the skeleton so as to guarantee maticatory proficiency at weaning and during the first post-weaning dry season (Godfrey *et al.*, 2001). At weaning, indrids are equipped with their first and (often) second

Table VI. Predicting Premolar and Molar Lengths at the P4/M1 Juncture[a]

Family	Is the length of P^4 better predicted by the length of the next most mesial premolar or by the first molar?	Is the length of M^1 better predicted by the length of the distalmost premolar or by M^2?	Is the length of P_4 better predicted by the length of the next most mesial premolar or by the first molar?	Is the length of M_1 better predicted by the length of the distalmost premolar or by M_2?
Palaeopropithecidae	premolar	molar	premolar	molar
Indridae	premolar	molar	premolar	molar
Archaeolemuridae	molar	molar	molar	molar
Cercopithecidae	molar	molar	molar	molar

[a] Multiple regression analyses were based on 52 palaeopropithecid maxillae, 52 palaeopropithecid mandibles, 169 indrid maxillae, 56 indrid mandibles, 87 archaeolemurid maxillae, 66 archaeolemurid mandibles, 145 cercopithecid maxillae, and 144 cercopithecid mandibles.

molars, as well as their adult incisors (including the toothcomb), and, sometimes, their adult posterior premolars. In every indrid species, the posterior premolars erupt shortly after weaning, if not before. Lemurid weanlings, in contrast, may lack their first molars, and have not yet shed any of their deciduous. The summed postcanine occlusal area (expressed as a proportion of adult postcanine occlusal area) is much higher at weaning in the indrids than in lemurids (Samonds *et al.*, 1999; Godfrey *et al.*, 2001). Indrids are seed-crunching folivore/frugivores, while lemurids tend to consume more fruit, and to discard or swallow seeds whole (Yamashita, 1996; Hemingway, 1996). The small size of indrid deciduous teeth does not compromise food processing at weaning *because these teeth are shed during or just prior to weaning*. One can assume that the relatively large deciduous premolars of the archaeolemurids served their owners well past weaning (just as the relatively large deciduous premolars of living lemurids do today). Given their extreme dental precocity, it seems likely that the Palaeopropithecidae had diminutive deciduous premolars that were shed early, as do extant indrids.

Growth Rates, Sexual Dimorphism, and Agonism in Extinct Indroids

Geographic patterns of adult body size variation may partly reflect absolute rates of skeletal growth and development. Closely related lemur taxa (subspecies or species) vary in body size along an ecogeographic gradient (Albrecht *et al.*, 1990). Differences in population means for metric traits such as skull length appear to be related to environmental parameters such as the length of the dry season and plant resource availability and/or quality (see Albrecht *et al.*, 1990; Godfrey *et al.*, 1990; Ravosa *et al.*, 1993). In contrast, body size gradients of many catarrhine taxa (with the exception of baboons living under severe ecological gradients) do not track vegetational clines (Albrecht *et al.*, 1990). The explanation for this difference may lie in the absolute rates and durations of postnatal growth and development. Slow, prolonged postnatal growth (as occurs in catarrhines) allows considerable flexibility in the rate of growth at any point prior to full skeletal maturation. Growth rates can decline under resource crunches only to bounce back when resources become available. Species with rapid postnatal growth rates and early cessation of growth may have limited flexibility, and net growth may correlate more strongly in these species with aspects of regional ecology. The existence, *in subfossil lemurs*, of ecogeographic size clines that strongly mirror patterns exhibited by extant lemur species (Albrecht *et al.*, 1990; Godfrey *et al.*, 1990) provides indirect support for the inference that, like living lemurs, extinct lemur species experienced rapid postnatal growth.

Leigh and Terranova (1998) explain the low levels or absence of sexual size dimorphism in extant lemurs as by-products of rapid postnatal growth and development. In comparison to like-sized anthropoids, lemurs tend to grow rapidly after birth. Among anthropoids, the growth trajectories of males and females gradually diverge. Sexual size dimorphism is generated by sexual differences in rates of growth, age at maturation, or both. Under rapid postnatal growth and development, however, bimaturism is severely limited if it occurs at all. Differences in reported weights of conspecific lemur males and females, when they exist, are based largely on seasonal weight fluctuations that are related to reproductive cycles (e.g., Kappeler, 1997).

Low values for body size, skull size, and canine dimorphism have been reported for many extant Malagasy prosimians (Tattersall, 1982; Kappeler, 1990, 1991, 1996b; Jenkins and Albrecht, 1991; Godfrey *et al.*, 1993). Within this group, skeletal growth rates correlate poorly with dental development. Thus, for example, *Varecia variegata* grows more rapidly than do like-sized indrids (Kappeler, 1996a), but it exhibits relatively retarded dental eruption (Eaglen, 1985). In comparison to other lemurs, the Indridae are unusual for their accelerated *dental* development but not somatic growth (Wright, 1999; Godfrey *et al.*, 2001). Nevertheless, somatic growth in the Indridae, like that of lemurids and other lemur families, is sufficiently rapid to obviate bimaturism and various manifestations of skeletal dimorphism.

Among the giant lemurs unimodality generally characterizes trait distributions for conspecifics from single sites, making Finite Mixture Analysis an ideal tool for the elucidation of dimorphism.* Finite Mixture Analysis can be used to model the maximum probable embedded bimodality in site-specific skull length distributions of extinct lemur species (Godfrey *et al.*, 1993). No assumption needs to be made regarding which of the sexes is larger. Instead, skull length dimorphism can be reported as a symmetric range around 1.0 (Table VII). The highest values represent maximum dimorphism under the assumption that males were larger than females; the lowest values represent maximum "reverse" dimorphism. It is clear that, like their smaller-bodied relatives, giant extinct lemurs display little or no skull length dimorphism. Table VII also reports FMA range estimates for canine dimorphism in giant extinct lemurs. Extinct lemur values are comparable to extant lemur values, and are considerably lower than those that characterize like-sized catarrhine primates.

Low canine and size dimorphism do not necessarily imply low levels of agonism, however (Plavcan and van Schaik, 1992; Plavcan *et al.*, 1995). Instead, the frequency and especially intensity of agonism is reflected in the *relative size* of the weapon teeth. Among extant lemurs species with relatively low levels of agonism (such as *Indri indri* and *Avahi laniger*) exhibit relatively short upper canines, while more highly aggressive species (such as *Lemur catta*) exhibit relatively tall upper canines. Because body mass values can only be estimated

*Only samples of *Archaeolemur* from the Ankarana and a few other localities violate unimodality for traits such as canine height, but these samples appear to comprise two species (see Simons, 1997).

Table VII. Sexual Size and Canine Dimorphism

Family	Genus and spices	Skull length dimorphism (♂ mean/ ♀ mean)	Maxillary canine height dimorphism (♂ mean/ ♀ mean)	Relative maxillary canine height (C^1 ht/M^1 mesiodistal length), for pooled-sex or separate-sex samples[a]
Palaeopropithecidae	*Palaeopropithecus maximus*	0.96–1.05[b]	0.93–1.08[b]	1.25
	Mesopropithecus pithecoides	0.94–1.06[b]	—	1.47
Archaeolemuridae	*Archaeolemur majori*	0.96–1.04[b]	—	1.43
	Archaeolemur edwardsi	0.96–1.04[b]	0.85–1.18[b]	1.43
	Hadropithecus stenognathus	—	—	0.82
Indridae	*Avahi laniger*	1.02	0.95	0.76
	Indri indri	1.01	0.92	0.91
	Propithecus verreauxi	0.97	0.96	1.08
	Propithecus diadema	0.99	0.99	1.26
Lemuridae	*Lemur catta*	1.00	1.05	2.25
	Eulemur fulvus	1.02	0.93	2.00
	Varecia variegata	1.01	0.96	1.78
Hylobatidae	*Hylobates lar*	1.01	1.16	♂ 2.80, ♀ 2.52
Cercopithecidae	*Trachypithecus obscura*	1.04	1.59	♂ 2.08, ♀ 1.33
	Semnopithecus entellus	1.10	1.99	♂ 2.42, ♀ 1.25
	Colobus guereza	1.10	1.35	♂ 2.54, ♀ 1.94
	Cercopithecus mitis	1.15	1.95	♂ 2.85, ♀ 1.55
	Macaca fascicularis	1.19	2.13	♂ 3.22, ♀ 1.58
	Papio cynocephalus	1.19	3.22	♂ 2.43, ♀ 0.83
	Mandrillus leucophaeus	1.43	3.43	♂ 3.13, ♀ 0.96

[a] Data for sexes displaying insignificant differences in mean canine height ($P > .05$) were pooled.
[b] Finite Mixture Analysis was used to estimate dimorphism limits for the extinct lemurs (see text). There were too few specimens of *Mesopropithecus pithecoides* to produce a reliable estimate, using Finite Mixture Analysis, for this species alone. However, a pooled-species sample comprising every known skull of *M. pithecoides*, *M. globiceps*, and *M. dolichobrachion* yields low range limits for skull size idmorphism; range limits for any single species of *Mesopropithecus* should be smaller. Extinct lemur skull samples comprise 4 *M. pithecoides* from Ampasambazimba, 13 *P. maximus* from Ampasambazimba, 15 *A. edwardsi* from Ampasambazimba, and 13 *A. majori* from the Bas Menarandra. Because dentitions are incomplete, sample sizes may be lower for individual indices. Samples of extant primates were drawn from single subspecies or from populations in single geographic regions, whenever possible.

for extinct species, we opted against using the regression procedure adopted by Plavcan *et al.* (1995) to assess relative upper canine height in extinct lemurs. Instead, we constructed a simple index that is directly measureable in fossils: Relative Upper Canine Height = C^1 height/M^1 mesiodistal length. Furthermore, because few of the skulls of extinct lemurs have intact upper canines, we used the ratios of site mean values for C^1 height and M^1 length (and not the means of these ratios calculated separately for individuals) to assess relative canine height in the extinct species. The relatively abundant isolated upper canines could thus be included in our sample.

Calculated in this manner, Relative Upper Canine Height displays a spectrum for extant lemurs similar to that obtained by Plavcan *et al.* (1995) using regression residuals.* Species with relatively low agonism, such as *Indri indri* (0.91) and *Avahi laniger* (0.76), have index values that are consistently lower than those exhibited by more aggressive extant species, such as the sifakas, and especially lemurids such as *Lemur catta* (2.25). Our extinct species' values also show variability, ranging from 0.82 for *Hadropithecus stenognathus* to 1.47 for *Mesopropithecus pithecoides*. *Palaeopropithecus* is most similar in this regard to *Propithecus*, and seems to have exhibited less intense agonism than *Mesopropithecus* or *Archaeolemur*.

Small sample sizes prevent us from using Finite Mixture Analysis to estimate the degree to which Relative Upper Canine Height may be bimodal in giant extinct lemur conspecifics at single sites. Few of the skulls of extinct lemur species bear intact upper canines *and* first maxillary molars. However, canine dimorphism and Relative Upper Canine Height dimorphism are highly correlated, and canine dimorphism is low in all of the giant lemurs tested. Upper canine height in *Palaeopropithecus maximus* from Ampasambazimba has a CV of only 5% (N = 8). This contrasts dramatically with pooled-sex CV's of 64% and 54% for *Mandrillus leucophaeus* and *Papio cynocephalus* respectively, the two most strongly dimorphic cercopithecoid species in our samples.

The dimorphism profile that we obtained for *Palaeopropithecus maximus* resembles that of *Propithecus diadema*. If Relative Upper Canine Height (in each sex) reflects the intensity of agonism for that sex, and if Skull Length and Canine Dimorphism reflect sex role differentiation in agonistic encounters, then *Palaeopropithecus maximus* displayed moderate agonism and males and females of this species differed little in their agonistic profiles. *Archaeolemur majori*, *A. edwardsi*, and *Mesopropithecus pithecoides* may have displayed slightly higher levels of agonism (and *Hadropithecus stenognathus* lower agonism), but, as in *Palaeopropithecus*, little sex role differentiation in agonistic profiles. The low estimates obtained for maximum skull-length dimorphism *in all extinct lemurs* suggest that these species probably exhibited more rapid postnatal growth than like-sized anthropoids (see Leigh and Terranova 1998).†

*This is, perhaps, unsurprising, as the relationship between maxillary canine height and maxillary first molar length does not deviate significantly from isometry (based on our sample of 429 individuals belonging to 32 species of primates ranging in body size from *Galagoides demidoff* to *Pongo pygmaeus*). Under such conditions, ratios and residuals should behave similarly.

†Current research on the microstructure of the teeth provides an independent means to explore the absolute pace of dental and skeletal development of the giant lemurs. Preliminary results for *Palaeopropithecus* (Samonds, unpubl. data) have corroborated our conclusions. Enamel deposition was very rapid in *Palaeopropithecus*.

Elasticity Analysis of Census and Skeletal Data: Building a Comparative Database

Our goal is to use elasticity analysis to model, for extinct species as well as extant, species-specific life-history strategies and the contributions of individuals in different life-cycle stages to population growth rates. We will begin with census and skeletal data for extant species, and then explore applications to giant extinct indroids.

Tables VIII and IX show the population growth-rate elasticities of life cycle stages in Verreaux' sifakas and crab-eating macaques calculated from census data taken from the literature (Jolly *et al.*, 1982; Wheatley *et al.*, 1996).*

Table VIII shows that sifaka population growth tends to depend most heavily on the survival rates and/or reproductive value of *adults*. Adult males and females influence the population growth rate roughly equally. Infants and juveniles together account for just over 10% of the variation in the population growth rate. Sifakas display high levels of infant and juvenile mortality (Wright, 1995, 1999). Adult investment in youngsters is minimal (Wright, 1999; Godfrey *et al.*, 2001), and reproduction itself may be long delayed (Richard *et al.*, 1991, 1993; Wright, 1999). At Beza Mahafaly, 3-year old *P. verreauxi* occasionally give birth in the wild, but half of the females do not reproduce before age 6, and mean age at first birth is 5 years. Wright (1995) reports age at first birth of 4 years for *P. diadema edwardsi* at Ranomafana; this is based on a sample of one. Females tend to leave their natal groups at around 4 years; therefore most do not reproduce until age 5 or later (Wright, pers. comm.). For *Indri*, first birth has been estimated at 7 to 9 years (Mittermeier *et al.*,

*Excellent census data reported by Richard *et al.* (1991, 1993) for Verreaux' sifakas at Beza Mahafaly were not used for elasticity analysis because the tables omit individuals younger than around 15 months. All reported individuals are dental adults.

Table VIII. Population Growth Rate Elasticity Matrix for Wild *Propithecus verreauxi* Based on 1970–1975 Census Data Collected at Berenty in Southern Madagascar (Jolly *et al.*, 1982)[a]

Stage	Adult male	Adult female	Juvenile	Infant
Adult male	0.1743	0.1940	0.0301	0.0184
Adult female	0.1904	0.2137	0.0332	0.0207
Juvenile	0.0301	0.0332	0.0052	0.0031
Infant	0.0184	0.0207	0.0031	0.0019
Cumulative elasticity	0.4167	0.4616	0.0716	0.0441

[a]Cumulative elasticities (column sums) are the relative contributons of each life cycle stage to population growth.

Table IX. Population Growth Rate Elasticity Matrix for Wild *Macaca fascicularis* Based on 1986–1990 Census Data Collected in Bali (Wheatley *et al.*, 1996)

Stage	Adult male	Adult female	Juvenile	Infant
Adult male	0.0001	0.0058	0.0046	0.0011
Adult female	0.0058	0.2575	0.1946	0.0480
Juvenile	0.0046	0.1946	0.1480	0.0360
Infant	0.0011	0.0480	0.0360	0.0090
Cumulative elasticity	0.0116	0.506	0.383	0.0942

1994). In contrast, lemurid females begin reproducing at an early age (often at 3 years).

The key to indrid population dynamics may be relatively low *adult* mortality under resource crunches. Adult mortality for sifakas at Beza Mahafaly during the disastrous drought of 1991–1992 was considerably lower than that of the sympatric ringtailed lemurs. During this period, Beza ringtail groups lost almost 50% of adult females and, through mortality and/or migration, 90% of adult males (Gould *et al.*, 1999). All life cycle stages were severely impacted by the drought. High dietary diversity normally helps ringtails to survive resource crunches, but severe crunches take a heavy toll, even among adults. Among the sympatric Beza sifakas adult mortality increased over background levels as well, but not nearly to the same extent as in the ringtails (Richard, pers. comm.). Ringtailed lemurs depend on rapid reproductive resilience to recover losses due to periodically high mortality (Gould *et al.*, 1999). It appears that sifaka populations have a different population maintenance strategy. Sifakas depend on the greater physiological capacity of *adults* to survive severe water shortages (plus a few successful reproductions under favorable environmental conditions). Relatively slow somatic growth (cf. that of other like-sized strepsirrhines) lessens the stress on lactating mothers increasing their probability of survival. Furthermore, given accelerated *dental* development sifaka youngsters become independent fibrous-food-processing machines as quickly as possible. Adult survival (not early reproduction) is critical to long-term population stability.

By contrast, analysis of the population census data on *Macaca fascicularis* (Wheatley *et al.*, 1996) reveals a greater dependence on juveniles and adult females (Table IX). Again, the remaining life-cycle stages (in this case infants and adult males) contribute less than 10% of the variation in population growth rate. Adult males are more "expendable" than adult females—at least in terms of their cumulative effect on the population growth rate. The success of adult females at nurturing their young is critically important to population maintenance in macaques, as is the survival of juveniles. Macaques maximize the benefits received from the life-cycle stages (juvenescence and adult females) that carry the highest reproductive value (Fisher, 1958; Wilson and Bossert, 1971).

Given our desire to apply elasticity analysis to subfossil assemblages, it behooves us to test the appropriateness of the assumptions that we must embrace in order to do so. Tables VIII and IX were each generated from census data collected at two successive points in time. Our subfossil assemblages do not afford us that luxury. Imposing the constraint of zero population growth allows us to treat one assemblage as we did the two consecutive censuses in the analyses above. Population growth rates sometimes fluctuate around zero (Jolly *et al.*, 1982; Dittus, 1977) and, at any point in time, most wild populations may be near enough to zero growth so that the effects of substituting this value for the observed change in time-successive samples is trivial. Assuming population stability "consecutive" samples can be constructed from a single sample via bootstrapping.

Computer-intensive sampling (repeated sampling with replacement) produces a series of sample "populations" from one or a limited number of original samples. The Central Limit Theorem predicts that this dataset ought to represent the "true" properties of the universe from which the original sample was drawn. Even with one original sample, this technique allows the survival rates of individuals entering any particular life-cycle stage to be calculated, along with the reproductive values of individuals in each life-cycle stage. One can derive estimates of variance that can be used to compare projected stable population distributions with the observed or standing population distributions, and projected estimates can be successively evaluated (Caswell, 1989; Noreen, 1989).

Table 10 shows the results of a repeat-analysis based on a single census of *Propithecus verreauxi*, under the assumption of zero population growth. Although the cumulative elasticities of life-cycle stages differ somewhat from those derived from actual successive censuses, the fundamental signals are the same.

The second condition is that skeletal data reflect census data. To test this condition, we applied cumulative elasticity analysis to life-cycle stages as

Table X. Another Analysis of Population Growth Rate Elasticities of *Propithecus verreauxi* Based on Census Data Collected at Berenty in 1970 Only (Jolly *et al.*, 1982) and Imposing Zero Population Growth

Stage	Adult male	Adult female	Juvenile	Infant
Adult male	0.2119	0.1853	0.0185	0.0438
Adult female	0.1853	0.1620	0.0159	0.0380
Juvenile	0.0185	0.0159	0.0016	0.0038
Infant	0.0438	0.0380	0.0038	0.0091
Cumulative elasticity	0.4594	0.4011	0.0398	0.0946

inferred from museum-housed skeletal "populations" of *Propithecus verreauxi verreauxi* and *Macaca fascicularis*. Our "population" of sifakas comprised the skeletons of individuals originally collected in the dry forests of southwest Madagascar and currently housed at several museums (including Harvard University's Museum of Comparative Zoology, the American Museum of Natural History, and others). Our macaque data were drawn entirely from the samples collected at Kinabatangan (Borneo) by the Asian Primate Expedition and now housed at the Museum of Comparative Zoology. Each individual was assigned a life-cycle stage. Dental data and cranial sutures were used to assign individuals to life-cycle stages.

A complicating factor is that the skulls of lemurs cannot be sexed on the basis of their size or morphology. To reflect this limitation, we collapsed our adult male and adult female categories for *Propithecus v. verreauxi* into a pooled-sex "adult" stage, ignoring museum-recorded sex designations. We also erected a "dental adult" stage for individuals lacking skeletal (or cranial) maturity, but displaying fully erupted adult dentitions. These individuals might have been recognized as subadults or older juveniles by field workers collecting actual census data. Our "juvenile" category corresponded to individuals displaying at least those teeth known to erupt by around the time of weaning. Weaning occurs at age 6 months in sifakas, after all of adult incisors and the first two permanent molars have erupted, and, often, after P4 has begun to erupt. Individuals assigned to our "infant" category were less dentally mature.

Our criteria for constructing skeletal life-cycle stages for crab-eating macaques differed somewhat, as it is indeed possible to sex skulls of adults fairly accurately on the basis of canine size and morphology. We thus retained the stages identified by field census-takers: "adult males," "adult females," "juveniles," and "infants." We used known dental eruption schedules (Smith *et al.*, 1994) to erect life-cycle stage boundaries. Our "adult female" category included some individuals lacking dental maturity, as reproductive maturation occurs long before females are dentally mature. Females were classified as adults if their upper canine and last molars were erupting. The appearance of the first molars marked the boundary between our "infant" and "juvenile" categories. The first molars begin to erupt by 1.25–1.50 years in *Macaca fascicularis*; weaning generally occurs shortly before this.

Tables XI and XII show the results of our elasticity analyses of the skeletal "populations" of sifakas and macaques. As in our earlier analysis, we used bootstrapping to create time depth under the assumption of zero population growth. The results reveal remarkable concordance across trials. While the elasticities derived from museum samples and from living population census data are by no means identical, they capture the same dominant signals. Sifakas have low relative abundances of immature individuals and the highest elasticities for adults; macaques have much higher relative abundances of immature individuals and the highest elasticities for juveniles and adult females. It appears that neither the constraint of zero population growth, nor

Table XI. Population Growth Rate Elasticity Matrices for Wild *Propithecus verreauxi verreauxi* by Stage, Based on Skeletal Data Collected at Several Museums and Imposing Zero Population Growth

Trial 1:

Stage	Full adult	Dental adult	Juvenile	Infant
Full adult	0.5917	0.1395	0.0076	0.0302
Dental adult	0.1395	0.0338	0.0019	0.0071
Juvenile	0.0076	0.0019	0.0001	0.0004
Infant	0.0302	0.0071	0.0004	0.0016
Cumulative elasticity	0.7690	0.1814	0.0100	0.0392

Trial 2, same museum data:

Stage	Full adult	Dental adult	Juvenile	Infant
Full adult	0.6193	0.1248	0.0126	0.0311
Dental adult	0.1248	0.0253	0.0026	0.0063
Juvenile	0.0126	0.0026	0.0257	0.0063
Infant	0.0311	0.0063	0.0063	0.0158
Cumulative elasticity	0.7864	0.1590	0.0160	0.0396

the use of skeletal rather than census data, affects the results dramatically. Pooling adult male and female elasticities for macaques would not yield a sifaka-like cumulative elasticity profile, because of the very large contribution from juvenile macaques.

The differences between sifaka and macaque elasticity profiles can be understood in terms of their ontogenetic profiles. *Propithecus verreauxi* and *Macaca fascicularis* differ in their weight gain and dental developmental trajectories, despite their similarities in adult female mass, gestation length,

Table XII. Population Growth Rate Elasticity Matrix for Wild *Macaca fascicularis* Based on Skeletal Data for the Population from the Kinabatangan R., Borneo, Museum of Comparative Zoology, Harvard University and Imposing Zero Population Growth

Stage	Adult male	Adult female	Juvenile	Infant
Adult male	0.0126	0.0313	0.0657	0.0010
Adult female	0.0313	0.0802	0.1680	0.0026
Juvenile	0.0657	0.1680	0.0026	0.0056
Infant	0.0010	0.0026	0.0056	0.0088
Cumulative elasticity	0.1107	0.2821	0.5918	0.0093

and age at first breeding for females (Table XIII). As is typical of all strepsirrhines when compared to like-sized *anthropoids* sifakas exhibit slow prenatal and rapid postnatal growth. Neonatal/maternal mass ratios are low, but by age one year, *Propithecus verreauxi* weigh considerably more than do year-old *Macaca fascicularis*. Dental acceleration in sifakas yields early acquisition of ecological (but not reproductive) adulthood. Maternal investment is time-limited; weaning occurs at least 6 months earlier in sifakas than in crab-eating macaques. A battery of largely permanent teeth enables sifaka weanlings to masticate hard seeds and fibrous foodstuffs at a relatively early age. This may be critical to their ability to survive under resource crunches. Adult sifakas are able to withstand drought due to their lack of physiological dependence on free-standing water. Under severe conditions, adults are far more likely to survive than youngsters. Rapid acquisition of ecological adulthood is reflected in the demographics of sifakas, with adults outnumbering infants and juveniles.

Crab-eating macaques, in contrast, exhibit the classic "slow but steady" life-history strategy typical of catarrhine primates (see Janson and van Schaik, 1993). Prenatal growth is rapid but postnatal growth is slow. Neonatal/maternal mass ratios are high. Weaning occurs after age one, but before the eruption of the first molar. Growth is prolonged; attainment of *ecological* as well as reproductive adulthood is delayed. For an extended period of time, juveniles live in a potentially hostile environment protected by adults. The demographic profile of macaques reflects their high investment in juvenile survival. Male macaques experience an increase in mortality as they enter adulthood and transfer out of their natal groups. (Male sifakas do not experience a comparable disadvantage, as both males and females transfer out of their natal groups.) The abundance of adult males can be relatively unimportant to population growth among species, such as *M. fascicularis*, for which a single male can have breeding access to a large number of females. Changes in adult male abundance can be expected to have a trivial effect on the reproductive value (or the expected remaining reproductive contribution) of the adult females.

Elasticity Analysis of Extinct Indroids

The success of our trial runs on skeletal populations of sifakas and macaques encouraged us to take the next step. We applied our technique to two species of giant extinct indroids, using only information gleaned from mandibles at single sites. We selected the subfossil site Tsirave (on the Mangoky River, Southwest Madagascar) because of the numerous well-preserved specimens of *Archaeolemur majori* that have been found there. We selected Ampasambazimba (Itasy Basin Central Madagascar) because of the number of specimens of *Palaeopropithecus maximus* found there. Nowhere else on Madagascar is *Palaeopropithecus* better represented.

Table XIII. Life History Characteristics of *Propithecus Verreauxi* and *Macaca fascicularis*[a]

Taxon	Gestation length (in days)	Adult ♀ mass (in kg)	Neonatal mass (in kg, pooled sexes)	Neonatal/Adult ♀ mass (as a percent)	Age at weaning (in years)	Age at ♀ first breeding (in years)	Age at completion of dental eruption (rounded to nearest year)
Propithecus verreauxi	154–162	3.62	0.100	2.8%	0.50	4.6	1.0
Macaca fascicularis	160–170	3.59	0.340	9.5%	1.17	3.9	7.0

[a]Sources: Richard (1987); Harvey *et al.* (1987); Ross (1991); Nowak (1991); B. H. Smith *et al.* (1994); R. J. Smith and Leigh (1998); R. J. Smith and Jungers (1997); and the records of the Duke University Primate Center.

Table XIV. Estimated Population Growth-Rate Elasticities for *Palaeopropithecus maximus* from Ampasambazimba (Central Madagascar) and *Archaeolemur majori* from Tsirave (Southwest Madagascar), Based on Mandibles in Subfossil Collections and Imposing Zero Population Growth

Palaeopropithecus maximus: Stage	Full adult	Dental adult	Juvenile	Infant
Full adult	0.5504	0.1877	0.0037	0
Dental adult	0.1877	0.0638	0.0013	0
Juvenile	0.0037	0.0013	0.0000	0
Infant	0	0	0	0
Cumulative elasticity	0.7418	0.2527	0.0050	0

Archaeolemur majori:

Stage	Full adult	Dental adult	Juvenile	Infant
Full adult	0.1897	0.0053	0.1913	0.0485
Dental adult	0.0053	0.0001	0.0053	0.0014
Juvenile	0.1913	0.0053	0.1925	0.0483
Infant	0.0485	0.0014	0.0483	0.0123
Culululative elasticity	0.4345	0.0121	0.4374	0.1105

Our sample of 16 partial or whole mandibles of *Archaeolemur majori* from Tsirave includes those of three infants (with deciduous dentitions but no M_1) six dentally immature individuals bearing at least some molars and perhaps some replacement teeth (but retaining some deciduous teeth), and seven individuals with their full adult dentition (including one with teeth in pristine condition and incomplete cranial suture closure). This contrasts with our sample of 20 partial or whole mandibles of *Palaeopropithecus maximus* from Ampasambazimba, all of which sport full adult mandibular dentitions. Five of these can be judged (on the basis of their shallow mandibular corpi, short premolar diastemata, and slight dental wear) to have belonged to immature individuals, despite their lack of deciduous teeth. Two have no diastema separating P_2 and P_4. One has very small corpus dimensions (e.g., a corpus height little over half of its projected full adult size) and can be judged on this basis to have belonged to a juvenile with incomplete eruption of its permanent maxillary dentition. The "standing" distribution for *Palaeopropithecus* from Ampasambazimba (15 full adults, four dental but not full adults, and one juvenile) is clearly different from that exhibited by *Archaeolemur majori* from Tsirave (six full adults, one dental but not full adult, six juveniles, and three infants).

Table XIV shows our cumulative elasticities for *Palaeopropithecus maximus* and *Archaeolemur majori*, based on these samples. The cumulative elasticities for *Palaeopropithecus* look very much like those of *Propithecus* while the cumulative elasticities for *Archaeolemur* do not. In some ways, they are more macaque-like. *Palaeopropithecus* seems to have placed a high selective premium on the survival and reproductive success of adults, while *Archaeolemur* seems to have placed a relatively higher selective premium on the survival and reproductive value of juveniles.

Developmental and Life-History Profiles of the Giant Extinct Indroids

Life-history strategies determine how populations respond to environmental perturbations. When those perturbations fall outside "expected" boundaries, populations may decline or disappear entirely. Those expectations depend on how individuals in different life-cycle stages are "supposed" to behave or interact: e.g., infants "should" be born at a particular time of year and weaned at another particular time of year; social or ecological independence "should" occur at a particular age, and so on. Such expectations describe species-specific life-history strategies. Particular life-cycle stages are never strictly comparable across taxa. By definition, weaning always separates "infancy" from "juvenescence," but it does not imply a uniform degree of ecological competence on the part of weanling (see, for example, Fragaszy and Bard, 1997; Godfrey *et al.*, 2001). Indrid life-history strategies and demographics differ from those of other lemurs (Jolly *et al.*, 1982; Sussman, 1991; Richard *et al.*, 1991; Wright, 1995; Jolly, 1998; Overdorff *et al.*, 1999); they differ even more strongly from those of cercopithecids such as macaques. The question is: To what extent can paleontologists ascertain variation in the ontogenies and life-history strategies of extinct species?

Palaeopropithecids and archaeolemurids were sympatric in a wide variety of habitats from the southern to the northern tips of Madagascar, and throughout the Central Highlands (Godfrey *et al.*, 1997, 1999). This means that their niche differences cannot be ascertained directly from geographic distributions; inferences regarding niche separation must be drawn from skeletal signals. Anatomical signals establish differences in diet and positional behavior (see Jungers *et al.*, this volume, Chapter 10). Our work allows us to begin to explore how niche separation may have been manifested in the behavioral ontogeny and population dynamics of these species.

Table XV summarizes some of our inferences regarding the adaptive strategies of *Palaeopropithecus* and *Archaeolemur*, and compares these with those of *Propithecus* and *Macaca*. Based on the data presented above, it seems likely that the palaeopropithecids shared with their closest extant relatives, the indrids, rapid postnatal growth, exceptionally rapid dental eruption, early acquisition of ecological adulthood, and an abbreviated period of dependence of infants on adults. The giant folivorous sloth lemurs, like their extant relatives, were undoubtedly constrained in their developmental schedules by the need to have an operational battery of cheek teeth to cope with the mastication of fibrous foods at weaning and during the first post-weaning dry

Table XV. Sifaka vs. Macaque Adaptive Strategies with Inferences for *Palaeopropithecus* and *Archaeolemur*

Trait	*Propithecus*	*Macaca*	*Palaeopropithecus*	*Archaeolemur*
Period of infance and ecological juvenescence	Relatively short	Relatively long	Relatively short	Intermediate. Relatively longer than in *Propithecus* or *Palaeopropithecus*, but not as prolonged as in *Macaca*.
Absolute pace of postnatal somatic growth	Fast	Slow	Fast	Fast
Bimaturism and skeletal body size dimorphism	Little or none	Moderate	Little or none	Little or none
Agonistic encounters	Moderate, more frequent and more intense than in *Avahi* or *Indri*. Sexes contribute roughly equally to agonistic encounters	High intensity, more intense and more frequent than in *Propithecus*. Male agonism > female agonism	Similar to *Propithecus*. Sexes contribute roughly equally to agonistic encounters	More intense than in *Propithecus*. Sexes contribute roughly equally to agonistic encounters
Life history strategy	Selective premium on adults of both sexes	Selective premium on juveniles and adult females	Selective premium on adults (of both sexes?)	Selective premium on juveniles as well as adults (of both sexes?)

season. Precocious eruption of the adult dentition in palaeopropithecids appears to have enabled them to develop dietary independence well before they had completed their somatic growth. Moderately high values for the Relative Upper Canine Height index, coupled with a lack of dimorphism in canine height, suggest moderate agonism with nearly equal engagement of males and females in agonistic encounters. The near-absence of dentally immature palaeopropithecids in the subfossil record lends support to our inference of abbreviated ecological dependency, and highlights the selective premium that sloth lemurs, like *Propithecus*, may have placed on adults of both sexes. Palaeopropithecid populations may have depended on the ability of adults to survive severe resource crunches caused by hurricanes or droughts. Younger and older individuals in species of *Archaeolemur* may have been more equally vulnerable to adverse conditions and may have depended (more like lemurids) on their dietary diversity to survive resource crunches.

Conclusions

In displaying ontogenetic relationships, adult trait correlation and life-cycle stage distributions that may or may not be manifested in extant species, extinct species can contribute much to our understanding of patterns of evolutionary change in ontogenetic pathways and their consequences for behavior and population dynamics. Indeed, Gould (1977) saw the fossil record as the ideal testing ground for the prevalence of such distinct patterns. Developmental information may be critical as well to the process of selecting appropriate extant reference populations. This is essential if researchers expect to use comparative data from extant species to draw behavioral inferences for extinct species.

Using skeletal clues, we have reconstructed aspects of the behavioral ontogeny, life-history strategies, and population dynamics of some extinct indroids. All of our behavioral inferences are based on skeletal signals. Some of these signals relate directly to the timing of dental vis-à-vis craniofacial growth and development: (1) advanced dental eruption in the skulls and mandibles of immature individuals; (2) high coefficients of variation of metric traits in dental adults from single sites. Others are indirect indicators of relative dental acceleration; they speak more to the unique consequences of rapid acquisition of dental adulthood in short-jawed strepsirrhines. The unique consequences of dental acceleration in the Indridae and Palaeopropithecidae include: (1) premolar loss; and (2) the appearance of a distinct "break" in the morphological gradient of the cheek teeth between the distal-most premolar and first molar (with concomitant modification of the adult dental correlation structure). Skeletal indicators of a rapid *absolute* pace of postnatal growth are manifested in both the archaeolemurids and the palaeopropithecids. These include: (1) little or no sexual dimorphism in skull length and canine height; and possibly (2) an ecogeographic pattern of body size variation among closely related species or subspecies. Finally, we argue that skeletal clues to life history dynamics can be manifested in the site-specific distributions of individuals belonging to different life cycle stages. Such comparisons have led us to hypothesize that palaeopropithecids placed a selective premium on the survival and reproductive value of adults while archaeolemurids placed a relatively greater selective premium on the survival and reproductive value of juveniles.

On the basis of the data reviewed here, we conclude that the palaeopropithecids were very like their closest extant relatives in aspects of their behavioral ontogeny and life-history strategies, but that the archaeolemurids were not. Extant indrids provide excellent models for the ontogeny and life history of *Palaeopropithecus* (despite their considerably smaller body size). Large-bodied anthropoids make decidedly poor models for both *Palaeopropithecus* and *Archaeolemur*. This should serve as a warning to anyone who would naively use body mass as a predictor of the behavior or life histories of extinct species.

Because giant extinct lemurs are found in abundance at numerous late Pleistocene or Holocene sites, excellent ontogenetic series are available for some of them. This is rare in the world of primate fossils. Some of the methods described here depend on the availability of good fossil samples. Nevertheless, even single immature specimens can provide information useful to the reconstruction of the behavioral ontogeny of extinct species. Our results bear testimony to the extent to which data critical to behavioral inference may be embedded in the fossil record, if one dares to probe deeply enough. They also demonstrate the utility of developmental data in affirming the appropriateness of particular extant species as models for behavioral reconstruction.

What is of particular interest to us here is how the discovery of unanticipated developmental relationships in *extinct* lemurs opened lines of inquiry into *extant* lemur biology that had previously gone untapped. It was the discovery of dental developmental acceleration in the Palaeopropithecidae that forced us not merely to reject an anthropoid reference model for palaeopropithecid biology and behavior, but also to explore in greater detail variation in the craniodental development of extant families of lemurs.

Finally, it was the dearth of young palaeopropithecids at fossil sites that prompted us to explore the utility of skeletal series in stage-based population studies. Elasticity analysis allows direct comparisons of species with very different patterns of somatic growth and skeletal maturation, or ages at reproductive maturation. It can be very useful in exploring the effects of different life-history strategies on population growth. Applied with care to the fossil record, researchers might be able to reconstruct aspects of the life-history strategies of extinct taxa, and expand the framework through which the advantages and disadvantages of different life-history strategies in the Order Primates can be interpreted.

ACKNOWLEDGMENTS

We thank Ken Glander for access to unpublished records of the Duke Primate Center. Skeletal data were collected at the following institutions: the Natural History Museum, London (BMNH); Museum of Comparative Zoology, Cambridge, MA; Université d'Antananarivo (UA); Duke University Primate Center (DUPC); the American Museum of Natural History, New York (AMNH); the Smithsonian Institution, Washington, D.C.; the Field Museum of Natural History, Chicago; the Rijksmuseum van Natuurlijke Historie, Leiden; and the Muséum National d'Histoire Naturelle, Paris. The research on dental development reported here was conducted in collaboration with a number of people, especially Karen Samonds and William Jungers. This research was carried out under the Protocol of Collaboration for paleontological research signed August 1, 1983, by officials of the University of Antananarivo and the Duke University Primate Center. The field research on the giant subfossil lemurs of Madagascar was headed by Elwyn Simons (Duke University Primate Center). Special thanks to the entire field crew, particularly Prithijit Chatrath and Donald DeBlieux. Special thanks, also, to Berthe Rakotosamimanana and to Gisèle Randria, who greatly facilitated research in Madagascar. Unpublished data on extant lemurs were generously provided by Claire Hemingway, Alison Richard, and Patricia Wright. Radiographs of young sifakas were produced with the generous assistance of Deryck Jones (Museum of Natural History London). The research was supported in part by NSF GER 9450175 to LRG and NSF SBR-9630350 to Elwyn Simons.

Appendix

Matrix Population Projection Methods

Leslie (1945) produced a matrix of values (constructed from age-specific survival and fertility rates) which he used to project the future size and age composition of a population. Leslie's matrix, **A**, is the "multiplicand" which premultiplies the population census, $\mathbf{n}_{(t)}$, and results in the projected age-distribution of a future census, $\mathbf{n}_{(t+1)}$. Summing $\mathbf{n}_{(t+1)}$ across all stages will give the projected population size at time $\mathbf{t+1}$. Algebraically, **A** transitions the age distribution at time **t** to one at time $\mathbf{t+1}$, namely $\mathbf{An}_{(t)} = \mathbf{n}_{(t+1)}$. The process of "transitioning" can be iterated many times allowing for projections about the population distribution among the age classes as well as population growth and age-class reproductive contributions.

Leslie matrices are an example of transition matrices—i.e., matrices that document how stages at time $\mathbf{t+1}$ are functions of stages at time **t**. A Leslie matrix has a typical set of nonzero elements; the survival probabilities are in the lower subdiagonal and the stage fertilities are in the first row. More complicated transition matrices having more nonzero elements are certainly possible. A four-stage example of a Leslie matrix is:

	Age Class 1	Age Class 2	Age Class 3	Age Class 4
Age Class 1	0	0.1	2	3
Age Class 2	0.4	0	0	0
Age Class 3	0	0.5	0	0
Age Class 4	0	0	0.7	0

Note the lack of symmetry. We should not expect a symmetric decomposition of these matrices, as we get in covariance matrices. Another way to visualize this matrix is to use a graph model. Here is the same example as a graph:

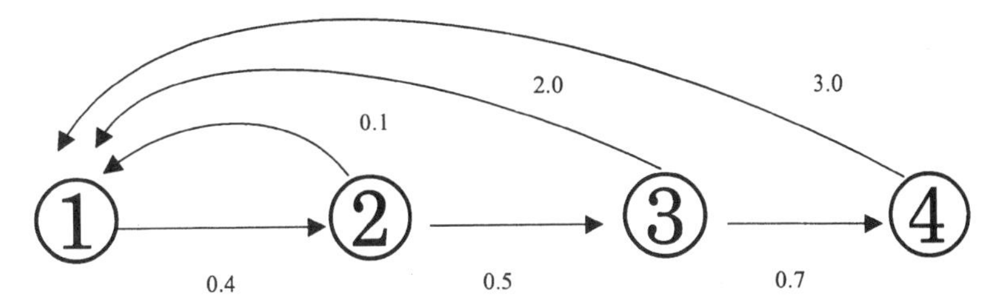

For either visualization, we have:

> probability of survival from age class 1 → age class 2 = 0.4,
> probability of survival from age class 2 → age class 3 = 0.8,
> probability of survival from age class 3 → age class 4 = 0.7,
> and individuals in the four age classes produce 0, 0.1, 2, and 3 offspring per projection interval.

Leslie (1945) demonstrated that the "dominant latent root" (or eigenvalue, λ) of the matrix characterizes the rate of increase in the population, $\mathbf{r}$, as the time interval, $\mathbf{t}$, increases: $\lambda = \mathbf{e}^{\mathbf{rt}}$. If population size does not change between samples, then $\mathbf{r} = \mathbf{0}$, and $\lambda = 1$.

Charlesworth (1980) illustrated the relationship of the matrix $\mathbf{A}$ with two important demographic variables—the stable age distribution and the reproductive value of the population described by $\mathbf{A}$. The long-term stable population distribution, $\mathbf{w}$, is represented by the right eigenvector associated with the dominant eigenvalue of the matrix, $\mathbf{Aw} = \lambda\mathbf{w}$. When $\mathbf{A}$ transitions the stable distribution, it simply gives back a rescaled version of the distribution. The stable population distribution is supported by reproductive contributions from each of its stable classes. The idealized or long-term reproductive value for each population class is calculated as: $\mathbf{v}'\mathbf{A} = \lambda\mathbf{v}'$.

Lefkovitch (1965) demonstrated the applicability of Leslie's method to the more general case of stage-classified organisms. Caswell (1989) illustrated more recent developments and applications using variables such as size, mass, or other stage-related features as "demographic" variables.

Eigenvalue (Growth-Rate) Sensitivity

The sensitivity of the intrinsic population growth rate (λ) to perturbations in the elements ($\mathbf{a}_{ij}$) of the projection matrix (**A**) can be shown to be proportional to the products of the corresponding reproductive value eigenvector times the corresponding life-cycle stage survival or abundance: $\partial\lambda/\partial\mathbf{a}_{ij} \propto \mathbf{v}_i\mathbf{w}_j$. A sensitivity is "the product of the *i*th element of the reproductive value vector and the *j*th element of the stable age distribution" (Caswell, 1989: 121). Larger sensitivity values indicate that a change in $\mathbf{a}_{ij}$ will influence the population growth rate more than those interactions with smaller products. Note that sensitivity measures only how *sensitive* λ is to changes in $\mathbf{a}_{ij}$, not the *amount* of change we can expect from changes in $\mathbf{a}_{ij}$.

The interpretation of sensitivities for any population requires some caution. It is important to remember that the product of the reproductive value of, say, juveniles with the survival of adult females will be different from the product of the reproductive value of adult females with the survival of juveniles (i.e., $\mathbf{v}_i\mathbf{w}_j \neq \mathbf{v}_j\mathbf{w}_i$). Furthermore, the data used to calculate **w** and **v** are measured on different scales (Caswell, 1989: 132). The survival vector must sum to 1. On the other hand, the reproductive value vector is scaled as a proportion of total reproductive output remaining at each stage in the life cycle, and therefore is not constrained to sum to any value. Sensitivities can be compared *within* but not across populations.

Using sensitivities, it is possible to draw a general profile of how much the population growth rate would change in response to changes in the survival or reproductive outcomes within various life-cycle stages. However, direct comparisons of sensitivities between species and even between populations of the same species must be done with care, because under different conditions, a greater *proportional* contribution to the population growth rate from either the survival or the reproductive value can be obscured in the calculation of the product.

Eigenvalue Elasticity

One solution to the difficulty of direct comparison among populations uses the concept of elasticity adapted from economic theory by Caswell and colleagues (Caswell, 1989; van Groenendael *et al.*, 1988). Elasticity is an estimate of the proportional change in λ due to a proportional change in $\mathbf{a}_{ij}$.

The formula for calculating the elasticity ($\mathbf{e}_{ij}$) of $\mathbf{A}$ with respect to $\mathbf{a}_{ij}$ is: $\mathbf{e}_{ij} = (\partial\lambda/\lambda)/(\partial\mathbf{a}_{ij}/\mathbf{a}_{ij})$. Elasticities can be summed for each stage and interpreted as the expected effect on the population growth rate (λ) of changes in the survival and reproduction profiles of particular life-cycle stages. A higher value for a column sum (or *cumulative elasticity*—the sum of the elasticities for any given life-cycle stage) indicates a greater impact on the population growth rate of changes in the survival or the reproductive value of individuals in that particular life-cycle stage. Because elasticities always sum to one, *direct* comparisons of the relative contributions of life-cycle stages to the ideal population growth rate are possible for different populations and for different species.

References

Acuña, J. D., and Muñoz, M. A. 1998. Mixture analysis in biology: Scope and limits. *J. Theor. Biol.* **191**:341–344.

Alberch, P. 1985. Problems with the interpretation of developmental sequences. *Syst. Zool.* **34**:46–58.

Albrecht, G. H., Jenkins, P. D., and Godfrey, L. R. 1990. Ecogeographic size variation among the living and subfossil prosimians of Madagascar. *Am. J. Primatol.* **22**:1–50.

Atchley, W. R. 1984. Ontogeny, timing of development, and genetic variance-covariance structure. *Am. Nat.* **123**:519–540.

Atchley, W. R., Plummer, A. A., and Riska, B. 1985. Genetics of mandible form in the mouse. *Genetics* **111**:555–577.

Benton, T. G., and Grant, A. 1999. Elasticity analysis as an important tool in evolutionary and population ecology. *Trends Ecol. Evol.* **14**:467–471.

Bromage, T. G., and Dean, M. C. 1985. Re-evaluation of the age at death of immature fossil hominids. *Nature* **317**:525–527.

Caswell, H. 1978. A general formula for the sensitivity of population growth rate to changes in life history parameters. *Theor. Popul. Biol.* **14**:215–230.

Caswell, H. 1989. *Matrix Population Models*. Sinauer, Sunderland, MA.

Caswell, H., and Twombly, S. 1989. Estimation of stage-specific demographic parameters for zooplankton populations: Methods based on stage-classified matrix projection models. In: L. McDonald, B. Manly, J. Lockwood, and J. Logan (eds.), *Estimation and Analysis of Insect Populations*, pp. 93–107. Springer-Verlag, New York.

Charlesworth, B. 1980. *Evolution in Age-Structured Populations*. Cambridge University Press, New York.

Cheverud, J. M. 1982. Phenotypic, genetic, and environmental morphological integration in the cranium. *Evolution* **36**:499–516.

Cheverud, J. M. 1995. Morphological integration in the saddle-back tamarin (*Saguinus fuscicollis*) cranium. *Am. Nat.* **145**:63–89.

Cheverud, J. M. 1996. Developmental integration and the evolution of pleiotropy. *Am. Zool.* **36**:44–50.

Churchill, S. E. 1996. Particulate versus integrated evolution of the upper body in Late Pleistocene humans: A test of two models. *Am. J. Phys. Anthropol.* **100**:559–583.

Cole, L. C. 1954. The population consequences of life history phenomena. *Q. Rev. Biol.* **29**:103–137.

Dean, M. C. 1987. Growth layers and incremental markings in hard tissue: A review of the lieterature and some preliminary observations about enamel structure in *Paranthropus boisei*. *J. Hum. Evol.* **16**:157–172.

Dittus, W. P. J. 1977. The social regulation of population density and age-sex specific sex distribution in the toque monkey. *Behaviour* **63:**281–322.

Eaglen, R. H. 1985. Behavioral correlates of tooth eruption in Madagascar lemurs. *Am. J. Phys. Anthropol.* **66:**307–315.

Fink, W. L., and Zelditch, M. L. 1995. Phylogenetic analysis of ontogenetic shape transformations: A reassessment of the piranha genus *Pygocentrus* (Teleostei). *Syst. Biol.* **44:**343–360.

Fisher, R. 1958. *The Genetical Theory of Natural Selection*. Dover, New York.

Flury, B. D., Airoldi, J. P., and Biber, J. P. 1992. Gender identification of water pipits (*Anthus spinoletta*) using mixtures of distributions. *J. Theor. Biol.* **158:**465–480.

Fragaszy, D. M., and Bard, K. 1997. Comparison of development and life history in *Pan* and *Cebus*. *Int. J. Primatol.* **18:**683–701.

Gilbert, S. F. 1997. *Developmental Biology*, 5th edition. Sinauer, Sunderland, MA.

Gilbert, S. F., Opitz, J. M., and Raff, R. A. 1996. Resynthesizing evolutionary and developmental biology. *Devel. Biol.* **173:**357–372.

Godfrey, L. R. 1993. Dental development in fossil lemurs: Phylogenetic and ecological interpretations. *Am. J. Phys. Anthropol.* Suppl. **16:**96.

Godfrey, L. R., Sutherland, M. R., Petto, A. J., and Boy, D. S. 1990. Size, space, and adaptation in some subfossil lemurs from Madagascar. *Am. J. Phys. Anthropol.* **81:**45–66.

Godfrey, L. R., Lyon, S. K., and Sutherland, M. R. 1993. Sexual dimorphism in large-bodied primates: The case of the subfossil lemurs. *Am. J. Phys. Anthropol.* **90:**315–334.

Godfrey, L. R., Jungers, W. L., Reed, K. E., Simons, E. L., and Chatrath, P. S. 1997. Subfossil lemurs: Inferences about past and present primate communities in Madagascar. In: S. M. Goodman and B. D. Patterson (eds.) *Natural Change and Human Impact in Madagascar*, pp. 218–256. Smithsonian Inst. Press, Washington DC.

Godfrey, L. R., Jungers, W. L., Wright, P. C., and Jernvall, J. 1998. Dental development in giant lemurs: Implications for their evolution and ecology. *Folia Primatol.* **69** (Suppl. 1): 398–399.

Godfrey, L. R., Jungers, W. L., Simons, E. L., Chatrath, P. S., and Rakotosamimanana, B. 1999. Past and present distributions of lemurs in Madagascar. In: B. Rakotosamimanana, H. Rasamimanana, J. Ganzhorn, and S. Goodman (eds.) *New Directions in Lemur Studies*, pp. 19–53. Kluwer Academic/Plenum Publishers, New York.

Godfrey, L. R., Samonds, K. E., Jungers, W. L., and Sutherland, M. R. 2001. Teeth, brains, and primate life histories. *Am. J. Phys. Anthropol.* **114:**192–214.

Gould, L., Sussman, R. W., Sauther, M. L. 1999. Natural disasters and primate populations: The effects of a 2-year drought on a naturally occurring population of ring-tailed lemurs (*Lemur catta*) in southwestern Madagascar. *Int. J. Primatol.* **20:**69–84.

Gould, S. J. 1977. *Ontogeny and Phylogeny*. Harvard University Press, Cambridge, MA.

Greenfield, L. O. 1996. Correlated response of homologous characteristics in the anthropoid anterior dentition. *J. Hum. Evol.* **31:**1–19.

Harvey, P. H., Martin, R. D., and Clutton-Brock, T. H. 1987. Life histories in comparative perspective. In: B. B. Smuts, D. L. Cheney, R. M. Seyfarth, R. W. Wrangham, and T. T. Struhsaker, Jr. (eds.) *Primate Societies*, pp. 181–196. Univ. of Chicago Press, Chicago.

Hemingway, C. A. 1996. Morphology and phenology of seeds and whole fruit eaten by Milne–Edwards' sifaka, *Propithecus diadema edwardsi*, in Ranomafana National Park, Madagascar. *Int. J. Primatol.* **17:**637–659.

Huxley, J. S. 1932. *Problems of Relative Growth*. Methuen & Co., London.

Janson, C., and van Schaik, C. 1993. Ecological risk aversion in juvenile primates: Slow and steady wins the race. In: M. E. Pereira and L. A. Fairbanks (eds.) *Juvenile Primates: Life History, Development and Behavior*, pp. 57–76. Oxford University Press New York.

Jenkins, P. D., and Albrecht, G. H. 1991. Sexual dimorphism and sex ratios in Madagascan prosimians. *Am. J. Primatol.* **24:**1–14.

Jolly, A. B. 1998. Pair-bonding, female aggression and the evolution of lemur societies. *Folia Primatol.* **69** (Suppl. 1):1–13.

Jolly, A. B., Oliver, W. L. R., and O'Connor, S. M. 1982. *Propithecus verreauxi* population and ranging at Berenty, Madagascar, 1975 and 1980. *Folia Primatol.* **39:**124–144.

Josephson, S. C., Juell, K. E., and Rogers, A. R. 1996. Estimating sexual dimorphism by method-of-moments. *Am. J. Phys. Anthropol.* **100:**191–206.

Jungers, W. L. 1980. Adaptive diversity in subfossil Malagasy prosimians. *Z. Morphol. Anthropol.* **71:**177–186.

Jungers, W. L., Godfrey, L. R., Simons, E. L., Chatrath, P. S., and Rakotosamimanana, B. 1991. Phylogenetic and functional affinities of *Babakotia* (Primates), a fossil lemur from northern Madagascar. *Proc. Natl. Acad. Sci.* USA **88:**9082–9086.

Jungers, W. L., Godfrey, L. R., Simons, E. L., and Chatrath, P. S. 1997. Phalangeal curvature and positional behavior in extinct sloth lemurs (Primates, Palaeopropithecidae). *Proc. Natl. Acad. Sci. USA* **94:**11998–12001.

Kappeler, P. M. 1990. The evolution of sexual size dimorphism in prosimian primates. *Am. J. Primatol.* **21:**201–214.

Kappeler, P. M. 1991. Patterns of sexual dimorphism in body weight among prosimian primates. *Folia Primatol.* **57:**132–146.

Kappeler, P. M. 1996a. Causes and consequences of life-history variation among strepsirhine primates. *Am. Nat.* **148:**868–891.

Kappeler, P. M., 1996b. Intrasexual selection and phylogenetic constraints in the evolution of sexual canine dimorphism in strepsirhine primates. *J. Evol. Biol.* **9:**43–65.

Kappeler, P. M., 1997. Intrasexual selection in *Mirza coquereli*: evidence for scramble competition polygyny in a solitary primate. *Behav. Ecol. Sociobiol.* **45:**115–127.

Lauder, G. V. 1981. Form and function: Structural analysis in evolutionary morphology. *Paleobiology* **7:**430–442.

Lefkovitch, L. P. 1965. The study of population growth in organisms grouped by stages. *Biometrics* **21:**1–18.

Leigh, S. R., and Terranova, C. J. 1998. Comparative perspectives on bimaturism, ontogeny, and dimorphism in lemurid primates. *Int. J. Primatol.* **19:**723–749.

Leslie, P. H. 1945. On the use of matrices in certain population mathematics. *Biometrika* **33:**183–212.

Leslie, P. H. 1948. Some further notes on the use of matrices in population mathematics. *Biometrika* **35:**213–245.

Luckett, W. P. 1984. Developmental evidence for toothcomb homology in the lemuriform primates *Propithecus* and *Lemur*. *Am. J. Phys. Anthropol.* **63:**187–188.

Lyman, R. L. 1994. *Vertebrate Taphonomy*. Cambridge University Press, Cambridge.

McKinney, M. L., and McNamara, K. J. 1991. *Heterochrony: The Evolution of Ontogeny*. Plenum Press, New York.

Meyers, D. M. and Wright P. C. 1993. Resource tracking: Food availability and *Propithecus* seasonal reproduction. In: P. M. Kappeler and J. U. Ganzhorn (eds.), *Lemur Social Systems and their Ecological Basis*, pp. 179–192. Plenum Press, New York.

Mittermeier, R. A., Tattersall, I., Konstant, B., Meyers, D. M., and Mast, R. B. (eds.). 1994. *Lemurs of Madagascar*. Conservation International, Washington DC.

Morbeck, M. E. 1997. Reading life history in teeth, bones, and fossils. In: M. E. Morbeck, A. Galloway, and A. L. Zihlman (eds.), *The Evolving Female: A Life-History Perspective*, pp. 117–131. Princeton University Press, Princeton, N.J.

Noreen, E. W. 1989. *Computer Intensive Methods for Testing Hypotheses: An Introduction*. John Wiley & Sons, New York.

Nowak, R. M. (ed.). 1991. Primates. In: *Walker's Mammals of the World*, 5th edition Vol. 1, pp. 400–514. Johns Hopkins Univ. Press, Baltimore.

Olson, E. C., and Miller, R. L. 1958. *Morphological Integration*. University of Chicago Press, Chicago.

Overdorff, D. J., Merenlender, A. M., Talata, P., Telo, A., and Forward, Z. A. 1999. Life history of *Eulemur fulvus rufus* from 1988–1997 in southeastern Madagascar. *Am. J. Phys. Anthropol.* **108:**295–310.

Pearson, K. 1894. Contribution to the mathematical theory of evolution. *Philos. Trans. R. Soc. A* **185:**71–110.

Pianka, E. R. 1970. On r- and K-selection. *Am. Nat.* **104:**592–597.

Pianka, E. R. 1971. r- and K-selection or b- and d-selection? *Am. Nat.* **105:**581–588.

Plavcan, J. M. 1994. Comparison of four simple methods for estimating sexual dimorphism in fossils. *Am. J. Phys. Anthropol.* **94:**465–476.

Plavcan, J. M. 1998. Correlated response, competition, and female canine size in primates. *Am. J. Phys. Anthropol.* **107:**401–416.

Plavcan, J. M., and van Schaik, C. P. 1992. Intrasexual competition and canine dimorphism in anthropoid primates. *Am. J. Phys. Anthropol.* **87:**461–477.

Plavcan, J. M., van Schaik, C. P., and Kappeler, P. M. 1995. Competition, coalitions and canine size in primates. *J. Hum. Evol.* **28:**245–276.

Ravosa, M. J., Meyers, D. M., and Glander, K. E. 1993. Relative growth of the limbs and trunk in sifakas: Heterochronic, ecological and functional considerations. *Am. J. Phys. Anthropol.* **92:**499–520.

Rehg, J. A. and Leigh, S. R. 1999. Estimating sexual dimorphism and size differences in the fossil record: A test of methods. *Am. J. Phys. Anthropol.* **110:**95–104.

Richard, A. F. 1987. Malagasy prosimians: female dominance. In: B. B. Smuts, D. L. Cheney, R. M. Seyfarth, R. W. Wrangham, and T. T. Struhsaker, Jr. (eds.) *Primate Societies*, pp 25–33. Univ. of Chicago Press, Chicago.

Richard, A. F., Rakotomanga, P., and Schwartz, M. 1991. Demography of *Propithecus verreauxi* at Beza Mahafaly, Madagascar: Sex ratio, survival, and fertility, 1984–1988. *Am. J. Phys. Anthropol.* **84:**307–322.

Richard, A. F., Rakotomanga, P., and Schwartz, M. 1993. Dispersal by *Propithecus verreauxi* at Beza Mahafaly, Madagascar: 1984–1991. *Am. J. Primatol.* **30:**1–20.

Robertson, C. A. and Fryer, J. G. 1969. Some descriptive properties of normal mixtures. *Scand. Actuar. J.* **52:**137–146.

Ross, C. 1991. Life history patterns of New World monkeys. *Int. J. Primatol.* **12:**481–502.

Samonds, K. E., Godfrey, L. R., Jungers, W. L., and Martin, L. B. 1999. Primate dental development and the reconstruction of life history strategies in subfossil lemurs. *Am. J. Phys. Anthropol. Suppl* **28:**238–239.

Sauer, J. R., and Slade, N. A. 1985. Mass-based demography of a hispid cotton rat (*Sigmodon hispidus*) population. *J. Mammal.* **66:**316–328.

Schwartz, J. H., and Tattersall, I. 1985. Evolutionary relationships of living lemurs and lorises (Mammalia, Primates) and their potential affinities with European Eocene Adapidae. *Anthropol. Pap. Am. Mus. Nat. Hist.* **60:**1–100.

Seachrist, L. 1996. Chewing up the fossil record. *Science* **271:**1056.

Shipman, P. 1981. *Life History of a Fossil: An Introduction to Taphonomy and Paleoecology.* Harvard University Press, Cambridge, MA.

Simons, C. V. M. 1997. Diet, dental eruption and dental variation in *Archaeolemur* specimens from Northwestern Madagascar. *Am. J. Phys. Anthropol. Suppl.* **24:**211–212.

Simons, E. L., Godfrey, L. R., Jungers, W. L., Chatrath, P. S., and Rakotosamimanana, B. 1992. A new giant subfossil lemur, *Babakotia*, and the evolution of the sloth lemurs. *Folia Primatol.* **58:**197–203.

Smith, B. H. 1992. Life history and the evolution of human maturation. *Evol. Anthropol.* **1:**134–142.

Smith, B. H., Crummett, T. L., and Brandt, K. L. 1994. Ages of eruption of primate teeth: A compendium for aging individuals and comparing life histories. *Yearb. Phys. Anthropol.* **37:**177–231.

Smith, R. J., and Jungers, W. L. 1997. Body mass in comparative primatology. *J. Hum. Evol.* **32:**523–559.

Smith, R. J., and Leigh, S. R. 1998. Sexual dimorphism in primate neonatal body mass. *J. Hum. Evol.* **34:**173–201.

Standing, H. F. 1908. On recently discovered subfossil primates from Madagascar. *Trans. Zool. Soc. London.* **18:**59–216.

Sussman, R. W. 1991. Demography and social organization of free-ranging *Lemur catta* in the Beza Mahafaly Reserve, Madagascar. *Am. J. Phys. Anthropol.* **84:**43–58.

Tanner, J. M. 1990. *Foetus into Man: Physical Growth from Conception to Maturity* 2nd edition. Harvard University Press, Cambridge MA.

Tattersall, I. 1982. *The Primates of Madagascar.* Columbia University Press, New York.

Thomas, R. D. K., and Reif, W.-E. 1993. The skeleton space: A finite set of organic designs. *Evolution* **47:**341–360.

Titterington, D. M., Smith, A. F. M., and Makov, V. E. 1985. *Statistical Analysis of Finite Mixture Distributions.* John Wiley & Sons, New York.

van Groenendael, J. de Kroon, H., and Caswell, H. 1988. Projection matrices in population biology. *Trends Ecol. Evol.* **3:**264–269.

Wheatley, B. P., Harya Putra, D. K., and Gonder, M. K. 1996. A comparison of wild and food-enhanced long-tailed macaques (*Macaca fascicularis*). In: J. E. Fa and D. G. Lindburg (eds.), *Evolution and Ecology of Macaque Societies*, pp. 182–206. Cambridge University Press, New York.

Wilson, E. O., and Bossert, W. 1971. *A Primer of Population Biology.* Sinauer, Sunderland, MA.

Wright, P. C. 1995. Demography and life history of free-ranging *Propithecus diadema edwardsi. Int. J. Primatol.* **16:**835–854.

Wright, P. C. 1999. Lemur traits and Madagascar ecology: Coping with an island environment. *Yearb. Phys. Anthropol.* **42:**31–72.

Yamashita, N. 1996. Seasonality and site specificity of mechanical dietary patterns in two Malagasy lemur families (Lemuridae and Indriidae). *Int. J. Primatol.* **17:**355–387.

Zelditch, M. L. 1988. Ontogenetic variation in patterns of phenotypic integration in the laboratory rat. *Evolution* **42**:28–41.

Zelditch, M. L. 1996. Introduction to the symposium: Historical patterns of developmental integration. *Am. Zool.* **36:**1–3.

A Comparative Approach to Reconstructing the Socioecology of Extinct Primates

5

CHARLES L. NUNN and CAREL P. VAN SCHAIK

Introduction

Some aspects of behavior can be reconstructed in the fossil record because structure and function (i.e., behavior) are directly linked. For instance, one may be able to reconstruct the diet of an extinct taxon based on a variety of morphological features that are known to have certain functions among extant species. Many chapters in this book follow this approach, which, for all its limitations (Kay and Cartmill, 1977; Kay, 1984), may produce the most reliable reconstructions of the behavior and ecology of extinct species.

Other aspects of behavior, including socioecological features such as group size and ranging patterns, have less direct links to morphology and are thus less easily reconstructed in extinct taxa. However, to obtain a deeper understanding of a species' lifestyle and the factors affecting its evolution,

---

CHARLES L. NUNN • Department of Biological Anthropology and Anatomy, Duke University, Durham, North Carolina 27708-0383. Present address: Section of Evolution and Ecology, University of California, Davis, California 95616. CAREL P. VAN SCHAIK • Department of Biological Anthropology and Anatomy, Duke University, Durham, North Carolina 27708-0383.

Reconstructing Behavior in the Primate Fossil Record, edited by Plavcan *et al.* Kluwer Academic/Plenum Publishers, New York, 2002.

some knowledge of socioecology is necessary. Socioecological variables of interest include feeding strategy and range use, group size, social relationships and mating system, between-group relations such as territoriality, and demographic features such as sex-biased dispersal or population density. While it is naive to think that we can ever reconstruct these features with the precision of extant species (especially given the appreciable intraspecific variation found in some of these traits in living species), even limited resolution of these features in a fossil taxon would be useful.

Cladistic techniques can be used to map the socioecological traits of extant species onto a phylogeny (Wrangham, 1987; Di Fiore and Rendall, 1994; see also Garland *et al.* 1997). However, the results of such studies depend critically on the phylogeny used to map the characters and on the accurate placement of the fossil taxon within this phylogeny. In addition, character reconstruction depends on the assumptions used to map the traits phylogenetically. The assumptions of unordered, equally weighted parsimony, the most commonly used method of character mapping, are often violated (Frumhoff and Reeve, 1994; Omland, 1997; Cunningham *et al.*, 1998). Errors in the phylogenetic hypothesis and violations of the assumptions of parsimony will lead to inaccurate behavioral reconstructions for fossil taxa, while the potential for such errors leads to uncertainty over behavioral reconstructions.

A second approach uses associations among traits in living species to infer behavioral traits in extinct ancestors, much like morphological traits are used to infer directly linked characters such as diet. This comparative approach to reconstructing socioecology, pioneered by Dunbar (1992a, 1993) for patterns in closely related taxa, extrapolates from detailed models in extant taxa that link socioecological variables to external variables (such as rainfall, temperature, or habitat type) and internal variables (such as body size). If the necessary input variables can be reconstructed in the fossil record, and if these estimates are not too far outside the range of modern values, then one can make inferences about the behavioral ecology of extinct representatives of a living taxon. Dunbar's (1992a, 1993) method is most applicable when an extinct taxon has a set of immediate living descendants. In this case, the input variables will be most readily available, and models based on living species will be most applicable. So far, this comparative approach has mainly been used to model group size, but there is no reason why it cannot be applied to other socioecological characters.

Many extinct taxa do not have living descendants which are sufficiently similar in size and ecology to warrant the use of these detailed models. In addition, the variables needed for these models are not available for all extant taxonomic groups, limiting the usefulness of this approach. Hence, it is worthwhile to explore a comparative approach that is based on more general associations between variables that are knowable in the fossil record, here called K variables, and those that are unknowable, here called U variables. For instance, if a particular diet (a K variable) is associated with other features of a species' behavioral ecology, such as home range size (a U variable), it might be reasonable to infer that a now extinct species with the same diet had a similar home range size. The soundness of any such conclusions will depend on the strength of the association in extant taxa and our understanding of its causality. Phylogenetic information, such as the likely taxonomic affiliation of the fossil, will improve these estimates.

The goal of this chapter is to evaluate the usefulness of this more general comparative approach to reconstructing the behavior of extinct primate species. To achieve this goal, we report on the statistical analysis of a data set containing socioecological variables for extant species, and we apply our results to several extinct primate taxa. For our analysis, the following behavioral features will be considered as the known (or knowable) K variables for the fossil taxon*: activity period (Kay and Cartmill, 1977), locomotor substrate (Kay and Williams, this volume, Chapter 9), habitat characteristics (an open or wooded environment, as determined from faunal and floral remains; Andrews and van Couvering, 1975; Andrews, 1996), diet (Kay, 1984), and body mass (Gingerich *et al.*, 1982). Because our knowledge of these variables in the fossil record is necessarily imprecise, we restrict ourselves to discrete categories of these variables. The U variables in our analysis include continuous variables describing foraging group size and population group size (see Clutton-Brock and Harvey, 1977a), home range size, day journey length, and the intensity of home range use (the defensibility, or "D," index; Mitani and Rodman, 1979). In a complementary analysis, Plavcan (this volume, Chapter 8) focuses on reconstructing aspects of the social system based on sexual dimorphism in dentition and body size.

Reconstructing the socioecology of extinct primates using our methods requires two steps. First, K variables should be identified for the fossil taxon in question. Although our methods do not require that all K variables be identified, the resolution of behavior in a fossil taxon will increase when more K variables can be assigned character states. The complete set of K variables captures basic lifestyle attributes of primates; we have therefore termed the set of possible combinations of K-variable character states "syndromes," where syndrome refers to a distinctive or characteristic pattern of behavior†. In addition, because not all K-variable combinations are equally likely, one can use information on patterns of K-variable character states in extant taxa to evaluate the likelihood of a combination of possible K-variable traits reconstructed for a fossil taxon. Thus, one can check the "internal consistency" of a fossil taxon's syndrome designation. Many of the relationships among the K-variable character states that we document below, as well as many of the relationships between K and U variables, are consistent with patterns found in previous comparative studies (e.g., Clutton-Brock and Harvey, 1977a,b).

After reconstructing K variables and checking their internal consistency, the next step is to estimate values for U variables in the extinct taxon. We

*We assume that K variables are inputs for the analyses that follow. Readers interested in issues regarding the reconstruction of individual K variables should refer to the sources provided.

†Our syndromes have similarities to the ecological categories of Clutton-Brock and Harvey (1977a). However, decisions regarding our categories were explicitly based on combination of K variables that have the potential to be reconstructed from appropriate fossil material.

present tables that make it possible to predict socioecological features of an extinct taxon when this taxon can be assigned to a particular syndrome. These tables take into account sample sizes and taxonomic patterns in living species. For cases where a fossil cannot be definitively placed in a syndrome, we also provide tables with single K variables. This reduces our resolution of behavior because fewer potentially confounding variables are controlled in the analysis; however, it might be the only approach since one can rarely identify all K variables for a fossil taxon. The tables provided in the Results section provide quantitative patterns; readers interested in more general patterns may wish to skip ahead to the Discussion, where we combine our results with a review of previous research on these associations in extant taxa. To strengthen the linkage between K and U variables (and thus our interpretations of behavior in extinct taxa), we also provide functional explanations for the statistical patterns documented here.

Before we begin the analysis, we discuss some methodological issues involved in reconstructing the behavior of a fossil taxon using this comparative approach.

Methodological Considerations

Four main problems complicate attempts to extrapolate from the known ($K_1, \ldots, K_n$) to the unknown ($U_1, \ldots, U_n$) features of behavioral ecology. The first is that the strength of the correlations between the known and the unknown traits may be strongly influenced by the hierarchical structure of relatedness that exists among the species in the sample. A conservative interpretation of this correlation is that the observed relationship is spurious, caused by descent from a common ancestor (see Harvey and Pagel, 1991). For these cases, the degrees of freedom may be overstated, leading to inflated Type I error rates in statistical tests of association (Felsenstein, 1985; Martins and Garland, 1991; Purvis *et al.*, 1994; Nunn, 1995). In order to correct for this possibility, several methods that incorporate phylogenetic information are used in our analyses when statistical significance levels are required (see Harvey and Pagel, 1991; Martins and Hansen, 1996).

The second problem is that strong correlations or associations may be due to the confounding effects of another trait which is correlated with the characters involved. When confounding variables are ignored, true patterns will be more difficult to discern. In addition, these factors may create spurious correlations and thus lead to incorrect reconstructions of behavior in the fossil record. Incidentally, the possibility of confounding factors is a major reason for limiting the scope of the present study to the order Primates. In many of our analyses, we have corrected for the effect of the most obvious confounding factor, namely, body mass; in addition, we examine patterns within particular evolutionary radiations (i.e., we incorporate general phylogenetic information).

A third problem involves limitations imposed by our focus on associations between K and U variables. Unlike the usual comparative study, where all the features of extant taxa are available for study, in this case we are limited to associations between features that are potentially knowable in the fossil record, such as diet, and those that are unknowable, such as home range size. We therefore have fewer characters available for study, and we are less able to account for the effects of multiple factors. Furthermore, the K variables used here are discrete, and these crude estimates may obscure variation that could be used to establish statistical relationships between variables. Thus, the patterns we document will probably explain less variation than is usually attained in comparative studies, and our resolution of behavior in the fossil record will be similarly compromised in that our estimates will have greater error. We elaborate on this issue further in the Discussion.

A final problem concerns how one K variable, body mass, is estimated in the fossil record (Smith, 1996). Body mass is often estimated using relationships in extant taxa between body mass and either dental dimensions (Gingerich *et al.*, 1982) or postcranial remains (Dagosto and Terranova, 1992). This body mass estimate might then be used to infer values for a U variable that is expected to scale with body mass, such as home range size. These inferences are based on the association between body mass and home range size in extant species, yet it is actually tooth size that is used in reconstructing the fossil taxon's home range size. Tooth size, as a proxy for body mass, will be less strongly linked to home range size, and this increased error should be accounted for in the analysis. The easiest method for dealing with this problem is to use a range of values for estimated body mass, preferably an estimate of potential error in the reconstructed body mass. However, this procedure may produce prediction intervals so wide that nothing meaningful can be inferred about the behavior of the fossil taxon (see Smith, 1996).

Methods

Taxonomy and Phylogeny

The first step in any comparative study involves compiling a database to be used for the tests. This requires a taxonomy to organize the data. Current methods also require explicit phylogenetic information in order to control for the nonindependence inherent in species-level data points (Felsenstein, 1985; Harvey and Pagel, 1991). We have adopted the composite estimate of primate phylogeny provided by Purvis (1995), which combines studies of different primate clades into one overall primate phylogeny. This scheme recognizes the 203 species of primates listed by Corbet and Hill (1991), and it has 160 out of 202 possible nodes resolved. In some cases, the literature reports on species not listed by Purvis (1995); these species were reassigned to species in Purvis's tree on the basis of specialized taxonomic monographs and by consultation with experts on the questionable taxonomic groups.

Some comparative methods require not only information on tree topology, but also information on branch lengths. Purvis (1995) provides branch length estimates for most of his phylogeny, but he was unable to provide estimates for some internode branches. In other cases, estimates of branch lengths conflict with the overall structure of the tree. The following rules were therefore implemented when entering branch lengths into the computer programs necessary for these analyses (see also Purvis 1995): (1) branches with negative values (because a lower node was given an older age than a preceding deeper node) were given branch lengths of 1.0; (2) ages of unknown nodes were calculated by interpolation, where information on preceding and following nodes were used; (3) branches within the tarsier clade were given values of 1.0, and the branch connecting this to the rest of the primates was figured given these values at lower nodes and the age of the node connecting them to the rest of the primates.

As acknowledged by Purvis (1995), the phylogeny used here probably contains errors in both topological structure and branch lengths, and this phylogeny is unlikely to be the final word on primate evolutionary relationships. However, a growing body of evidence suggests that while phylogenetic error can cause statistical problems (inflated Type I error rates and reduced statistical power), these statistical errors are less severe than those that result when phylogenetic relationships are ignored (Martins and Garland, 1991; Gittleman and Luh, 1992, 1994; Purvis *et al.*, 1994; Nunn, 1995). To control for the effect of phylogeny in some analyses of continuous characters, we use the method of independent contrasts (Felsenstein, 1985). Contrasts were calculated using the computer program CAIC (Purvis and Rambaut 1995).

Choice of Variables

Our K variables serve as predictor variables in this analysis. For categorical variables, we aim at the smallest possible number of character states. While this limits the resolution of our analysis, the relationship between morphology and function, especially in fossil species, is usually not so precise that a fine-grained classification can be supported. Moreover, such fine-grained classifications are likely to be limited to certain taxonomic groups, and thus cannot be readily extrapolated to other taxa.

Tables I and II summarize the knowable (K) and unknowable (U) variables used in our analyses. Appendix A provides the rules used to assign character states, and the values assigned to each species are provided in Appendix B. Multiple references are common for each cell in Appendix B. We therefore do

Table I. Knowable ("K") Variables

Variable	Character states/units	Codes[a]
Activity period	Nocturnal	N
	Diurnal	D
Locomotor substrate	Arboreal	A
	Terrestrial	T
Habitat	Wooded	W
	Open	O
Diet	Frugivore	Fr
	Folivore	Fo
	Gummivore	G
	Insectivore	I
Body size	Kilograms	

[a] Criteria for deciding character states are provided in Appendix A.

not provide references here, which in any case are constantly outdated by new information.

Two potentially knowable variables were not included in our study: locomotor mode (e.g., vertically clinging and leaping, quadrupedal, and suspensory) and relative brain size (corrected for body mass). Locomotor mode depends primarily on taxonomic affiliation and body mass (Fleagle, 1984); hence, this trait is not as useful as the other characters, which are potentially more labile and thus more informative. Regarding relative brain size, a number of associations have been proposed that might be of interest for this chapter (Clutton-Brock and Harvey, 1980; Gibson, 1986; Sawaguchi, 1990; Barton and Purvis, 1994). However, uncertainty surrounds the generality of these relationships, perhaps because overall brain size is not a useful measure when specific functions are considered (Barton and Purvis, 1994; Barton *et al.*, 1995). A more successful approach is to examine the relationships between social or ecological variables with the relative size of specific brain structures (e.g., Dunbar, 1992b; Barton *et al.*, 1995). However, because the fossil record does not usually provide such detailed information, we will not examine correlations involving brain size here.

Table II. Unknowable ("U") Variables

Variable	Units
Population group size	Mean number of individuals
Foraging group size	Mean number of individuals
Home range size	Hectares (ha)
Day journey length	Meters (m)
Defensibility-index (D-index)	Number of diameters of home range crossed per day

Intraspecific variation is not easily incorporated using existing comparative methods. For discrete K variables, we incorporated intraspecific variation by using multiple syndrome designations for a species when enough information on U variables was available to make this worthwhile (only two cases: open and closed habitats in *Papio anubis* and *Pan troglodytes*).

For the continuous U characters, we used means when more than one value was available. To avoid over-representation of better-studied populations, averages were taken for each population or study site, and the mean of these population averages was assigned to the species. Consideration was also given to the possible bias imposed when social groups within a population are studied for different periods of time in longitudinal studies. For these cases, we incorporated information on group identity in our master database, and we calculated weighted averages that avoid over-representation of groups that have been studied for different periods of time at long-term field sites. Thus, our approach in this chapter is to sample as many populations as possible. Some references provided only a range for a given species rather than a statistic of location, such as the mean, median, or mode. In these cases, we used the midpoint of this range as an estimate of the mean for that population.

For statistical analyses, all continuous variables (body size and the U variables) were log-transformed. This improved the fit of all variables to normality, with the effect that all continuous characters except foraging group size were not significantly different from a normal distribution after the data were log-transformed. No other transformations were found to improve the normality of the foraging group size data, which was highly skewed to the right due to the high representation of prosimian species that are solitary.

Body mass is treated as the predictor variable in estimating values for U variables, making regression an appropriate statistical method for analyzing patterns involving body mass. We assumed that measurement error in body size is less than the error in our U variables, so that least-squares regression (rather than reduced major axis regression) is appropriate (see also Harvey and Pagel 1991; Charnov, 1993).

Results

Distribution and Analysis of the "K" Variable Syndromes

Syndromes in Living Species

One goal of this chapter is to identify syndromes of traits that are functionally incompatible, and to distinguish these syndromes from combinations that are functionally possible, but not observed among living species. We first document the pattern of K-variable character states in living taxa, with special attention to absent syndromes. This information is important, because if a fossil is identified with a syndrome not found in living species, our ability to infer the behavior of this fossil taxon is more limited. Failure to match a fossil taxon to a syndrome in living species raises additional questions regarding our inferences of the K variables for the fossil material, especially when we have strong theoretical reasons for doubting the existence of such a combination of traits. Hence, these patterns serve as an internal check on the reconstruction of K variables for a fossil taxon.

One hundred and eighty of the 203 species in Purvis's (1995) tree could be assigned to one of 32 possible combinations of K variables (four K variables with 2, 2, 2, and 4 character states, respectively; see Table I). Of these 32 possible syndromes, only 12 syndromes (38% of total possible combinations) were actually found to exist among extant taxa, and only 9 of these (28%) are represented by 3 or more taxa (see Tables III and Table IV).

Certain patterns occur among the K variables which allow identification of syndromes using fewer than the four K variables examined here (see also Clutton-Brock and Harvey, 1977a). These conventions, provided in Table III (column labeled "Syndrome"), will be used throughout this chapter to reinforce the patterns that exist among the K-variable character states. The first pattern relates to substrate use and habitat: arboreal species obviously require wooded environments in order to have appropriate substrates for locomotion, so that arboreality is always associated with wooded habitats. Hence, once it is decided that a fossil taxon was arboreal, only two other characters are needed to identify its syndrome (i.e., activity period and diet). In contrast, terrestrial species can live in open or wooded habitats, and species that are terrestrial therefore require all four K variables to differentiate the possible syndromes (activity period, diet, substrate, and habitat). If a reconstruction indicates that a fossil taxon possessed an impossible combination of traits (e.g., arboreal animals living in an open environment), at least one of the K variables must have been reconstructed incorrectly.

Other simplified syndromes result from combinations of K variables that are missing in primates, but are nonetheless possible (Table IV). Thus, no nocturnal primate species are terrestrial, although this combination cannot be precluded on functional grounds. Because nocturnal primate species are always arboreal, they must therefore also live in wooded habitats. Hence, nocturnal species can be identified with only two characters: activity period and diet (Table III). The lack of nocturnal terrestrial species can be accounted for with a constraint-based explanation. A taxon would most likely make the transition to a nocturnal-terrestrial niche by one of two routes: a nocturnal-arboreal species can become terrestrial, or a diurnal-terrestrial species can become nocturnal. However, either of these transitions are unlikely in primates. Arboreal primate species tend to have slower life histories than terrestrial species (Ross, 1988), and animals with slow life histories may be unlikely to invade the terrestrial niche where unavoidable mortality for such small creatures must be much higher than in the canopy. On the other hand, diurnal-terrestrial species may be unable to invade the nocturnal-terrestrial

Table III. Syndromes Present

Activity period	Substrate	Habitat	Diet	Syndrome	N	Radiations[a]	Genera included[b]
N[c]	A	W	I	N-I	14	3	
					4	*TR*	*Tarsius*
					8	*LO*	*Arctocebus, Euoticus, Galago, Galagoides, Loris, Otolemur*
					2	*LM*	*Allocebus, Daubentonia*
N	A	W	G	N-G	3	2	
					1	*LM*	*Phaner*
					2	*LO*	*Euoticus, Otolemur*
N	A	W	FR	N-Fr	10	3	
					3	*LO*	*Galagoides, Nycticebus, Perodicticus*
					5	*LM*	*Cheirogaleus, Microcebus, Mirza*
					2	*NW*	*Aotus*
N	A	W	FO	N-Fo	2	1	
					2	*LM*	*Avahi, Lepilemur*
D	A	W	I	D-I	1	1	
					1	*NW*	*Saimiri*
D	A	W	G	D-G	2	1	
					2	*NW*	*Callithrix, Cebuella*
D	A	W	FR	D-A-Fr	83	4	
					8	*LM*	*Eulemur, Propithecus, Varecia*
					40	*NW*	*Ateles, Cacajao, Callicebus, Callimico, Callithrix, Cebus, Chiropotes, Lagothrix, Leontopithecus, Pithecia, Saguinus, Saimiri*

					27	*OW*	*Cercocebus, Cercopithecus, Colobus, Macaca, Miopithecus, Presbytis, Pygathrix*
					8	*AP*	*Hylobates, Pongo*
D	A	W	FO	D-A-Fo	34	4	
					5	*LM*	*Hapalemur, Indri, Propithecus*
					6	*NW*	*Alouatta, Brachyteles*
					21	*OW*	*Colobus, Nasalis, Presbytis, Procolobus, Pygathrix, Simias*
					2	*AP*	*Hylobates*
D	T	W	FR	D-T-W-Fr	20	3	
					1	*LM*	*Lemur*
					17	*OW*	*Allenopithecus, Cercocebus, Cercopithecus, Macaca, Mandrillus, Papio*
					2	*AP*	*Pan*
D	T	W	FO	D-T-W-Fo	4	2	
					3	*OW*	*Macaca, Presbytis, Pygathrix*
					1	*AP*	*Gorilla*
D	T	O	FR	D-T-O-Fr	5	2	
					4	*OW*	*Cercopithecus, Erythrocebus, Papio*
					1	*AP*	*Pan*
D	T	O	FO	D-T-O-Fo	3	1	
					3	*OW*	*Papio, Theropithecus*

[a] LO, lorisids; LM, lemurs; TR, Tarsius; NW, New World monkeys; OW, Old World monkeys; AP, apes.
[b] Some genera have species in more than one syndrome.
[c] Syndrome codes given in Table I.

Table IV. Observed, Possible, and Impossible K-Variable Syndromes in Primates[a]

	Arboreal—wooded	Arboreal—open	Terrestrial—wooded	Terrestrial—open
Nocturnal-	Frugivore	Frugivore	Frugivore	Frugivore
	Folivore	Folivore	Foliore	Folivore
	Insectivore	Insectivore	Insectivore	Insectivore
	Gummivore	Gummivore	Gummivore	Gummivore
Diurnal-	Frugivore	Frugivore	Frugivore	Frugivore
	Folivore	Folivore	Folivore	Folivore
	Insectivore	Insectivore	Insectivore	Insectivore
	Gummivore	Gummivore	Gummivore	Gummivore

[a] Unshaded cells, syndrome observed in extant primates; lightly shaded cell ( ), syndrome possible but not observed in primates; shaded cell ( ), syndrome not observed because the set of traits is functionally incompatible (i.e., strict arboreality is not possible in an open environment).

niche because diurnal animals are constrained in their diurnality by visual adaptations that might make the transition to nocturnality exceedingly difficult.

Likewise, insectivores and gummivores are found in some diurnal and nocturnal arboreal primates, but no terrestrial species eat primarily gum or insects. This means that diurnal gummivores and insectivores can also be identified with just two K variables (activity period and diet) because they must be arboreal and therefore live in wooded habitats. The absence of terrestrial gummivores and insectivores may be due to two factors. First, their food sources (insects, tree gums) are usually only available above the ground or best accessed there, requiring at least a partly arboreal life style. Second, these dietary specialists are small, and small animals would face much higher mortality on the ground. Thus, transitions to the terrestrial niche are unlikely here as well.

When reconstructions indicate that a fossil taxon possessed any suite of K variables not observed in extant taxa, caution is required before reconstructing U variables using the charts presented below. Although our intuition indicates that such niches are difficult for primates to occupy, these syndromes may have existed at one point in the evolutionary history of primates; after all, some of these missing syndromes are currently occupied by other mammalian taxa, and adaptive patterns in primates may have shifted over time. If we can confidently assign a fossil taxon to a particular syndrome not observed among extant species, U variable ranges for single K variables (see below) could be combined to estimate the past behavior of the fossil taxon.

Taxonomic Distribution of Syndromes

Several patterns are also present in the taxonomic distribution of syndromes (Table V). First, D-A-Fr and D-A-Fo are the most common syndromes,

Table V. Diversity among Radiations in Observed K-Variable Syndromes

Syndrome	Tarsiers (TR)	Lorisids (LO)	Lemurs[a] (LM)	New World Monkeys (NW)	Old World Monkeys (OW)	Apes (AP)
N-I	X	X	X			
N-G		X	X			
N-Fr		X	X	X		
N-Fo			X			
D-G				X		
D-I				X		
D-A-Fr			X	X	X	X
D-A-Fo			X	X	X	X
D-T-W-Fr			X		X	X
D-T-O-Fr					X	X
D-T-W-Fo					X	X
D-T-O-Fo					X	
# syndromes observed	1	3	7	5	6	5
# spp in the radiation[b]	4	13	24	51	75	14
avg # spp per syndrome	4	4.33	3.43	10.2	12.5	2.8

[a] Does not include subfossil taxa.
[b] Calculations only include species that can be assigned to one of the syndromes (i.e., those with no missing information for any of the K variables).

found in 84 and 34 species of primates, respectively. These two syndromes are found in all major primate radiations except the lorisids and tarsiers. Second, New World primates do not contain any terrestrial species, as is well known, and they do not occupy as many ecological niches as species in other radiations (see also Fleagle and Reed, 1996; Kappeler and Heymann, 1996). However, some extinct New World primates may have been terrestrial (Kay and Williams, this volume, Chapter 9), suggesting that patterns in the past may have differed from those in the present. Third, the great ape species tend to be classified into different syndromes, with the orangutan D-A-Fr, the gorilla D-T-W-Fo, and the two species of chimpanzee either D-T-W-Fr or D-T-O-Fr (taking into account intraspecific variation in habitat use of *Pan troglodytes*). Finally, other relationships between syndromes and taxonomy are present in Table V, with D-G and D-I limited to the New World, terrestriality observed only in *Lemur catta* and in Old World monkeys and apes, and nocturnality confined almost exclusively to prosimians. Among nocturnal species, frugivores and insectivores predominate, with 10 and 14 species respectively (Table III).

Syndrome diversity also differs among the major primate radiations. Table V provides the number of syndromes and the average number of species for each syndrome in the six primate radiations. The actual syndromes are distributed unevenly among the different radiations, although each radiation has roughly the same number of syndromes (between 5 and 7 syndromes per radiation in anthropoids and lemurs, but only 3 in lorisids). In addition, the average number of taxa per syndrome varies greatly among the primate radiations, with noticeably more species per syndrome in New World and Old World monkeys. Cases where particular syndromes have more species may indicate higher net speciation rates for these syndromes (e.g., Old World monkeys; see Purvis *et al.*, 1995).

A Phylogenetic Analysis of Syndromes

We need characters that are to some degree independent, because if two characters always evolve perfectly in tandem, it would be better to reduce the number of K variables to simplify the analysis. To investigate evolutionary relationships among the K variables, the syndromes found in extant primate taxa were mapped onto Purvis's (1995) phylogeny. Sensitivity of the results to the assumptions of parsimony (e.g., Omland, 1997) and use of maximum likelihood methods (Pagel 1994; Schluter *et al.*, 1998) are two ways of dealing with error in reconstructing ancestral states. For now, however, we assume that the assumptions of parsimony hold; nevertheless, we examine phylogenetic sensitivity by creating 100 trees in which the polytomies in Purvis's (1995) phylogeny are randomly resolved, and then calculating the statistics of interest on these multiple phylogenetic trees.

To examine phylogenetic patterns, a transition matrix was set up in MacClade (Maddison and Maddison, 1992) that weighted transitions between syndromes according to the number of K variables required to change in a transition from any one syndrome to another. We assumed that the number of "steps" (used in calculating treelengths) in a transition from one syndrome to another was equal to the number of K variables that must change for a transition of that type to take place. For example, a shift from D-A-Fo to D-T-O-Fo would involve two transitions (arboreal to terrestrial, and wooded to open), and would thus count as two steps when calculating treelengths. Diet was treated as an unordered character, so that shifts from any dietary category to another counted as a single step. The results of this "syndrome mapping approach" match almost exactly syndromes reconstructed by individual character reconstructions.

On most internode branches on the phylogeny, no change was inferred (the diagonal in the matrix in Table VI). The mean treelength was calculated as 55.51 (range: 54–56), which is only slightly higher than the average number of branches on which any changes occur (53.08 mean reconstructed changes, range: 50–56). Therefore, on the majority of branches for which one or more changes are reconstructed, the change involves only a single K variable (an average of 1.046 changes per branch, looking only at branches where any changes are reconstructed). Only four of the transitions from one syndrome to another involved a change in two K variables (Table VI), and no transitions required three or more K-variable character state changes. Hence, K variables

Table VI. Possible and Observed Evolutionary Transitions among Syndromes[a,b]

	N-I	N-G	N-Fr	N-Fo	D-I	D-G	D-A-Fr	D-A-Fo	D-T-W-Fr	D-T-W-Fo	D-T-O-Fr	D-T-O-Fo
N-I	22.9	2.3	0.9									
N-G	0.3	0.3										
N-Fr	3.8	0.7	23.9	1.1			2.1					
N-Fo	1.0			3.8			0.1	0.9				
D-I												
D-G						1.5	0.5					
D-A-Fr			1.0		1.0	1.5	165.3	5.1	8.3	0.8		
D-A-Fo							6.3	67.9		2.0		
D-T-W-Fr							4.7		29.8	1.2	3.0	1.0
D-T-W-Fo												
D-T-O-Fr							0.2		1.7		7.4	1.5
D-T-O-Fo											0.5	1.5

[a] Transitions from "row" syndromes to "column" syndromes.
[b] Shaded cells indicate an observed transition that counts as two steps. Cells in the diagonal indicate branches on which no change was inferred.

Table VII. Indices Calculated for Individual K Variables and Syndromes

Character	Number of character states	Tree length (steps)	Consistency Index (CI)	Retention Index (RI)
All characters (Ensemble Index)	—	53	0.113	0.638
Activity period	2	4	0.250	0.900
Locomotor substrate	2	16	0.067	0.576
Habitat	2	6	0.167	0.286
Diet	4	28	0.107	0.583

are not tightly linked (i.e., they do not change perfectly in tandem), and individual K variables should therefore be informative.

Some other interesting patterns were revealed by this exercise. For example, there were 11.1 cases of an arboreal species becoming terrestrial, but only 4.9 cases of a terrestrial species becoming arboreal. All of the latter cases involved Old World frugivores (e.g., the macaques, mangabeys, and guenons). On 3.1 branches a nocturnal species switched to diurnality, but on only one branch did a diurnal species become nocturnal (*Aotus*). *Tarsius* is inferred to have had a nocturnal ancestor when examined phylogenetically despite morphological evidence indicating that nocturnality is secondarily derived in this genus (Cartmill, 1980).

To measure patterns in the evolution of K-variable character states, several additional statistics were calculated for individual K variables using MacClade, including treelengths (see above), the consistency index (CI), and the retention index (RI) (see Farris, 1989). The CI measures homoplasy, and is lower when more cases of convergent evolution occur. The RI measures the evolutionary conservatism of a trait: higher values for a trait indicate that the trait tends to be retained after it has evolved (Farris, 1989; Di Fiore and Rendall, 1994). RI's and CI's were calculated on the polytomous Purvis (1995) phylogeny for individual K variables assuming that character states are unordered. The results show that activity period has a higher RI and CI than the other characters (Table VII), indicating that transitions between diurnality and nocturnality are rare in the evolutionary history of primates. Diet and locomotor substrate have moderately high RI's, while habitat has the lowest RI among the individual K variables, suggesting that this K variable is the most evolutionarily labile.

Reconstructing the Behavior of Extinct Taxa

After evaluating the internal consistency of the reconstructed K variables for a fossil taxon, the next step is to use the associations among K and U variables in extant taxa to predict U variables for the fossil. In this section we document variation in U variables for each of the syndromes in living species. For fossil taxa in which not all K variables can be identified, we also provide estimates for each of the K variables separately (i.e., not in the context of other K variables). This latter analysis is expected to give wider ranges for the U variables than the syndrome analysis. We attempt to narrow the ranges of our estimates in fossils by incorporating lineage effects (variation among radiations) and by including body mass as a covariate.

Our general approach is to identify a range of values from extant species that can be used as a reasonable estimate for extinct species. However, taking a simple range from extant taxa would not be justified, because when more extant examples of a given syndrome exist, our estimates of range will likely become larger simply because we are sampling more data points and, possibly, a greater number of potentially confounding factors. We therefore calculated ranges based on the mean $\pm$ 2 standard deviations (roughly encompassing the range in which 95% of all observations are expected) using log-transformed data. The results are presented in Table VIII (these results are back-transformed; thus, ranges are asymmetrical around the mean). We also provide sample sizes, which differ among the U variables because of different numbers of taxa (see Table III) and biases in collection (e.g., fewer studies of ranging patterns in nocturnal species). Obviously, special care should be taken when sample sizes are small.

In many cases, a fossil taxon cannot be readily associated with a particular syndrome. This is true when not all the K variables can be identified for a fossil, perhaps because the fossilized remains are not sufficiently complete. Syndrome identification is also less certain when features of the fossil give conflicting predictions for K variables, or when a fossil is classified to a syndrome not filled by any living species. For these cases, Table VIII will not be useful. Instead, Table IX should be used, which documents the relationship between individual K and U variables. These ranges can be combined to narrow the estimate for the fossil taxon's behavior when more than a single K variable is known. As an example, if we wish to estimate the D-index of a fossil taxon reconstructed to be a terrestrial folivore, we would combine the ranges for terrestrial (0.18–3.85) and folivorous (0.20–3.95) taxa to come up with an estimated D-index of 0.20–3.85. Because this second analysis cannot control for as many confounding factors as the syndrome analysis, it provides wider ranges as estimates for U. For example, if we knew that our terrestrial-folivore was most often found in a wooded environment, we could combine that information with our analysis of K variables that showed terrestrial species are always diurnal, and then classify this species to the D-T-W-Fo syndrome. We could thus use ranges in Table VIII, which would reduce our upper limit for this taxon's D-index to 1.83.

These tables show that ranges are extremely wide within syndromes, so that ranges often overlap greatly between syndromes. In some cases, a difference greater than an order of magnitude exists between the lowest and

Table VIII. Means and Ranges Based on 2 Standard Deviations for Syndromes[a]

Syndrome	N	Radiations[b]	Body size	N	Population group size	N	Foraging group size	N	Home range size	N	Day journey	N	D-Index	N
N-I	14	3	0.2 [0.0–1.6]	13	2.6 [1.3–5.2]	12	1.0 [1.0–1.0]	14	5.9 [0.3–119.1]	8	1666 [1172–2369]	5	4.36 [1.06–17.98]	5
	4	TR	0.1 [0.1–0.3]	3	2.6 [1.6–4.2]	3	1.0 [1.0–1.0]	4	2.0 [0.3–14.9]	2	1800	1	5.47	1
	8	LO	0.2 [0.0–1.0]	8	2.7 [1.4–5.2]	7	1.0 [1.0–1.0]	8	5.1 [0.6–40.2]	5	1632 [1007–2644]	3	5.85 [2.24–15.29]	3
	2	LM	0.5 [0.5–53.0]	2	2.4 [0.6–9.8]	2	1.0 [1.0–1.0]	2	103	1	1641	1	1.43	1
N-G	3	2	0.5 [0.1–2.5]	3	3.3 [2.0–5.3]	3	1.0 [1.0–1.0]	3	8.5	1	1250	1	3.80	1
	2	LO	0.6 [0.1–4.7]	2	3.7 [3.1–4.5]	2	1.0	2	8.5	1	1250	1	3.80	1
	1	LM	0.3	1	2.5	1	1	1						
N-Fr	10	3	0.3 [0–3.2]	10	2.2 [0.8–5.7]	9	1.3 [0.4–3.7]	10	6.4 [0.8–51.1]	6	473 [182–1231]	2	2.26	1
	3	LO	0.7 [0.2–3.2]	3	1.6 [0.4–5.8]	2	1.0 [1.0–1.0]	3	20.4 [10.3–40.5]	2				
	5	LM	0.2 [0–1.1]	5	2.1 [0.9–4.9]	5	1.0 [1.0–1.0]	5	2.7 [1.2–6.1]	3				
	2	NW	0.8 [0.7–1.1]	2	3.5 [3.0–4.1]	2	3.5 [3.0–4.1]	2	8.3	1	473 [182–1231]	2	2.26	1
N-Fo	2	LM	0.9 [0.4–2.5]	2	1.7 [0.6–4.9]	2	1.7 [0.6–4.9]	2	0.9 [0.1–8.7]	2	457	1	3.42	1
D-G	2	NW	0.2 [0.1–0.5]	1	8.0 [7.6–8.5]	1	8.0 [7.6–8.5]	1	2.8 [0.1–67.7]	1	469 [121–1820]	1	2.51 [2.13–2.97]	1
D-I	1	NW	0.8	2	45	2	45	2	128.7	2	1755	2	2.14	2
D-A-Fr	83	4	2.4 [0.3–18.1]	73	9.5 [1.9–48.9]	77	8.4 [1.7–42.3]	78	54.9 [3.8–787.8]	59	1334 [492–3618]	51	1.61 [0.52–5.00]	50
	8	LM	2.5 [1.3–4.6]	8	5.4 [2.3–13.1]	8	4.5 [1.4–14.5]	8	11.7 [0.4–324.8]	6	596 [422–843]	5	1.61 [0.32–7.99]	5
	40	NW	1.2 [0.2–9.5]	34	8.3 [1.8–38.4]	36	7.1 [1.7–28.6]	37	60.7 [3.4–1087.0]	27	1694 [699–4106]	24	1.99 [0.76–5.19]	23
	27	OW	4.1 [1.5–11.1]	25	18.3 [7.1–47.6]	26	16.7 [5.9–47.6]	26	90.7 [19.2–429.2]	18	1306 [575–2967]	14	1.21 [0.48–3.06]	14
	8	AP	8.0 [1.7–37.5]	6	3.2 [1.6–6.5]	7	3.2 [1.6–6.5]	7	40.2 [7.6–212.1]	8	1116 [578–2155]	8	1.43 [0.38–5.39]	8
D-A-Fo	34	4	5.4 [1.8–16.9]	30	9.3 [1.6–55.7]	33	8.4 [2.8–25.6]	33	39.4 [2.7–568.4]	29	652 [250–1703]	20	0.92 [0.26–3.27]	18
	5	LM	2.3 [0.4–14.1]	5	4.8 [2.7–8.4]	5	4.8 [2.7–8.4]	5	32.4 [2.8–375.7]	4	555 [345–892]	3	1.31 [1.17–1.46]	2
	6	NW	5.9 [3.9–8.9]	6	8.8 [3.0–26.0]	6	7.9 [3.9–16.2]	6	33.3 [2.1–521.4]	5	461 [169–1259]	5	0.75 [0.14–4.03]	5
	21	OW	6.5 [3.9–10.8]	17	12.0 [1.6–89.8]	20	10.5 [3.5–31.0]	20	41.7 [2.2–790.4]	18	772 [307–1938]	11	0.91 [0.26–3.12]	10
	2	AP	7.8 [3.3–18.4]	2	4.6 [3.1–6.6]	2	4.6 [3.1–6.6]	2	52.7 [12.6–220.7]	2	946	1	1.36	1
D-T-W-Fr	20	3	7.4 [1.7–31.9]	19	26.8 [5.3–135.8]	20	22.8 [3.8–134.7]	20	262.5 [6.6–10507.1]	15	1438 [458–4510]	8	0.97 [0.17–5.63]	7
	1	LM	2.7	1	15.6	1	15.6	1	15.2	1	957	1	3.11	1
	17	OW	6.5 [2.4–17.9]	16	26.9 [5.0–144.7]	17	26.9 [5.0–144.7]	17	267.9 [9.8–7330.7]	13	1413 [418–4780]	6	0.98 [0.26–3.68]	5
	2	AP	33.2 [33.2–33.2]	2	43.4	1	6.6 [4.5–9.7]	2	3475	1	2400	1	0.28	1
D-T-W-Fo	4	2	18.3 [2.0–171.0]	4	37.2 [6.1–225.2]	4	21.3 [9.4–48.2]	4	564.6 [11.8–26953.4]	4	956 [423–2162]	3	0.33 [0.06–1.83]	3
	3	OW	10.5 [9.5–11.5]	3	49.4 [9.0–272.7]	3	23.5 [9.8–56.3]	3	341.8 [6.0–19543.2]	3	1008 [327–3103]	2	0.47 [0.08–2.62]	2
	1	AP	97.7	1	15.8	1	15.8	1	2544.5	1	861	1	0.16	1
D-T-O-Fr	5	2	10.5 [1.8–62.6]	5	40.4 [19.5–83.8]	5	26.3 [4.1–170.8]	5	1250.7 [41.4–37794.4]	5	3524 [1092–11372]	5	0.93 [0.42–2.07]	5
	4	OW	7.9 [1.9–32.8]	4	38.8 [17.2–87.6]	4	38.8 [17.2–87.5]	4	1147.1 [23.0–57261.0]	4	3493 [903–13501]	4	0.94 [0.37–2.35]	4
	1	AP	33.2	1	47.6	1	5.6	1	1767.3	1	3650	1	0.92	1
D-T-O-Fo	3	OW	11.8 [8.4–16.7]	3	74.2 [17.8–309.9]	3	46.3 [23.2–92.4]	3	907.8 [64.6–12755.6]	3	4193 [481–36544]	3	1.16 [0.58–2.29]	3

[a] Values in table are: mean [mean + 2SD, mean − 2SD]. Values are back-transformed from log-transformed data.
[b] LO, lorisids; LM, lemurs; TR, Tarsius; NW, New World monkeys; OW, Old World monkeys; AP, apes.

highest values for the ranges in these tables (e.g., home range size, especially in Table IX). Because these ranges are so wide, it appears that basic socioecological variables, such as group size, cannot be inferred with great resolution using this approach. We therefore considered other potentially confounding effects which could narrow our estimates of behavior in fossil taxa.

Narrowing the Ranges: Taxonomic Affiliation

Several studies have shown that taxonomic affiliation affects the values of socioecological variables in primates (e.g., Di Fiore and Rendall, 1994; Fleagle and Reed, 1996; Kappeler and Heymann, 1996). This may be true in our analysis if species within radiations share confounding factors that are not taken into account with our necessarily simplistic syndrome categories (e.g., interspecific competition or gross habitat similarities; see Kappeler and Heymann, 1996). By looking within radiations, we are in essence taking into account some of these potentially confounding factors. This basically crude phylogenetic approach should therefore increase the resolution of U variable estimates.

While in most cases looking within radiations narrows the estimates (see Tables VIII and IX), many of these ranges still remain quite wide. For example, among species classified as D-A-Fr, the mean foraging group size differs substantially among radiations, yet ranges overlap completely, spanning in two radiations greater than one order of magnitude (i.e., lemurs: 1.4–14.5; New World monkeys: 1.7–28.6; ANOVA cannot be performed because phylogenetic effects make calculating the degrees of freedom difficult; Garland *et al.*, 1993; Nunn and Smith 1998). However, while these ranges do not provide the precision desired, ranges calculated within radiations are superior to those in which this taxonomic information is ignored (e.g., for all D-A-Fr, the range is 1.7–42.3, which is wider than any ranges within radiations).

Narrowing the Ranges: The Effect of Body Mass

Body mass is another potentially serious confounding factor which can be incorporated into the analysis. Table X provides the slopes of phylogenetically corrected regressions of U variables on body mass. In all cases for which significant results were obtained, the intercept did not differ significantly from 0, so we used the slope of regressions with the intercept forced to 0 as the estimated relationship between the U variable and body mass. Phylogenetic tests were conducted only in cases where the syndrome-radiation was represented by 5 or more extant species, and where the relationship between body mass and the U variable is significant at $P < 0.10$ in analyses that did not control for the effects of phylogeny. We provide slopes for all syndromes and radiations in which contrasts were calculated, but we provide P-values only for those cases

Table IX. Means ±2 Standard Deviations, for Single K Variables[a]

Variable	N	Radiations[b]	Body size	N	Population group size	N	Foraging group size	N	Home range size	N	Day journey	N	D-Index	N
Nocturnal	31	All	0.3 [0.0–2.4]	29	2.5 [1.1–5.5]	26	1.1 [0.6–2.2]	31	5.0 [0.3–74.4]	17	1057 [286–3901]	9	3.83 [1.19–12.30]	8
	15	LO	0.3 [0.1–2.0]	14	2.6 [1.1–6.2]	11	1.0 [1.0–1.0]	15	7.6 [1.0–58.2]	8	1527 [949–2457)	4	5.25 [2.15–12.85]	4
	10	LM	0.3 [0.0–3.9]	10	2.1 [0.9–4.9]	10	1.1 [0.6–2.0]	10	3.4 [0.1–137.0]	6	866 [142–5279]	2	2.21 [0.65–7.59]	2
	4	TR	0.1 [0.1–0.3]	3	2.6 [1.6–4.2]	3	1.0 [1.0–1.0]	4	2.0 [0.3–14.9]	2	1800	1	5.47	1
	2	NW	0.8 [0.7–1.1]	2	3.5 [3.0–4.1]	2	3.5 [3.0–4.1]	2	8.3	1	473 [182–1231]	2	2.26	1
Diurnal	159	All	3.7 [0.4–33.5]	143	12.4 [1.8–85.9]	146	10.7 [1.9–61.6]	148	77.9 [2.3–2657.9]	119	1221 [306–4863]	94	1.27 [0.32–5.05]	90
	14	LM	2.4 [0.8–7.4]	14	5.6 [2.2–14.3]	14	5.0 [1.6–15.5]	14	17.4 [1.0–310.8]	11	614 [379–994]	9	1.66 [0.44–6.26]	8
	51	NW	1.4 [0.1–14.3]	45	8.6 [1.9–38.2]	46	7.4 [1.9–29.5]	47	47.7 [2.1–1098.0]	35	1277 [315–5182]	32	1.73 [0.48–6.17]	31
	80	OW	5.8 [2.0–16.6]	72	20.3 [3.5–117.1]	74	18.0 [4.2–78.3]	74	127.1 [4.3–3792.1]	60	1356[315–5837]	41	1.02 [0.34–3.10]	39
	14	AP	14.0 [1.8–110.8]	12	6.1 [0.7–52.2]	12	4.5 [1.5–13.2]	13	108.7 [2.4–5014.8]	13	1267 [475–3382]	12	1.00 [0.16–6.17]	12
Terrestrial	34	All	8.9 [1.8–44.3]	33	31.8 [6.3–160.0]	32	24.0 [4.7–123.6]	33	450.6 [12.1–16844.5]	27	2021 [392–10407]	19	0.82 [0.18–3.85]	18
	1	LM	2.7	1	15.6	1	15.6	1	15.2	1	957	1	3.11	1
	29	OW	7.4 [2.6–21.0]	28	32.6 [6.1–174.3]	28	28.6 [6.5–125.6]	28	417.6 [13.6–12848.7]	23	2137 [382–11954]	15	0.90 [0.28–2.91]	14
	4	AP	43.5 [14.8–128.0]	4	32.0 [9.4–108.6]	3	7.9 [3.0–20.7]	4	2500.1 [1270.6–4919.3]	3	1961 [444–8670]	3	0.35 [0.05–2.02]	3
Arboreal	155	All	1.8 [0.1–27.7]	138	7.4 [1.1–48.3]	140	5.5 [0.6–47.1]	146	32.8 [1.3–803.4]	109	1073 [324–3548]	84	1.57 [0.40–6.19]	80
	15	LO	0.3 [0.1–2.0]	14	2.6 [1.1–6.2]	11	1.0 [1.0–1.0]	15	7.6 [1.0–58.2]	8	1527 [949–2457]	4	5.25 [2.15–12.85]	4
	23	LM	1.0 [0.1–16.1]	23	3.5 [1.1–11.6]	23	2.5 [0.5–12.9]	23	9.5 [0.3–338.2]	16	629 [293–1351]	10	1.65 [0.46–5.89]	9
	4	TR	0.1 [0.1–0.3]	3	2.6 [1.6–4.2]	3	1.0 [1.0–1.0]	4	2.0 [0.3–14.9]	2	1800	1	5.47	1
	53	NW	1.4 [0.1–13.5]	47	8.3 [1.8–37.2)	48	7.2 [1.8–28.7]	49	45.5 [2.0–1055.9]	36	1205 [283–5123]	34	1.74 [0.50–6.11]	32
	50	OW	5.0 [2.0–12.6)	43	15.3 [3.2–71.7]	46	13.6 [4.3–43.2]	46	60.7 [5.4–682.5]	37	1043 [390–2786]	26	1.09 [0.37–3.21]	25
	10	AP	7.9 [2.1–30.6]	8	3.5 [1.8–7.0]	9	3.5 [1.8–7.0]	9	42.4 [8.9–201.8]	10	1095 [586–2047]	9	1.43 [0.41–4.92)	9
Open	8	All	11.0 [2.8–43.1]	8	50.8 [16.4–157.6]	8	32.6 [6.8–156.9]	8	1109.1 [57.6–21339.7]	8	3761 [866–16331]	8	1.01 [0.48–2.11]	8
	7	OW	9.4 [3.1–28.7]	7	51.3 [15.1–173.9]	7	41.9 [20.3–86.4]	7	1037.7 (43.7–24654.1)	7	3777 [773–18445]	7	1.03 [0.46–2.26)	7
	1	AP	33.2	1	47.6	1	5.6	1	1767.3	1	3650	1	0.92	1

Wooded	182	All	2.3 [0.1–38.1]	164	9.0 [1.1–70.1]	164	6.7 [0.7–66.7]	171	45.8 [1.2–1723.4]	128	1095 [335–3583]	95	1.43 [0.31–6.57]	90
	15	LO	0.3 [0.1–2.0]	14	2.6 [1.1–6.2]	11	1.0 [1.0–1.0]	15	7.6 [1.0–58.2]	8	1527 [949–2457]	4	5.25 [2.15–12.85]	4
	24	LM	1.0 [0.1–16.3]	24	3.7 [1.0–14.0]	24	2.7 [0.5–15.9]	24	9.8 [0.3–312.7]	17	653 [303–1409]	11	1.76 [0.50–6.22]	10
	4	TR	0.1 [0.1–0.3]	3	2.6 [1.6–4.2]	3	1.0 [1.0–1.0]	4	2.0 [0.3–14.9]	2	1800	1	5.47	1
	53	NW	1.4 [0.1–13.5]	47	8.3 [1.8–37.2]	48	7.2 [1.8–28.7]	49	45.5 [2.0–1055.9]	36	1205 [283–5123]	34	1.74 [0.50–6.11]	32
	73	OW	5.5 [2.0–14.9]	65	18.5 [3.4–100.0]	67	16.5 [4.0–68.0]	67	96.3 [4.6–2012.0]	53	1098 [394–3062]	34	1.02 [0.31–3.32]	32
	13	AP	12.9 [1.6–105.0]	11	5.0 [0.8–30.5]	11	4.4 [1.4–13.5]	12	86.2 [2.4–3149.9]	12	1151 [540–2451]	11	1.01 [0.15–6.80]	11
Frugivore	118	All	2.6 [0.2–32.7]	107	10.8 [1.4–83.1]	110	8.9 [1.0–78.6]	113	74.7 [2.0–2744.7]	85	1404 [426–4622]	66	1.46 [0.42–6.07]	63
	3	LO	0.7 [0.2–3.2]	3	1.6 [0.4–5.8)	2	1.0 [1.0–1.0]	3	20.4 [10.3–40.5]	2				
	14	LM	0.9 [0.0–18.7]	14	4.1 [1.0–17.7]	14	2.9 [0.4–20.4]	14	7.8 [0.4–140.6]	10	645 [393–1059)	6	1.80 [0.39–8.30]	6
	42	NW	1.2 [0.2–8.9]	36	7.9 [1.7–37.0]	38	6.8 [1.7–27.6]	39	56.6 [3.0–1058.3]	28	1536 [505–4674]	26	2.00 [0.78–5.12]	24
	48	OW	5.2 [1.7–16.0]	45	22.5 [5.9–84.9]	47	21.3 [5.3–86.6]	47	181.2 [8.6–3808.7]	35	1570 [465–5298]	24	1.10 [0.41–3.00]	23
	11	AP	12.8 [2.0–84.1]	9	5.8 [0.5–64.6]	9	4.0 [1.7–9.4]	10	91.7 [2.1–4039.7]	10	1356 [486–3784]	10	1.16 [0.24–5.57]	10
Folivore	43	All	6.0 (1.1–32.1]	39	11.4 [1.1–112.6]	42	9.6 [1.9–49.2]	42	54.7 [1.0–3000.3]	38	826 [168–4063]	27	0.88 [0.20–3.95]	25
	7	LM	1.8 [0.3–10.4]	7	3.6 [1.1–11.6]	7	3.6 [1.1–11.6]	7	9.8 [0.1–713.0]	6	529 [343–816]	4	1.80 [0.59–5.49]	3
	6	NW	5.9 [3.9–8.9]	6	8.8 [3.0–26.0]	6	7.9 [3.9–16.2]	6	33.3 [2.1–521.4]	5	461 [169–1259]	5	0.75 [0.14–4.03)	5
	27	OW	7.5 [3.9–14.5]	23	17.4 [1.7–180.0]	26	13.6 [3.2–57.7]	26	79.7 [1.9–3332.8]	24	1096 [190–6337]	16	0.87 [0.25–3.04]	15
	3	AP	18.2 [0.9–355.9]	3	6.9 [1.6–29.7]	3	6.9 [1.6–29.7]	3	192.0 [2.0–18890.2]	3	902 [789–1032]	2	0.47 [0.02–9.40]	2
Insectivore	15	All	0.2 [0.0–1.8]	14	3.3 [0.6–18.0]	13	1.3 [0.2–9.2]	15	8.3 [0.3–270.8]	9	1681 [1223–2309]	6	3.87 [0.96–15.60]	6
	8	LO	0.2 [0.0–1.0]	8	2.7 [1.4–5.2]	7	1.0 [1.0–1.0]	8	5.1 [0.6–40.2]	5	1632 [1007–2644]	3	5.85 [2.24–15.29]	3
	2	LM	0.5 [0.0–53.0]	2	2.4 [0.6–9.8]	2	1.0 [1.0–1.0]	2	103	1	1641	1	1.43	1
	4	TR	0.1 [0.1–0.3]	3	2.6 [1.6–4.2]	3	[1.0, 1.0–1.0]	4	2.0 [0.3–14.9]	2	1800	1	5.47	1
	1	NW	0.8	1	45	1	45	1	128.7	1	1755	1	2.14	1
Gummivore	5	All	0.3 [0.1–1.8]	5	4.7 [1.7–13.3]	5	2.3 [0.2–22.6]	5	4.0 [0.3–54.3]	3	650 [147–2868]	3	2.88 [1.76–4.72]	3
	2	LO	0.6 [0.1–4.7]	2	3.7 [3.1–4.5]	2	1.0 [1.0–1.0]	2	8.5	1	1250	1	3.8	1
	1	LM	0.3	1	2.5	1	1	1						
	2	NW	0.2 [0.1–0.5]	2	8.0 [7.6–8.5]	2	8.0 [7.6–8.5]	2	2.8 [0.1–67.7]	2	469 [121–1820]	2	2.51 [2.13–2.97]	2

[a] Values in table are: mean [mean + 2SD, mean-2SD]. Values are back-transformed from log-transformed data.
[b] LO, Lorisids; LM, lemurs; TR, Tarsius; NW, New World monkeys; OW, Old World monkeys; AP, apes.

Table X. Association between Body Size and U Variables in the Different Syndromes[a]

Syndrome	N	Radiation[b]	Population group size	Foraging group size	Home range size	Day journey	D-Index
N-G	3	*LO*					
N-1	14	3	−0.18, P = 0.0942	all have FGS = 1	0.57	—	−0.22, P = 0.893
	4	*TR*					
	8	*LO*					
	2	*LM*					
N-Fr	10	3	−0.20, P = 0.0744	all have FGS = 1	—		
	3	*LO*					
	5	*LM*	—	all have FGS = 1			
	2	*NW*					
N-Fo	2	*LM*					
D-G	2	*NW*					
D-I	1	*NW*					
D-A-Fr	83	4	−0.0317	—	**1.53, p = 0.0004**	—	**−0.60, p = 0.0014**
	8	*LM*	—	—	—	—	—
	40	*NW*	0.30	—	**1.40, P = 0.0090**	—	−0.30
	27	*OW*	−0.08	—	—	−0.22	**−0.98, p = 0.0074**
	8	*AP*	−0.41	−0.41	0.61	−0.57	−0.76

D-A-Fo	34	4	0.13	—	—	—	—
	5	*LM*	—	—			
	6	*NW*	—	—	only two contrasts	—	—
	21	*OW*	—	—	2.60	—	−0.77
	2	*AP*					
D-T-W-Fr	20	3	0.61, p = 0.0521	—	0.74	**0.66, p = 0.0426**	—
	1	*LM*					
	17	*OW*	0.65, p = 0.0692	0.65, p = 0.0692	0.59	0.73, p = 0.0822	—
	2	*AP*					
D-T-W-Fo	4	2					
	3	*OW*					
	1	*AP*					
D-T-O-Fr	5	2	0.30	—	—	—	—
	4	*OW*					
	1	*AP*					
D-T-O-Fo	3	*OW*					

[a] Slopes of the least-squares regression line, forced through the origin (a = 0), all analyses conducted on log-transformed data. Blank cells indicate cases where not enough data points were available to perform the analysis in nonphylogenic tests ($n < 5$). A dashed line (—) indicates cases where $n > 5$, but p-values in norphylogenetic results are greater than 0.10, so that phylogenetic tests were not conducted. P-values are provided only if less than 0.10, with significant results shown in bold.

[b] LO, Lorisids; LM, lemurs; TR, Tarsius; NW, New World monkeys; OW, Old World monkeys; AP, apes.

that are significant ($P \leqslant 0.05$) or close to significant ($P < 0.10$). A line in a cell in Table X indicates that a syndrome is represented by 5 or more species, but $P \geqslant 0.10$ in nonphylogenetic tests, and so contrasts were not calculated. Regressions of foraging group size on body mass were not calculated in N-I and N-Fr because most extant taxa feed solitarily.

The majority of the cells in Table X do not contain the requisite 5 or more species (only 67 of the 165 total cells meet this criterion), so we did not calculate nonphylogenetic regressions for most of the possible syndromes and radiations within syndromes. Of the 67 cells with 5 or more species, the relationship between body mass and a given U variable was significant in the nonphylogenetic tests in nearly half of the cases ($n = 30$). Many of the syndromes have very few data points, however, so the absence of statistically significant results may be due to low statistical power (see Cohen 1988; Thomas and Juanes 1996). When phylogeny is taken into account using independent contrasts, only 11 radiation-syndromes show a relationship between body size and U variables with P-values less than 0.10; of these, only 5 are statistically significant at $P < 0.05$. Four of these significant results are found in the best represented syndrome (D-A-Fr), again suggesting that statistical relationships are present but discernible only when sample sizes are large.

Despite these statistical limitations, several patterns emerge.* The relationship between body mass and group size is negative in nocturnal species, but positive in diurnal terrestrial ones. No clear pattern between these two variables can be seen in D-A-Fr and D-A-Fo, with some radiations having a positive slope, some radiations a negative slope, but none having a slope statistically different from 0 at $P < 0.10$.

A fairly consistent relationship is observed between body mass and home range size, with all cases that meet our criteria for calculating independent contrasts showing a positive slope (all slopes in the nonphylogenetic analyses were also positive). In some cases, the association between body mass and home range size is highly significant (e.g., D-A-Fr all species, and D-A-Fr New World monkeys), and the slopes are the largest found among the U variables (in the nonphylogenetic analysis, a slope of 5.53 was calculated for D-A-Fo, New World, but only 2 contrasts are possible here because of unresolved relationships among the species of howler monkeys in Purvis' phylogeny).

The relationship between day journey length and body mass is ambiguous. Among D-T-W-Fr this trend is positive ($P < 0.10$), while in D-A-Fr, the trend is negative, but not significantly so. The D-index, in contrast, is negatively related to body mass in all phylogenetic tests, with some results statistically significant in phylogenetic tests (D-A-Fr, all species and Old World).

Thus, some U variables scale with body mass in extant taxa. When these associations can be identified, they can be used to narrow the estimated ranges

* In the Discussion, we consider more general trends and the theoretical linkage between body size and U variables.

of U behavioral traits in extinct taxa. Before the relationship between body mass and U variables can be used, however, the error involved in estimating body mass should be incorporated in the analysis (Smith, 1996; see above). Body mass is usually estimated through associations between body mass and morphological characters in extant taxa, including characters such as molar tooth area (Gingerich *et al.*, 1982) or postcranial dimensions (Dagosto and Terranova, 1992). These morphological features estimate body mass with some error, which can often be quantified. For example, Gingerich *et al.* (1982) showed that body mass based on M_1 dimensions for *Aegyptopithecus zeuxis* is predicted to be 6.04 kg. However, 95% confidence limits provide a range from 5.62 to 6.48 kg. Estimates based on other morphological features can lead to even wider confidence limits (see Gingerich *et al.*, 1982).

Therefore, when body mass is used to refine estimates of behavior in the fossil record, two sources of error should be incorporated: the error in measuring body mass for a fossil taxon, and the error in the relationship between body mass and the U variable among extant species. Error in estimated body mass can be incorporated by using confidence limits to find a reasonable range of estimated body masses (Gingerich *et al.*, 1982). We suggest that error in the relationship between U and body mass be incorporated by using prediction limits* on the regression of the U variable on body mass. Figure 1 demonstrates this general approach. In the case of a positive relationship between a U variable and body mass, the lower estimate of body mass (L1) is used to predict the lower prediction limit (P1), while the higher estimate of body mass (L2) is used to predict the upper prediction limit (P2). The corresponding prediction interval for the U variable is shown on the Y axis of Fig. 1. The variables are reversed in the case of a negative relationship between U and body mass, with L1 used to predict P2, and L2 used to predict P1. This gives wider predicted ranges for the U variable than simply using a single estimate of body mass, and this procedure therefore provides more conservative reconstructions (i.e., wider estimates) of behavior in the fossil record.

To examine the extent to which incorporating the effect of body mass narrows estimates of U variables in fossil taxa, we calculated prediction intervals for cases in which a significant association ($P < 0.05$) was found between a U variable and body mass after taking phylogeny into account (only five cases; see Table X). These relationships (with the regression coefficients calculated as in Table X and forced on the log-transformed data) are plotted in Fig. 2 (a–e). Table XI provides the parameters and formulas needed to predict exact upper and lower limits for the five U variables that are significant in phylogenetic tests when estimates of body mass are available. In general, body mass regressions can narrow the ranges of U variables in fossil taxa, although the potential improvement in resolution is not great.

* Prediction limits on a regression line are used in predicting Y from X (i.e., U from K). This differs from confidence limits on a regression line, which represent probabilities of the location of the line itself, and are usually narrower than the prediction limits (see Sokal and Rohlf, 1995).

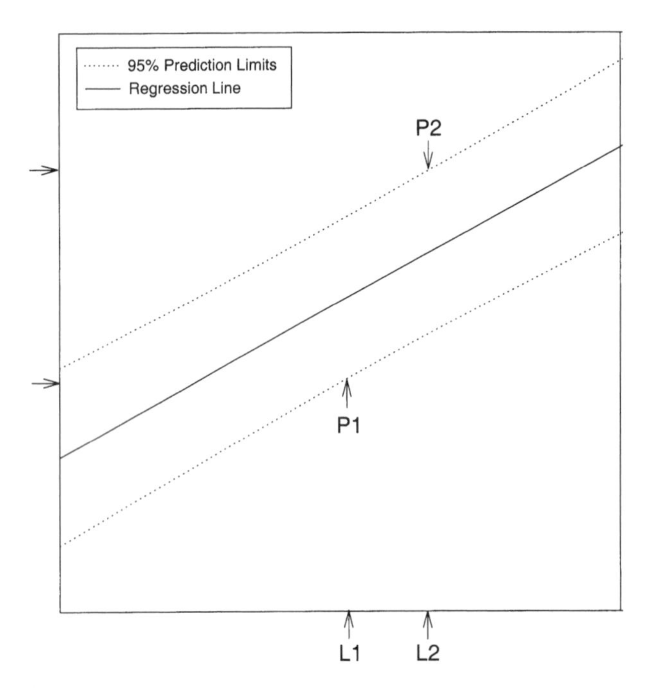

Fig. 1. Using body size to narrow reconstructions of behavior in the fossil record. Reconstructions of U variables based on body size should incorporate the error involved in estimating body size for the fossil taxon, in addition to error in the relationship between U and body size in extant taxa. In the present case, the slope of the regression line is positive. Thus, the lower estimate of body size for the fossil taxon (L1) is used in the formulas (see Table XI) to calculate the lower prediction limit (P1), and the upper estimate of body size (L2) is used to calculate the upper prediction limit (P2).

Worked Examples

In order to explore the usefulness of the approach taken here, we investigate the ability of our methods to reconstruct the socioecology of several fossil taxa. First, we examine how the syndrome categories apply to actual fossil taxa. In most cases, the syndromes present in extant taxa can also be identified in extinct taxa. Furthermore, syndromes missing in living species are not found to be common in the past. For example, a number of fossil taxa (*Aegyptopithecus*, *Apidium*, *Oreopithecus*) have been classified as D-A-Fr or D-A-Fo (Fleagle and Kay, 1985), which are the most common syndromes in extant species (Table III). Still others are interpreted as terrestrial (either wholly or partly), and either frugivorous (*Proconsul*) or folivorous (*Mesopithecus*) (Fleagle and Kay, 1985). However, no fossil primates have been reconstructed as terrestrial and nocturnal (R. Kay, pers. comm.). Some diurnal gum-eaters probably existed in the past (e.g., *Lagonimico conclucatus*, Kay 1994). Finally, insectivory probably also has deep phylogenetic roots (e.g., Williams and Covert, 1994) and is present in many diurnal taxa as a major dietary component (see information on individual species in Smuts *et al.*, 1987).

However, when the focus narrows to radiations, some anomalies appear. For example, based on patterns of molar wear and crown height in catarrhines, Kay and Williams (this volume, Chapter 9) argue that the New World taxon *Branisella boliviana* was at least partly terrestrial. However, no extant New World primate species fills a terrestrial niche (see Table V). In addition, *Branisella* is thought to weigh 721–759 grams (Kay and Williams, this volume, Chapter 9; see also Gingerich *et al.*, 1982, who give a range of 1005–1310 grams). This range falls outside the range of body masses for the terrestrial niche in Tables VIII and IX. In this case, then, the unusual combination of K traits calls for a close examination of the evidence for these character states. Additional information, especially in the form of postcranial remains, may be

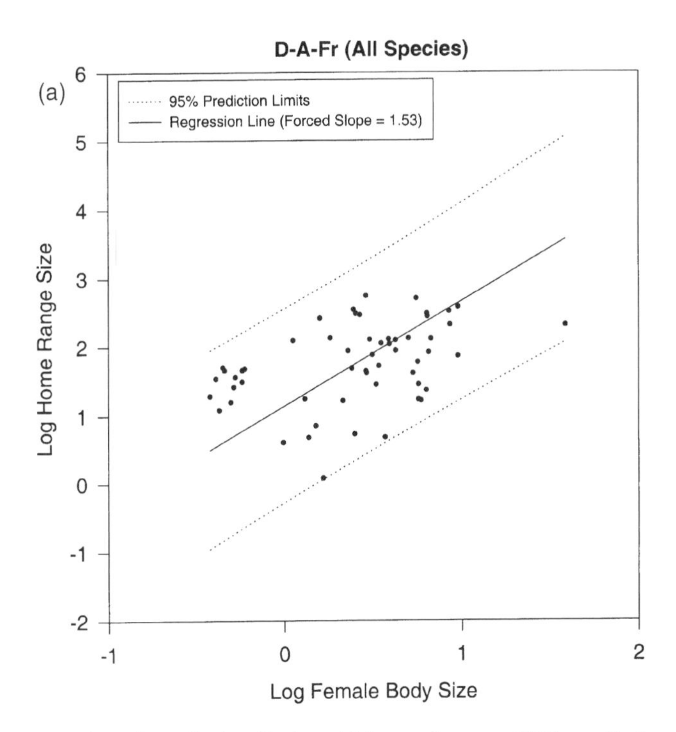

Fig. 2a–e. Examples of prediction limits within syndromes. 95% prediction limits were calculated for cases within syndromes where a significant association between body size and a U variable could be documented after taking phylogeny into effect. The slope of the line was calculated using independent contrasts on log-transformed data, and this slope was forced on the log-transformed data with the Y-intercept free to vary.

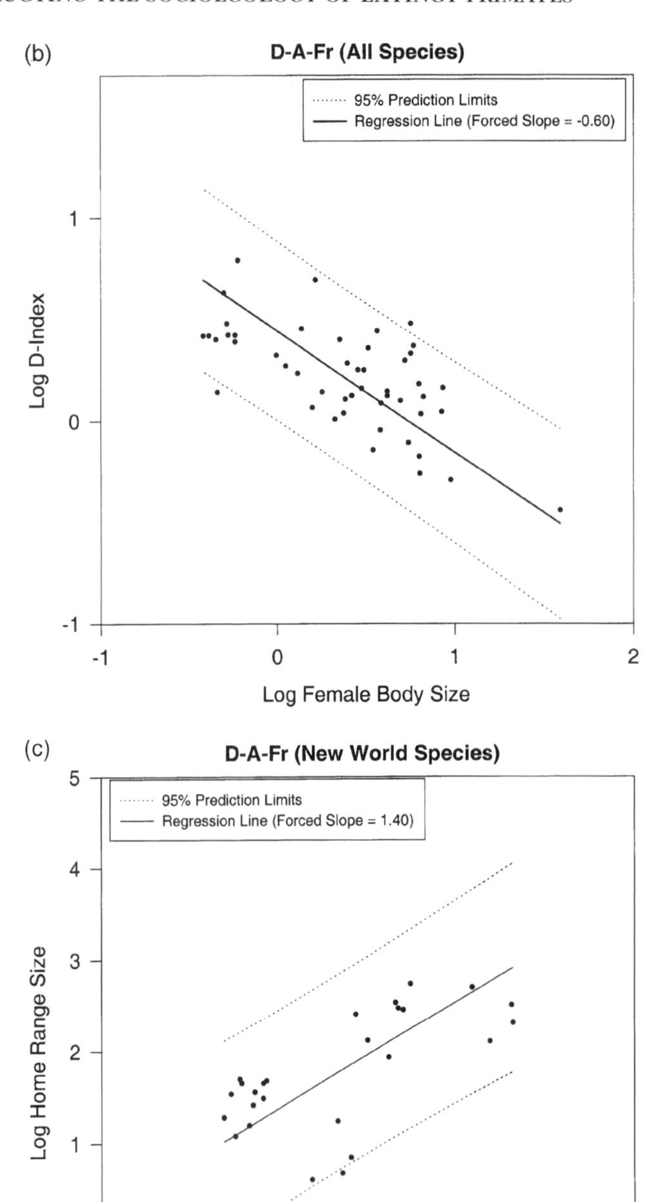

Fig. 2. Continued.

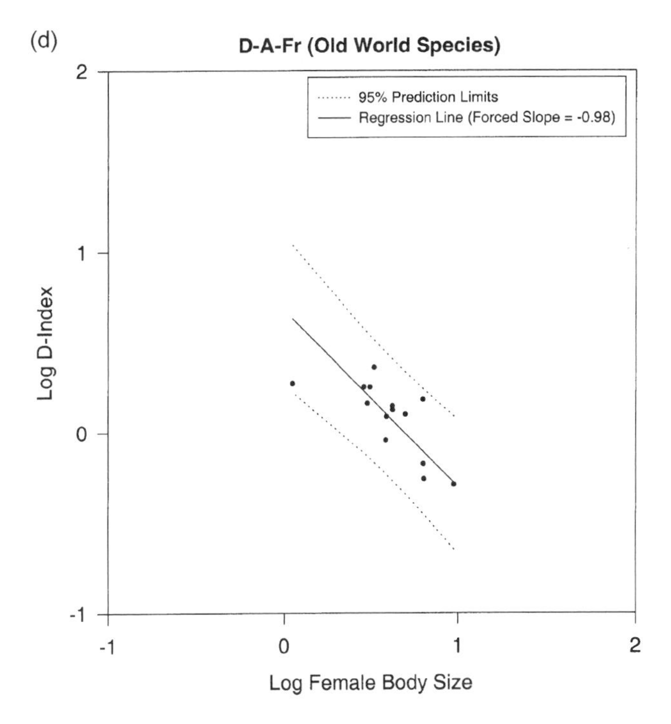

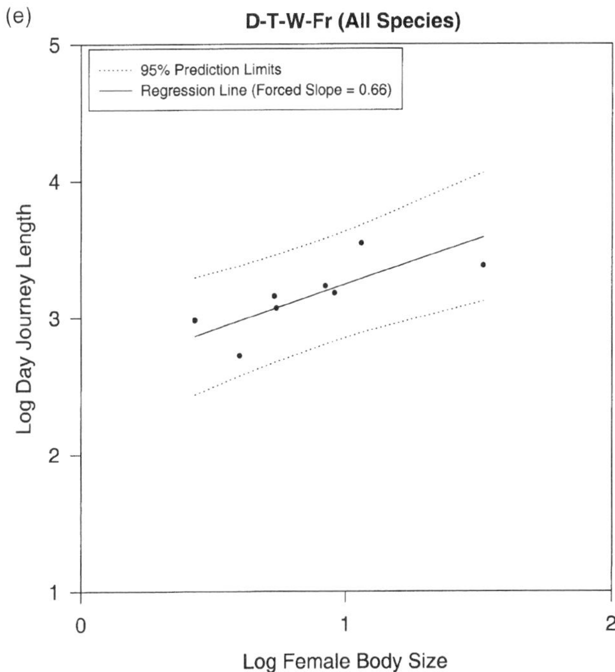

Fig. 2. Continued.

Table XI. Calculation of Prediction Limits for Significant Associations between U Variables and Body Size

Syndrome (Radiation)—U variables	a	b	$S^2_{\bar{Y} \cdot X}$	n	$\bar{X}$	$\sum x^2$
D-A-Fr (All Species)—Home Range Size	1.14	1.53	0.4849	56	0.3933	10.7577
D-A-Fr (All Species)—D-Index	0.44	−0.60	0.0463	48	0.3882	9.6353
D-A-Fr (New World)—Home Range Size	1.61	1.40	0.1257	26	0.1131	4.7436
D-A-Fr (Old World)—D-Index	0.68	−0.98	0.0060	14	0.6083	0.6324
D-T-W-Fr (All Species)—Day Journey Length	2.58	0.66	0.0249	8	0.8699	0.7776

Formulas[a]

$$P_1 = \hat{Y}_i - t_{a[n-1]}\hat{s}_{\bar{Y}}$$

$$P_2 = \hat{Y}_i + t_{a[n-1]}\hat{s}_{\bar{Y}}$$

where

$$\hat{Y}_i = a + bX_i$$

$$\hat{S}_{\bar{Y}} = \sqrt{S^2_{\bar{Y} \cdot X}\left[1 + \frac{1}{n} + \frac{(X_i - \bar{X})^2}{\sum x^2}\right]}$$

[a] We use standard statistical notation for confidence and prediction limits on regression lines. Thus, X_i refers to a given estimate of body weight for a fossil, and $\hat{Y}_i$ refers to the estimated value of the U variable, as determined from the regression formula. To determine which value of body size (lower confidence limit L_1 or upper confidence limit L_2) to use as X_i in calculating $\hat{s}_{\bar{Y}}$, see text and Fig. 4.

necessary to resolve the issue of substrate use in this fossil taxon. In the meantime, analyses of *Branisella's* socioecology should also consider the possibility that this species is arboreal.

After examining the internal consistency of the K variables, the next step is to use Tables VIII–XI to estimate U variables for the fossil taxon. In some cases, such as *Branisella*, this will not be straightforward: because its syndrome is not present in the New World, ranges cannot be narrowed taxonomically using Table VIII. However, using the information in Table IX for Terrestrial-All species, we infer that *Branisella* foraged in large groups (mean = 24) with a large home range (450.6 ha), a long daily path length (2021 m), and a low D-Index (0.82). As seen in Table IX, however, the ranges for these U variables always span more than one order of magnitude, reflecting the uncertainty inherent in these estimates. If we instead assume that *Branisella* was arboreal, we would infer a much smaller foraging group size (5.5), home range (32.8 ha), and day journey length (1073 m), and a larger D-index (1.57) (from Arboreal-All species in Table IX; taxonomic information not incorporated to facilitate comparison with terrestrial patterns). Thus, one ends up with different interpretations depending on the

Table XII. Reconstructing Socioecological Variables for *Aegyptopithecus*

Variable	D-A-Fr	D-A-Fr, taxonomy	D-A-Fr, body size[a]
Population group size	9.5 (1.9–48.9)	18.3 (7.1–47.6)	—
Foraging group size	8.4 (1.7–42.3)	16.7 (5.9–47.6)	—
Home range size	54.9 ha (3.8–787.8)	90.7 ha (19.2–429.2)	216 ha (7.4–6300)
Day journey length	1334 m (492–3618)	1306 m (575–2967)	—
D-index	1.61 (0.52–5.00)	1.21 (0.48–3.06)	0.82 (0.51–1.32)

[a] Body size rantge used in formulas: estimated body size = 6.04, 95% confidence interval 5.62 to 6.48 (from Gingerich *et al.*, 1982). Relationship between body size and home range size comes from the pattern across all primates, and therefore does not control for taxonomy. Relationship between body size and the D-index comes from the pattern found in Old World monkeys, and therefore controls for both taxonomy and body size.

inference of the relevant K variables, even though many of these ranges overlap.

In other cases, a fossil taxon can be assigned to one of the syndromes in Table VIII, and taxonomy and body size can be used to narrow estimates for the fossil taxon. For example, *Aegyptopithecus* inhabited the Old World, and is interpreted as D-A-Fr (Fleagle and Kay, 1985). Table XII provides estimates for group size, home range size, day journey length, and the D-index for this fossil taxon. Using only syndrome information, ranges are quite large, in one case spanning two orders of magnitude (home range size). These ranges can be narrowed using taxonomic information (i.e., the radiation in which the fossil is most likely to be placed, in this case Old World monkeys; Fleagle, 1988) and body size (taken from Gingerich *et al.*, 1982). This reduces the ranges somewhat, but they still remain quite wide. Body size reduces the range for the D-index substantially, possibly because slopes were calculated within the Old World monkey radiation (see Table X). On the other hand, the range for estimated home range size based on body size is actually wider than either of the previous estimates, although statistics here are based on patterns in all species, and not within just the Old World monkeys. Nevertheless, it does suggest that based on body size, *Aegyptopithecus* had a larger home range than might be inferred without this information. At this group size, multiple males were probably present (Andelman, 1986), and extant species with a D-index of greater than 1 are typically territorial (Mitani and Rodman, 1979). In conclusion, we infer that *Aegyptopithecus* lived in medium-sized, multimale groups that were moderately territorial, perhaps similar to extant *Cercocebus* (Waser, 1976).

Discussion

In reconstructing behavior in the fossil record, we face an obvious problem: behavior does not fossilize. However, it is often possible to reconstruct some behavioral traits on the basis of fairly reliable functional relationships with anatomical features. We therefore explored how a more general comparative approach might be taken in the case of socioecological variables. In particular, we attempted to predict unknowable socioecological variables (U variables) from combinations of variables that are more directly knowable from the fossil record (K variables). We also examined ways to improve resolution using body mass estimates and general phylogenetic information.

We have not attempted to predict aspects of the social system beyond group size (e.g., social relationships), because these features are not likely to be simple correlates of the crude K variables that can be obtained from fossil specimens. At least for now, aspects of the mating system and male–male competition are best predicted by the degree of sexual dimorphism in canine size and body mass (Plavcan, this volume, Chapter 8). Thus, when reliable dimorphism estimates are available, more detailed reconstructions may be possible.

While in some cases our approach has been successful, in other cases ranges remain too wide to be generally useful. Nevertheless, some general trends can be identified that are useful for comparing particular fossils, or for making broad statements about the general socioecology of a fossil taxon. These general patterns are summarized in Table XIII and discussed in more detail in the next section.

General Patterns among K and U Variables in Extant Primates

This review is based on two sources. First, we survey previous comparative studies, including the most comprehensive study to date (Clutton-Brock and Harvey, 1977a,b). It should be noted, however, that many of the statistics generated in these previous studies did not fully consider the effects of phylogeny, in many cases because phylogenetic comparative methods, or the necessary phylogenetic information, were not yet available.

The second source of information are the results presented in the above tables and figures. Our results make use of an expanded database which includes more species and field studies than earlier work. The influence of potentially confounding K variables was controlled by comparing the mean values of syndromes that differ in the variable of interest (i.e., using Table VIII). We did not evaluate these patterns statistically, mainly because these calculations involve complicated phylogenetic analyses beyond the scope of the present chapter. Future studies aimed at accounting for extant variation should correct for phylogenetic nonindependence and control for the effects of multiple, potentially confounding variables.

Table XIII. General Patterns between K and U Variables in Extant Taxa

Variable	Character state	Group size[a]	Home range size	Day journey length	D-index	Body mass	Relationship to other K variables
Activity period	Nocturnal	lower	lower	slightly lower	higher	lower	arboreal and wooded, tend to be insectivorous
	Diurnal	higher	higher	higher	lower	higher	variable
Substrate use	Arboreal	lower	lower	lower	higher	lower	wooded habitat
	Terrestrial	higher	higher	higher	lower	higher	diurnal
Habitat	Wooded	lower	lower	lower	no clear relationship	lower	—
	Open	higher	higher	higher	no clear relationship	higher	diurnal and terrestrial
Diet	Gummivore	lower	very low	lower	very high	very low	arboreal, wooded
	Insectivore	lower	very low	higher	very high	very low	arboreal, wooded
	Frugivore	higher	higher	higher	higher	lower	—
	Folivore	higher	lower	lower	lower	higher	usually diurnal
Body mass	—	no clear relationship	positive correlation	no clear relationship	negative correlation	—	—

When examining the general associations among K and U variables, we also discuss the possible factors that underlie these patterns. Such information provides additional support for the postulated patterns by considering the theoretical linkage among the traits. In almost all cases, however, these factors require additional research, sometimes in detailed behavioral studies rather than comparative ones. We therefore hope that this review identifies areas for future research.

Activity Period

Activity period has a profound effect on life history and socioecology. Relative to diurnal species, nocturnal species tend to be smaller in body mass, are always arboreal, forage alone or in pairs rather than larger groups, live in smaller home ranges, and tend to have lower population densities and biomasses (Clutton-Brock and Harvey, 1977a,b). The results from above confirm some of these patterns (Table VIII), and also show that nocturnal species have slightly smaller day journey lengths, but higher D-indices (Fig. 3).

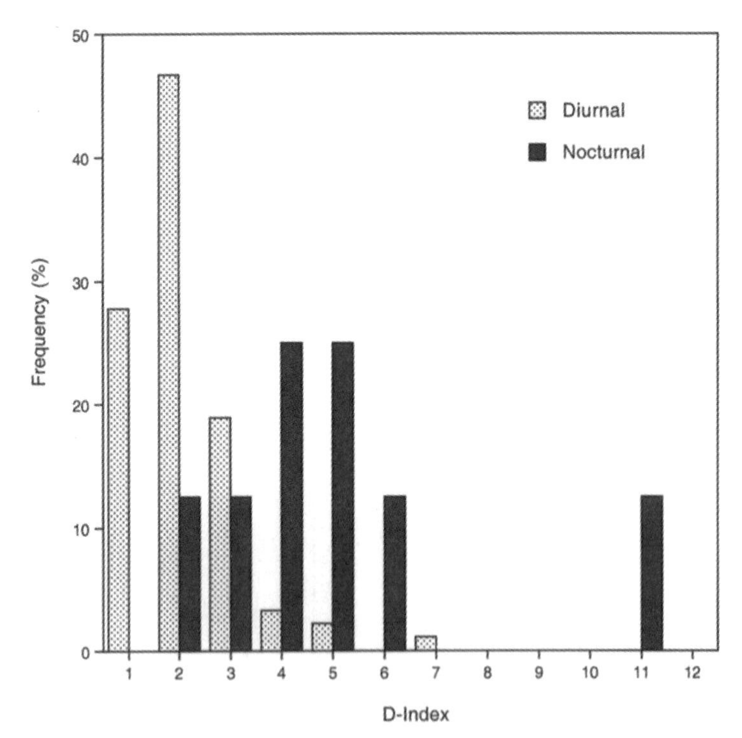

Fig. 3. The intensity of range use in diurnal and nocturnal primates. Nocturnal primates tend to use their ranges more intensely (as measured by the D-index; Mitani and Rodman, 1979).

This means that nocturnal species use their home ranges more intensely, perhaps because their locomotion is facilitated by familiarity with local arboreal pathways.

Several factors may contribute to the generally solitary lifestyle of nocturnal species. First, solitary foraging may aid ambush hunting of active prey; if the animals were to hunt in a group, stalks might more often be interrupted, thereby reducing hunting success (Clutton-Brock and Harvey, 1977b). However, this cannot hold for all species, because highly folivorous (e.g., *Lepilemur mustelinus*) or highly frugivorous (e.g., *Nycticebus coucang*) nocturnal species also occur. Second, predation risk may be a major factor. Nocturnal predators that hunt by sound (e.g., owls) may be attracted to animals that use contact calls to coordinate travel as a dispersed group. Because it is more difficult to distinguish between sounds of predator and conspecific movements at night in dense forest understories, false alarm rates would likely be high. Moreover, given the low acuity associated with nocturnal vision, collective visual vigilance may be less effective.

With one exception (*Varecia*; see van Schaik and Kappeler, 1996), all primates that deposit their young in nests are nocturnal, perhaps because it is difficult for a diurnal animal to have a nest with altricial young. However, the opposite is not true: some nocturnal primates carry their young. In summary, then, relative to diurnal species, nocturnal species are small, solitary and arboreal, live at relatively low densities in intensively used ranges, and are more likely to be insectivores and to use nests.

Locomotor Substrate

Substrate use is an important influence on primate socioecology. Relative to arboreal species, terrestrial animals tend to be larger, live in larger groups, have larger home ranges, cover longer distances each day, and live at lower densities (Clutton-Brock and Harvey, 1977a). Terrestrial species also tend to have "faster" life histories than expected for their body masses (Ross, 1988; Harvey *et al*., 1989). They are also more likely to live in multilevel societies, where foraging units come together to sleep on a common refuge (e.g., hamadryas baboons), or forage together when ecological opportunities allow (e.g., mandrills, pigtailed macaques; Table VIII). It is not clear to what extent the effects ascribed to locomotor substrate in previous studies are due to the correlated effects of habitat, in that terrestrial primate taxa are often assumed to inhabit open environments. Here, we controlled for this by comparing forest-living species that differ in substrate use. Nonetheless, our tables support the above patterns. In addition, terrestrial species were found to have smaller D-indices.

Species that are primarily terrestrial may live in larger groups for two related reasons. First, they are at greater risk of predation because they face a greater variety of predator species, and escape may be more difficult because of shorter detection ranges. Second, they have lower energy costs, in that locomotion in the vertical dimension requires more energy than horizontal travel (see Schmidt-Nielsen, 1984). Hence, terrestrial taxa may overcome the costs of group living more easily. Because adults in terrestrial taxa tend to weigh more, and because group sizes tend to be larger, foraging groups of terrestrial species probably have greater energy requirements. Thus, terrestrial species probably require larger home ranges and day journey lengths to meet their energetic demands, which results in lower population densities and D-indices.

Habitat

The effect of habitat is best studied by comparing the terrestrial forest-living animals with those inhabiting open environments. The results in Table VIII confirm that species inhabiting open environments are larger in body size, live in larger groups with larger home ranges, and travel further per day. Habitat has no clear effect on range use intensity. Two of the cases in Table VIII represent intraspecific patterns, in that populations of olive baboons and common chimpanzees can inhabit either forests or savannas.

In general, then, the effects of living in an open habitat are in the same direction as those of being terrestrial. Increased group size and body mass are probably related to an increased threat of predation. First, an open environment probably increases predation risk due to the increased distance to refuges (i.e., high trees or cliffs). Second, species living in open environments might gain more than other diurnal taxa through collective vigilance in larger groups because they can potentially spot predators from further away, and therefore have more time to give alarm calls and take necessary precautions.

Home range size and day journey length are expected to be larger in an open environment for three reasons. First, body mass and group size are larger, and individuals thus face greater energy requirements. Second, longer travel distances may be caused by the lower food density in open habitats, which can be related to lower rainfall compared to wooded habitats. Third, the relative absence of vertical climbing makes travel more efficient, and thus allows larger animals in open environments to travel more extensively.

Diet

Clutton-Brock and Harvey (1977a,b) distinguished between frugivorous, folivorous, and insectivorous species. Most of their conclusions concerned differences between frugivores and folivores, because the effect of insectivory is confounded by the effects of nocturnality and small body size. They found that frugivores tend to be smaller in body mass, live in larger groups and larger home ranges, and travel farther per day. Our preliminary results show mixed support for these generalizations (Table VIII): diet (frugivory versus folivory) did not affect group size; the home ranges of frugivores were only marginally larger than those of folivores; but frugivores did tend to be smaller and did tend to travel more (and thus used their ranges more intensively). Compared to frugivores and folivores, the even smaller-bodied insectivorous and gummivorous primates have smaller home ranges that they use more intensely. Obviously, these analyses need to be repeated with modern comparative techniques using finer dietary characterizations, supplemented by intraspecific studies.

These different findings are difficult to discuss because of the many potentially confounding factors that exist among the relevant variables. However, some of the more prominent effects can be explained relatively straightforwardly. First, folivores tend to have larger body sizes than related frugivores, probably because this increases the efficiency of digesting a poor quality diet (Kay, 1984). Second, folivores are less mobile because their food tends to be more abundant, and perhaps because they require more time resting to digest their lower-quality diet. Third, frugivores use their ranges more intensively, and therefore have a greater potential for territoriality.

Previous analyses suggested that frugivores live in larger groups than folivores, possibly because a frugivorous diet constrains maximum daily path length less, and frugivores are thus better able to increase their day journey lengths. However, this relationship was weak in previous studies (Clutton-Brock and Harvey, 1977a) and absent in this one. A more refined analysis that looks at sister taxa or intraspecific variation may be helpful in this regard, as there are probably many confounding variables involved.

Diet also affects other variables that might be of interest to paleontologists. Thus, folivores have smaller brains for their body masses than frugivores (Clutton-Brock and Harvey, 1980; Aiello and Wheeler, 1995), perhaps implying reduced cognitive abilities. Diet may also affect life history. Chapman *et al.* (1990) suggested that, after the influence of body mass is removed statistically, the more insectivorous species have larger litters. Leigh (1994) showed that folivores also have faster life histories, perhaps because of the increased mortality from plant toxins. In general, biomass decreases from folivores through frugivores to insectivores, in accordance with overall food abundance (Hladik, 1975). In summary then, while insectivory and gummivory have strong socioecological correlates, the socioecological contrasts between frugivores and folivores are probably more complicated.

Body Mass

Body mass affects many aspects of primate socioecology. Larger species spend more time feeding and have higher dietary diversity (Clutton-Brock and Harvey, 1977b), and they live in larger groups and at lower densities (ibid.; cf. Robinson and Janson, 1987). Our results (Table X) confirm some of these patterns: larger animals live in larger groups (Fig. 4), have larger home ranges,

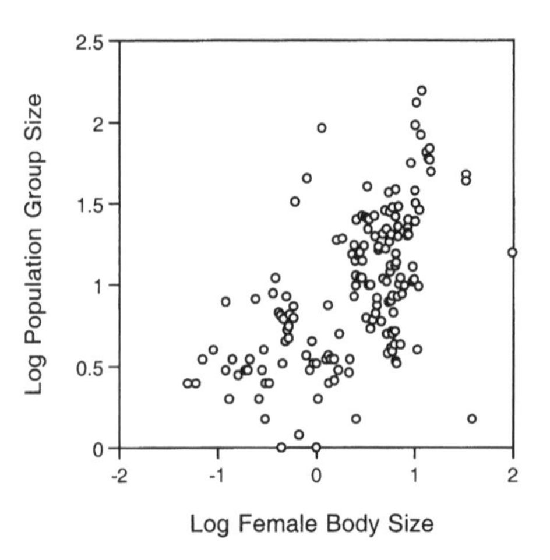

Fig. 4. The relationship between body size and population group size. In analyses across primate species and without controlling for phylogeny, population group size is positively related to female body size. A similar pattern is observed using foraging group size.

and have smaller D-indices. However, when looking within syndromes and radiations, there is no clear association between body size and either group size or day journey length.

The basis of the relationship between body mass and group size in primates remains a long-standing question. These two variables may be influenced by other variables, including the effects of nocturnality, predation avoidance strategies, and terrestriality. First, nocturnal species are probably forced to live in small ranges (see above), which limits foraging and population group sizes (i.e., sleeping groups). Second, small animals rely more on crypsis or on hiding to avoid predation. Only diurnal species may gain reduced predation risk from gregariousness (see above), thus making gregariousness unprofitable for nocturnal taxa. Among diurnal species, small animals also tend to follow a more cryptic strategy, while large-bodied species tend to follow a more active avoidance strategy (Janson and Goldsmith, 1995). Third, terrestrial species can probably reach a greater body size than arboreal species (Clutton-Brock and Harvey, 1977a), which might be constrained to a lower body mass because of their dependence on tree limbs for locomotion. Also, terrestrial species may be at greater risk for predation than arboreal species (Cheney and Wrangham, 1987), and they counter this threat by living in larger groups. Thus, body size systematically interacts with several confounding variables, most of which probably involve antipredation strategies. As a result, looking within syndromes and radiations to control for confounding effects eliminates the relationship between body size and group size observed across all species.

A positive association also exists between home range size and body size. This pattern is expected because, all else equal, larger species require larger home ranges in order to meet their energetic requirements. However, theory suggests that home range size should scale with body mass raised to the 0.75 (Kleiber, 1961; Clutton-Brock and Harvey, 1984), yet many of our regressions indicate a steeper slope. One reason for this steeper slope is that our analyses ignore the effect of group size. The relevant test is thus a regression of home range size on foraging group mass (calculated using group composition and expected masses of different classes of individuals; Clutton-Brock and Harvey, 1977a). Calculating foraging group mass, however, requires knowledge of variables that are not knowable in the fossil record (i.e., group composition), and this analysis is therefore beyond the scope of the present chapter. Differences in resource distribution also may be important factors in accounting for variance in home range size, so that species with different diets might be expected to show different relationships between home range size and body mass (Table X; see also McNab, 1963; Milton and May, 1976).

Across all primates, a regression of contrasts in the D-index on contrasts in body mass gives a negative slope ($b = -0.335$, $p = 0.10$), although this relationship is not significant unless two outliers (in colobines and macaques) are removed ($p = 0.0036$) (Fig. 5). Thus, smaller animals tend to use their ranges more intensely (even though they must feed more selectively on food

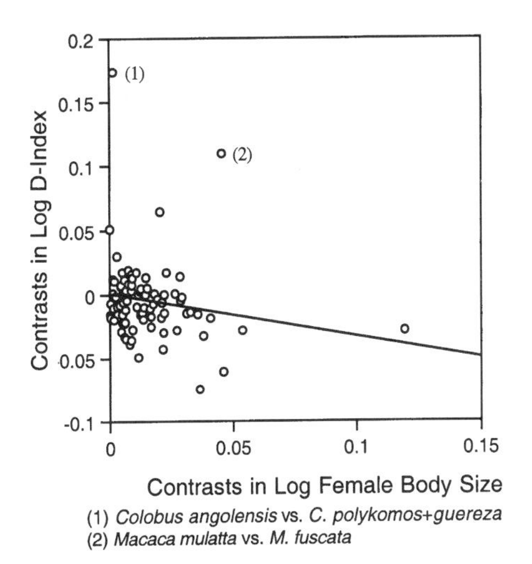

Fig. 5. The relationship between range use intensity (the D-index) and female body size. Smaller-bodied primate species tend to use their ranges more intensely. Values on this graph are independent contrasts, as calculated by CAIC (Purvis and Rambaut, 1995). The slope of the regression line is significantly different from 0 when the outlying data points labeled (1) and (2) are excluded from the analysis.

items that tend to be less abundant). Body mass may affect the D-index through its effects on home range size and day journey length: the positive relationship between home range size and body mass and the lack of a strong relationship between day journey length and body mass leads to a negative relationship between the D-index and body mass. Thus, body mass may be an important influence on range use and between-group relations.

Body mass may also affect aspects of social behavior, for instance time spent grooming (but only in platyrrhines: Dunbar 1991). Finally, body mass profoundly affects life history. Larger species tend to have lower birth rates, slower development, and longer life spans (e.g., Harvey *et al.*, 1989), perhaps because they are subjected to lower unavoidable mortality, such as predation (Cheney and Wrangham, 1987). Body mass is also positively correlated with sexual dimorphism in body and canine size (Mitani *et al.*, 1996; Plavcan, this volume, Chapter 8). In summary, then, a more detailed analysis of the effects of body mass on socioecological variables within syndromes shows that larger primate species tend to have larger home ranges and smaller D-indices, while other effects are less prominent.

Conclusions

We have shown in this chapter that correlations exist among features that are knowable in the fossil record, such as body mass or diet, and features that are unknowable, such as foraging group size and home range size. However, these unknowable features cannot be reconstructed with the accuracy of the knowable features. Thus, our conclusions have turned out to be more qualitative than quantitative (Table XIII).

The goal of reconstructing behavior in the fossil record imposes serious constraints on the comparative patterns documented here in that our predictor variables are limited to characters that are knowable in the fossil record. Thus, in cases where multiple variables might be needed to explain variation in some other character, we are limited to using only the knowable variables in our multivariate analyses. Had our goal been to redo the analyses of Clutton-Brock and Harvey (1977a,b), our ability to explain variation in Y variables would doubtless have been greater. To see why, again consider the variables that influence home range size. The home range size of a gregarious animal is a function of the energetic needs of the group in which it finds itself during foraging, which is a function of the composition (number of individuals of each class) of its foraging group and the masses of each of the age-sex classes in these groups. When foraging group size and female body mass are examined in simple regression analyses, foraging group size accounts for 43.0% of the variation in home range size, while female body mass accounts for 45.1% of the variation in home range size.* However, if multiple regression is used, with both foraging group size and female body mass as the independent variables, 57.6% of the variation in home range size is accounted for. This is still not the ideal analysis, which would regress home range size on foraging group mass (calculated using group composition and corresponding body masses; see above) while controlling for diet, habitat, and substrate. More complicated analyses are therefore expected to account for even more of the variation in home range size and other variables.

Improved socioecological reconstructions may be possible when a more complete model of primate socioecology is developed. A truly strategic model (*sensu* Tooby and DeVore, 1987) would link behavioral ecology and social systems of primates to external variables (defined relative to the animal), such as predation risk and the abundance and distribution of food, and internal variables, such as life history features and reproductive biology (and hence the potential for sexual coercion; Smuts and Smuts 1993). In addition, this model must incorporate the effects of "social rules" (e.g., the collective action problem; Hawkes 1992; Nunn, 2000) and irreversible features that act as phylogenetic constraints (Antonovics and van Tienderen, 1991). Such a strategic model is still a distant goal (Wrangham, 1987; van Schaik, 1996), but should be our priority if we hope to achieve dramatic improvements in the resolution of reconstructions.

Finally, reconstructions will always be uncertain, no matter how sophisticated the methods, because their accuracy cannot be checked directly. However, if multiple independent lines of evidence converge on similar conclusions, our confidence in the reconstructions will increase. Hence, a diversity of approaches should be used. For example, some improvement can be obtained by combining our approach with attempts to reconstruct character evolution using phylogenetic techniques (Di Fiore and Rendall 1994; Garland *et al.*, 1997).

ACKNOWLEDGMENTS

Steve Heulett assisted in searching the primary literature to build the database. Steve also provided helpful discussion and comments, as did Cliff Cunningham, Rich Kay, Michael Plavcan, Holger Preuschoft, Marcelo Sanchez, Karen Strier, Maria van Noordwijk, Blythe Williams, and three anonymous reviewers. CLN was supported by an NSF Graduate Student Fellowship and NSF Dissertation Improvement Grant (#SBR-9711806), and CvS by a Forschungspreis of the Alexander von Humboldt Foundation.

* Based on r^2 from least-squares regression. Phylogeny is not controlled for here, as we are simply determining how much of the variation can be accounted for by different variables among the same set of species.

Appendix

A. Rules for Assigning Character States to Species

In compiling our information, we started with solid critical reviews, including those in Smuts *et al.* (1987), Gautier-Hion *et al.* (1988), Rylands (1993), and Davies and Oates (1994). Rowe (1996) was used to identify additional references that were not picked up by the earlier reviews. For review material, primary sources were consulted to confirm questionable or conflicting values. In addition, we systematically searched five primate-related journals from 1984 to 1995 (*American Journal of Primatology*, *International Journal of Primatology*, *American Journal of Physical Anthropology*, *Primates*, and *Folia Primatologica*).

Knowable "K" Variables

Body Mass

It is often possible to reconstruct body size of extinct species using associations between body size and dentition (Gingerich *et al.*, 1982) or postcranial remains (Dagosto and Terranova, 1992) in extant taxa. Body size is perhaps the most important ecological variable because it affects virtually all aspects of a species' physiology (Schmidt-Nielson, 1984), social behavior (Clutton-Brock and Harvey, 1984), and life history (Harvey *et al.*, 1989). In this study, we used female body mass because females, as the primary care-givers to infants, are more dependent on access to resources and avoidance of predation for their reproductive success, making females the "ecological" sex (see Emlen and Oring, 1977; Wrangham, 1979). We used several references in compiling our body mass data. First, we took body mass information from an early version of the information provided in Plavcan and van Schaik (1997). In general, we used the same data as these authors. If the information was not available in Plavcan and van Schaik (1997), we used Kappeler and Heymann (1996). For the remainder of the species, we used Rowe (1996) to identify sources for body mass and, where these sources were based on free-ranging animals in natural environments, we used this information (only 6 cases). From these 3 sources, a total of 172 species could be assigned a body mass estimate.

Activity Pattern

Among the extant species, nocturnal, diurnal, and cathemeral (Tattersall, 1987) activity periods are recognized. However, there is uncertainty over the status of cathemerality and its adaptive function (van Schaik and Kappeler, 1996), so that reconstructed socioecological variables may be misleading. Nonetheless, because cathemeral species (e.g., *Eulemur fulvus*) are behaviorally closer to diurnal species, cathemeral animals are classified here as diurnal, and we retain a dichotomous diurnal–nocturnal character assignment.

Locomotor Substrate

Terrestrial locomotion can have important consequences for socioecological variables, such as group size, perhaps through its related effect on predation risk. Even though many species use both the ground and the vegetation as locomotor substrate to some extent, we have adopted a simple two-way classification, terrestrial vs. aboreal, on the basis of the habitual substrate for travel in undisturbed habitats. As an operational definition of habitual terrestrial travel, species that spend approximately 20% or more of their time on the ground were classified as terrestrial. Thus, *Macaca fascicularis*, while spending some time on the ground, does not habitually travel long distances on the ground and is therefore classified as arboreal, while *Macaca nemestrina* is classified as terrestrial even though it spends much time in the trees (Crockett and Wilson, 1980).

Habitat

The comparison between terrestrially and arboreally traveling species confounds the effects of locomotor substrate and that of habitat, in that terrestrial species are often considered to occupy a more open habitat (e.g., Clutton-Brock and Harvey, 1977a). Thus, a characterization of habitat (open versus wooded) was included in our database, as culled from sources on general ecology of different field sites. A habitat is considered closed whenever there is a significant presence of vertical structures (i.e., woody plants) which would allow for arboreal locomotion and foraging; thus, forests of all kinds (woodland, savanna-woodland and closed thornscrub) are considered wooded habitats. This does not mean, however, that some arboreal foraging does not occur in open habitats (e.g., Kummer, 1968). Intraspecific variation in habitat use is taken into account in *Pan troglodytes* and *Papio anubis*, each of which is assigned to two syndromes.

Diet

Our characterizations of diet are categorical rather than continuous, partly because we lack the ability to make fine-grained, continuous estimates of diet in the fossil record for all fossilized material, and partly to reduce the number of categories used in order to maintain a clear link with morphology. In addition, morphological correlates used to infer percentages of dietary categories may have differed in the past (Fleagle, 1988), making quantitative inferences especially dubious. Our categorization uses four classes: fauna (mainly insects but also including vertebrates), reproductive parts (mainly fruits but also including flowers and seeds), vegetative parts (leaves and shoots), and exudates (gum). Diet for each species was characterized as a single category, namely, the one that had the greatest proportion of feeding time devoted to it. Where there was intraspecific variation in percentage time eating different dietary items, the raw data of different studies were averaged to obtain a species average, which was then used to classify the species into one of the four categories. Diet is often variable within species in our database, although this variation was rarely sufficient to change the classification of a species from one dietary category to another, and when it was, there was usually insufficient information available on U variables to make including this intraspecific variation worthwhile.

Unknown "U" Variables

Group Size

In most cases, primates form clearly defined social groups that are relatively stable over time. In some species, however, one can discern multiple levels of grouping. First, animals may form temporary parties or subgroups during foraging (e.g., *Macaca fascicularis*: van Schaik and van Noordwijk, 1988), sometimes even to the extent that one rarely if ever finds all members of one "group" or community together (e.g., *Pan troglodytes*: Wrangham, 1977). Second, groups may come together to sleep in larger units (e.g., *Rhinopithecus brelichi*: Bleisch *et al.*, 1993; *Papio hamadryas*: Kummer, 1968). Following Clutton-Brock and Harvey (1977a), we have distinguished between two levels of grouping: the *foraging group*, being the animals that forage together as a unit, and the community or *population group*, being the animals that share a common range or at least come together frequently, usually to sleep together, and among which foraging units have highly overlapping ranges. For most of the analyses, the foraging group size and population group size are identical.

As noted in the Methods, we occasionally used the midpoint of a range in estimating U variables. Means calculated from ranges may be systematically underestimated if solitary, perhaps dispersing individuals are included as the lower group size in a primary reference. We therefore avoided calculating means in this way for species when the lower range was given as 1 in the primary or secondary literature.

Home Range Size and Day Journey Length

Home range sizes and day journey lengths were taken from the published literature, and intraspecific variation was averaged using the procedures discussed in the text. Attempts were made to favor estimates of home range and day journey length based on systematic, long-term observations of stable social groups over the entire yearly cycle, rather than brief surveys.

Range Use (D-Index)

We measured the intensity of home range use with Mitani and Rodman's (1979) D-index. This measure was calculated for study groups in which both home range size and day range length were available. When more than a single D-index could be calculated for a species, these values were averaged, and this mean value used in the analyses. Following the procedures of Mitani and Rodman (1979), the D-index is calculated using the formula $D = d/(4A/p)^{0.5}$, where d = day journey length (in km) and A = the home range size (in km^2). Although D will be overstated when individuals recross paths, it captures the potential for range use, and thus for territoriality.

Appendix B

Knowable "K" Variables (Syndromes)[a]

Species	Radiation[b]	Female body size	Activity period	Substrate	Habitat	Diet
Lemur catta	LM	2.68	D	T	W	FR
Eulemur coronatus	LM	1.69	D	A	W	FR
Eulemur fulvus	LM	2.4	D	A	W	FR
Eulemur macaco	LM	2.49	D	A	W	FR
Eulemur mongoz	LM	1.66	D	A	W	FR
Eulemur rubriventer	LM	2.14	D	A	W	FR
Hapalemur aureus	LM	1.5	D	A	W	FO
Hapelemur griseus	LM	0.89	D	A	W	FO
Hapelemur simus	LM	1.3	D	A	W	FO
Varecia variegata	LM	3.51	D	A	W	FR
Lepilemur mustelinus	LM	0.66	N	A	W	FO
Microcebus murinus	LM	0.06	N	A	W	FR
Microcebus rufus	LM	0.05	N	A	W	FR
Microcebus coquereli	LM	0.26	N	A	W	FR
Cheirogaleus medius	LM	0.28	N	A	W	FR
Cheirogaleus major	LM	0.44	N	A	W	FR
Allocebus trichotis	LM	0.09	N	A	W	I
Phaner furcifer	LM	0.33	N	A	W	G
Avahi laniger	LM	1.32	N	A	W	FO
Propithecus diadema	LM	5.9	D	A	W	FO
Propithecus tattersalli	LM	3.17	D	A	W	FR
Propithecus verreauxi	LM	3.7	D	A	W	FR
Indri indri	LM	6.25	D	A	W	FO
Daubentonia madagascarensis	LM	2.52	N	A	W	I
Loris tardigradis	LO	0.19	N	A	W	I
Nycticebus coucang	LO	1.2	N	A	W	FR
Nycticebus pygmaeus	LO	0.38	N	A	W	
Arctocebus calabarensis	LO	0.3	N	A	W	I

(Cont.)

Species	Radiation[b]	Female body size	Activity period	Substrate	Habitat	Diet
Perodicticus potto	LO	0.99	N	A	W	FR
Galagoides alleni	LO	0.3	N	A	W	FR
Galagoides demidoff	LO	0.07	N	A	W	I
Galagoides zanaibaricus	LO	0.14	N	A	W	I
Otolemur crassicaudatus	LO	1.24	N	A	W	G
Otolemur garnetti	LO	1.03	N	A	W	I
Euoticus elegantulus	LO	0.29	N	A	W	G
Euoticus inustus	LO	0.21	N	A	W	I
Galago granti	LO		N	A	W	
Galago moholi	LO	0.16	N	A	W	I
Galago senegalensis	LO	0.21	N	A	W	I
Tarsius bancanus	TR	0.13	N	A	W	I
Tarsius syrichta	TR	0.12	N	A	W	I
Tarsius pumilus	TR		N	A	W	I
Tarsius spectrum	TR	0.2	N	A	W	I
Callithrix argentata	NW	0.36	D	A	W	FR
Callithrix humeralifer	NW	0.38	D	A	W	FR
Callithrix jacchus	NW	0.24	D	A	W	G
Cebuella pygmaea	NW	0.12	D	A	W	G
Saguinus bicolor	NW	0.43	D	A	W	FR
Saguinus fuscicollis	NW	0.41	D	A	W	FR
Saguinus imperator	NW	0.45	D	A	W	FR
Saguinus inustus	NW	0.8	D	A	W	
Saguinus labiatus	NW	0.52	D	A	W	FR
Saguinus leucopus	NW	0.49	D	A	W	FR
Saguinus midas	NW	0.52	D	A	W	FR
Saguinus mystax	NW	0.58	D	A	W	FR
Saguinus nigricollis	NW	0.46	D	A	W	FR
Saguinus oedipus	NW	0.5	D	A	W	FR
Saguinus tripartitus	NW		D	A	W	FR
Leontopithecus chrysomelas	NW	0.53	D	A	W	FR
Leontopithecus chrysopygus	NW		D	A	W	FR
Leontopithecus rosalia	NW	0.48	D	A	W	FR
Callimico goeldi	NW	0.58	D	A	W	FR
Cebus albifrons	NW	1.81	D	A	W	FR
Cebus apella	NW	2.45	D	A	W	FR
Cebus capucinus	NW	2.28	D	A	W	FR
Cebus olivaceous	NW	1.59	D	A	W	FR
Aotus trivirgatus	NW	0.92	N	A	W	FR
Aotus azarae	NW	0.78	N	A	W	FR
Callicebus brunneus	NW	0.85	D	A	W	FR
Callicebus personatus	NW	1.38	D	A	W	FR
Callicebus moloch	NW	0.99	D	A	W	FR
Callicebus torquatus	NW	1.31	D	A	W	FR
Saimiri oestedi	NW	0.6	D	A	W	FR
Saimiri sciureus	NW	0.79	D	A	W	I

(Cont.)

Species	Radiation[b]	Female body size	Activity period	Substrate	Habitat	Diet
Pithecia albicans	NW		D	A	W	FR
Pithecia hirsuta	NW		D	A	W	FR
Pithecia monachus	NW	2.17	D	A	W	FR
Pithecia pithecia	NW	1.51	D	A	W	FR
Cacajao calvus	NW	2.88	D	A	W	FR
Cacajao melanocephalus	NW	2.74	D	A	W	FR
Cacajao rubicundus	NW		D	A	W	FR
Chiropotes albinasus	NW	2.52	D	A	W	FR
Chiropotes satanas	NW	2.66	D	A	W	FR
Alouatta belzebul	NW	5.18	D	A	W	
Alouatta caraya	NW	5.41	D	A	W	FO
Alouatta fusca	NW	4.53	D	A	W	FO
Alouatta palliata	NW	5.6	D	A	W	FO
Alouatta pigra	NW	6.29	D	A	W	FO
Alouatta seniculus	NW	6.02	D	A	W	FO
Ateles belzebuth	NW	8.47	D	A	W	FR
Ateles fusciceps	NW	8.8	D	A	W	FR
Ateles geoffroyi	NW	6.7	D	A	W	FR
Ateles paniscus	NW	8.59	D	A	W	FR
Brachyteles arachnoides	NW	8.38	D	A	W	FO
Lathothrix flavicauda	NW		D	A	W	FR
Lathothrix lagothricha	NW	5.54	D	A	W	FR
Macaca arctoides	OW	8.4	D	T	W	FR
Macaca assamensis	OW	6.7	D	A	W	FR
Macaca cyclopis	OW	4.95	D	T	W	FR
Macaca fascicularis	OW	3.12	D	A	W	FR
Macaca fuscata	OW	9.1	D	T	W	FR
Macaca maurus	OW	5.1	D	A	W	FR
Macaca mulatta	OW	5.37	D	T	W	FR
Macaca nemestrina	OW	6.35	D	A	W	FR
Macaca nigra	OW	4.69	D	T	W	FR
Macaca ochreata	OW		D	A	W	FR
Macaca radiata	OW	3.85	D	A	W	FR
Macaca silenus	OW	5	D	A	W	FR
Macaca sinica	OW	3.4	D	A	W	FR
Macaca sylvanus	OW	10	D	T	W	FR
Macaca thibetana	OW	10.1	D	T	W	FO
Macaca tonkeana	OW	8.64	D	T	W	
Cercocebus albigena	OW	6.4	D	A	W	FR
Cercocebus atterimus	OW	5.64	D	A	W	FR
Cercocebus galeritus	OW	5.47	D	T	W	FR
Cercocebus torquatus	OW	6.33	D	T	W	FR
Papio anubis (o)	OW	14.1	D	T	O	FO
Papio anubis (w)	OW	14.1	D	T	W	FR
Papio cynocephalus	OW	13.61	D	T	O	FR
Papio hamadryas	OW	9.98	D	T	O	FO
Papio papio	OW	13	D	T	W	FR
Papio ursinus	OW	14.52	D	T	O	FR
Mandrillus leucophaeus	OW	10	D	T	W	FR

Cont.)

Species	Radiation[b]	Female body size	Activity period	Substrate	Habitat	Diet
Mandrillus sphinx	OW	11.5	D	T	W	FR
Theropithecus gelada	OW	11.7	D	T	O	FO
Cercopithecus aethiops	OW	3.26	D	T	O	FR
Cercopithecus ascanius	OW	3.3	D	A	W	FR
Cercopithecus campbelli	OW	2.7	D	T	W	FR
Cercopithecus cephus	OW	2.88	D	A	W	FR
Cercopithecus denti	OW		D	A	W	
Cercopithecus diana	OW	3.9	D	A	W	FR
Cercopithecus dryas	OW	2.25	D		W	
Cercopithecus erythogaster	OW	2.4	D	A	W	FR
Cercopithecus erythrotis	OW		D	A	W	FR
Cercopithecus hamlyni	OW	3.36	D	A	W	FR
Cercopithecus lhoesti	OW	4.7	D	T	W	FR
Cercopithecus mitis	OW	4.23	D	A	W	FR
Cercopithecus mona	OW	2.5	D	A	W	FR
Cercopithecus neglectus	OW	3.96	D	T	W	FR
Cercopithecus nictitans	OW	4.22	D	A	W	FR
Cercopithecus petaurista	OW	2.9	D	A	W	FR
Cercopithecus pogonias	OW	3.02	D	A	W	FR
Cercopithecus preussi	OW		D	T	W	FR
Cercopithecus solatus	OW	3.5	D	T	W	
Cercopithecus wolfi	OW	2.76	D	A	W	FR
Miopithecus talapoin	OW	1.12	D	A	W	FR
Allenopithecus nirgoviridis	OW	3.25	D	T	W	FR
Erythrocebus patas	OW	5.9	D	T	O	FR
Presbytis aurata	OW		D	A	W	FO
Presbytis comata	OW	6.66	D	A	W	FO
Presbytis cristata	OW	5.7	D	A	W	FO
Presbytis entellus	OW	11.05	D	T	W	FO
Presbytis francoisi	OW		D	A	W	FO
Presbytis frontata	OW	5.66	D	A	W	
Presbytis geei	OW	9.5	D	A	W	FR
Presbytis johnii	OW	10.9	D	A	W	FO
Presbytis melalophos	OW	6.32	D	A	W	FR
Presbytis obscura	OW	6.47	D	A	W	FO
Presbytis phayrei	OW	6.9	D	A	W	FO
Presbytis pileatus	OW	5.14	D	A	W	FO
Presbytis potenziani	OW	6.4	D	A	W	FO
Presbytis rubicunda	OW	5.7	D	A	W	FO
Presbytis senex	OW	5.9	D	A	W	FO
Pygathrix avunculus	OW	8.5	D	A	W	FR
Pygathrix brelichi	OW		D	A	W	FO
Pygathrix roxellanae	OW	10.3	D	T	W	FO

(Cont.)

Species	Radiation[b]	Female body size	Activity period	Substrate	Habitat	Diet
Pygathrix nemaeus	OW	4.1	D	A	W	FO
Nasalis larvatus	OW	9.87	D	A	W	FO
Simias concolor	OW	7.1	D	A	W	FO
Colobus angolensis	OW	7.4	D	A	W	FO
Colobus badius	OW	6.77	D	A	W	FO
Colobus guereza	OW	7.83	D	A	W	FO
Colobus kirkii	OW		D	A	W	FO
Colobus polykomos	OW	7.13	D	A	W	FO
Colobus satanas	OW	9.5	D	A	W	FR
Procolobus verus	OW	4.1	D	A	W	FO
Hylobates agillis	AP	5.7	D	A	W	FR
Hylobates concolor	AP	5.8	D	A	W	FO
Hylobates hoolock	AP	6.5	D	A	W	FR
Hylobates klossii	AP	5.9	D	A	W	FR
Hylobates lar	AP	5.3	D	A	W	FR
Hylobates moloch	AP	5.7	D	A	W	FR
Hylobates muelleri	AP		D	A	W	FR
Hylobates pileatus	AP		D	A	W	FR
Hylobates syndactylus	AP	10.6	D	A	W	FO
Pongo pygmaeus	AP	38.5	D	A	W	FR
Pan troglodytes (o)	AP	33.2	D	T	O	FR
Pan troglodytes (w)	AP	33.2	D	T	W	FR
Pan paniscus	AP	33.2	D	T	W	FR
Glorilla gorilla	AP	97.7	D	T	W	FO

[a]Values log-transformed for the analyses. Blank cells indicate that data were not available.
[b]LM, lemurs; LO, lorisids; NW, New World monkeys; OW, Old World monkeys; AP, apes.

Unknown "U" Variables[a]

Species	Population group size	Foraging group size	Home range size	Day journey length	D-index
Lemur catta	15.6	15.6	15.2	956.8	3.1074
Eulemur coronatus	5.0	2			
Eulemur fulvus	8.5	8.5	47.9	549.8	1.0909
Eulemur macaco	9.9	9.9	5.3		
Eulemur mongoz	3.0	3	1.2	610	4.9350
Eulemur rubriventer	2.9	2.9	16.3	463.7	1.0185
Hapalemur aureus	3.5	3.5	80		
Hapelemur griseus	4.5	4.5	7.7	425	1.3573
Hapelemur simus	7.5	7.5	100		
Varecia variegata	5.4	3.1	110.8	718	0.7159
Lepilemur mustelinus	1.2	1.2	0.4		
Microcebus murinus	2.5	1	1.8		
Microcebus rufus	2.5	1			
Microcebus coquereli	2.0	1	2.8		

(Cont.)

Species	Population group size	Foraging group size	Home range size	Day journey length	D-index
Cheirogaleuse medius	3.0	1	4		
Cheirogaleuse major	1.0	1			
Allocebus trichotis	4.0	1			
Phaner furcifer	2.5	1			
Avahi laniger	2.5	2.5	2	457	3.4229
Propithecus diadema	5.0	5		670	
Propithecus tattersalli	6.3	6.3			
Propithecus verreauxi	6.1	6.1	4.7	675	2.7667
Indri indri	4.3	4.3	17.9	600	1.2568
Daubentonia madagascarensis	1.5	1	103	1640.8	1.4330
Loris tardigradis	3.0	1			
Nycticebus courcang		1			
Nycticebus pygmaeus		1			
Arctocebus calabarensis	1.5	1			
Perodicticus potto	1.0	1	16		
Galagoides alleni	2.5	1	26		
Galagoides demidoff	3.5	1	1.3		
Galagoides zanzibaricus	3.5	1	2.3	1738.8	10.1606
Otolemur crassicaudatus	3.5	1	8.5	1250	3.7997
Otolemur garnetti	2.0	1	14.6	2000	4.6387
Euoticus elegantulus	4.0	1			
Euoticus inustus		1			
Galago granti		1			
Galago moholi	2.8	1	11.1		
Galago senegalensis	3.5	1	6.8	1250	4.2443
Tarsius bancanus	2.0	1	4.1	1800	5.4715
Tarsius syrichta	3.0	1			
Tarsius pumilus		1			
Tarsius spectrum	3.0	1	1		
Callithrix argentata	8.9	8.9			
Callithrix humeralifer	11.0	11	19.1	1485.3	2.6235
Callithrix jacchus	8.2	8.2	8.6	757	2.3681
Cebuella pygmaea	7.9	7.9	0.9	290	2.6650
Saguinus bicolor	6.5	6.5	12		
Saguinus fuscicollis	6.8	6.8	34.3	1612	2.6303
Saguinus imperator	3.3	3.3	50	1420	2.5169
Saguinus inustus					
Saguinus labiatus	5.6	5.6	26	1487	3.0101
Saguinus leucopus	8.5	8.5			
Saguinus midas	4.7	4.7			
Saguinus mystax	6.3	6.3	30.9	1903	2.4564
Saguinus nigricollis	6.2	6.2	45	1000	1.3841
Saguinus oedipus	5.3	5.3	15.7	1893	4.2726
Saguinus tripartitus					
Leontopithecus chrysomelas	6.6	6.6	36	1796	2.6528
Leontopithecus chrysopygus	3.5	3.5			
Leontopithecus rosalia	4.5	4.5		1496	
Callimico goeldii	7.4	7.4	45	2000	2.6422

(Cont.)

Species	Population group size	Foraging group size	Home range size	Day journey length	D-index
Cebus albifrons	19.2	19.2	132.5	1800	1.3858
Cebus apella	14.0	14	344	2110	1.2769
Cebus capucinus	15.4	15.4	86.3	2000	2.5066
Cebus olivaceous	18.8	18.8	257	2100	1.1609
Aotus trivirgatus	3.3	3.3	8.3	663.5	2.2610
Aotus azarae	3.7	3.7		337.4	
Callicebus brunneus	3.0	3			
Callicebus personatus	3.5	3.5	4.7	695	2.8411
Callicebus moloch	3.3	3.3	4	642.5	2.0993
Callicebus torquatus	3.7	3.7	17.4	819.5	1.7129
Saimiri torquatus	32.3	32.3	47.8	3350	6.1726
Saimiri sciureus	45.0	45	128.7	1755	2.1424
Pithecia albicans	4.6	4.6	172.4		
Pithecia hirsuta	4.5	4.5			
Pithecia monachus	3.5	3.5			
Pithecia pithecia	2.6	2.6	7		
Cacajao calvus	26.5	26.5	550		
Cacajao melanocephalus	15.8	15.8			
Cacajao rubicundus					
Chiropotes albinasus	25.0	25	300	3750	1.9187
Chiropotes satanas	15.0	15	287.5	3200	1.3307
Alouatta belzebul	5.0	5			
Alouatta caraya	7.9	7.9			
Alouatta fusca	6.0	6	6	495	1.8428
Alouatta palliata	12.0	12	38.9	319.9	0.7829
Alouatta pigra	5.2	5.2	71.3	250	0.1982
Alouatta seniculus	6.8	6.8	12.7	586.6	1.2096
Ateles belzebuth	20.8	4.1	324	2250	1.1078
Ateles fusciceps					
Ateles geoffroyi	22.7	3.9	128.7	1675	1.3086
Ateles paniscus	20.3	3.1	206	2356.8	1.4538
Brachyteles arachnoides	22.5	12.3	193.6	894.3	0.6972
Lathothrix flavicauda		9.1			
Lathothrix lagothricha	28.0	12.1	503.8	1539.5	0.7798
Macaca arctoides	20.0	20		1700	
Macaca assamensis	19.8	19.8			
Macaca cyclopis	28.5	28.5	109.8		
Macaca fascicularis	26.0	26	75	1700	1.7762
Macaca fuscata	56.2	56.2	133.1	1505	1.1560
Macaca maurus	22.0	22			
Macaca mulatta	36.8	36.8	319.7	1428	0.3268
Macaca nemestrina	38.5	15	300	1500	0.6647
Macaca nigra	20.5	20.5	217		
Macaca ochreata					
Macaca radiata	26.5	26.5	127.5	846.7	0.9008
Macaca silenus	16.6	16.6	131	1625	1.2582
Macaca sinica	25.2	25.2	51.8		
Macaca sylvanus	24.5	24.5	288.5		
Macaca thibetana	31.7	31.7	300		

(Cont.)

Species	Population group size	Foraging group size	Home range size	Day journey length	D-index
Macaca tonkeana					
Cercocebus albigena	15.5	15.5	272.8	1193.1	0.5476
Cercocebus atterimus	13.1	10.3	59		
Cercocebus galeritus	18.3	18.3	117.5	1174	0.9598
Cercocebus torquatus	26.2	26.2	247.4		
Papio anubis (o)	69	69	1260.5	5400	0.9988
Papio anubis (w)	58.9	58.9			
Papio cynocephalus	60.1	60.1	4564	5900	1.0655
Papio hamadryas	37.8	37.8	2800	10666.7	1.7083
Papio papio	65.3	65.3	1295		
Papio ursinus	49.7	49.6	1347.5	5585	1.1346
Mandrillus leucophaeus	96.5	96.5	4500		
Mandrillus sphinx	84.0	84	2216.7	3500	1.3872
Theropithecus gelada	156.9	38.1	212	1280	0.9054
Cercopithecus aethicops	25.6	25.6	68.8	1372.5	1.3338
Cercopithecus ascanius	22.0	22	27.8	1501.9	2.2755
Cercopithecus campbelli	15.8	15.8	40		
Cercopithecus cephus	11.0	11	43.5	1196	1.7799
Cercopithecus denti			37	1217	1.7731
Cercopithecus diana	19.8	19.8	107.6	1892	1.2196
Cercopithecus dryas					
Cercopithecus erythogaster	17.5	5			
Cercopithecus erythrotis	24.5	24.5			
Cercopithecus hamlyni	10.0	10			
Cercopithecus lhoesti	10.9	10.9	850		
Cercopithecus mitis	17.1	17.1	86.5	1053.7	1.3279
Cercopithecus mona	11.4	11.4			
Cercopithecus neglectus	6.7	6.7	9.8	530	1.8283
Cercopithecus nictitans	16.3	16.3	123.8	1684.3	1.3930
Cercopithecus petaurista	14.0	14	41		
Cercopithecus pogonias	17.4	17.4	125.8	1751.5	1.4446
Cercopithecus preussi	5.0	5			
Cercopithecus solatus	10.0	10			
Cercopithecus wolfi	11.1	11.1			
Miopithecus talapoin	92.6	92.6	122.3	2311.5	1.8623
Allenopithecus nirgoviridis	40.0	40			
Erythrocebus patas	29.7	29.7	4092.5	3290	0.4774
Presbytis aurata	9.0	9	6.4		
Presbytis comata	8.5	8.5	36.3	595.5	0.9376
Presbytis cristata	20.4	20.4	14.4	440	1.1385
Presbytis entellus	28.8	28.8	48.4	677.3	0.8592
Presbytis francoisi	17.1	17.1		2000	
Presbytis frontata					
Presbytis geei	12.9	12.9	375		
Presbytis johnii	9.8	9.8	161.6	500	0.3485
Presbytis melalophos	13.1	13.1	23	851.8	1.5141
Presbytis obscura	13.7	13.7	33	754.5	1.1648

(Cont.)

Species	Population group size	Foraging group size	Home range size	Day journey length	D-index
Presbytis phayrei	10.1	10.1	75		
Presbytis pileatus	10.5	10.5	21		
Presbytis potenziani	3.4	3.4	15.9		
Presbytis rubicunda	8.0	8	64.5	819	0.9669
Presbytis senex	8.6	8.6	6		
Pygathrix avunculus	25.0	25			
Pygathrix brelichi	522.0	32.65	1000		
Pygathrix roxellanae	132.3	14.25	2750	1500	0.2535
Pygathrix nemaeus	7.5	7.5			
Nasalis larvatus	10.8	10.8	433.4	1150	0.3397
Simias concolor	4.3	4.3	14.4		
Colobus angolensis	8.8	8.8	400	1250	0.5539
Colobus badius	30.2	30.2	53.3	538	1.5636
Colobus guereza	9.9	9.9	9.5	554.5	2.0250
Colobus kirkii					
Colobus polykomos	11.0	11	24	832	1.5051
Colobus satanas	10.5	10.5	72	484.5	0.5091
Procolobus verus	8.3	8.3	27.8		
Hylobates agilis	4.1	4.1	28	1276	2.1398
Hylobates concolor	5.2	5.2	87.5		
Hylobates hoolock	3.3	3.3	81.7	1100	1.0785
Hylobates klossii	3.9	3.9	16.3	1514	2.3357
Hylobates lar	3.8	3.8	40.8	1569.4	1.9735
Hylobates moloch			17	1400	3.0092
Hylobates muelleri	3.4	3.4	36	1030	1.2555
Hylobates pileatus	3.7	3.7	36	833	1.2304
Hylobates syndactylus	4.0	4	31.8	946.2	1.3624
Pongo pygmaeus	1.5	1.5	203.3	598.8	0.3616
Pan troglodytes (o)	47.6	5.6	1767.3	3650	0.9161
Pan troglodytes (w)		5.8			
Pan paniscus	43.4	7.6	3475	2400	0.2793
Gorilla gorilla	15.8	15.8	2544.5	860.8	0.1647

[a]Values log-transformed for the analyses. Blank cells indicate that data were not available.

References

Aiello, L. C. and Wheeler, P. 1995. The expensive tissue hypothesis. *Curr. Anthropol.* **36:**199–221.

Andelman, S. J. 1986. Ecological and social determinants of cercopithecine mating patterns. In: D. I. Rubenstein and R. W. Wrangham (eds.), *Ecological Aspects of Social Evolution: Birds and Mammals*, pp. 201–216. Princeton University Press, Princeton, N.J.

Andrews, P. 1996. Palaeoecology and hominoid palaeoenvironments. *Biol. Rev.* **71:**257–300.

Andrews, P., and van Couvering, J. A. H. 1975. Palaeoenvironments in the East African Miocene. In: F. Szalay (ed.), *Contributions to Primatology*. pp. 62–103. Karger, Basel.

Antonovics, J., and van Tienderen, P. H. 1991. Ontoecogenophyloconstraints? The chaos of constraint terminology. *Trends Ecol. Evol.* **6:**166–167.

Barton, R. A., and Purvis, A. 1994. Primate brains and ecology: looking beneath the surface. In: B. Thierry, J. R. Anderson, J. J. Roeder, and N. Herrenschmidt (eds.), *Current Primatology*, pp. 1–9. Universite Louis Pasteur, Strasbourg.

Barton, R. A., Purvis, A., and Harvey, P. H. 1995. Evolutionary radiation of visual and olfactory brain systems in primates, bats and insectivores. *Philos. Trans. R. Soc. London, Ser. B* **348:**381–392.

Bleisch, W., Cheng Ao-Song, Ren Xiao-Dong, and Xie Jia-Hua 1993. Preliminary results from a field study of wild guizhou snub-nosed monkeys (*Rhinopithecus brelichi*). *Folia Primatol.* **60:**72–82.

Cartmill, M. 1980. Morphology, function, and evolution of the anthropoid postorbital septum. In: R. L. Ciochon and A. B. Chiarelli (eds.), *Evolutionary Biology of New World Monkeys and Continental Drift*, pp. 243–274. Plenum Press, New York.

Chapman, C. A., Walker, S., and Lefebvre, L. 1990. Reproductive strategies of primates: the influence of body size and diet on litter size. *Primates* **31:**1–13.

Charnov, E. L. 1993. *Life History Invariants*. Oxford University Press, Oxford.

Cheney, D. L., and Wrangham, R. W. 1987. Predation. In: B. B. Smuts, D. L. Cheney, R. M. Seyfarth, R. W. Wrangham, and T. T. Struhsaker (eds.), *Primate Societies*, pp. 227–239. Chicago University Press, Chicago.

Clutton-Brock, T. H., and Harvey, P. H., 1977a. Primate ecology and social organization. *J. Zool.* **183:**1–39.

Clutton-Brock, T. H., and Harvey, P. H. 1977b. Species differences in feeding and ranging behaviour in primates. In: T. H. Clutton-Brock (ed.), *Primate Ecology*, pp. 557–579. Academic Press, London.

Clutton-Brock, T. H., and Harvey, P. H. 1980. Primates, brains and ecology. *J Zool.* **190:**309–323.

Clutton-Brock, T. H., and Harvey, P. H. 1984. Comparative approaches to investigating adaptation. In: J. R. Krebs and N. B. Davies (eds.), *Behavioural Ecology*, pp. 7–29. Blackwell, Oxford.

Cohen, J. 1988. *Statistical Power Analysis for the Behavioral Sciences*. Erlbaum Associates, Hillsdale, NJ.

Corbet, G. B., and Hill, J. E. 1991. *A World List of Mammalian Species*. Oxford University Press, Oxford.

Crockett, C. M., and Wilson, W. L. 1980. The ecological separation of *Macaca nemestrina* and *Macaca fascicularis* in Sumatra. In: D. G. Lindburg (ed.), *The Macaques: Studies in Ecology, Behavior and Evolution*, pp. 148–181. Van Nostrand Reinhold, New York.

Cunningham, C., Omland, K., and Oakley, T. 1998. Reconstructing ancestral character states. *Trends Ecol. Evol.* **13:**361–366.

Dagosto, M., and Terranova, C. J. 1992. Estimating the body size of Eocene primates: a comparison of results from dental and postcranial variables. *Int. J. Primatol.* **13:**307–344.

Davies, A. G., and Oates, J. F. 1994. *Colobine Monkeys: Their Ecology, Behaviour, and Evolution*. Cambridge University Press, Cambridge.

Di Fiori, A., and Rendall, D. 1994. Evolution of social organization: a reappraisal for primates by using phylogenetic methods. *Proc. Natl. Acad. Sci. USA* **91:**9941–9945.

Dunbar, R. I. M. 1991. Functional significance of social grooming in primates. *Folia Primatol.* **57:**121–131.

Dunbar, R. I. M. 1992a. Behavioural ecology of the extinct papionines. *J. Hum. Evol.* **22:**407–421.

Dunbar, R. I. M. 1992b. Neocortex size as a constraint on group size in primates. *J. Hum. Evol.* **20:**469–493.

Dunbar, R. I. M. 1993. Socioecology of the extinct theropiths: a modelling approach. In:N. G. Jablonski (ed.), *Theropithecus: The Rise and Fall of a Primate Genus*, pp. 465–486. Cambridge University Press, Cambridge.

Emlen, S. T., and Oring, L. W. 1977. Ecology, sexual selection, and the evolution of mating systems. *Science* **197:**215–223.

Farris, J. S. 1989. The retention index and the rescaled consistency index. *Cladistics* **5:**417–419.

Felsenstein, J. 1985. Phylogenies and the comparative method. *Am. Nat.* **125:**1–15.

Fleagle, J. G. 1984. Primate locomotion and diet In: D. J. Chivers, B. A. Wood, and A. Bilsborough (eds.), *Food Acquisition and Processing in Primates*. Plenum Press, New York.

Fleagle, J. G. 1988. *Primate Adaptation and Evolution*. Academic Press, New York.

Fleagle, J. G., and Kay, 1985. The paleobiology of catarrhines. In: E. Delson (ed.), *Ancestors: The Hard Evidence*, pp. 23–36. A. R. Liss, New York.

Fleagle, J. G., and Reed, K. E. 1996. Comparing primate communities: a multivariate approach. *J. Hum. Evol.* **30:**489–510.

Frumhoff, P. C., and Reeve, H. K. 1994. Using phylogenies to test hypotheses of adaptation: a critique of some current proposals. *Evolution* **48:**172–180.

Garland, T., Jr., Dickerman, A. W., Janis, C. M., and Jones, J. A. 1993. Phylogenetic analysis of covariance by computer simulation. *Syst. Biol.* **42:**265–292.

Garland, T., Jr., Martin, K. L. M., and Díaz-Uriarte, R. 1997. Reconstructing ancestral trait values using squared-change parsimony: plasma osmolarity at the origin of amniotes. In: S. S. Sumida and K. L. M. Martin (eds.), *Amniote Origins*, pp. 425–501. Academic Press, San Diego, CA.

Gautier-Hion, A., Bourlière, F., Guatier, J.-P., and Kingdon, J. 1988. *A Primate Radiation: Evolutionary Biology of the African Guenons*. Cambridge University Press, Cambridge.

Gibson, K. R. 1986. Cognition, brain size and the extraction of embedded food resources. In: G. Else and P. C. Lee (eds.), *Primate Ontogeny, Cognitive and Social Behavior*, pp. 93–105. Cambridge University Press, Cambridge.

Gingerich, P. D., Smith, B. H., and Rosenberg, K. 1982. Allometric scaling in the dentition of primates and prediction of body weight from tooth size in fossils. *Am. J. Phys. Anthropol.* **58:**81–100.

Gittleman, J. L., and Luh, 1992. On comparing comparative methods. *Annu. Rev. Ecol. Syst.* **23:**383–404.

Gittleman, J. L., and Luh, H. K. 1994. Phylogeny, evolutionary models, and comparative methods: a simulation study. In: P. Eggleton and R. I. Vane-Wright (eds.), *Phylogenetics and Ecology*, pp. 103–122. Academic Press, London.

Harvey, P. H., and Pagel, M. D. 1991. *The Comparative Method in Evolutionary Biology*. Oxford University Press, Oxford.

Harvey, P. H., Read, A. F., and Promislow, D. E. L. 1989. Life history variation in placental mammals: unifying the data with theory. In: P. H. H. Partridge (ed.), *Oxford Surveys in Evolutionary Biology*, pp. 13–31. Oxford University Press, Oxford.

Hawkes, K. 1992. Sharing and collective action. In: E. A. Smith and B. Winterhalder (eds.), *Evolutionary Ecology and Human Behavior*, pp. 269–300. Aldine de Gruyter, New York.

Hladik, C. M. 1975. Ecology, diet, and social patterning in Old and New World primates. In: R. H. Tuttle (ed.), *Socioecology and Psychology of Primates*, pp. 3–35. Mouton, The Hague.

Janson, C. H., and Goldsmith, M. 1995. Predicting group size in primates: foraging costs and predation risk. *Behav. Ecol.* **6:**326–336.

Kappeler, P. M., and Heymann, E. W. 1996. Nonconvergence in the evolution of primate life history and socio-ecology. *Biol. J. Linn. Soc.* **59:**297–326.

Kay, R. F. 1984. On the use of anatomical features to infer foraging behavior in extinct primates. In: P. Rodman, and J. Cant (eds.), *Adaptations for Foraging in Nonhuman Primates*. Columbia University Press, New York.

Kay, R. F. 1994. "Giant" tamarin from the Miocene of Columbia. *Am. J. Phys. Anthropol.* **95:**333–353.

Kay, R. F., and Cartmill, M. 1977. Cranial morphology and adaptations of *Palaecthon nacimienti* and other Paromomyidae (Plesiadapoidea, ?Primates), with a description of a new genus and species. *J. Hum. Evol.* **6:**19–53.

Kay, R. F., and Williams, B. A. this volume. The adaptations of *Branisella boliviana*, the earliest South American monkey.

Kleiber, M. 1961. *The Fire of Life: An Introduction to Animal Energetics*. John Wiley, New York.

Kummer, H. 1968. *Social Organization of Hamadryas Baboons*. University of Chicago Press, Chicago.

Leigh, S. R. 1994. Ontogenetic correlates of diet in anthropoid primates. *Am. J. Phys. Anthropol.* **94:** 499–522.

Maddison, W. P., and Maddison, D. R. 1992. *MacClade. Analysis of Phylogeny and Character Evolution*. Sinauer Associates, Inc., Sunderland, MA.

Martins, E. P., and Garland, T., Jr. 1991. Phylogenetic analyses of the correlated evolution of continuous characters: a simulation study. *Evolution* **45:**534–557.

Martins, E. P., and Hansen, T. F. 1996. The statistical analysis of interspecific data: a review and evaluation of phylogenetic comparative methods. In: E. Martins (ed.), *Phylogenies and the Comparative Method in Animal Behavior*, pp. 22–75. Oxford University Press, New York.

McNab, B. K. 1963. Bioenergetics and the determination of home range size. *Am. Nat.* **97:**133–140.

Milton, K., and May, M. L. 1976. Body weight, diet and home range area in primates. *Nature* **259:**459–462.

Mitani, J. C., and Rodman, P. S. 1979. Territoriality: the relation of ranging pattern and home range size to defendability, with an analysis of territoriality among primate species. *Behav. Ecol. Sociobiol.* **5:**241–251.

Mitani, J. C., Gros-Louis, J., and Richards, A. F. 1996. Sexual dimorphism, the operational sex ratio, and the intensity of male competition in polygynous primates. *Am. Nat.* **147:**966–980.

Nunn, C. L. 1995. A simulation test of Smith's "degrees of freedom" correction for comparative studies. *Am. J. Phys. Anthropol.* **98:**355–367.

Nunn, C. L. (2000). Collective action, free-riders, and male extragroup conflict. In: P. Kappeler (ed.), *Primate Males*, pp. 192–204. Cambridge University Press, Cambridge.

Nunn, C. L., and Smith, K. K. 1998. Statistical analyses of developmental sequences: the craniofacial region in marsupial and placental mammals. *Am. Nat.* **152:**82–101.

Omland, K. 1997. Examining two standard assumptions of ancestral reconstructions: repeated loss of dichromatism in dabbling ducks (Anatini). *Evolution*, **51:**1636–1646.

Pagel, M. 1994. Detecting correlated evolution on phylogenies: a general method for the comparative analysis of discrete characters. *Proc. R. Soc. London, Ser. B* **255:**37–45.

Plavcan, J. M. this volume. Reconstructing social behavior from dimorphism in the fossil record.

Plavcan, J. M., and van Schaik, C. P. 1997. Intrasexual competition and body weight dimorphism in anthropoid primates. *Am. J. Phys. Anthropol.* **103:**37–68.

Purvis, A. 1995. A composite estimate of primate phylogeny. *Philos. Trans. R. Soc. London, Ser. B* **348:**405–421.

Purvis, A., and Rambaut, A. 1995. Comparative analysis by independent contrasts (CAIC): an Apple Macintosh application for analysing comparative data. *Computer Applications in the Biosciences (CABIOS)* **11:**247–251.

Purvis, A., Gittleman, J. L., and Luh, H. 1994. Truth or consequences: effects of phylogenetic accuracy on two comparative methods. *J. Theor. Biol.* **167:**293–300.

Purvis, A., Nee, S., and Harvey, P. H. 1995. Macroevolutionary inferences from primate phylogeny. *Proc. R. Soc. London, Ser. B* **260:**329–333.

Robinson, J. G., and Janson, C. H. 1987. Capuchins, squirrel monkeys, and atelines: socioecological convergence with Old World primates. In: B. B. Smuts, D. L. Cheney, R. M. Seyfarth, R. W. Wrangham, and T. T. Struhsaker (eds.), *Primate Societies*. Chicago University Press, Chicago.

Ross, C. 1988. The intrinsic rate of natural increase and reproductive effort in primates. *J. Zool.* **214:**199–219.

Rowe, N. 1996. *The Pictorial Guide to the Living Primates*. Pogonias Press, East Hampton, NY.

Rylands, A. B. 1993. *Marmosets and Tamarins: Systematics, Behaviour, and Ecology*. Oxford University Press, Oxford.

Sawaguchi, T. 1990. Relative brain size, stratification, and social structure in anthropoids. *Primates* **31:**257–272.

Schluter, D., Price, T., Mooers, A. O., and Ludwig, D. 1998. Likelihood of ancestor states in adaptive radiation. *Evolution* **51:**1699–1711.

Schmidt-Nielsen, K. 1984. *Scaling: Why is Animal Size So Important?* Cambridge University Press, Cambridge.

Smith, R. J. 1996. Biology and body size in human evolution. *Curr. Anthropol.* **37:**451–481.

Smuts, B. B. and Smuts, R. W. 1993. Male aggression and sexual coercion of females in nonhuman primates and other mammals: evidence and theoretical implications. *Adv. Study Behav.* **22:**1–63.

Smuts, B. B., Cheney, D. L., Seyfarth, R. M., Wrangham, R. W., and Struhsaker, T. T. 1987. *Primate Societies.* Chicago University Press, Chicago.

Sokal, R. R., and Rohlf, F. J. 1995. *Biometry.* W.H. Freeman and Company, New York.

Tattersall, I. 1987. Cathemeral activity in primates: a definition. *Folia Primatol.* **49:**200–202.

Thomas, L., and Juanes, F. 1996. The importance of statistical power analysis: an example from Animal Behaviour. *Anim. Behav.* **52:**856–859.

Tooby, J., and DeVore, I. 1987. The reconstruction of hominid behavioral evolution through strategic modeling. In: W. G. Kinzey (ed.), *The Evolution of Human Behavior: Primate Models,* pp. 183–237. State University of New York Press, Albany, NY.

van Schaik, C. P. 1996. Social evolution in primates: the role of ecological factors and male behavior. *Proc. Br. Acad.* **88:**9–31.

van Schaik, C. P., and Kappeler, P. M. 1996. The social systems of gregarious lemurs: lack of convergence or evolutionary disequilibrium? *Ethology* **102:**915–941.

van Schaik, C. P., and van Noordwijk, M. A. 1988. Scramble and contest in feeding competition among female long-tailed macaques (*Macaca fascicularis*). *Behaviour* **105:**77–98.

Waser, P. M. 1976. *Cercocebus albigena*: site attachment, avoidance, and intergroup spacing. *Am. Nat.* **110:**911–935.

Williams, B. A., and Covert, H. H. 1994. New Early Eocene anaptomorphine primate (Omomyidae) from the Washakie Basin, Wyoming, with comments on the phylogeny and paleobiology of anaptomorphines. *Am. J. Phys. Anthropol.* **93:**323–340.

Wrangham, R. W. 1977. Feeding behaviour of chimpanzees in Gombe National Park, Tanzania. In: T. H. Clutton-Brock (ed.), *Primate Ecology: Studies of Feeding and Ranging Behaviour in Lemurs, Monkeys and Apes,* pp. 503–538. Academic Press, London.

Wrangham, R. W. 1979. On the evolution of ape social systems. *Soc. Sci. Inf.* **18:**335–368.

Wrangham, R. W. 1987. The significance of African apes for reconstructing human social evolution. In: W. G. Kinzey (ed.), *The Evolution of Human Behavior: Primate Models,* pp. 51–71. State University of New York Press, Albany, NY.

The Use of Paleocommunity and Taphonomic Studies in Reconstructing Primate Behavior 6

KAYE E. REED

Introduction

Behavioral ecology is concerned with not only describing animal behavior but placing that description within an evolutionary context, that is, how or why did a particular behavior evolve (Krebs and Davies, 1981). The "how" is often supplied by studies of primate morphology and comparisons with behaviors such as feeding ecology, activity pattern, substrate use, social organization, etc. The "why" is often only available through scenarios or with the help of contextual information. Community paleoecology and taphonomic studies are an essential part of answering why particular primate behavior may have evolved because the contextual evidence that they provide can furnish insights about changes in primate interactions with other primates, mammals, vegetation, and climate.

Paleoecology is the study of ancient organisms, their environment, and the interactions of those organisms with one another using ecological theory

KAYE E. REED • Institute of Human Origins and Department of Anthropology, Arizona State University, Tempe, Arizona 85281-4101 and Bernard Price Institute for Palaeontological Research, University of the Witwatersrand, Johannesburg, South Africa.

Reconstructing Behavior in the Primate Fossil Record, edited by Plavcan *et al.* Kluwer Academic/Plenum Publishers, New York, 2002.

and methodology as is possible (Dodd and Stanton, 1990). Thus, paleontological studies centering on morphological analysis of one particular taxon are only one component of paleoecology, but one that usually provides extensive functional data, and thus behavioral correlates of fossil primates. Comparative morphological studies attempt to provide reconstructions of fossil primate locomotion, trophic behavior, sexual dimorphism, and activity pattern (see other chapters). Evidence to reconstruct social structure, dominance hierarchy, competition, and dispersal patterns is somewhat more remote as morphological studies only provide secondary lines of evidence to these behaviors. For example, canine dimorphism is usually suggestive of either single-male or multi-male social groups, but as Plavcan and van Schaik have shown (1997) this is not always straightforward. Unless an entire group of fossil primates was preserved *in situ*, scenarios of social organization, based on canine dimorphism, remain inferential.

The goal of this chapter is to describe how paleocommunity and taphonomic studies can expand knowledge of primate adaptations and behavior. Using abiotic and biotic information about communities in which ancient primates existed can bolster information recovered from morphological studies of these primates. Examining this contextual information across time could show ecological patterns that may have influenced primate behavior. Primate behavior as interpreted through morphological studies can then be compared to ecological patterns through time providing data that may answer why particular behaviors evolved. For example, vegetation changes from forest to more open regions could possibly be followed by changes in primate feeding behavior or substrate use. Patterns in changes of the feeding behaviors of other animals may influence primate reliance on similar resources, etc.

Ecologists often do not study complete biological communities because it is almost impossible to study all of the organisms in a meaningful way. In paleo-assemblages, especially, it is impossible to reconstruct the interrelationships of microbes, bacteria, plants, insects, reptiles, mammals, etc. There is nothing wrong with working with a single aspect of a community and ecologists have shown this to be effective methodologically (e.g., Morris *et al.*, 1989; Strong *et al.*, 1984). Often studying a single aspect of a community is referred to as analyzing guilds, i.e., terrestrial vertebrate guilds (Brown and Bowers, 1984) or primate guilds, but is just as often referred to as "land bird communities" (Jarvinen and Haila, 1984), "community structure of coral-associated crustaceans" (Abele, 1984), or "primate communities" (Terborgh and van Schaik, 1987). The choice of an organismal group or groups for community analysis depends on the questions to be answered, and for paleoecology, questions that can be answered about fossil assemblages.

Many studies of communities are based on data that show that the communities have an underlying structure that can be determined. Observations and research on extant communities have shown that community composition or structure varies depending on vegetation and climate. Abiotic factors such as sunlight, soil, rainfall, seasonality, temperature, and evapotranspiration in various combinations influence the type or architecture of vegetation that grows in particular geographic areas. I use vegetation architecture to refer to vertical stratification levels, such that forests have a highly stratified architecture while grasslands have little vertical stratification for large mammals. In turn, animals are adapted to the vegetation that surrounds them as well as the abiotic factors (Odum, 1971). Thus, community composition is not random, but based on abiotic features that support particular plant life which in turn supports particular animal life (Roughgarden, 1989). Species found in communities change across geographic regions, but within communities of the same vegetation architecture or type, there are plant and animal species that are ecological equivalents (Odum, 1971). The principles of community paleoecology are based on this being true not only across geographic space, but across time as well. Thus, ecological processes that cause community cohesiveness exist in the same way in the geological past.

An example of these principles is found in extant African mammal communities. These mammal communities exist in various vegetation types (forests, woodlands, deserts, etc.). For each community the percentages of arboreal, aquatic, fruigvorous, grazing, etc. adaptations were calculated. The percentages of these adaptations are consistent within vegetation type and significantly different among vegetation types, illustrating that the ecological adaptations in each community show a particular pattern (Reed, 1997, 1998). Thus, large mammalian communities have predictable ecological compositions or structures based on adaptations that depend upon vegetation, although individual species within them many differ. Because the distribution of mammalian adaptations is filled by ecologically equivalent species, community structure such as this can be compared with other communities across geographic areas and time. Finally, as these adaptations are significantly different among vegetation types they can be used to reconstruct vegetation in the past. Thus, fossil assemblages of mammals can reconstruct vegetation type, and provide information about how community composition has altered over time. The fossil record is the only place where patterns of long-term changes in the structure of communities can be examined from an evolutionary perspective (Wing *et al.*, 1992). The resultant information provided by these data can give insights into primate diversity, competition, replacement, displacement, and immigration or emigration over time within a larger ecological framework. Community context affects the behavior of primate species as it involves primate interactions with the vegetation, other mammals, and other primates.

Recent taphonomic studies have focussed on gaining ecological information from biased fossil assemblages (Wilson, 1988). Information gain is contrasted with prior studies that concentrated on information that was lost during the period from death to recovery of fossil material. Information gain about accumulating agents whether biotic (e.g., predators) or abiotic (e.g., seasonal floods), for example, can help to interpret fossil assemblages in context. Some fossil assemblages created by carnivores not only supply data about the carnivore's hunting patterns, but can give insight into the behavior of the prey, which assuredly are sometimes primates.

The Comparative Method and Actualistic Studies in Paleoecology

The comparative method, in conjunction with studies of functional morphology, is used to discover the likely biological role of certain structures of fossil primates. A primatologist can study the behavior of an ape and know what it eats, how it moves, and where it lives. The only thing a paleontologist can interpret about the adaptations of a taxon comes from the remains of its bones and teeth (Kay, 1984). That there is a relationship between behavior and morphology, and the assumption that this relationship is uniform throughout time is the theoretical basis for believing that the ecological adaptations of a fossil taxon can be reconstructed (Fleagle, 1979).

Comparative or analogous methodology is also used to link extant and extinct communities (Wing *et al.*, 1992). However, it is not individual species that are compared, but aspects of the communities. Analyses are devised in which structural or other patterns in extant communities are compared to those in the past. Ecological patterns in communities can include, but are not limited to, stratification, activity, trophic, locomotor, size, reproductive, social, etc. Some patterns are more viable for fossil research than others. An aspect of community structure that is comparable over time and space must be selected. Andrews *et al.*, (1979) recognized that the percentages of trophic, locomotor, and body size adaptations of taxa within communities work well for this purpose. They used these characteristics of extant and extinct mammal communities in order to compare these patterns and reconstruct environments. Damuth (1992) refers to this type of methodology as "taxon-free" analysis because phylogeny and historical circumstances are removed from analogies between communities. Theoretically, one could compare an extant rain forest community with its high percentage of species devoted to arboreal locomotion with a Jurassic dinosaur community. However, each specific region and time period should be meticulously analyzed before sweeping comparisons are made.

The main problem is designing tests that will eventually make valid analogies for interpretation of the fossil record. First, we need to know how primate morphology relates to particular behaviors and ecological adaptations, e.g., trophic preferences, locomotor patterns, activity patterns, etc. Second, we would like to know the ecological background of these behaviors, that is, how primates interact with other organisms, compete or share resources, what is preying on primates, and the habitat in which they lived. It has been shown that these factors influence primate behavior (e.g., Barton *et al.*, 1993; Iwamoto *et al.*, 1996; Cowlishaw, 1994, 1997; Wahungu, 1998; Zinner and Pelaez, 1999).

The ecological background in many fossil primate studies has consisted of recreating the taxonomic composition of a fossil assemblage, and depicting community structure by species abundances, species diversity, equitability (or dominance), and by trophic relationships (Shotwell, 1955; Bishop, 1968; Andrews and Van Couvering, 1975; Van Couvering and Van Couvering, 1976; Nesbit-Evans *et al.*, 1981; Pickford and Andrews, 1981; Andrews *et al.*, 1981; Rose 1981; Shipman *et al.*, 1981; Shipman, 1982; Hill, 1985a, 1985b; Leakey and Walker, 1985; Shipman, 1986). Taxonomic composition, while important, is inappropriate to use for ecological background since taxonomic composition shifts through time and space and is therefore difficult to compare across these boundaries. Species abundances and diversities, although essentially comparable across time and space because of their taxon-free nature, often reflect predator preferences and other taphonomic biases in fossil assemblages, rather than give meaningful information about the ecological composition of a community. On the other hand, these data are important in identifying primate behavior patterns with regard to predators.

Dodd and Stanton (1990) suggest that paleoecology depends upon a good stratigraphic framework, good taxonomy, and an ecological background. Ecological background should consist of information about how living organisms exist in their communities and how that information might be recovered from fossil assemblages. Although it is not the only way, actualistic studies provide an important link between the past and present.

The term actualistic study is derived from the term "actuopaleontology" (Dodd and Stanton, 1990) and is used to refer to a study of modern organisms or modern death assemblages using a paleontological perspective. The study or experiment is created with a hypothesis that can be tested using information from extant data. The results of the test can then be used to make an analogy with some aspect of the fossil record. This is simply reasoning through analogy as the extant experiment examines processes that result in a discernable pattern that can be compared to those found in the fossil record.

A broad-scaled example is one that reconstructs environments through mammalian community structure. I analyzed modern African mammal communities and discovered that certain locomotor and trophic adaptations are significantly different among types of vegetation, e.g., forests, woodlands, bushlands, etc. (Reed, 1996, 1997, 1998). I hypothesized that these data would be discernable in the fossil record, but first tested the community data with extant African death assemblages. That is, information about the adaptations of species assemblages recovered from hyena dens, leopard and lion kills, porcupine dens, and transect data were compared to trophic and locomotor structure of macromammal communities to see if taphonomy affected the representation of these adaptations. These death assemblages were a sample of the living community from which they were derived, and similar to what might be recovered from a fossil assemblage. The death assemblages were able to accurately reflect many of the ecological adaptations percentages, and as such, the community composition. African Pliocene fossil assemblages were then examined to reconstruct vegetation and to analyze shifts in community structures (Reed, 1997, 1998). In addition to accurate habitat reconstructions, these studies and others (e.g., Brain, 1981) have furnished information about carnivore behavior that may influence primate behavior.

A taphonomic example of an actualistic study is based on an analysis of extant carnivore behavior. Blumenschine (1989) proposed that numbers of bones left on the modern landscape are able to predict the degree of scavenging competition in different areas of the modern ecosystem. He tested this hypothesis first by relating that the number of bones, the percentages of skeletal parts, and the completeness of surviving bones were different in individual East African communities. He then correlated these results with a scale of scavenging opportunities in each area: season, type of ecosystem and habitat, size of carcass, and initial consumer species. The results of this test can be used as a taphonomic model to estimate the degree of competition for carcasses in paleocommunities. Blumenschine tested an ecological *process* (scavenging) that left a *pattern* (skeletal remains) on the landscape. This pattern can be extrapolated to another community of a similar type in an ancient landscape. This study was applied to learning more about early hominin behavior.

Actualistic studies of this nature are extremely important in that they can help explain interactions of paleospecies and their behavior. There is always the problem of equafinality, but continuing to examine more complex processes and patterns in living communities will allow this problem to be addressed from multiple lines of evidence (Gifford-Gonzalez, 1991). Many small hypotheses and tests on modern communities can help build a storehouse of knowledge that can be used to analyze past primate communities and behavior.

In summary, to be able to reconstruct primate paleoecology (from habitats primates live in through various ecological behaviors), extant ecological patterns and processes must be found to act as a basis for examining past communities. These patterns can then be tested using actualistic studies and the comparative method to reconstruct contextual and behavioral information about primates in the past.

Fossil Assemblages

The Makapansgat Limeworks is a well-known fossil locality as the early hominin, *Australopithecus africanus*, has been recovered from two members of this South African cave site (Dart, 1952). In addition to this hominin, several other primate species have also been recovered from each member. I will first detail how large mammal communities are useful for vegetation reconstruction. Reconstruction of vegetation is one of the major uses of paleocommunity studies in primate behavior as it provides an assessment of the environment in which primates lived and evolved. Then I will compare the Makapansgat fossil primate assemblages with extant primate communities showing how both community and taphonomic data can be used to (1) suggest primate behavior, (2) serve as a basis for producing further hypotheses about primate behavior, and (3) place behavioral descriptions generated by morphological studies of these primates into evolutionary context (Krebs and Davies, 1981).

Inferring Environments from Mammal Morphology

I used an analysis of the ecological structure of the community to reconstruct the vegetation because this method helps eliminate the confounding factors of biogeography and chronology. This taxon-free method uses the percentages of various ecological adaptations of mammalian species in each community rather than their taxonomic identities and abundances to represent community structure. Vegetation reconstructions based on this type of analysis are likely to be accurate despite taphonomic biases that might occur when species abundances are used to infer environments (Behrensmeyer, 1991). These biases are caused by errors in counting or estimating abundances stemming from trampling, fluvial dispersion, and fragmentation. The ecological structure of communities is also comparable across time and geographic area (Andrews *et al.*, 1979; Damuth, 1992). I used only macromammal communities as these eliminated problems encountered when basing estimates of past vegetation on one animal or group of animals, e.g., bovids.

Fleming (1973) and Andrews *et al.*, (1979) showed that mammalian ecological adaptations form different patterns when communities exist in various types of vegetation. Both studies assigned ecological adaptations to each mammal for a variety of communities in three categories: trophic, locomotor, and body size. They produced and compared histograms of these categorical percentages among communities. In contrast, I examined the percentage of each adaptation (e.g., arboreal locomotion) across vegetation types and tested for significant differences among them (Fig. 1).

Communities of large mammal species were compiled from 32 National Parks, Game Reserves, and specific vegetation regions (Swynnerton, 1958; Lamprey, 1962; Child, 1964; Vesey-Fitzgerald, 1964; Sheppe and Osborne, 1971; Kingdon, 1974a, 1974b, 1974c, 1977, 1979, 1982a, 1982b; Rautenbach, 1978; Smithers, 1978; Behrensmeyer *et al.*, 1979; Perera, 1982; Emmons *et al.*, 1983; Happold, 1989; Skinner and Smithers, 1990). Each large mammal species (186 species) from these 32 areas and seven vegetation types was coded for its locomotor and trophic behavioral adaptations based on observations recorded in the literature (Kingdon, 1974a, 1974b, 1974c, 1977, 1979, 1982a, 1982b; Estes, 1991; Skinner and Smithers, 1990). These adaptations were broadly defined so that morphological analyses could eventually be used to infer adaptations for fossil taxa (Table I).

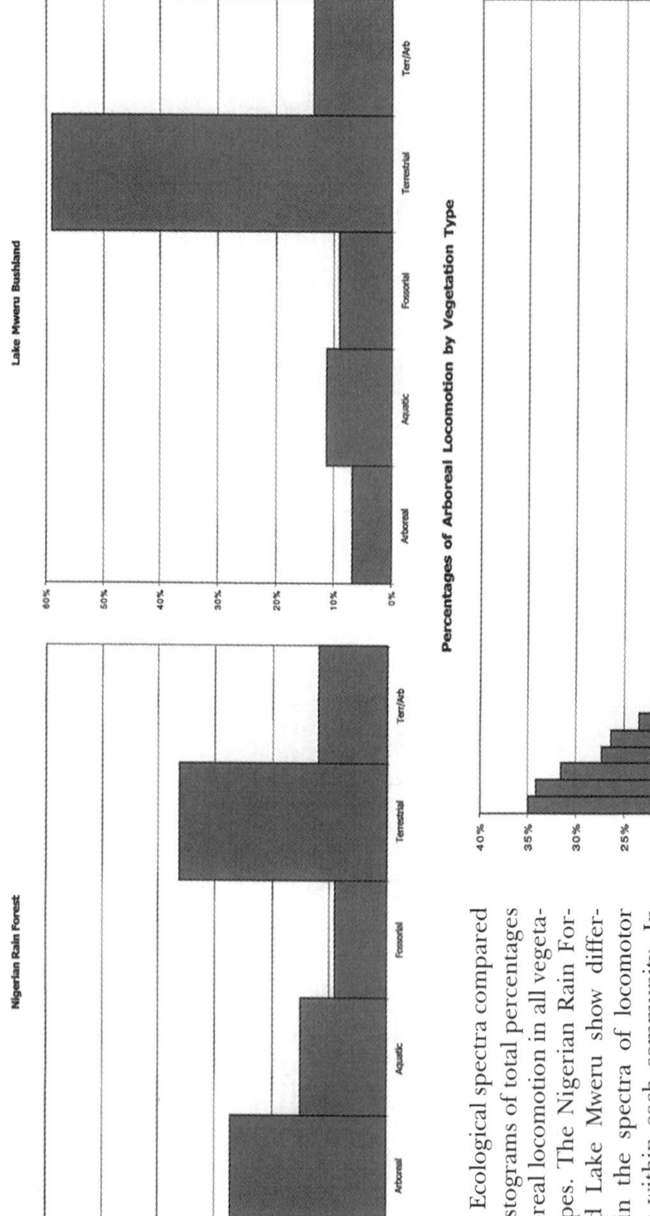

Fig. 1. Ecological spectra compared with histograms of total percentages of arboreal locomotion in all vegetation types. The Nigerian Rain Forest and Lake Mweru show differences in the spectra of locomotor pattern within each community. In contrast, in the column graph showing percentages of arboreal locomotion the contrast between vegetation types is immediately visible.

Table I. Adaptations Assigned to Mammals in Extant Communities

Trophic adaptation			Example
Carnivore	Meat		*Panthera leo* (lion)
	Meat/bone		*Crocuta crocuta* (hyena)
	Meat/invertebrates		*Mungos mungo* (mongoose)
Insectivore	Insects, invertebrates		*Manis tricuspis* (pangolin)
Herbivore	Grazer	Grass	*Connochaetes taurinus* (wildebeest)
	Fresh-grass grazer	Edaphic grasses	*Kobus kob* (kob)
	Browser	Leaves	*Tragelaphus strepsiceros* (kudu)
	Mixed feeder	Grass and leaves	*Aepyceros melampus* (impala)
	Roots/Bulbs		*Hystrix africaeaustralis* (porcupine)
Frugivore	Fruit/Leaves/Insects	In various combinations	*Papio ursinus* (Chacma baboon)
Omnivore	Three or more of the above with no preference		*Melivora capensis* (honey badger)
Locomotor adaptation			
Arboreal	Moves and feeds in trees most of the time		*Galago moholi* (bushbaby)
Aquatic	Feeds or locomotes in water 50% of the time		*Hippopotamus amphibius* (hippo)
Fossorial	Lives or acquires food underground by digging		*Orycteropus afer* (aardvark)
Terrestrial	Moves and feeds on the ground		*Equus burchelli* (zebra)
Terrestrial/arboreal	Moves and feeds on the ground and in trees		*Panthera pardus* (leopard)

Vegetation types were also broadly defined, for example, the "forest" category includes lowland rain, tropical temperate, and montane forests (Reed, 1998; Table II). Mammal adaptations from fossil assemblages are more likely to be able to predict broad vegetation types rather than microhabitats, e.g., primary and secondary forests.

Locomotor adaptations of communities that were found to predict vegetation include: (1) the percentage of species that were arboreal in each community, and (2) the percentage of species exhibiting aquatic locomotion in each community. The percentages of these mammalian adaptations were dependent upon the vegetation type and climate such that, as the number of trees decreased, such as from a forest to open woodland, the percentages of arboreal locomotion decreased. Greater amounts of water within a vegetation type were reflected in higher percentages of aquatic locomotion. The percentages of arboreal locomotion in each community were also partly attributable to mean

Table II. Localities, Vegetation Type, and Adaptation Percentages[a]

Locality	Rain	D	A	AQ	F	T	TA	TC	TF	B	C	CB	CI	FG	FI	FL	G	I	MF	OM	R
E of Niger	1596	1	27.3	15.2	9.1	36.4	12.1	9.1	36.4	3.0	9.1	0.0	0.0	9.1	6.1	30.3	0.0	12.1	15.2	12.1	3.0
Congo Basin	1800	1	26.4	13.2	9.4	28.3	22.6	26.4	43.4	3.8	15.1	0.0	11.3	1.9	5.7	37.7	0.0	5.7	7.5	7.5	3.8
Knysna	1016	1	23.5	5.9	11.8	35.3	23.5	29.4	41.2	0.0	11.8	0.0	17.6	0.0	11.8	29.4	0.0	0.0	5.9	11.8	11.8
Killimanjaro	1050	1	31.6	0.0	0.0	63.2	5.3	10.5	42.1	15.8	10.5	0.0	0.0	0.0	10.5	31.6	0.0	0.0	15.8	10.5	5.3
W of Niger	1600	1	37.5	12.5	9.4	28.1	12.5	6.3	40.6	9.4	6.3	0.0	0.0	6.3	15.6	25.0	0.0	6.3	12.5	12.5	6.3
E of Cross	1550	1	35.0	17.5	10.0	27.5	10.0	15.0	47.5	2.5	12.5	0.0	2.5	5.0	15.0	32.5	0.0	5.0	10.0	10.0	5.0
Makakou	1800	1	34.1	7.3	2.4	36.6	19.5	12.2	56.1	2.4	9.8	0.0	2.4	7.3	19.5	36.6	0.0	0.0	9.8	7.3	4.9
Rwenzori NP	900	2	15.7	9.8	11.8	45.1	17.7	31.4	27.5	2.0	15.7	3.9	11.8	11.8	7.8	19.6	3.9	3.9	9.8	7.8	2.0
Guinea Woodland	1000	2	11.9	8.5	10.2	52.5	16.9	32.2	20.3	5.1	18.6	3.4	10.2	11.9	1.7	18.6	5.1	5.1	11.9	6.8	1.7
W Lunga NP	875	2	6.8	11.4	9.1	59.1	13.6	31.8	18.2	4.6	11.4	2.3	18.2	6.8	2.3	15.9	11.4	0.0	15.9	9.1	2.3
Natal Woodland	875	2	7.5	7.5	10.0	57.5	17.5	35.0	22.5	5.7	13.2	3.8	11.3	9.4	3.8	15.1	9.4	3.8	17.0	5.7	1.9
Serengeti NP	750	3	5.3	8.0	10.7	62.7	13.3	29.3	17.3	4.0	12.0	4.0	13.3	6.7	2.7	14.7	10.7	6.7	17.3	6.7	1.3
Lake Mweru	750	3	5.7	8.6	5.7	71.4	8.6	28.6	22.9	5.7	11.4	5.7	11.4	14.3	2.9	20.0	11.4	0.0	14.3	2.9	0.0
Serengeti Bush	803	3	6.3	4.7	14.1	59.4	15.6	31.3	17.2	4.7	15.6	4.7	10.9	4.7	3.1	14.1	12.5	6.3	17.2	4.7	1.6
Rukwa Valley	700	3	5.8	7.7	9.6	63.5	13.5	28.9	19.2	5.8	13.5	3.9	11.5	9.6	3.9	13.5	9.6	3.9	17.3	5.8	1.9
Kafue NP	821	4	5.5	10.9	14.5	52.7	16.4	34.5	12.7	5.5	20.0	3.6	10.9	12.7	3.6	9.1	10.9	5.5	9.1	7.3	1.8
Kruger NP	675	4	3.2	7.9	12.7	61.9	14.3	34.9	12.7	7.9	14.3	4.8	15.9	6.3	3.2	9.5	12.7	4.8	14.3	4.8	1.6
Linyanti Swamp	650	4	2.9	14.3	2.9	68.6	11.4	28.6	11.4	5.7	11.4	2.9	14.3	17.1	2.9	8.6	17.1	0.0	17.1	2.9	0.0
SS Woodland	650	4	4.8	6.0	12.0	66.3	10.8	30.1	13.3	6.0	12.0	3.6	14.5	6.0	2.4	10.8	16.9	6.0	16.9	3.6	1.2
Sudan	689	4	3.9	5.9	13.7	62.7	13.7	39.2	11.8	5.9	21.6	5.9	11.8	7.8	2.0	9.8	7.8	3.9	13.7	7.8	2.0
SW Arid	400	5	1.7	6.7	18.3	61.7	11.7	33.3	8.3	6.7	16.7	5.0	11.7	3.3	1.7	6.7	23.3	8.3	13.3	1.7	1.7
Kalahari TV	450	5	0.0	0.0	33.3	60.0	6.7	33.3	6.7	0.0	13.3	6.7	13.3	0.0	0.0	6.7	13.3	20.0	20.0	0.0	6.7
Sahel	450	5	0.0	9.7	16.1	51.6	22.6	41.9	6.5	0.0	25.8	3.2	12.9	9.7	0.0	6.5	6.5	6.5	16.1	9.7	3.2
Chobe	650	5	1.8	8.9	17.9	57.1	14.3	28.6	7.1	7.1	14.3	3.6	10.7	10.7	1.8	5.4	16.1	8.9	14.3	5.4	1.8
Amboseli NP	510	5	2.2	4.3	15.2	60.9	17.4	23.9	8.7	10.9	15.2	4.3	8.7	6.5	2.2	6.5	15.2	6.5	15.2	6.5	2.2
Tarangire NP	600	5	2.1	2.1	16.7	68.8	10.4	27.1	8.3	10.4	12.5	6.3	8.3	4.2	0.0	8.3	12.5	10.4	16.7	8.3	2.1
Okavango	600	5	1.9	11.1	18.5	53.7	14.8	33.3	7.4	5.6	14.8	3.7	14.8	9.3	1.9	5.6	16.7	9.3	11.1	5.6	1.9
Kafue Flats	821	6	0.0	13.5	2.7	70.3	13.5	37.8	5.4	5.4	21.6	5.4	10.8	16.2	0.0	5.4	16.2	0.0	13.5	5.4	0.0
Serengeti Plains	500	6	0.0	0.0	15.8	84.2	0.0	31.6	0.0	0.0	15.8	15.8	0.0	0.0	0.0	0.0	26.3	10.5	21.1	10.5	0.0
SS Grassland	500	6	0.0	9.8	17.1	58.5	14.6	36.6	4.9	2.4	17.1	4.9	17.1	2.4	0.0	4.9	24.4	9.8	14.6	0.0	2.4
Namib Desert	125	7	0.0	0.0	22.2	50.0	27.8	50.0	11.1	0.0	27.8	11.1	16.7	0.0	0.0	11.1	11.1	5.6	11.1	0.0	5.6
Makapansgat 3			5.3	3.5	7.0	64.6	17.5	26.3	19.3	21.1	15.7	3.5	5.3	5.3	0.0	15.7	15.8	3.5	8.8	1.8	5.3
Makapansgat 4			5.9	2.9	5.9	70.6	14.7	17.6	20.6	20.6	2.9	5.9	8.8	5.9	0.0	20.6	11.8	0.0	14.7	2.9	5.9

[a] Rain = mean annual rainfall, DF = discriminant function code, A = arboreal locomotion, AQ = Aquatic locomotion, F = fossorial locomotion, T = terrestrial locomotion, TA = terrestrial/arboreal locomotion, TC = total carnivory, TF = total frugivory, B = leaves, C = meat, CB = meat/bone, CI = meat/invertebrates, FG = fresh grass, FI = fruit/insects, FL = fruit/leaves, G = grass, I = insects, MF = grass/leaves, OM = omnivore, R = roots/tubers. DF Codes: 1 = forests, 2 = closed woodlands and closed woodland/bushland, transition, 3 = bushlands, 4 = open woodlands, 5 = shrublands, 6 = grasslands, 7 = desert. Makapansgat percentages differ from previous publications due to newly recovered specimens and carbon isotope data.

annual rainfall (Fig. 2). Because the amount of rainfall received influences the abundance of trees and other plants (Archibold, 1995), this adaptation can also be used to roughly estimate the amount of tree cover, and thus vertical stratification of the community.

Trophic adaptations also delineate vegetation. The best trophic predictor is the percentage of frugivores in a community (frugivores including those mammals that supplement with insects or leaves). Percentages of frugivory in communities are also partly attributable to mean annual rainfall (Fig. 2). When the percentages of frugivory are plotted against the percentages of arboreal locomotion, most extant vegetation types are separated from one another (Fig. 3). Other trophic adaptations that show significant differences among vegetation types include percentages of grazing, fresh grass grazing, and meat-bone eating (Table III).

After showing that percentages of mammalian community adaptations could reflect or even predict the vegetation from which they derived (Reed, 1997, 1998), I tested the method on extant African death assemblages to see if accurate vegetation reconstructions could be made using biased assemblages of mammals. African death assemblages, like fossil assemblages, are biased by accumulating agent and other factors, such as type of predator, predator choice, scavenged removal of specimens, trampling, fragmentation, etc. However, unlike fossil assemblages the reconstruction of vegetation based on adaptations of species from death assemblages can be compared with extant mammal communities and vegetation for accuracy. Death assemblage data were taken from published accounts of mammals that produce bone assemblages, and a surface bone collection transect (Kruuk, 1972; Schaller, 1976; Behrensmeyer *et al.*, 1979; Brain, 1981; Skinner *et al.*, 1986; Henschel and Skinner, 1990; Kitchner, 1991). This was an actualistic study as the death assemblages were stand-ins for fossil assemblages. The percentage of mammalian adaptations from these death assemblages accurately predicted the vegetation type from which they were derived, despite small sample sizes and predator or accumulation biases (Fig. 4). There were some anomalies, however. For example, remains of small arboreal mammals were not recovered from hyena dens although they were present in the extant community. This is likely because smaller animals can be completely consumed by these predators, and their bones are eliminated elsewhere. Thus, there is a signature pattern for hyena assemblages (Klein and Cruz-Uribe, 1984). One of the reasons for using the large mammal community and percentages of adaptations within them to analyze fossil assemblages is that it is immediately noticeable if entire adaptations are missing from a deposit. For example, several members in the Shungura Formation, Omo, Ethiopia have no carnivores associated with them. No one would propose that these ancient communities did not have carnivores. These biases can be examined in the context of ecological structure to begin to look for reasons why various adaptive categories are missing from assemblages. In any case, the death assemblages had smaller species sample sizes than any of the fossil assemblages, but had enough representative adaptations to successfully predict the vegetation from which they were derived.

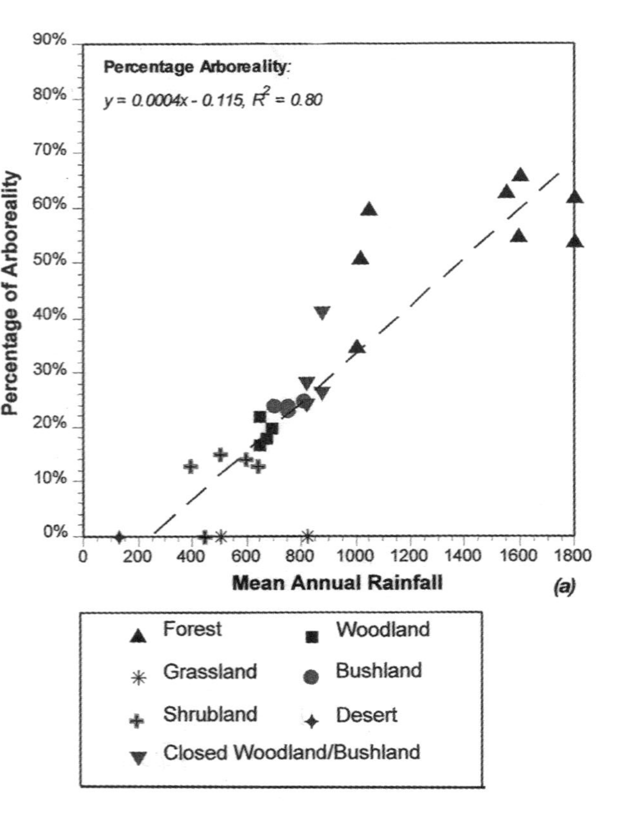

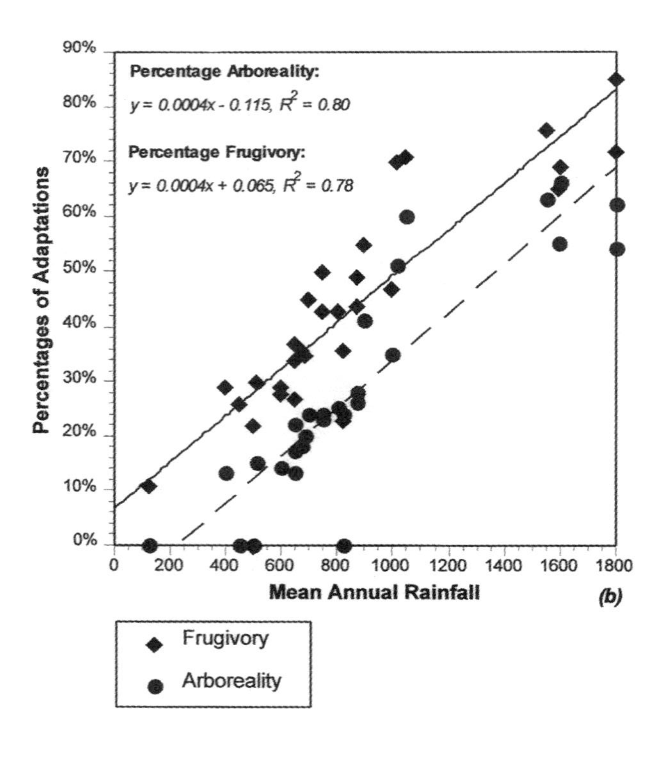

Fig. 2. Regressions of percentages of arboreality and frugivory in extant African communities against mean annual rainfall. (a) This regression shows that percentages of arboreal locomotion increase as the vegetation types change from desert to rain forest reflecting mean annual rainfall increases. (b) Percentages of frugivorous mammals also increase with mean annual rainfall, and the slope of the increase is nearly identical to that of the percentages of arboreal locomotion.

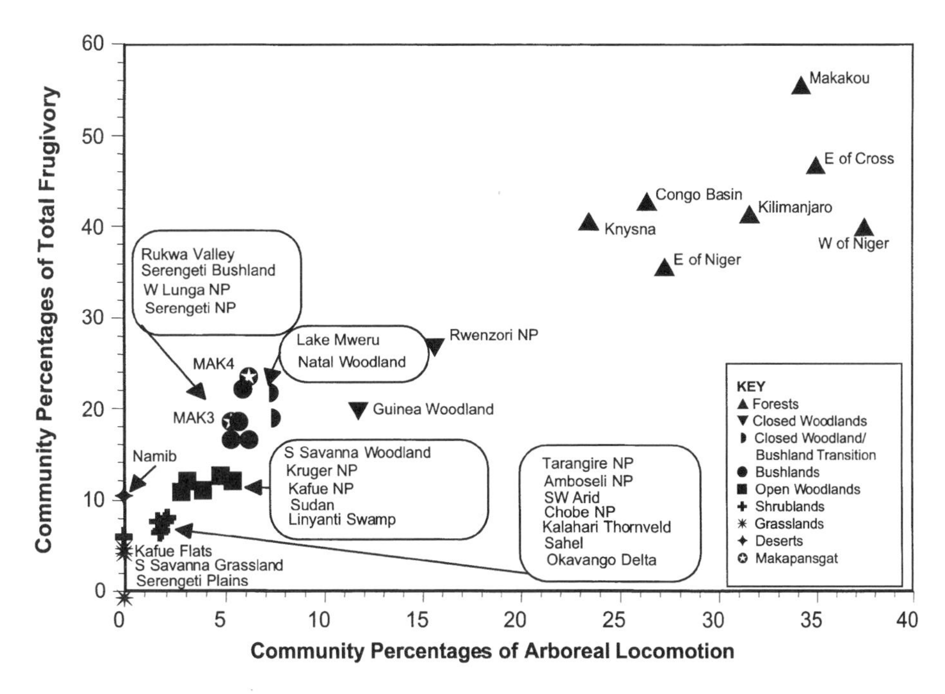

Fig. 3. Percentages of arboreal locomotion plotted against percentages of frugivory in extant African communities. Vegetation types are grouped together on this graph showing that the percentages of these two adaptations within communities serve to indicate the vegetation from which they were derived.

Makapansgat

The Makapansgat Valley is located in Northern Province, South Africa, about 325 km northeast of Johannesburg. The fossil assemblages discussed here derive from the Limeworks Cave. Breccia from the Limeworks has been mined for fossils since 1945 and over 50,000 mammalian specimens have been recovered. Collection assemblages have been designated as Members 1–5 with Member 1 consisting of material from the oldest basal travertine of unknown age, and Member 5 that is Pleistocene in age. Previous research has determined that the Member 3 (M3) assemblage (?3.1–2.9 m.y.) was accumulated by hyenas and porcupines (Maguire *et al.*, 1980), while the Member 4 (M4) assemblage (?3.0–2.8 m.y.) was likely created by leopards (Maguire *et al.*, 1980) and large birds of prey (Reed, 1996). Predators often select particular animals (Brain, 1981) and assemblages could reflect a biased sample of a living community if abundances of specimens were used to reconstruct environments or examine community structure. However, vegetation reconstructions based on the presence of taxa can be accurate despite these taphonomic problems. This is because numerous (e.g., carnivore preferences) and rare species are counted equally.

Table III. Mean and Range of Significant Adaptations for Each Vegetation Type[a]

		A	C	CB	F	G	I	OM	R	T	TC	TF
Forests	Mean	0.308	0.107	0.000	0.074	0.000	0.041	0.103	0.054	0.365	0.156	0.439
n = 7	Std. Deviation	0.051	0.028	0.000	0.044	0.000	0.045	0.021	0.031	0.125	0.089	0.063
	Range	0.140	0.088	0.000	0.118	0.000	0.121	0.052	0.093	0.357	0.232	0.197
Closed Wood	Mean	0.107	0.148	0.038	0.099	0.071[c]	0.037	0.065	0.019	0.540	0.330	0.225[c]
Transition	Std. Deviation	0.039	0.030	0.003	0.016	0.030	0.013	0.010	0.001	0.068	0.016	0.035
n = 4	Range	0.082	0.069	0.005	0.039	0.059	0.031	0.022	0.003	0.157	0.036	0.078
Bushlands	Mean	0.058	0.123	0.040	0.100	0.098[b,c]	0.033	0.063	0.014	0.657[b]	0.295	0.192[c]
n = 4	Std. Deviation	0.004	0.009	0.013	0.034	0.017	0.028	0.029	0.011	0.079	0.012	0.026
	Range	0.009	0.020	0.032	0.083	0.039	0.067	0.071	0.025	0.178	0.027	0.057
Open Wood	Mean	0.040	0.159	0.042	0.112	0.131[b]	0.040	0.053	0.013	0.624[b]	0.335	0.124
n = 5	Std. Deviation	0.011	0.046	0.012	0.047	0.040	0.024	0.022	0.008	0.061	0.042	0.008
	Range	0.026	0.101	0.030	0.117	0.093	0.060	0.050	0.020	0.158	0.106	0.018
Shrubland	Mean	0.014	0.161	0.047	0.194	0.148	0.100	0.053	0.028	0.591	0.316	0.076
n = 7	Std. Deviation	0.009	0.045	0.013	0.062	0.051	0.046	0.034	0.018	0.057	0.058	0.009
	Range	0.022	0.133	0.034	0.181	0.169	0.135	0.097	0.050	0.171	0.180	0.022
Grassland	Mean	0.000	0.173	0.087	0.119	0.223	0.068	0.053	0.008	0.701	0.353	0.034
n = 3	Std. Deviation	0.000	0.037	0.062	0.080	0.054	0.059	0.053	0.014	0.130	0.033	0.030
	Range	0.000	0.070	0.109	0.144	0.101	0.105	0.105	0.024	0.257	0.063	0.054
	Variance	0.000	0.001	0.004	0.006	0.003	0.003	0.003	0.000	0.017	0.001	0.001
Kruskal-Wallis		27.675	12.633	18.583	17.726	23.353	11.220	11.434	16.299	15.056	16.341	27.748
df = 5, p =		*0.000*	*0.027*	*0.002*	*0.003*	*0.000*	*0.047*	*0.043*	*0.006*	*0.010*	*0.006*	*0.000*

[a] Kruskal-Wallis showed significant differences between groups for all vegetation types at $X^2_{0.01[5]} = 150.9$. Mann-Whitney *U*-tests between each pair of communities showed significant differences at $p < 0.05$ for arboreal and terrestrial locomotion, and for grazing and total frugivory except for no significant difference between bushland/open woodland and [c]no significant difference between closed woodland/bushlands. All other adaptations could discern between a pair or two of vegetation types, but were not significant for many between-pair comparisons. Abbreviations: A, arboreal; C, meat-eating; CB, meat/bone-eating; F, fossorial; G, grazing; I, insectivory; OM, omnivory; R, root/bulb-eating; T, terrestrial; TC, meat & meat/invertebrate & meat/bone-eating; TF, fruit & fruit/insect & fruit/leaf-eating.

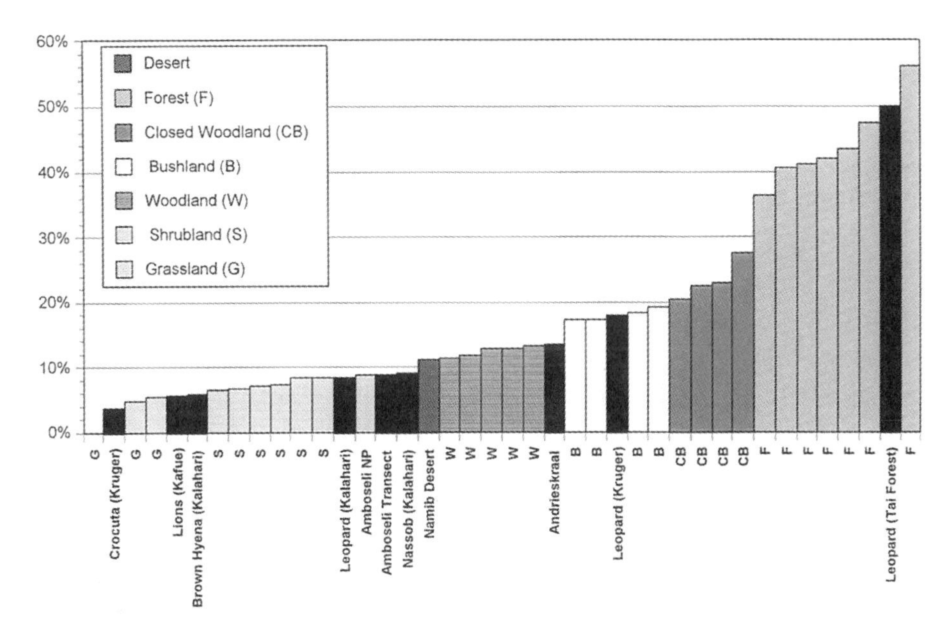

Fig. 4. Percentages of frugivores recovered from death assemblages interspersed with extant African vegetation types. All death assemblages are in black and labeled as to type of assemblage and locality. Differences in the Kruger assemblages relate to predators. Lions hunt in the more open regions of Kruger National Park, hence the assemblage is positioned between grasslands and shrublands. Leopards, on the other hand, usually hunt in riverine forests and bushlands at Kruger, thus they are positioned with other bushlands. The bone transect material from Amboseli National Park is positioned next to that locality on the column graph. The Kalahari is a thornveld (shrubland) vegetation, not a desert.

Sorting, Identification, and Measurement

The mammalian fossils from Makapansgat were collected and prepared by a variety of researchers over about 50 years. Most of the fossil material is stored at the Bernard Price Institute (BPI), University of the Witwatersrand, South Africa. Hominin and many of the cercopithecine fossils are stored in the Department of Anatomy and Human Biology of the same university. Depending on the mammalian order, different measurements were taken of each taxon to arrive at predictions of trophic and locomotor adaptations.

Indices created by Van Valkenburgh (1988) allowed differentiation of carnivores into meat-eaters, meat/bone-eaters, and meat/invertebrate-eaters. Four of these indices were used in this study: (1) upper canine shape, (2) premolar shape of the largest premolar (the largest premolar defined by area; P_4 for all but hyenas in which P_3 is the largest), (3) relative premolar size, and (4) relative blade length of the lower carnassial, M_1. Postcranial assignments for carnivores were based on the work of Lewis (1995).

Two methods were used to classify trophic preference of herbivores. First, hypsodonty indices (HI) derived by Janis (1988) were calculated by taking the M3 height and dividing it by the M3 width. Height measurements were taken on unworn teeth and width measurements taken at the occlusal surface. In general, mammals in the grazer category have the highest, or most hypsodont, teeth (mean = 5.18) while omnivores have the lowest (mean = 1.16) (Janis, 1988). Because Janis discovered that habitat preference sometimes affected HI, mandibular indices and carbon isotope data (Sponheimer *et al.*, 1999) were also used to reconstruct trophic preferences of bovids. The collection contains many specimens of each species, and often the index means for species could be used for trophic analysis, rather than predicting trophic behavior through only one specimen. All ungulates were assigned to terrestrial locomotion.

Benefit and McCrossin (1990) have published data on the trophic adaptation of some South African papionins, based on shearing crests [measurements devised by Kay (1984)] and wear pattern. The Makapansgat material was examined according to their methodology and trophic adaptations for these fossil monkeys were predicted (McBratney and Reed, 1999). Locomotor assignments for primates were based on the work of Fleagle (Bown *et al.*, 1982) and Ciochon (1993).

Some of the fossil animals, such as the aardvark (*Orycteropus* cf. *O. afer*), were assigned trophic and/or locomotor adaptations based on similarity to their modern congeners. Because aardvark teeth are unique, it is likely that they were used for the same function by living and fossil species. In addition, the humerus of the fossil aardvark exhibited morphology that was indicative of digging. Unfortunately, until morphological comparative samples are available for all extant trophic groups, taxonomic analogy and qualitative morphological comparisons must occasionally be used as a surrogate for quantitative and tested morphological analysis.

Environmental Reconstructions

Each fossil mammal from each member from the Makapansgat assemblage was assigned a locomotor and trophic category based on craniodental and limb morphology. The percentages of each adaptation present in the large mammal faunas of Members 3 and 4 were calculated. Proportions of adaptations from Makapansgat were then compared with those from the sample of extant communities (Table II). Fossil assemblages were assigned to a vegetation category through a discriminant function analysis (DFA). These analyses not only provided the vegetation reconstruction for each Member, but indicated the ecological community structure as well (Table IV; Fig. 5). Reconstructed habitats for M3 and M4 using the large mammal assemblage suggest that primates existed in bushland with riparian woodland or forest with adjacent edaphic grasslands (Reed, 1997, 1998).

Table IV. Discriminant Function Analysis for Vegetation Types Using Adaptations of Mammalian Communities

(a) Structure matrix (pooled within groups correlations between functions and variables)

Variable	Function 1	Function 2
A	−0.359[a]	0.150
TF	−0.315[a]	0.184
R	−0.080	−0.078
F	0.056	−0.041
I	0.032	−0.077
TA	−0.019	0.073
T	0.115	0.039
MF	0.054	0.015
G	0.282	0.027
OM	−0.055	0.017
B	−0.002	0.059
TC	0.114	0.104
FG	0.022	0.138
AQ	−0.018	0.057
Eigenvalue	139.75	13.83
% Variance	89.4	8.8
Significance	$p<0.001$	$p<0.002$

[a] Largest absolute correlation between variable and all discriminant functions. Abbreviations as in Appendix 1.

(b) F-test between groups (df = 14,11)

Category	CW	B	OW	S	G
Forests	29.629 $p<0.001$	34.677 $p<0.001$	53.567 $p<0.001$	83.570 $p<0.001$	59.672 $p<0.001$
Closed woodland		**1.273** $\boldsymbol{p<0.349}$	3.339 $p<0.026$	13.243 $p<0.001$	14.325 $p<0.001$
Bushland			**1.525** $\boldsymbol{p<0.244}$	7.634 $p<0.001$	9.501 $p<0.001$
Open woodland				4.302 $p<0.010$	5.922 $p<0.003$
Shrubland					**2.003** $\boldsymbol{p<0.126}$

CW = Closed Woodland, B = Bushland, OW = Open Woodland, S = Shrubland, G = Grassland.

Primates as Indicators of Environment

The large mammal community from Makapansgat M3 is represented by 57 species of which eight are primates and seven of those are cercopithecoids (Table V). There have been 33 species recovered from M4 of which six are primates and five are cercopithecids (Table V). In support of the vegetation reconstructions mentioned above, the number of primate species has been shown to indicate mean annual rainfall, such that the greater the number of

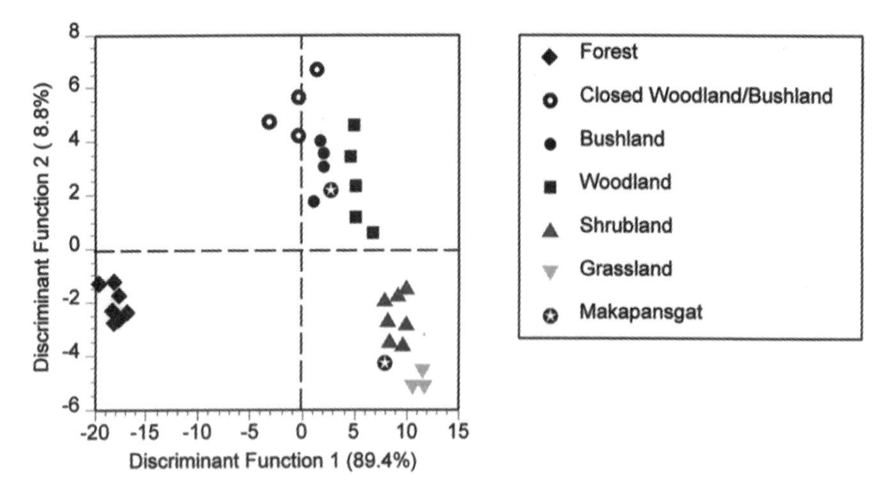

Fig. 5. Discriminant function plot of first two functions using all adaptations of extant communities and assigning a vegetation type to the Makapansgat fossil assemblages.

primate species the greater the annual rainfall (Reed and Fleagle, 1995). Although Kay *et al.*, (1997) have shown that this relationship is not linear in South American primate communities after about 3000 mm of rainfall is reached, bushlands receive considerably less rainfall than this and therefore the linear relationship holds for these communities. In African communities, primate species numbering five or greater indicate mean annual rainfall amounts consistent with closed woodlands and/or bushland that also have riparian forests within the overall vegetation. As this supports the reconstructions based on the Makapansgat large mammal communities, subsequent comparisons of M3 and M4 primate assemblages will be contrasted with extant African bushland and woodland communities.

Community Comparisons

Ecological Adaptations

Makapansgat community structures are similar to, but slightly different from, bushland and woodland extant communities. Figure 6 shows similarities and differences in percentages of each adaptation in M3 and M4 compared with the percentage means of adaptations from extant bushland and woodland habitats. Differences in the Makapansgat communities are more interesting than similarities. There are more browsing animals, meat-eaters (M3), fruit/leaf eaters, and root and bulb consumers, but fewer meat-eaters (M4), fresh grass grazers, carnivore/insectivores, and frugivore/insectivores.

Table V. Species Recovered from Makapansgat Members 3 and 4[a]

Order	Family	Species	Trophic[b]	Locomotor[b]	M3	M4
Artiodactyla	Bovidae	*Aepyceros* sp.	B	T	X	
Artiodactyla	Bovidae	*Alcelaphini* sp.	G	T	X	X
Artiodactyla	Bovidae	*Boselaphini* sp.	B	T	X	
Artiodactyla	Bovidae	*Cephalophus* cf. *C. natalensis*	FL	T	X	
Artiodactyla	Bovidae	*Gazella gracilior*	MF	T	X	
Artiodactyla	Bovidae	*Gazella vanhoepeni*	B	T	X	X
Artiodactyla	Bovidae	*Hippotragus cookei*	G	T	X	
Artiodactyla	Bovidae	Large impala species	MF	T	X	
Artiodactyla	Bovidae	*Makapania broomi*	MF	T	X	X
Artiodactyla	Bovidae	*Megalotragus* sp.	G	T	X	
Artiodactyla	Bovidae	*Neotragini* sp.	B	T	X	X
Artiodactyla	Bovidae	*Oreotragus oreotragus*	B	T	X	
Artiodactyla	Bovidae	*Parmularius braini*	G	T	X	
Artiodactyla	Bovidae	*Parmularius* sp. nov.	G	T	X	
Artiodactyla	Bovidae	*Redunca darti*	FG	T	X	X
Artiodactyla	Bovidae	*Simatherium* cf. *S. kohllarseni*	MF	T	X	X
Artiodactyla	Bovidae	*Tragelaphus pricei*	B	T	X	X
Artiodactyla	Bovidae	*Tragelaphus* sp.	B	T	X	X
Artiodactyla	Bovidae	*Tragelapus* sp. aff. *T. angasi*	B	T	X	X
Artiodactyla	Giraffidae	*Giraffa* cf. *G. jumae*	B	T	X	X
Artiodactyla	Giraffidae	*Sivatherium marusium*	B	T	X	
Artiodactyla	Hippopotamidae	*Hippopotamus amphibius*	FG	AQ	X	
Artiodactyla	Suidae	*Notochoerus capensis*	G	T	X	X
Artiodactyla	Suidae	*Potamochoeroides shawi*	OM	T	X	X
Carnivora	Canidae	*Canis* cf. *C. adustus*	Cl	T	X	
Carnivora	Canidae	*Canis* cf. *C. mesomelas*	Cl	T	X	X
Carnivora	Canidae	*Vulpes chama*	Cl	F	X	X
Carnivora	Felidae	*Acinonyx* cf. *A. jubatus*	C	T	X	
Carnivora	Felidae	*Dinofelis* sp.	C	TA	X	X
Carnivora	Felidae	*Felis* cf. *F. lybica*	C	TA	X	
Carnivora	Felidae	*Felis serval*	C	TA	X	
Carnivora	Felidae	*Felis* sp. nov.	C	TA	X	
Carnivora	Felidae	*Homotherium* sp.	C	TA	X	
Carnivora	Felidae	*Lynx* cf. *L. caracal*	C	T	X	
Carnivora	Felidae	*Panthera* cf. *P. pardus*	C	TA	X	
Carnivora	Hyaenidae	*Crocuta* sp.	CB	T		X
Carnivora	Hyaenidae	*Hyaena makapani*	CB	T	X	X
Carnivora	Hyaenidae	*Pachycrocuta* sp.	CB	T	X	
Carnivora	Mustelidae	Otter sp.	C	AQ	X	
Carnivora	Viverridae	*Crossarchus* sp.	Cl	T		X
Carnivora	Viverridae	*Galerella* sp.	C	T	X	
Hyracoidea	Procaviidae	*Gigantohyrax maguirei*	MF	T		X
Hyracoidea	Procaviidae	*Procavia antiquus*	MF	T		X
Hyracoidea	Procaviidae	*Procavia* sp.	MF	T	X	
Hyracoidea	Procaviidae	*Procavia transvaalensis*	MF	T		X

(Cont.)

Table V. *Continued*

Order	Family	Species	Trophic[b]	Locomotor[b]	M3	M4
Lagomorpha	Leporiidae	*Lepus* sp.	G	T	X	X
Perissodactyla	Chalicotheridae	*Ancylotherium hennegi*	B	T	X	
Perissodactyla	Equidae	*Hipparion* cf. *H. lybicum*	G	T	X	
Perissodactyla	Rhinocerotidae	*Ceratotherium simum*	G	T	X	X
Perissodactyla	Rhinocerotidae	*Diceros bicornis*	B	T	X	
Primates	Cercopithecidae	*Cercocebus* sp.	FL	?	X	X
Primates	Cercopithecidae	*Cercopithecoides williamsi*	FL	TA	X	X
Primates	Cercopithecidae	*Colobinae gen et sp indet*	FL	?	X	
Primates	Cercopithecidae	*Parapapio broomi*	FL	TA	X	X
Primates	Cercopithecidae	*Parapapio jonesi*	FL	A	X	X
Primates	Cercopithecidae	*Parapapio whitei*	FL	TA	X	X
Primates	Cercopithecidae	*Theropithecus darti*	FLG	TA	X	X
Primates	Hominidae	*Australopithecus africanus*	FL	T	X	X
Proboscidea	Gomphotheriidae	*Anancus kenyensis*	B	T		X
Rodentia	Hystricidae	*Hystrix africaeaustralis*	R	F	X	X
Rodentia	Hystricidae	*Hystrix makapanensis*	R	F	X	X
Rodentia	Hystricidae	*Xenohystrix crassidens*	R	F	X	
Rodentia	Thryonomidae	*Thryonomys* sp.	FG	AQ		X
Tubulidentata	Orycteropidae	*Orycteropus* cf. *O. afer*	I	F	X	

[a]A = arboreal locomotion, AQ = Aquatic locomotion, F = fossorial locomotion, T = terrestrial locomotion, TA = terrestrial/arboreal locomotion, B = leaves, C = meat, CB = meat/bone, Cl = meat/invertebrates, FG = fresh grass, FI = fruit/insects, FL = fruit/leaves, G = grass, I = insects, MF = grass/leaves, OM = omnivore, R = roots/tubers; x = present.

[b]Trophic and locomotor assignments are based on information found in the text.

Browsing animals are those that consume C_3 or dicot plants. Extant bushland and woodland communities consist of 2%–12% large mammals exhibiting a browsing adaptation. Makapansgat M3 has 21.1% and M4 has 20.6% browsing animals in their communities. The most similar extant sites to M3 are those that are montane: the forest atop Kilimanjaro (15.8%) and the forest–moorland region of Aberdare National Park (12.2%). Greater percentages of browsing mammals exist in many African Pliocene localities that have been reconstructed as bushland or woodland vegetation (Reed, 1997). These browsing mammals consist of large-sized species that no longer exist in extant communities: four giraffoids, a chalicothere, a browsing elephant, as well as many browsing bovids. The Makapansgat assemblages are unique in only having a single giraffe species (*Giraffa* cf. *G. jumae*), but there are several more browsing antelope. In fact, the dominant gazelle (*Gazella vanhoepeni*) and a fossil impala species (*Aepyceros* sp.) are both reconstructed as browsers, although their extant congeners are mixed feeding animals (Sponheimer *et al.*, 1999).

More meat-eating carnivores have been recovered from M3 than found in extant communities. These mammals include several large felid species that have no extant counterparts (saber-toothed cats). More fruit-and-leaf eating

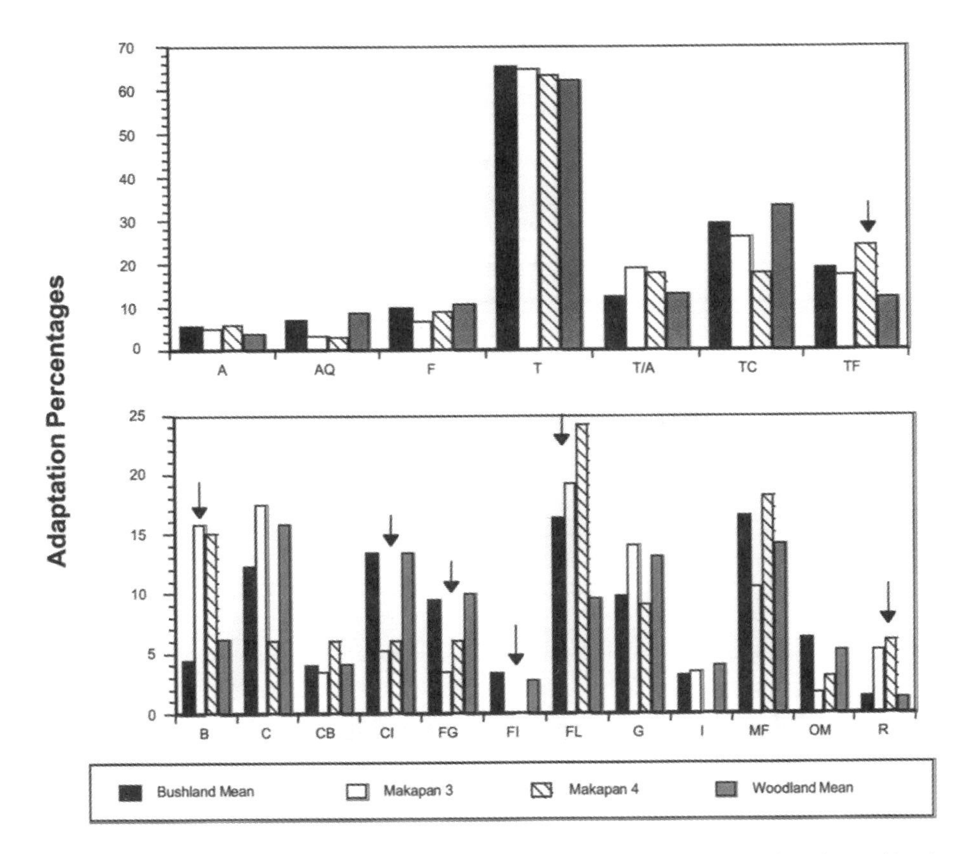

Fig. 6. Comparisons of adaptations between extant means of bushland and woodland communities with Makapansgat M3 and M4. Bushland and woodland localities are listed in Table II. Arrows indicate differences between extant and fossil assemblages discussed in the text.

species have been recovered from both members, and this has to do with a greater representation of baboon species, which will be discussed later.

Root and bulb eating mammals include only porcupine species, of which there are three at Makapansgat, including the giant *Xenohystrix crassidens*. These porcupines were assigned to the root and bulb category because that is what extant African porcupines eat. The teeth are virtually identically in all three species, but different in size. Further work may show that they ate something different, but it appears that the number of these adaptations is unique compared with extant bushlands and woodlands.

Fresh grass grazers are abundant in extant eastern African bushlands and woodlands that have wetlands, swamps, or flood plains for at least part of the year. There are two fresh grass-grazing species from each member at Makapansgat that simply indicate that although there are edaphic grasslands, they were not as extensive as those that support a greater diversity of animals.

M4 has fewer meat-eating felids than one might expect from their representation in M3 or from extant communities. This is probably due to accumulator bias in M4. Most of the fossils recovered from this member were victims of leopards or large birds of prey. These taphonomic aspects will be discussed below.

Finally, carnivore/insectivores and frugivore/insectivores are generally small-bodied animals in extant communities, for example, genets, mongooses, and guenons. These small-sized mammals range from about 500 g to about 5 kg. As will be discussed below, the possibility exists that they were not accumulated in the Makapansgat deposits due to predator selection for larger prey.

The community contexts for the extinct primate species at Makapansgat are similar to extant bushland and woodland communities. The major differences that could affect primate species behaviors are the presence of more browsing and root/bulb eating animals, and the fact that there are more primate species reconstructed as fruit and leaf eaters when compared with extant communities.

There could be several causes for the presence of more browsing animals in Pliocene localities. First, the productivity of the ecosystem could have been higher three million years ago. That is, bushland vegetation architecture could have been the same as extant communities, but the underlying productivity could have been greater. Productivity is the measurement of energy transfer: from sunlight to plant parts in primary production, and from leaves to feeding material for other consumers in secondary production. Productivity has been measured in a variety of ways in extant communities, including litterfall (Kay *et al.*, 1997), standing crop biomass, biomass production, or amount of light absorbed by plants, and fixation of carbon dioxide (Johnson *et al.*, 1996). Increased biodiversity is often a result of greater community productivity (Johnson *et al.*, 1996; Kay *et al.*, 1997), thus increased numbers of browsing species in the Pliocene could reflect this phenomenon.

Second, recent historical circumstances could be influencing the percentage of these browsing species. Bushlands, woodlands, and eventually grasslands developed and expanded over Africa from the Miocene forward in time, and seasons were becoming more pronounced (Singh, 1988). There were higher percentages of browsing animals in the Miocene (Van Couvering and Van Couvering, 1976). Thus, the high percentages of browsing species at the mid-Pliocene Makapansgat could be the result of relatively recent historical events. Browsing animals in Pliocene communities are more plentiful than in extant communities. Lineages that possess browsing species (e.g., bovids and giraffids) from this epoch contain congeners that have gradually changing adaptations to other sources of food (bovids) or have several congeners or species that go extinct (giraffids) by the end of the Pleistocene. Percentages of mixed feeders and grazers in bushland, woodland, and shrubland communities are higher in Pleistocene localities than they are in Pliocene localities, indicating an overall community shift dependent upon changing environments. These mixed feeders and grazers include genera that were browsers in the Pliocene (Reed, 1997; Sponheimer *et al.*, 1999). This scenario is supported at Makapansgat by two bovids that are reconstructed as browsers, but whose extant congeners are mixed feeders (preferring grass in one case).

A third suggestion for the increased percentages of browsing mammals is interrelated with the presence of megamammals. Other than bovid species, many other browsers recovered from Makapansgat are larger than extant mammals and, for whatever reason, many of these giants became extinct during the Plio-Pleistocene. Thus, the relationship of browsing animals with a changing environment and/or increased productivity may be moot if it is the disappearance of megamammals that is crucial to these percentage differences in extant and extinct communities. Finally, the differences between the extant communities and these fossil assemblages could simply be stochastic.

The ecological reconstructions of bushland with riverine forest and flood plains in the near vicinity are based on knowledge of extant African ecosystems. There are, however, differences that, while not affecting the reconstruction of overall vegetation, indicate that the primate community and ecosystem could have been structured somewhat differently in the past. We can therefore look at primate community structure, and primate behavior within that context.

Mammals that Share Niche Space with Primates

The space that is occupied by each species in a community is called its ecological niche. It is assumed that every creature differs from every other in some aspect of that niche. For example, animals that eat the same food often have different body sizes, or animals similar in trophic adaptation and body size may differ in locomotor adaptation. There are also numerous niche aspects that cannot be discerned for fossil or extant species that separate them from other species (Hutchinson, 1978).

In extant African bushland communities, many primates utilize frugivore/folivore trophic niches that are also occupied by small bovid species. Bovids that eat fruit, flowers and/or leaves include: *Madoqua kirkii* (dik-dik), *Nesotragus moschatus* (suni), *Cephalophus monticola* or *C. natalensis* (small duikers), and *Sylvicapra grimmia* (bush duiker). There may be two or three in any bushland habitat. These bovids are terrestrial and eat fallen fruit, flowers, and leaves. Primates that might share these resources in extant bushlands and woodlands include *Papio* species and *Chlorocebus aethiops*. *Erythrocebus patas* might share food resources with bush duikers in more open areas. However, it is rare for all three of these primates to be present in a single community in extant bushlands.

Extant savanna baboons are also known to consume roots and tubers along with fruits, grass seeds, and other herbaceous material (Fleagle, 1998). Other larger mammals that eat roots and tubers are few, but include the two porcupine species present in eastern (*Hystrix cristata*) and southern (*H. africaeaustralis*) Africa and possibly the bush pig (*Potomochoerus porcus*).

A slightly different pattern is evident in the fossil assemblages. There are two bovid species whose dental morphology indicates a possible adaptation to fruit and leaves: a fossil neotragine (dwarf antelope) recovered from M3 and M4, and a *Cephalophus* species (duiker) recovered from M3. However, these two bovids are likely sharing resources with at least four primate species. In addition, there are three porcupine species from the fossil assemblage whose teeth indicate they were likely eating very similar substances to the extant porcupines. Thus, there are fewer fruit and leaf consuming bovids, but more fruit and leaf-consuming primate species; species that are possibly foraging for roots and tubers are greatly increased in the fossil localities.

Primate Taxonomic Comparisons and Extant Behavior Summaries

Sympatric primates in extant African woodland and bushland communities, usually consist of two prosimians, at least one colobine, a guenon, a vervet, and one papionin (Table VI). At Makapansgat M3, there are no prosimians, two colobines, and five papionins. M4 has only one colobine, but is otherwise the same. This taxonomic composition of primates is not unique in the African Pliocene, but is unknown in extant communities. Percentages of primates in relation to other mammals, and percentages of infraorder and tribal representations of primates against total primates are depicted in Table VI. There is a definite distinction between the extant communities and the fossil ones.

Prosimians or small *Cercopithecus* species have not been recovered from either of the fossil assemblages. It is possible that diurnal fruit and insect-eating primates did not exist in this community as none have been recovered. However, *Galago* species, at least, have been recovered from other Pliocene fossil assemblages (e.g., Laetoli). The lack of prosimians in this deposit might be a taphonomic bias because none of the accumulators of these fossil assemblages preyed upon small, mostly nocturnal primates (see below). Nevertheless, there are more anthropoid species recovered from each fossil assemblage than found in any extant bushland/woodland community.

Colobines account for 14% to 17% of the primate species in extant bushland communities; M3 contains 28.6% colobine species. This is simply a difference between one and two species. In extant communities, this could be a recent, human-induced factor as several closed woodland/bushland communities often have two colobine species. Having two colobines is a common Pliocene phenomenon with most south and east African fossil assemblages bearing two colobines: one large (e.g., *Paracolobus*) and one somewhat smaller

Table VI. Primate Species in Fossil and Extant Communities and Percentages of Primate Higher Order Representations[a]

Species	M3	M4	West Lunga NP	Natal Lowlands
	Parapapio jones	*P. jonesi*	*Papio cynocephalus*	*P. ursinus*
	P. whitei	*P. whitei*	*Cercopithecus aethiops*	*C. aethiops*
	P. broomi	*P. broomi*	*Cercopithecus mitas*	*C. mitas*
	Theropithecus darti	*T. darti*	*Otolemur crassicaudatus*	*O. crassicaudatus*
	Ceropithecoides williamsi	*C. williamsi*		*G. molholi*
	Cercocebus sp.	*Cercocebus* sp		
	Colobinae sp.			
	Lake Mweru	**Serengeti Bush**	**Serengeti NP**	**Rukwa Valley**
	P. cynocephalus	*P. ursinus*	*P. ursinus*	*P. cynocephalus*
	Colobus guereza	*C. guereza*	*C. guereza*	*C. aethiops*
	C. aethiops	*C. aethiops*	*C. aethiops*	*C. mitas*
	O. crassicaudatus	*C. mitas*	*C. mitas*	*O. crassicaudatus*
	E. patas	*E. patas*	*E. patas*	*G. senegalensis*
	O. crassicaudatus	*O. crassicaudatus*	*O. crassicaudatus*	
	G. senegalensis	*G. senegalensis*	*G. senegalensis*	
Percentages	**M3**	**M4**	**West Lunga NP**	**Natal Lowlands**
Primates	12.28%	18.18%	8.8%	12.5%
Prosimians	0%	0%	2.22%	5.0%
Anthropoids	12.28%	18.18%	6.67%	7.5%
Papionins	8.88%	15.15%	2.22%	2.5%
Cercopithcini	0%	0%	4.44%	2.5%
Colobini	1.75%	3.03%		0%
	Lake Mweru	**Serengeti Bush**	**Serengeti NP**	**Rukwa Valley**
Primates	11.43%	10.94%	9.34%	9.62%
Prosimians	2.86%	3.13%	2.66%	3.85%
Anthropoids	8.57%	7.81%	6.67%	5.77%
Papionins	2.86%	1.56%	1.33%	1.92%
Cercopithicins	2.86%	1.56%	1.33%	3.85%
Colobinins	2.86%	1.56%	1.33%	0%

[a] Primate species in fossil and extant communities and percentages of primate higher order representations. KEY: M3,Member 3 Makapansgat; M4, Member 4 Makapansgat; West Lunga NP, West Lunga National Park, Zambia; Natal Lowlands, Closed woodland/bushland regions in KwaZulu-Natal, South Africa; Lake Mweru, Lake Mweru National Park, Botswana; Serengeti Bush, Serengeti National Park, bushland regions only, Tanzania; Serengeti National Park, all regions of park, Tanzania; Rukwa Valley, Rukwa Valley, Kenya.

(e.g., *Cercopithecoides, Rhinocolobus*). Extant African colobine social organization tends to be single male groups (Oates, 1977). Colobines eat leaves, fruits, and seeds, and are arboreal compared with extant papionins. They are somewhat sexually dimorphic and range in body mass from about 4.2 kg to 13.5 kg (Fleagle, 1998). Fossil species differ from this pattern and are discussed below.

Papionins contribute to a great taxonomic difference between primates in extant and Makapansgat communities. In most extant, and indeed, most later

Pleistocene communities (forest through grassland) there is only one or at most two papionin species. Usually this consists of one baboon or drill species and a mangabey, although two mangabeys are often found together in forests. In the Ituri Forest, there are two mangabeys and a baboon species (Lanjaow, 1987), as greater numbers of papionin species usually occur in extant forest communities. Five papionin species have been recovered from the Makapansgat fossil assemblages.

Extant papionin social organization is variable, and all the species are sexually dimorphic. Body mass ranges from 5.6 kg to about 30 kg. The "savanna" baboons live in habitats ranging from forests to grasslands in large, multi-male troops, with the males emigrating to other troops. They forage terrestrially and arboreally, eating fruit, roots, insects, gums, and some fauna (Fleagle, 1998), and always sleep in trees or cliffs. *P. hamadryas* groups are small, single-male foraging groups that come together with other small groups to sleep at night. Geladas live in the highlands of Ethiopia and forage in small, single-male groups. They join in large groups at certain times of the year. They eat mostly grass seeds, roots, and sometimes fruit. Extant *Cercocebus* species tend to prefer swampy areas of forests and eat fruits, hard nuts, seeds, and invertebrates, all of which they forage for on the ground. They also have groups that can be small, or come together in larger troops (Fleagle, 1998). Again, reconstructed behaviors of the fossil species differ from these patterns.

Finally, a fossil hominin has been recovered from both members at Makapansgat. Although bipedal hominins are found in extant bushlands, *Australopithecus africanus* is assumed to have differed greatly in behavioral ecology from extant hunters and gatherers.

Behavioral Ecology of Primates in Extant and Makapansgat Communities

I will compare the overall behavioral ecology of Old World monkeys found in extant African communities with fossil monkeys from Makapansgat; that is, I will compare the percentages of various trophic and locomotor behavior, as well as body mass. Calculations were made among primates only (Table VII).

Primate Substrate and Locomotor Comparisons

Extant communities have higher percentages of arboreal primates. This is partially the result of the lack of prosimians in the Makapansgat communities. If prosimians are not included in the extant calculations, then the percentages of arboreal primates are more similar among communities. One small ulna

Table VII. Percentages of Primate Adaptations in Fossil and Extant Communities[a]

Adaptations	M3	M4	West Lunga NP	Natal Lowlands	Lake Mweru	S Bushland	Serengeti NP	Rukwa Valley
Fruit and leaves	57.14%	66.67%	50%	40%	50%	28.57%	28.57%	40%
Fruit and Grass Seeds	14.3%	16.67%	25%	20%	0%	14.29%	14.29%	0%
Fruit and Insects	0%	0%	25%	40%	25%	42.86%	42.86%	60%
Leaves and Fruit	28.57%	16.67%	0%	0%	25%	14.29%	14.29%	0%
Arboreal	42.86%	33.33%	50%	60%	25%	57.14%	57.14%	60%
Ter/Arboreal	57.14%	66.67%	50%	40%	25%	28.57%	28.57%	40%
Terrestrial	0%	0%	0%	0%	0%	14.29%	14.29%	0%

[a]The distribution of percentages is based on the total number of primate species in each community. Localities as in Table VI.

from the Makapansgat collection, possibly belonging to either *Cercocebus* or *Parapapio jonesi*, has arboreal characteristics, thus at least one of these species is arboreal. The colobine species are a completely different story. The large colobine species that is not *Cercopithecoides williamsi* is only represented by three dental specimens. Large size does not indicate terrestriality, however, as large *Paracolobus* species are reconstructed as arboreal. Therefore, the substrate and locomotor pattern of this species is unknown. *C. williamsi* was likely utilizing both the ground and the trees for foraging (Ciochon, 1993). This is in contrast to east African *Cercopithecoides* sp. that are considered hyper-terrestrial (Birchette 1988; Fleagle, 1998). *C. williamsi* is a quadrupedal walker or runner with no leaping ability indicated in its postcranial anatomy (Ciochon, 1993).

Most of the papionins are terrestrial/arboreal. This substrate classification means that these monkeys foraged both on the ground and in trees, or traveled between feeding patches on the ground. Ciochon (1993) and others have suggested that *Parapapio* species are quite similar to extant *Papio* postcranially, thus indicating that these species likely moved terrestrially and climbed trees. Lucas and Teaford (1994) suggest it is possible that *Parapapio* species could have been more arboreal based on expected vs. observed microwear characteristics of the teeth.

Krentz (1993) suggests that *Theropithecus darti* at Hadar likely had arboreal tendencies, but also notes that the morphological defining characters are not distinct in this species. Ciochon (1993) suggests that *T. darti* at Makapansgat is also more arboreal than the living gelada. Therefore, papionin species from Makapansgat appear to be similar to extant papionins in substrate use and locomotor pattern. There appear to be no strictly terrestrial primates from the Makapansgat fossil assemblages, but only one primate that is terrestrial from any extant locality in this study (*Erythrocebus patas*).

Most of the primate species recovered from Makapansgat were likely foraging in trees and terrestrially, and traveling between food patches terrestrially as well. In addition, most of the fossil species appear to be quadrupedal walkers and runners, whether on the ground or in the trees. None of the postcranial material examined had evidence of leaping or climbing morphology (Ciochon, 1993). *A. africanus* was a biped that probably climbed trees to forage or to escape predation. Although the number of colobine species is greater in these fossil communities, the lack of arboreal colobine species makes these fossil communities unique in substrate and locomotor structure.

Primate Trophic Comparisons

Primate species in extant bushland communities eat 28% to 50% fruit and leaves. M3 and M4 contain 57% and 67% primates, respectively, that are categorized as fruit-and-leaf eaters. Four Makapansgat primate species probably ate primarily fruit and leaves, and three of those are closely related species. According to McBratney and Reed (1999) *P. jonesi* is estimated to eat

58% fruit and 28% leaves, *P. whitei* is estimated to eat 53% fruit and 33% leaves, and *P. broomi* 68% and 13%. These are fairly similar percentages of fruits, although *P. broomi* (n = 5) appears to be more frugivorous. In any case, these primates appear to be eating very similar vegetation. Ungar and Teaford (1996) note that *Parapapio* sp. from East Africa have less occlusal pitting and suggest that (1) microwear indicates that they had fewer small, hard objects in their diets, and (2) nonocclusal microwear indicates that they ingested grit, perhaps indicating terrestrial foraging or more airborne dust in Pliocene fossil localities. Lucas and Teaford (1994) suggest that *Parapapio* species ate less roots and tubers (less pitting on the occlusal surfaces), but otherwise their diet was similar. Perhaps the three porcupine species at Makapansgat had free reign over the hard roots and tubers. Jablonski (1993) suggests that *Theropithecus darti* ate grass seeds. Shearing crests and wear patterns, however, also suggest high percentages of fruit in the diet of *T. darti* (McBratney and Reed, 1999). *T. darti* probably ate fruit, leaves, and grass seeds, and was most similar to *E. patas* in extant bushland communities.

Shearing crests of the fossil colobine species suggest that they ate more fruit that extant species (McBratney and Reed, 1999). Ungar and Teaford (1996) suggest that *Cercopithecoides* (East Africa) might also have been foraging on the ground because of more nonocclusal grit as seen in microwear. Finally, *A. africanus* was probably eating tough herbivorous foods as indicated by its masticatory system and microwear, which shows little evidence of pitting caused by grit (Grine, 1986).

Trophically, these fossil primate communities are not structurally similar to extant bushland communities. The ability to reconstruct "partitioned" diets of these similar taxa has proved to be rather difficult. An overall impression, however, is that most of the species in the fossil assemblage were eating fruits, vegetation, seeds, etc. in possibly different amounts than related species in extant communities.

Primate Body Size Comparisons

One of the most interesting differences between the Makapansgat primate guilds and those from extant communities is the range of body sizes (Table VIII). In extant bushland communities, the largest primate is always a *Papio* species, with extreme sexual size dimorphism. Females average about 15 kg while males can be as large as 31kg (Smith and Jungers, 1997). Extant African colobines and other cercopithecines range from 3 kg to 12 kg (Fleagle, 1998). However, the primate species from M3 and M4 range from 13 kg to 45 kg. I have estimated the fossil primate species body masses from the area of m_1 (Gingerich *et al.*, 1982). Use of dental variables to predict body size may over- or underestimate the mass, however. To test the accuracy of estimating the fossil body sizes with the area of M1, I estimated the mass of 103 *Papio ursinus* specimens, 45 females and 68 males (data from Eisenhart, 1974). The male

Table VIII. Estimated Body Sizes of Primates in M3 and M4 Makapansgat

Species	Sex	N	M1 area	Body mass
Theropithecus darti	M	2	109.41	39 kg
	F	1	95.34	32 kg
Parapapio jonesi	M	1	78.22	24 kg
	F	3	64.5	18 kg
P. broomi	M	5	72.95	21 kg
	F	2	56.2	14 kg
P. whitei	M	1	80.56	25 kg
Cercocebus sp.	M	1	61.34	16.5 kg
	F	1	53.29	13.4 kg
Cercopithecoides williamsi	M	2	69.6	20 kg
	F	2	62.2	17 kg
Colobinae gen. et. sp. indet.			153.67 (m3)	30 kg

mean for *P. ursinus* was within a kilogram of the *P. ursinus* mean reported by Smith and Jungers (1997). The female mean, however, was 10 kg greater than the reported mean of 14.8 kg. It is evident that while the M1 estimates of the fossil male papionins are in the correct ballpark, the female estimates cannot be used. Nevertheless, using the male estimates, the maximum sizes of these primates, as a group, exceed those of the primate communities in extant woodlands and bushlands.

Postcranial material is not associated with cranial material at this cave site, so it is impossible to use joint surface predictors to estimate body mass. Estimated body mass of *T. darti* at Makapansgat, using M1 area, is greater than that of *A. africanus* as estimated using postcranial material (Fleagle, 1998). Krentz (1993) estimated *T. darti* body mass from postcrania of the Hadar specimens to be15 kg. Eck (1993) suggests that *T. darti* at Makapansgat is about 4% larger than those specimen from Hadar. If this were used to suggest the body mass of *T. darti* at Makapansgat, it would be approximately 17 kg. This smaller body size for the *T. darti* specimens at Makapansgat is unlikely, as specimens are some of the largest in these assemblages.

Primates in the body size ranges of *Chlorocebus aethiops*, *Cercopithecus mitis*, or *Colobus guereza* have not been recovered from these fossil assemblages. As body size is used to infer much about behavior, it is important to understand why, as a group, primate species at Makapansgat are larger than primates from modern bushland/woodland communities. The obvious reason is that there are no *Cercopithecus* species in the fossil assemblages. The species are mostly papionins that have body masses and canine dimorphism similar to extant papionins. Fossil colobines from this and other Pliocene localities are also considerably larger than extant ones. The estimated masses of *Cercopithecoides* and *Paracolobus* are as much as 30 kg. Finally, *A. africanus* males are estimated to have had a mass of about 40.8 kg (Fleagle, 1998).

Primate Interactions and Behavior Based on Accumulating Agents

Makapansgat Member 3

Primates recovered from M3 were preyed upon by *Hyaena makapani* (Maguire *et al.*, 1980; Reed, 1996). Primates make up about 1.4% of the specimens of M3 when both cranial and postcranial material are considered (Reed, 1996). Primates are a limited contribution to the total assemblage compared with artiodactyls, which are 86% of the total species number. There are two probable reasons for fewer primates. First, they were possibly adept at avoiding hyena predation, because they could climb trees to avoid this predator. Second, the average-sized prey for these hyenas was about 100 kg (Reed, 1996). Consequently, primates could have been too small for hyenas to hunt or scavenge on a regular basis.

Representation of the primate adults and juveniles and sex ratios can contribute to inferences about primate behavior. Of the 47 minimum number of primate individuals recovered, 40 are adults and only seven are juveniles (Table IX). Superficially, this appears to be a taphonomic bias because hyenas may have been able to completely consume juvenile individuals. However, 50% to 60% of *C. williamsi*, *P. whitei*, and *T. darti* remains are male individuals. These are all large primate species and are probably the most terrestrial. The preponderance of hyena male victims may indicate that the social organization of these primates was such that males defended the group. Cowlishaw (1994) has observed interactions of *Papio* with felid predators and noted that male baboons are more likely to fight back than females. Although the males can sometimes win an encounter, the predators can and will dominate the males. In fact, leopards appear to prefer the large males. Iwamoto *et al.*, (1996) observed *Theropithecus gelada* males jointly attacking both dogs and a leopard.

Alternatives to the male defense scenario are few. First, females could have been subject to more complete hyena consumption as they are smaller than the males. Second, males could have formed subgroups that were more vulnerable to attack. Extant male, primate groups are usually composed of young adult males. However, while the ages, based on tooth wear, of the Makapansgat M3 male specimens vary from young adults to mature individuals, it is more likely that the males attacked hyenas in protection of the troop and lost the encounters.

Makapansgat Member 4

In contrast to the low percentages of primates in M3, primate specimens constitute 51% of the total M4 assemblage. This member was accumulated by large birds of prey and by leopards (Maguire *et al.*, 1985; Reed, 1996). Primates could have been the preferred prey of eagle-like birds. Black eagles

Table IX. Distribution of Primate Specimens Recovered from Makapansgat M3 and M4[a]

Makapansgat M3 Species Distribution

Species	NISP	MNI	ADULT	Juv	F	M	I
Cercocebus sp.	4	4	2	2	1	0	2
C. williamsi	17	9	8	1	3	5	1
Colobinae sp.	3	2	2			1	1
P. broomi	53	15	13	2	2	4	7
P. Jonesi	22	5	4	1	2	1	2
P. whitei	7	6	5	1		6	
T. darti	27	6	6		2	4	
Total	133	47	40	7	10	21	13

Distribution Percentages by Category

Species	NISP%	MNI%	ADULT%	JUV%	F%	M%	I%
Cercocebus sp.	3%	6.5%	50%	50%	25%	0%	75%
C. williamsi	12.8%	19.6%	89%	11%	33%	56%	11%
Colobinae sp.	2.26%	4.4%	100%	0%	0%	50%	50%
P. broomi	39.9%	32.6%	87%	13%	13%	27%	47%
P. jonesi	16.5%	10.8%	80%	20%	40%	20%	40%
P. whitei	5.26%	13%	83%	17%	0%	100%	0%
T. darti	20.3%	13%	100%	0%	40%	60%	0%
Total overall distribution		100%	85.1%	14.8%	21.3%	44.7%	27.6%

Makapansgat M4 Species Distribution

Species	NISP	MNI	ADULT	JUV	F	M	I
Cercocebus sp.	2	2	2		2		
C. williamsi	17	7	5	2	3	3	1
P. broomi	23	10	7	3	4	1	5
P. Jonesi	10	5	5		4	1	
P. whitei	10	6	6		1	5	
T. darti	24	14	10	4	8	5	1
Total	86	44	35	9	22	15	7

Distribution Percentages by Category

Species	NISP%	MNI%	ADULT%	JUV%	F%	M%	I%
Cercocebus sp.	2.3%	4.6%	100%	0%	100%	0%	0%
C. williamsi	19.7%	15.9%	71%	29%	42.8%	42.8%	14.3%
P. broomi	50%	22.7%	70%	30%	40%	10%	50%
P. jonesi	11.6%	11.4%	100%	0%	80%	20%	40%
P. whitei	11.6%	13.6%	100%	0%	20%	80%	0%
T. darti	27.9%	31.8%	71%	28.6%	57%	36%	7%
Total overall distribution		100%	79.5%	20.5%	50%	34%	16%

[a] Distribution of primate specimens recovered from Makapansgat M3 and M4. Percentages of adults, juveniles (JUV), females (F), males (M), and indeterminate (I) are based on total MNI (minimum numbers of individuals). NISP is number of identifiable specimens.

in southern Africa today focus on a preferred species. The presentation is such that there are many more adults than juvenile specimens (Table IX). However, most of the adult specimens are females, and most of the male specimens are juveniles. I suggest that most of these specimens are female and juvenile males, because they weighed less than adult males and were thus easier for eagles to carry. Females and juveniles could have also been easier for the eagles to hunt in some way that is related to their social organization. This taphonomic bias also supports extreme sexual size dimorphism in these fossil papionins. When the primate species foraged on the ground, they may have had a wide group spread. This might have made them easier targets from the air. When troops of primates foraged in trees, it is possible that the females and juveniles were at a higher height, either because they weighed less and could get out to where they were more exposed, or because the males were defending the base of the trees.

Integration of Community Comparisons

Primates are a product of their historical circumstances and their adaptations to surrounding environments. In addition, despite their historical circumstances and phylogenies, they also exist together as sympatric species dividing a common vegetation type and habitat. Sympatric primate species have partitioned the resources that they utilize in their environments for several reasons. One reason is to reduce competition with other primates and other mammals. Another reason may stem from what each species is obligated to do during a lean trophic period, such as the reduction of fruiting plants in the dry seasons of African bushlands. Many studies of primate species have shown that very similar species do not overlap in range, or effectively partition resources. Resource partitioning could occur through primates eating different plant parts of the same plant species as in the three species of *Hapalemur* (Wright, 1988). Partitioning could also occur by procuring similar foods in different ways that successfully allows access to different patches of foods (Horovitz and Meyer, 1997). Closely related species are often separated in ecological space from one another by the amounts of various food groups that they eat, or by different locomotor patterns (Fleagle and Reed, 1996).

Primates from the Makapansgat fossil assemblages probably possessed some behaviors that were similar to those exhibited by primates in extant bushland communities. *Parapapio* species were probably similar to extant *Papio* species and foraged in closed to open areas, on both the ground and in trees. It is also probable that they had similar social organizations to extant savanna baboons. *Cercocebus* sp. and/or *P. jonesi* likely inhabited more riparian regions. *C. williamsi* and *T. darti* perhaps had behavioral adaptations that are unique and therefore have no modern counterparts.

Cercocebus and *Lophocebus* species now occur only in forest environments despite the fact that one group is strictly arboreal and the other utilizes more terrestrial locomotion. In addition, the *Cercocebus* group is suggested to be most closely related to *Mandrillus* and have unique adaptations for foraging on forest floors (Fleagle and McGraw, 1999). Eisenhart (1974) suggested that *Cercocebus* species from Makapansgat preferred the gallery forest of this ancient community. As there are very few fossil specimens of *Cercocebus* compared with the other papionins, it is plausible that this species preferred the riparian forest. Fewer specimens of a species from this assemblage may indicate that they did live in the riparian environment as otter and hippo specimens from this assemblage are also few in number. This suggests that the accumulators were not scavenging or hunting much in these riparian regions. Extant hyena species are not often found in riverine forests. As *Cercocebus* species are more closely related to *Mandrillus*, this might indicate that the fossil *Cercocebus* were foraging on this forest floor, thus avoiding eagles or other birds of prey that hunt in the tops of the trees.

Parapapio species are the most abundant primates in Member 3. They probably inhabited the bushland, as well as the riparian area. It is probable that the M3 hyenas hunted and scavenged more frequently in the bushland and associated woodland. As *Cercocebus* and the large colobine species are the least numerous in these assemblages (Table IX), it follows that they existed predominantly in the forest while the *Parapapio* species did not. Even though *Parapapio jonesi* was possibly arboreal, the other *Parapapio* species may have been in competition such that they partitioned the fruits and leaves similar to *Hapalemur* species (Wright, 1988). Another possibility is that there were greater abundances of fruiting plants in Pliocene bushlands that could support more papionin species. Still another alternative is that there are only two species of *Parapapio* (McKee, 1991) and they have split their niches such that one is mostly arboreal and the other travels more on the ground.

Finally, *T. darti* was the largest papionin species from these assemblages, although smaller than some later *Theropithecus* species. Shearing crest data indicate that this species probably ate some fruit, but also leaves and probably grass seeds. Therefore, it probably foraged in open areas or possibly in the edaphic grasslands inhabited by *R. darti*, the reduncine bovid. Leakey (1993) has suggested that later *Theropithecus* lineages exploited these types of grass-lands.

Colobine species may have partitioned niches with both locomotor and body size differences. *C. williamsi*, the smaller species, likely foraged for leaves and fruits in the more open regions. The behaviors of the large colobine are unknown at this time, but its limited representation suggests that this species existed in the riparian region where hyenas scavenged or hunted infrequently.

A. africanus traveled bipedally, but could have climbed trees to gather food and avoid predation. The masticatory system suggests an adaptation to tough plant foods. This hominin exhibits considerable sexual dimorphism, but reconstruction of their social behavior using comparisons with extant primates is problematic (Plavcan and van Schaik, 1997). It has also been suggested that *A. africanus* also inhabited the forested areas close to the river (Rayner *et al.*,

1993) and, as support for this suggestion, their presence is also rare in these assemblages.

By examining the primates as a community, the morphological limitations of behavioral reconstructions become apparent. There are several species reconstructed as eating fruit and leaves in the fossil communities. When they are compared with the niche occupation of the Serengeti bushland, there appear to be more species sharing the same trophic niches at M3 (Figs. 7 and 8). Perhaps McKee (1991) is correct suggesting that large *P. jonesi* and small *P. whitei* have been erroneously classified as a mid-sized species, *P. broomi*. This might provide greater niche separation between the two remaining species. *A. africanus* overlaps with *P. whitei* in Fig. 8, but the early hominin had a different mode of locomotion, and thus foraging behavior from *P. whitei*. However, other researchers suggest that these *Parapapio* represent three distinct species and, as such, they must have partitioned their trophic resources more finely than morphological evidence indicates. Thus, comparison of fossil primate communities with extant ones allows identification of conceivable niche separation that is not possible reconstructing behaviors of single species.

There are two major differences between these early South African primate communities and extant bushland communities. First, there are more papionin species from the Makapansgat assemblages than are found in extant bushland communities. Would greater productivity promote greater primate species diversity? Kay *et al.*, (1997) have suggested that there is greater primate biodiversity in South American rain forests with greater productivity. While they measure productivity as litterfall, others have substituted mean annual rainfall because rainfall increases plant growth. Too much rainfall likely causes soil leaching and less productivity, but this occurs at very high rain forest levels of rainfall (Kay *et al.*, 1997). The problem of too much rainfall would not be a factor for a bushland community that usually has under 1000 mm of mean annual rainfall. Increased productivity in bushland communities might occur if wet and dry seasons were not as pronounced as they are in Africa today. In eastern Africa, there is one long and one short dry season interspersed with rainy seasons. In southern Africa, the dry season is six months, followed by rainfall of six months. It is possible that in the past, more evenly dispersed rainfall patterns produced consistently abundant leafing and fruiting cycles. In other words, increased productivity in bushlands might be the result of staggered fruiting and leafing availability within an environment of more even and abundant rainfall.

Second, the primates within these communities are larger, as a group, than primate communities in extant bushlands. Popp (1983) suggests that body masses of papionin species increase as mean annual rainfall increases. Mean annual rainfall is used in his study as a surrogate for primary plant productivity. The estimated body sizes of the papionin males from Makapansgat fall into the mean annual rainfall range of 635–1500 mm according to Table 1 of Popp (1983). However, the extant bushland mean annual rainfall ranges used in this study are from 700–875 mm. Thus, several large-sized

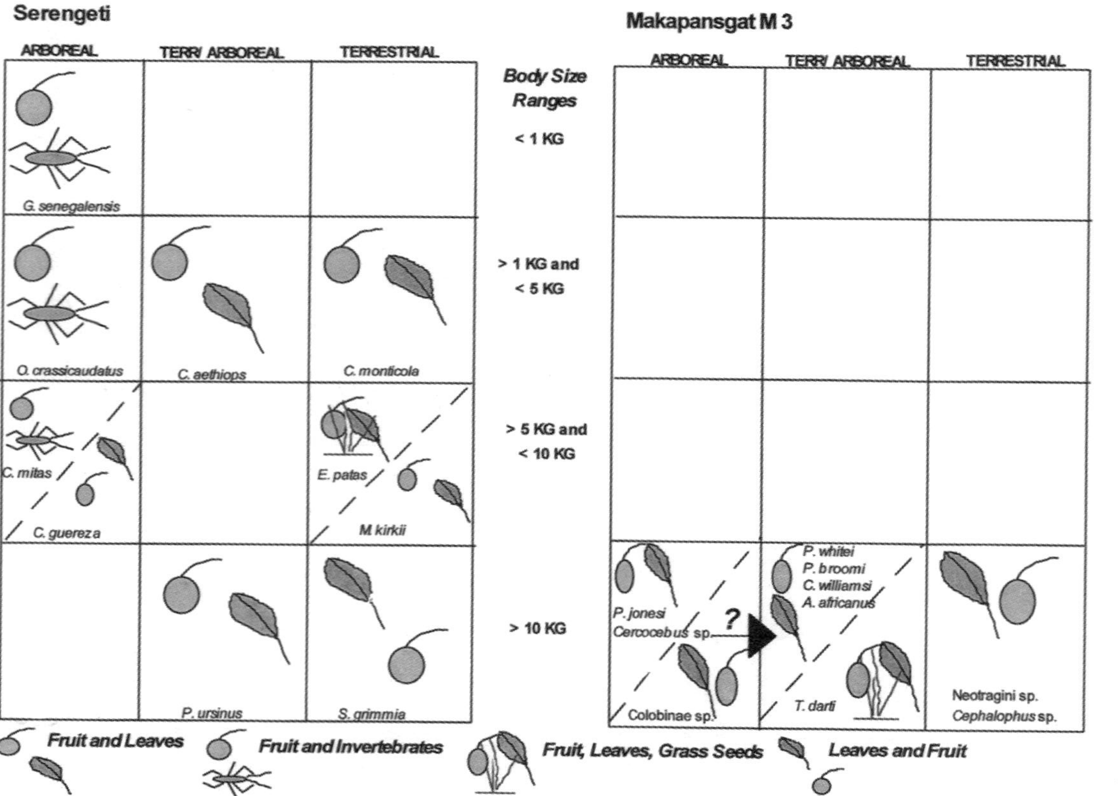

Fig. 7. Comparison of Serengeti Bushland and Makapansgat M3 ecological niches held by primate and bovid frugivore/folivores. Columns are grouped into three substrates: arboreal, indicating animals that seldom forage on the ground; terr/arboreal, indicating animals that forage on the ground and in trees; and terrestrial indicating animals that rarely, if ever, are found in trees. Rows are grouped into body size categories. Symbols within the boxes indicate the trophic behavior of the animals found in that substrate and body size category. All of the mammals are separated by one of the three categories in the Serengeti model, that is, there is no niche overlap. However, note that all of the Makapansgat primates are found in the >10 kg body size category and there appears to be niche overlap.

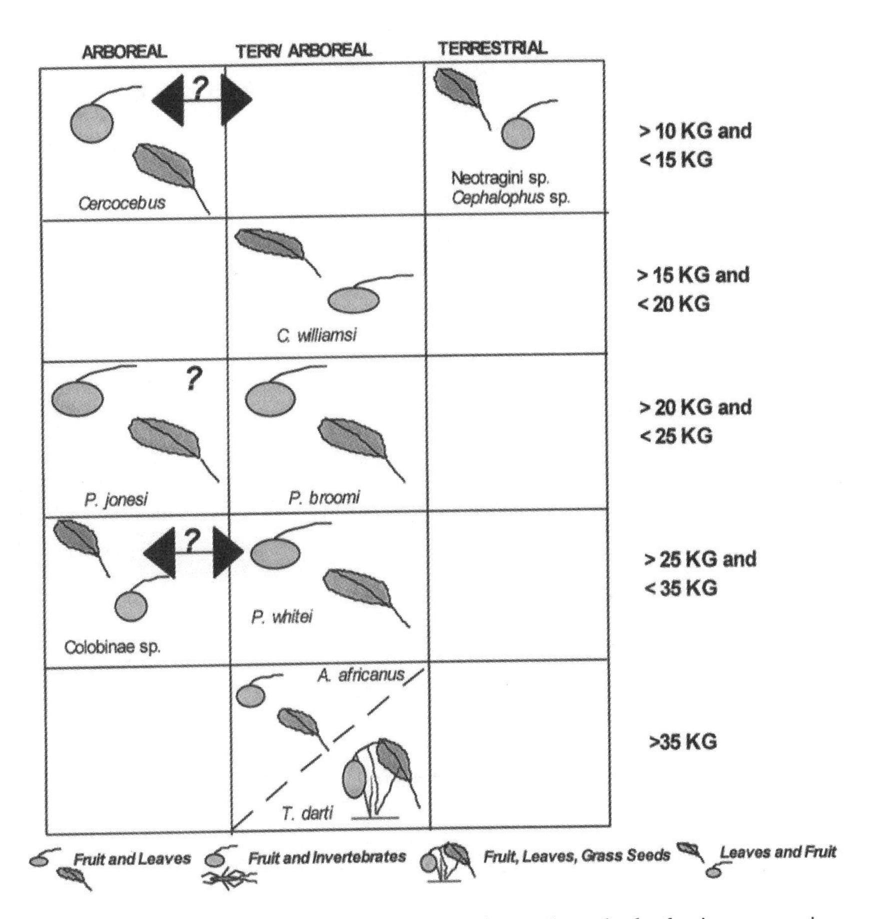

Fig. 8. Makapansgat M3 ecological niche separation. When the body size categories are expanded for M3, niche separation becomes apparent. Question marks and arrows indicate uncertain substrate usage, but only *P. jonesi* would overlap with *P. broomi* if their substrates coincided.

papionin species may indicate that bushland productivity was indeed higher three million years ago.

As previously mentioned, there are 21% and 20% browsing species from Makapansgat M3 and M4 assemblages compared with 2% to 6% browsing species in extant bushland communities. Accordingly, there are more browsing species, as well as primate species in this ancient bushland. One way to analyze the hypothesis that ancient Pliocene communities were more productive, is to examine the biodiversity and size differences among browsing and frugivorous animals within many bushland areas. The seasonal distribution of rainfall, amount of rainfall, soil nutrients, etc. could be compared among extant communities with species numbers and body sizes in order to see if a pattern exists. That is, are there increased body sizes or primate diversity when the abiotic factors mentioned above increase or are more nutrient rich? This pattern could then be used to compare with fossil assemblages.

Finally, it is also possible that the Makapansgat paleocommunity is simply unlike any extant community (Reed, 1998). Pliocene localities might be replete with papionin species simply because this is a radiation of a new lineage into new habitats. That is, when new primate species appear in the fossil record, are there always many more sympatric species? Comparisons of other primate radiations within other time periods and other paleocommunities will enlighten on this issue.

Environments became more xeric and open in the Pleistocene (Behrensmeyer *et al.*, 1997; Reed 1997), and hence less productive. There are many large-sized and/or browsing mammals in Pliocene communities that eventually disappear in the Pleistocene. As primate species found in fossil assemblages changed throughout the Plio-Pleistocene, the communities in which they existed also changed. Several species of *Parapapio* in fossil assemblages are often replaced by a single *Papio* species in Pleistocene fossil assemblages. *Theropithecus* species also change adaptations, and thus behavior, over time in eastern and southern African fossil sites. These changes in adaptations over time can be used to generate hypotheses about primate evolution and changes in primate behavior.

Evolution, Taphonomy, and Community Paleoecology

Examination of fossil primates from a community perspective gives us information on vegetation type and other environmental factors, interactions with other species, and niche structure that is not readily apparent by reconstructing behavior of individual primates. This contextual information also shows independent evidence of ecological changes with which various primate lineages can be compared. Thus, various primate behaviors can be studied within the context of changing environments, interactions, and niche structure so that the "why" of a particular behavior can possibly be discovered. Taphonomic studies also contribute to information about primate behavior. For example, ancient birds of prey appeared to prefer primate species at M4 times at Makapansgat. This predator behavior suggests possible foraging strategies of these papionins.

Some of the information provided by community studies serves to produce hypotheses regarding primate behavioral ecology. For example, only by looking at the entire community is it apparent that there is a different pattern of primate body sizes within ancient bushland communities. Some possible reasons for this phenomenon include increased productivity and/or historical circumstances. Various alternative hypotheses can be tested using data from extant communities, e.g., productivity measures vs. species diversity and body sizes, in order to compare evident patterns with fossil assemblages.

Are Pliocene localities replete with papionin species simply because this is a radiation of a new lineage into new habitats, or because of a fundamental difference in habitat or productivity? Papionins went through a radiation in the Pliocene at a time when the environment was changing. Although extant "savanna" baboons appear to be well adapted to life in fairly open, xeric regions, their fossil relatives, who appear to have similar behaviors, apparently did not do so well in those environments. Paleoecological community studies have the ability to address questions regarding the extinction of certain lineages by comparing and contrasting the biotic and abiotic factors that surrounded, and probably contributed to, their demise.

Again, these questions can be answered using actualistic studies and information on other fossil assemblages in a community perspective. Despite the fact that paleocommunity analyses and taphonomic studies do not reconstruct primate behavior, *per se*, the context that they provide helps trace the course of behavior and evolution in primates.

Acknowledgments

Thanks are due to everyone in South Africa who allowed access to the primate collections: P. V. Tobias, B. Rubidge, and F. Thackeray. I thank M. Plavcan for asking me to participate in this volume (at least I think I do). Peter Ungar, Michele Goldsmith, Charlie Lockwood, and Brandeis McBratney discussed various aspects of primate behavior with me, and made valuable points and suggestions. Two anonymous reviewers also made great suggestions. This research was supported by the National Science Foundation, Leakey Foundation, Association of University Women, and the Institute of Human Origins.

References

Abele, M. 1984. In: D. R. Strong, Jr., D. Simberloff, L. G. Abele, and A. B. Thistle (eds.), *Ecological Communities: Conceptual Issues and Evidence*, pp. 123–137. Princeton University Press, Princeton.

Andrews, P. and Van Couvering, J. A. H. 1975. Palaeoenvironments in the East African Miocene. In F. Szalay, (ed.), *Approaches to Primate Paleobiology, Contributions to Primatology*, Volume 5, pp. 62–103. Karger, Basel.

Andrews, P., Lord, J. M., and Nesbit-Evans, E. M. 1979. Patterns of ecological diversity in fossil and modern mammalian faunas. *Biol. J. Linn. Soc.* **11**:177–205.

Andrews, P., Meyer, G. E., Pileam, D. R., Van Couvering, J. A. and Van Couvering, J. A. H. 1981. The Miocene fossil beds of Maboko Island, Kenya: Geology, age, taphonomy, and paleontology. *J. Hum. Evol.* **10**:35–48.

Archibold, O. W. 1995. *Ecology of World Vegetation*. Chapman and Hall, New York, 510 pp.

Barton, R.A., Whiten, A., Byrne, R. W., and English, M. 1993. Chemical-composition of baboon plant foods: implications for the interpretation of intraspecific and interspecific differences in diet. *Folia Primatol.* **61**:1–20.

Behrensmeyer, A. K. 1991. Terrestrial vertebrate accumulations. In: P. A. Allison and D. E. G. Briggs (eds.), *Taphonomy: Releasing the Data Locked in the Fossil Record.* Plenum Press, New York.

Behrensmeyer, A. K., Western, D., and Dechant Boaz, D. E. 1979. New perspectives in vertebrate paleoecology from a recent bone assemblage. *Paleobiology* **5:**12–21.

Behrensmeyer, A. K., Todd, N. E., Potts, R., and McBrinn, G. E. 1997. Late Pliocene faunal turnover in the Turkana Basin, Kenya and Ethiopia. *Science* **278:**1589–1594.

Benefit, B. R. and McCrossin, M. L. 1990. Diet, species diversity, and distribution of African fossil baboons. *Kroeber Anthropol. Soc. Papers* **71–72:**79–93.

Birchette, M. G., Jr. 1988. *The Postcranial Skeleton of Paracolobus Chemeroni.* Ph.D. Dissertation, Harvard University.

Bishop, W. W. 1968. The Later Tertiary in East Africa: Volcanics, sediments, and faunal inventory. In: W. W. Bishop and J. D. Clark (eds.), *Background to Evolution in Africa,* pp. 31–56. University of Chicago Press, Chicago.

Blumenschine, R. J. 1989. A landscape taphonomic model of the scale of prehistoric scavenging opportunities. *J. Hum. Evol.* **18:**345–372.

Bown, T. M, Kraus, M. J., Wing, S. L., Fleagle, J. G., Tiffany, B., Simons, E. L., and Vondra, C. F. 1982. The Fayum forest revisited. *J. Hum. Evol.* **11:**603–632.

Brain, C. K. 1981. *The Hunters or the Hunted?: An Introduction to African Cave Taphonomy.* University of Chicago Press, Chicago.

Brown, J. H. and Bowers, M. A. 1984. Patterns and processes in three guilds of terrestrial vertebrates. In: D. R. Strong, Jr., D. Simberloff, L. G. Abele, and A. B. Thistle (eds.), *Ecological Communities: Conceptual Issues and Evidence,* pp. 282–296. Princeton University Press, Princeton.

Child, G. S. 1964. Some notes on the mammals of Kilimanjaro. *Tanganyika Notes and Records* **64:**77–89.

Ciochon, R. L. 1993. *Evolution of the Cercopithecid Forelimb: Phylogenetic Implications from Morphometric Analyses.* Geological Sciences, Vol. 48. University of California Press, Berkeley.

Cowlishaw, G. 1994. Vulnerability to predation in baboon populations. *Behaviour.* **131:**293–304.

Cowlishaw, G. 1997. Refuge use and predation risk in a desert baboon population. *Anim. Behav.* **54:**241–253.

Damuth, J. D. 1992. Taxon-free characterization of animal communities. In: A. K. Behrensmeyer, J. D. Damuth, W. A. DiMichele, R. Potts, H-D. Sues, and S. L. Wing (eds.), *Terrestrial Ecosystems through Time,* pp. 183–204. University of Chicago Press, Chicago.

Dart, R. A. 1952. Faunal and climatic fluctuations in Makapansgat Valley: their relation to the geologic age and Promethean stutus of Australopithecus. In: L. S. B. Leakey, and S. Cole (eds.), *Proceedings of the 1st Pan African Congress on Prehistory, Nairobi, 1947,* pp. 96–106.

Dodd, J. R. and Stanton, R. J. 1990. *Paleoecology: Concepts and Applications.* 2d. John Wiley and Sons, New York.

Eck, G. G. 1993. *Theropithecus darti* from the Hadar Formation, Ethiopia. In: N. G. Jablonski (ed), *Theropithecus: The Rise and Fall of a Primate Genus,* pp. 15–84. Cambridge University Press, Cambridge.

Eisenhart, W. L. 1974. *The Fossil Cercopithecoids of Makapansgat and Sterkfontein.* Unpublished BA thesis, Harvard College.

Emmons, L. H., Gautier-Hion, A., and Dubost, G. 1983. Community structure of the frugivorous-folivorous mammals of Gabon. *J. Zool. London.* **199:**209–222.

Estes, R. D. 1991. *The Behavior Guide to African Mammals.* University of California Press, Berkeley.

Fleagle, J. G. 1979. Primate positional behavior and anatomy: naturalistic and experimental approaches. In: M. E. Morbeck, H. Preuschoft, and N. Gomberg (eds.), *Environment, Behavior, and Morphology: Dynamic Interactions in Primates,* pp. 313–325. Gustav Fischer, New York.

Fleagle, J. G. 1998. *Primate Adaptation and Evolution,* 2nd ed. Plenum Press, New York, 455 pp.

Fleagle, J. G. and McGraw, W. S. 1999. Skeletal and dental morphology supports diphyletic origin of baboons and mandrills. *Proc. Natl. Acad. Sci. USA* **96:** 1157–1161.

Fleagle, J. G. and Reed, K. E. 1996. Comparing primate communities: a multivariate approach. *J. Hum. Evol.* **30:**489–510.

Fleming, T. H. 1973. Numbers of mammal species in North and Central American forest communities. *Ecology* **54:**555–563.
Gifford-Gonzales, D. 1991. Bones are not enough: analogues, knowledge, and interpretive strategies in zooarchaeology. *J. Anthropol. Archaeol.* **10:**215–254.
Gingerich, P. D., Smith, B. H., and Rosenberg, K. 1982. Allometric scaling in the dentition of primates and prediction of body weight from tooth size in fossil. *Am. J. Phys. Anthropol.* **58:**81–100.
Grine, F. E. 1986. Dental evidence for dietary differences in *Australopithecus* and *Paranthropus: a quantitative assesment of permanent molar microwear. J. Hum. Evol.* **15:**783–822.
Happold, D. C. D. 1989. *The Mammals of Nigeria.* Clarendon Press, Oxford.
Henschel, J. R. and Skinner, J. D. 1990. The diet of spotted hyenas *Crocuta crocuta*: a case report. *Afr. J. Ecol.* **28:** 69–82.
Hill, A. 1985a. Early Hominid from Baringo, Kenya. *Nature* **315:**222–224.
Hill, A. 1985b. Les variations de la faune Miocene recent et du Pliocene d'Afrique de L'est (1). *L'Anthropologie (Paris)* **89:**275–279.
Horovitz, I., and Meyer, A. 1997. Evolutionary trends in the ecology of New World monkeys inferred from a combined phylogenetic analysis of nuclear, mitochondrial, and morphological data. In: T. J. Givnish and K. J. Sytsnice (eds.), *Molecular Evolution and Adaptive Radiation*, pp. 189–224. Cambridge University Press, Cambridge.
Hutchinson, G. E. 1978. *An Introduction to Population Ecology.* Yale University Press, New Haven.
Iwamoto, T., Mori, A., Kawai, M., and Bekele, A. 1996. Antipredator behavior of gelada baboons. *Primates.* **37:**389–397.
Jablonski, N. G. 1993. Evolution of the masticatory apparatus in *Theropithecus.* In: N. G. Jablonski (ed.), *Theropithecus: The Rise and Fall of a Primate Genus*, pp. 299–330. Cambridge University Press, Cambridge.
Janis, C. M. 1988. An estimation of tooth volume and hypsodonty indices in ungulate mammals and the correlation of these factors with dietary preference. In: D. E. Russel, J.-P. Santoro, and D. Sigogneau-Russell (eds.), *Teeth Revisited.* Mem. Mus. Natn. Hist. Nat., Paris (serie C) **53:**367–387.
Jarvinen, O. and Haila, Y. 1984. Assembly of land bird communities on northern islands: a quatitative analysis of insular impoverishment. In: D. R. Strong, Jr., D. Simberloff, L. G. Abele, and A. B. Thistle (eds.), *Ecological Communities: Conceptual Issues and Evidence*, pp. 138–150. Princeton University Press, Princeton.
Johnson, K. H., Vogt, K. A., Clark, H. J., Schmitz, O. J., and Vogt, D. J. 1996. Biodiversity and the productivity and stability of ecosystems. *TREE* **11:**372–377.
Kay, R. F. 1984. On the use of anatomical features to infer foraging behavior in extinct primates. In: P. S. Rodman and I. G. Cant (eds.), *Adaptations for Foraging in Nonhuman Primates*, pp. 21–53. Columbia University Press, New York.
Kay, R. F., Madden, R. H., van Schaik, C., and Higdon, D. 1997. Primate species richness is determined by plant productivity: implications for conservation. *Proc. Natl. Acad. Sci. USA.* **94:**13023–13027.
Kingdon, J. 1974a. *East African Mammals. Volume I.* University of Chicago Press, Chicago.
Kingdon, J. 1974b. *East African Mammals. Volume IIA.* University of Chicago Press, Chicago.
Kingdon, J. 1974c. *East African Mammals. Volume IIB.* University of Chicago Press, Chicago.
Kingdon, J. 1977. *East African Mammals. Volume IIIA.* University of Chicago Press, Chicago.
Kingdon, J. 1979. *East African Mammals. Volume IIIB.* University of Chicago Press, Chicago.
Kingdon, J. 1982a. *East African Mammals. Volume IIIC.* University of Chicago Press, Chicago.
Kingdon, J. 1982b. *East African Mammals. Volume IIID.* University of Chicago Press, Chicago.
Kitchner, A. 1991. *The Natural History of the Wild Cats.* Comstock Publishing Associates, Ithaca.
Klein, R.G., and Curz-Uribe, K. 1984. *The Analysis of Animal Bones from Archaeological Sites.* University of Chicago Press, Chicago, 266 pp.
Krebs, J. R. and Davies, N. B. 1981. *An Introduction to Behavioural Ecology.* Sinauer Associates, Sunderland.
Krentz, H.B. 1993. Postcranial anatomy of extant and extinct species of *Theropithecus.* In: N. G. Jablonski (ed.), *Theropithecus: The Rise and Fall of A Primate Genus*, pp. 383–423. Cambridge University Press, Cambridge.

Kruuk, H. 1972. *The Spotted Hyena: A Study of Predation and Social Behavior.* University of Chicago Press, Chicago.

Lamprey, H. F. 1962. The Tarangire Game Reserve. *Tanganyika Notes and Records.* **60:**10–22.

Lanjouw, A. 1987. *Data Review on the Central Congo Swamp and Floodplain Forest Ecosystem.* Unpublished Report of the Royal Tropical Institute Rural Development Program.

Leakey, M. G. 1993. Evolution of *Theropithecus* in the Turkana Basin. In: N. G. Jablonski, (ed.), *Theropithecus: The Rise and Fall of a Primate Genus,* pp. 85–156. Cambridge University Press, Cambridge.

Leakey, R. E. F., and Walker, A. 1985. New Higher Primates from the early Miocene of Buluk, Kenya. *Nature* **318:**173–175.

Lewis, M. E. 1995. *Plio-Pleistocene Guilds: Implications for Hominid Paleoecology.* PhD Dissertation, SUNY—Stony Brook, UMI.

Lucas, P., and Teaford, M. 1994. Functional morphology of colobine teeth. In: A. G. Davies, and J. F. Oates (eds.), *Colobine Monkeys; Their Ecology, Behavior, and Evolution,* pp. 173–204. Cambridge University Press, Cambridge.

Maguire, J. M., Pemberton, D., and Collett, M. H. 1980. The Makapansgat limeworks grey breccia: hominids, hyenas, or hillwash? *Palaeontol. Afr.* **23:**75–98.

Maguire, J. M., Schrenk, F., and Stanistreet, I. G. 1985. The lithostratigraphy of the Makapansgat Limeworks Australopithecine site: some matters arising. *Annu. Geol. Surv. S. Afr.* **19:**31–51.

McBratney, B. M., and Reed, K. E. 1999. The ecological diversity of paleoprimate communities from South and East African hominin-bearing localities. *Am. J. Phys. Anthropol. Supp.* **28**:194.

McKee, J. 1991. Paleo-ecology of the Sterkfontein hominids: a review and synthesis. *Palaeontol. Afr.* **28:**41–51.

Morris, D. W., Abramsky, Z., Fox, B. J., and Willig, M. R. 1989. *Patterns in the Structure of Ecological Communities.* Texas Tech University Press, Lubbock, TX, 266 pp.

Nesbit-Evans, E. M., Van Couvering, J. A. H. and Andrews, P. 1981. Paleoecology of Miocene sites in Western Kenya. *J. Hum. Evol.* **10:**99–116.

Oates, J. 1977. The guereza and its food. In: T. H. Clutton-Brock, (ed.), *Primate Ecology: Studies of Feeding and Ranging Behavior in Lemurs, Monkeys, and Apes,* pp. 276–323. Academic Press, New York.

Odum, E. P. 1971. *Fundamentals of Ecology.* 3rd edition. Saunders Publishing Co., Philadelphia.

Perera, N. P. 1982. Ecological considerations in the management of the wetlands of Zambia. In: B. Gopal, R. E. Turner, R. G. Wetzel, and F. F. Whighan (eds.), *Wetlands Ecology and Management,* pp. 21–30. UNESCO, Paris.

Pickford, M., and Andrews, P. 1981. The Tinderet Miocene Sequence in Kenya. *J. Hum. Evol.* **10:**11–33.

Plavcan, M., and van Schaik, C. 1997. Interpreting hominid behavior on the basis of sexual dimorphism. *J. Hum. Evol.* **32:**345–374.

Popp, J. L. 1983. Ecological determinism in the life histories of baboons. *Primates* **24:**198–210.

Rautenbach, I. L. 1978. A numerical re-appraisal of southern African biotic zones. *Bull. Carnegie Mus. Nat. Hist.* **6:**175–187.

Rayner, R. J., Moon, B., and Masters, J. C. 1993. The Makapansgat australopithcine environment. *J. Hum. Evol.* **24:**219–231.

Reed, K. E. 1996. *The Paleoecology of Makapansgat and other African Plio-Pleistocene Hominid localities.* Ph.D. Dissertation, UMI.

Reed, K. E. 1997. Early hominid evolution and ecological change through the African Plio-Pleistocene. *J. Hum. Evol.* **32:**289–322.

Reed, K. E. 1998. Using large mammal communities to examine ecological and taxonomic structure and predict vegetation in extant and extinct assemblages. *Paleobiology* **24:**384–408.

Reed, K. E., and Fleagle, J. G. 1995. Geographic and climate control of primate diversity. *Proc. Natl. Acad. Sci. USA* **92:**7874–7876.

Rose, K. D. 1981. Composition and species diversity in Paleocene and Eocene mammal assemblages: an empirical study. *J. Vert. Paleontol.* **1:**367–388.

Roughgarden, J. 1989. The structure and assembly of communities. In: J. Roughgarden, R. May, and S. Levin (eds.), *Perspectives in Ecological Theory,* pp. 203–226. Princeton University Press, Princeton.

Schaller, G. B. 1976. *The Serengeti Lion*. University of Chicago Press, Chicago.

Sheppe, W., and T. Osborne 1971. Patterns of use of a flood plain by Zambian mammals. *Ecol. Monogr.* **41:**179–205.

Shipman, P. 1982. Reconstructing the Paleoecology and Taphonomic History of *Ramapithecus wickeri* at Fort Ternan Kenya. Columbia, MO: Curators of the University of Missouri, Museum Brief #26.

Shipman, P. 1986. Paleoecology of Fort Ternan reconsidered. *J. Hum. Evol.* **15:**193–204.

Shipman, P., Walker, A., Van Couvering, J. A., Hooker, P. A., and Miller. J. A. 1981. The Fort Ternan hominoid site, Kenya: Geology, age, taphonomy and paleoecology. *J. Hum. Evol.* **10:**49–72.

Shotwell, J. A. 1955. An approach to the paleoecology of mammals. *Ecology* **36:** 327–337.

Singh, G. 1988. History of aridland vegetation and climate: a global perspective. *Biol. Rev.* **63:**159–195.

Skinner, J. D., and Smithers, R. H. N. 1990. *The Mammals of the Southern African Subregion*. University of Pretoria Press, Pretoria.

Skinner, J. D., Henschel, J. R. and van Jaarsveld, A. S. 1986. Bone collecting habits of spotted hyaenas (*Crocuta crocuta*) in the Kruger National Park. *S. Afr. J. Zool.* **2:**33–38.

Smith, R. J., and Jungers, W. L. 1997. Body mass in comparative primatology. *J. Hum. Evol.* **32:**523–559.

Smithers, R. H. N. 1978. *A Checklist of the Mammals of Botswana*. Trustees of the National Museum of Rhodesia, Salisbury.

Sponheimer, M., Reed, K. E., and Lee-Thorp, J. 1999. Combining isotopic and ecomorphological data to refine bovid paleodietary reconstruction: a case study from the Makapansgat Limeworks hominin locality. *J. Hum. Evol.*

Strong, D. R. Jr., Simberloff, D., Abele, L. G., and Thistle, A. B. 1984. *Ecological Communities: Conceptual Issues and Evidence*. Princeton University Press, Princeton, 614 pp.

Swynnerton, G. H. 1958. Fauna of the Serengeti National Park. *Mammalia* **22:**435–450.

Terborgh, J., and van Schaik, C. P. 1987. Convergence and nonconvergence in primate communities. In: J. H. R. Gee and P. S. Giller (eds.), *Organization of Communities: Past and Present*, pp. 205–226. Blackwell Scientific Publications, Oxford.

Ungar, P. S., and Teaford, M. F. 1996. Preliminary examination of non-occlusal dental microwear in anthropoids: implications for the study of fossil primates. *Am. J. Phys. Anthropol.* **100:**90–113.

Van Couvering, J. A. H., and Van Couvering, J. A. 1976. Early Miocene mammal fossils from East Africa: Aspects of geology, faunistics, and paleoecology. In: G. L. Isaac and E. McCown (eds.), *Human Origins: Louis Leakey and the East African Evidence*, pp. 155–207. Staples Press, Menlo Park, CA.

Van Valkenburgh, B. 1988. Trophic diversity in past and present guilds of large predatory mammals. *Paleobiology* **14:**155–173.

Vesey-Fitzgerald, D. F. 1964. Mammals of the Rukwa Valley. *Tanganyika Notes and Records* **62:**61–72.

Wahungu, G. M. 1998. Diet and habitat overlap in two sympatric primate species, the Tana crested mangabey (*Cercocebus galeritus*) and yellow baboon (*Papio cynocephalus*). *Afr. J. Ecol.* **36:**159–173.

Wilson, M. V. H. 1988. Taphonomic processes: information loss and information gain. *Geosci. Canada* **15:**131–148.

Wing, S. L., Sues, H.-D., Potts, R., DiMichele, W. A., and Behrensmeyer, A. K. 1992. Evolutionary paleoecology. In: A. K. Behrensmeyer, J. D. Damuth, W. A. DiMichele, R. Potts, H.-D. Sues, and S. L. Wing (eds.), *Terrestrial Ecosystems through Time*, pp. 1–13. University of Chicago Press, Chicago.

Wright, P. C. 1988. A lemur's last stand. *Anim. Kingdom* **91:**12–25.

Zinner, D., and Pelaez, F. 1999. Verreaux's eagles (*Aquila verreauxi*) as potential predators of Hamadryas baboons (*Papio hamadryas hamadryas*) in Eritrea. *Am. J. Primatol.* **47:**61–66.

Reconstructing the Diets of Fossil Primates

7

PETER UNGAR

Introduction

Feeding adaptations are of great interest to primatologists whether they study living or fossil species. Diet both underlies many of the behavioral and ecological differences that separate extant taxa, and plays an important role in defining ecological niche, with all its implications for the ecology and evolution of extinct forms. While we can directly observe most living primates to see what they eat, deductions about the diets of fossils must be based on indirect evidence. Most recent work in this domain has employed the comparative method. Researchers attempt to relate anatomical evidence to diet in living primates to form a baseline for the inference of feeding behaviors from the remains of fossil forms. The idea is simple enough. We observe the manifestation of some feature in a fossil, and look for a corresponding condition in living primates. If every time we observe that trait in living primates it functions in a given way, we assume it functioned that way for the fossil form.

Both adaptive and nonadaptive evidence have been used to infer diets of fossil primates. The adaptive evidence studied has been dominated by analyses of the sizes and shapes of jaws and teeth, and the thickness and structure of tooth enamel. Nonadaptive evidence is that pertaining to the actual foods eaten by the individual whose remains are being studied. Most of this work has

---

PETER UNGAR • Department of Anthropology, University of Arkansas, Fayetteville, Arkansas 72701.

Reconstructing Behavior in the Primate Fossil Record, edited by Plavcan *et al.* Kluwer Academic/Plenum Publishers, New York, 2002.

involved the examination of tooth wear and mineralized tissue chemistry. Workers have set out to discover relationships among diet and these lines of evidence in a wide range of living primates. Consistent associations have allowed workers to infer (or perhaps more precisely, to predict) diet from teeth in fossil forms (Anthony and Kay, 1993). This paper surveys these, the more common approaches to how we relate bones and teeth to diet in primates. This is not meant to be a comprehensive review, but merely to provide an entry to the literature. Studies of anatomical evidence for diet in primates have contributed much to our understanding of fossil primate feeding behaviors and, when taken together, can provide important perspectives on dietary adaptations and diversity throughout the history of our order.

Adaptive Signals for Diet in Primates

Dental Allometry

Paleobiologists have considered the functional implications of relative tooth sizes of adult primates since the early 1970's. Debates stem from Robinson's (1954) observation that *Paranthropus* had smaller incisors and larger molars than did *Australopithecus*. He reasoned that these differences reflected functional specializations. Robinson related large molars in the "robust" australopithecines to an herbivorous diet that required grinding large quantities of tough vegetable foods. In contrast, he related the relatively larger front teeth of "gracile" australopithecines to a more omnivorous diet requiring extensive incisal preparation of meat and other foods during ingestion. Groves and Napier (1968) provided evidence for this by computing incisor to molar row length ratios for living apes—chimpanzees had the largest incisor-to-molar index and gorillas had the lowest. They suggested that chimpanzees need stronger incisors for husking fruits, while gorillas eat coarse vegetable foods that require considerable mastication but little incisal preparation.

Unfortunately, we cannot tell whether incisor-to-molar size ratio differences reflect selection on incisors, molars, or both. Therefore, most subsequent workers have chosen to focus on either incisors or molars, and to consider tooth size with respect to body size.

Cheek Tooth Allometry

Most of the discussion about cheek tooth allometry has revolved around debates concerning whether molar size varies isometrically with body size, or whether it is positively allometric and proportional to basal metabolic rate. Gould (1971), and Pilbeam and Gould (1974; 1975) reasoned that cheek tooth size should be proportional to energy requirements because a larger occlusal surface might be used to process more of a given type of food in a given amount of time. They noted that among a broad range of mammals (including primates, rodents, suids, and cervids), molar surface areas scale at about 0.75 power of body mass, the same rate that metabolism increases with body mass (Kleiber, 1961). They argued, based on weight estimates of the day, that early hominid molars were likewise "metabolically scaled" and therefore "positive allometry of tooth area in australopithecines affords no evidence for differences in diets or behavior" (Pilbeam and Gould, 1974).

Kay (1975, 1978) countered by noting that relationships between body size and occlusal area should be assessed by comparing animals with similar diets. Because different foods need to be chewed to different degrees, a sample representing a mix of diets would confound attempts to determine the relationship between tooth area and the rate at which food can be processed. Kay examined molar size in a number of primate taxa identified as frugivores, folivores, and insectivores and found that *within* each of these categories, primate postcanine tooth surface area varied isometrically, not with positive allometry. Therefore, if australopithecine molar area was positively allometric, larger forms would likely have eaten different foods than would have smaller ones.

These results were corroborated in independent tests (Corruccini and Henderson, 1978; Goldstein *et al.*, 1978). While molar size varies isometrically within diet categories, it is positively allometric across such categories. Why might this be so? This may be an effect of the idea that larger mammals eat abundant foods with low energy content, while smaller taxa tend to eat foods with the higher energy and protein content (Kay, 1975; Lucas, 1980; Rensberger, 1973). Larger primates would therefore require more tooth surface area for processing greater quantities of lower energy foods.

Some subsequent workers (Walker, 1981; Demes and Creel, 1988) have loaned support to metabolic or functional equivalence models for the early hominids. These authors have argued that since "robust" australopithecines had both relatively larger molars and attachment sites for the muscles of mastication, their bite force per unit tooth area might have remained constant. If so, the more megadont hominids could have been processing greater quantities of the same food items, a case of functional equivalence. Others, however (Wood, 1981; Wood and Abbott, 1983), have maintained that functional equivalence models are unlikely. Jungers (1988) and McHenry (1988, 1992) have independently reconstructed australopithecine body weights using estimates based on various aspects of skeletal size and both found that "gracile" and "robust" australopithecines from South Africa were essentially of equivalent body weight. Therefore, a metabolic equivalence model to explain australopithecine dental allometry can no longer be supported.

While theory dictates that larger primates should have relatively larger teeth to process greater quantities, low quality foods, specific associations between tooth size, and broad diet category have not held for primates. While this idea might explain why folivorous noncercopithecoid primates have relatively larger molars than frugivores, it is not clear why insectivorous noncercopithecoids would have relatively larger molars than frugivores (Kay, 1973; Strait, 1993a; Anapol and Lee, 1994). Further, it certainly does not explain why frugivorous Old World monkeys have larger molars than folivorous cercopithecoids (Lucas, 1980; Kay, 1977a). As such, inferences of broad diet category from molar size in higher primates are problematic. Kay and Cartmill (1977) noted that in order for a morphological trait to be useful for predicting a behavior, it must be associated with that behavior in all animals that possess the trait. Therefore, molar size, at least in the absence of good phylogenetic control, cannot be considered useful for predicting diet in fossil primates.

Incisor Allometry

Following Robinson's (1954, 1963) studies, many researchers have also considered relative incisor size to be indicative of diet. Jolly (1970a) speculated that small incisors relative to molars in australopithecines could be associated with terrestrial seed-eating, as seen in *Theropithecus* today. While this idea has been the subject of some controversy (Dunbar, 1976), Jolly's efforts have stimulated considerable research on relative incisor size (Fig. 1) in a wide

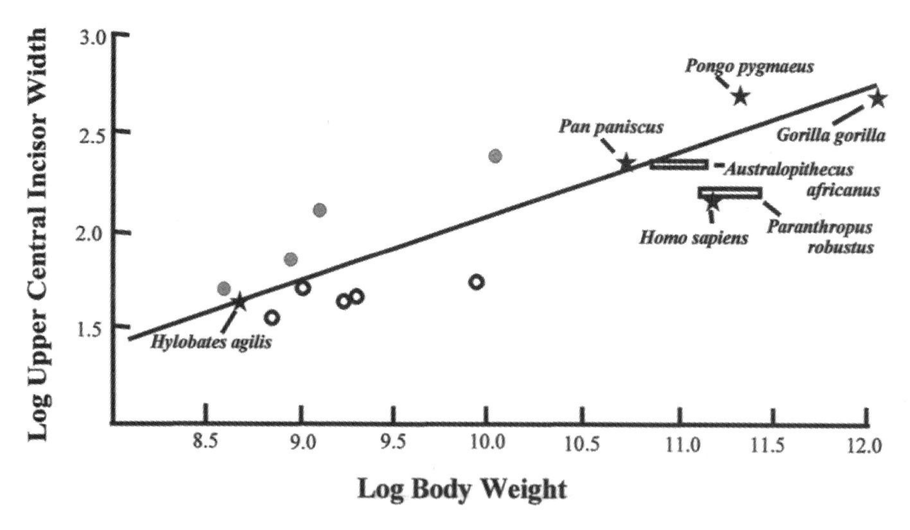

Fig. 1. Regressions of maxillary central incisor mesiodistal widths (mm) versus body weights (g) for selected male extant catarrhine taxa, *A. africanus* and *P. robustus*. Closed circles = colobines, open circles = cercopithecines, stars = hominoids. Rectangles for fossil taxa designate presumptive male body size ranges [based on data from Fleagle (1988) and Swindler (1976)]. The regression line is based on extant species representing several catarrhine genera (Ungar and Grine, 1991).

variety of living and fossil primates including strepsirrhines (Eaglen, 1986; Jolly, 1970b; Kay and Hylander, 1978), platyrrhines (Eaglen, 1984; Kinzey, 1974), cercopithecoids (Goldstein *et al.*, 1978; Hylander, 1975), and hominoids (Conroy, 1972; Groves, 1970; Simons and Ettel, 1980; Simons and Pilbeam, 1972; Ungar and Grine, 1991).

Most incisor allometry work from the mid-1970's onward considered incisor size relative to body size (rather than with respect to cheek tooth size). Hylander (1975) examined maxillary incisor row lengths in 57 anthropoid species and found that those with relatively larger incisors tend to consume larger, tougher fruits, while those with smaller incisors tend to feed on smaller foods or those that require less extensive incisal preparation, such as leaves or berries. He reasoned that broader incisor rows provide more working area to prepare larger foods (e.g., fruits), and a greater surface area to extend the life of the tooth given wear resulting from frequent use.

Others have since extended this line of reasoning to suggest direct associations between large incisors and fruits, and between small incisors and leaves (Anthony and Kay, 1993; Fleagle, 1988; Goldstein *et al.*, 1978). For example, Goldstein and coauthors (1978) verified that cercopithecine frugivores and "omnivores" have wider incisors than do colobine folivores, and that incisor size is actually a better predictor of diet than is molar area. Eaglen (1984) conducted a similar study for platyrhines. He found that New World monkeys with relatively larger incisors tend to consume foods that require greater incisal preparation than those with smaller front teeth (see Rosenberger [1992] for further discussion).

Eaglen warned, however, that platyrrhines as a group tend to have smaller incisors than do catarrhines, so that incisor size comparisons should be made bearing in mind phylogenetic effects. The same holds for comparisons with strepsirhines who, as a group, have even smaller incisors. In fact, Eaglen (1986) found that anterior tooth size does not relate well at all to broad dietary differences in strepsirhines. On the other hand, some aspects of front tooth size do reflect phyletic differences among strepsirhines. So, again, phylogeny can play an important role in determining relative incisor size independent of tooth function.

Relationships between diet and anterior tooth size do not necessarily hold when comparing specific catarrhine taxa either. For example, incisor size differences in sympatric Sumatran monkeys and apes cannot be explained by differences in broad food type preferences (i.e., frugivory, folivory), food item sizes, or degree of habitual incisor use alone. Gibbons have absolutely and relatively much smaller incisors than do sympatric orangutans even though the latter have been reported to spend more than four times as much of their feeding time eating leaves (Ungar, 1996a). While there may be compelling evidence for a tendency among *closely related* anthropoids for those species with larger incisors to use these teeth more often in food processing (Fig. 1, see Ungar and Grine, 1991), inferences of diet or ingestive behaviors from incisor size alone should be approached with caution, particularly when comparing distantly related taxa or those with uncertain phyletic affinities.

As a case in point, Groves and Napier (1968) argued that *Aegyptopithecus* and *Dryopithecus* (as defined at the time) had relatively small incisors compared with their molars, suggesting a gorilla-like folivorous adaptation. Similarly, Harrison (1982, 1993) argued on the basis of incisor size that most of the Early Miocene catarrhine taxa he studied (i.e., *Dendropithecus*, *Proconsul*, *Rangwapithecus*) were folivorous. Indeed, early catarrhines as a group have relatively small incisors compared with extant hominoids (Kay and Ungar, 1997). Perhaps then the relatively smaller incisors in these forms reflects phylogenetic history rather than diet *per se*. Again, platyrrhines as a group have smaller incisors than do extant cercopithecoids, but the former do not tend to be more folivorous than the latter. Therefore, incisor size alone may not be all that useful for inferring diet in fossil primates, especially when comparing distantly related forms, or those with uncertain phyletic affinities (Eaglen, 1984).

In sum, while diet and ingestive behaviors probably influence tooth size, phylogenetic history plays an important role that can make teasing function from phylogeny extremely difficult. This is especially true where relationships between tooth size and diet are opposite in different groups (e.g., molar sizes in hominoids versus cercopithecoids). Fortunately, studies of dental morphology have been rather more effective in distinguishing diets among extant primates.

Dental Morphology

Most prosimians have elongate, mesiodistally narrowed lower front teeth that form a tooth comb. Their upper front teeth tend to be diminutive, with a diastema between the central incisors. In contrast, most anthropoids have broader, more spatulate incisors. These differences have been related to specializations for grooming in the former group and a shift toward incisal biting during ingestion in the latter (i.e., Eaglen, 1984, 1986; Kay and Hiiemae, 1974; Kay and Hylander, 1978; Rosenberger and Strasser, 1985). There are also differences in incisor shape within each of these higher-level taxa that undoubtedly relate to differences in tooth use. Gorillas, for example, have short stubby incisors with very thick cingula compared with other hominoids (Walker, 1973). In contrast, baboons have long and curved, labiolingually narrow incisors virtually devoid of any cingulum. To this point, there have been very few studies of the functional implications of such differences.

In contrast, many workers have considered the functional implications of primate *molar* form. Gregory (1922) was among the first to consider dental functional anatomy, suggesting that primate molar shape evolved to improve mechanical efficiency for chewing. Several subsequent workers (e.g., Simpson, 1933; Crompton and Sita-Lumsden, 1970) have viewed teeth as guides for masticatory movement.

Kay and Hiiemae (1974), for example, associated specific dental morphologies with shearing, crushing, and grinding [though Lucas and Teaford (1994) suggest alternative terms—see below]. Food is sheared between the leading edges of crown crests. Shearing blades are generally reciprocally concave to minimize contact area. In contrast, food is crushed between planar surfaces on teeth. Finally, grinding involves both shear and crush components. In this case, two smooth "mortar and pestle"-like surfaces are occluded and moved across one another. Kay and Hiiemae applied this model to the fossil record, and argued for a progressive transformation from shearing to crushing and grinding from the middle Paleocene paromomyid *Palenochtha* to the Eocene adapid *Pelycodus* (=*Cantius*) to the Oligocene catarrhine *Aegyptopithecus*. These authors suggested that this marks a shift from insectivory to frugivory.

Other workers have associated specific crown morphologies with shearing, crushing, and grinding as well. For example, Rosenberger and Kinzey (1976) noted features associated with shearing in the more insectivorous *Callithrix* and the more folivorous *Alouatta*, but rounded or flattened cusps associated with crushing in more frugivorous *Cebus* and *Pithecia*. Further, Seligsohn and Szalay (1978) suggested that the occlusal morphologies of *Lepilemur* and *Hapalemur* are consistent with an emphasis on cutting edges for leaf-eating by the former, and puncturing/point penetration for the preparation of bamboo stems by the latter.

In sum, reciprocally concave, highly crested teeth have been associated with insectivorous and folivorous diets, while rounder/flatter cusped teeth have been associated with a more frugivorous diet. Kinzey (1978) warned, however, that these general food categories are simplifications, and that dental morphology might reflect not just primary specializations, but also critical though less often eaten food resources.

Similarities between the molars of leaf- and insect-eaters reflect the fact that plant fiber and insect chitin form resistant, almost two-dimensional sheets and rods more efficiently broken by shearing than by crushing. Moreover, chitin and leaf cellulose are both structurally complex and present similar challenges to the digestive system. In contrast, frugivores consume more crushable three-dimensional fruits and nuts, which often contain proteins and easy to digest simple sugars. Because shearing crest length correlates inversely with chewed particle size and more finely ground particles are digested more completely, it makes sense that a diet of leaves or insects should select for longer crested teeth (Kay, 1984; Kay and Sheine, 1979; Sheine, 1979; Sheine and Kay, 1977). This is consistent with findings that colobines have finer ground foods in their stomachs than do cercopithecoids (Walker and Murray, 1975).

So, how can we distinguish folivores from insectivores? Kay (1984) has demonstrated that among living primates, insectivores tend not to range above about 350 g while folivores tend to exceed 700 g. Frugivores overlap these ranges. Therefore, in combination, shearing crest lengths and body weights can distinguish frugivores, folivores, and insectivores.

Molar Shearing Quotient Studies

Kay (1978, 1984; Kay and Covert, 1984; Kay and Hylander, 1978) devised a "shearing quotient" (SQ) as a measure of relative shear potential of a tooth. First, the lengths of mesiodistally running crests connecting the main cusps of lower second molars are summed. A least-squares regression line is fit to frugivorous extant species in logarithmic space with mesiodistal occlusal surface length as the independent variable and summed crest length as the dependent variable. Frugivorous species alone are used to control for allometric changes in animals with similar adaptations. SQ's are computed as deviations from the frugivore regression (i.e., as measures of differences between observed and "expected" shearing crest lengths for a given tooth length). The higher the SQ, the relatively longer the crests. Results of Kay's studies confirm that folivores and insectivores do indeed have higher SQ's than frugivores (Fig. 2). Further, among frugivores, hard-object specialists have even lower SQ's than soft-fruit eaters, indicating that the former have extremely blunt teeth designed for crushing (Anthony and Kay, 1993; Meldrum and Kay,

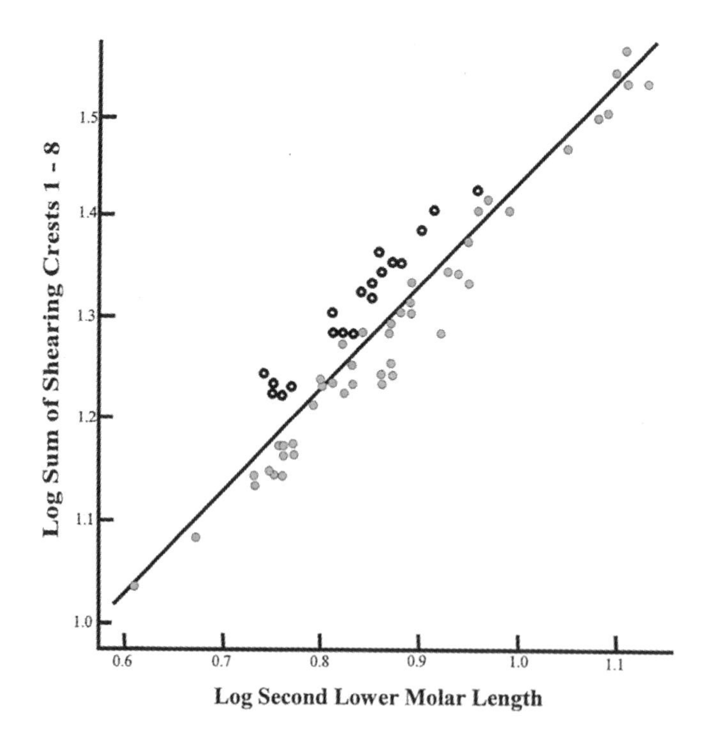

Fig. 2. Regression of summed shearing crest length (mm) versus second lower molar (mm) length for cercopithecoids. Darker circles = colobines, lighter circles = cercopithecines. Cercopithecine data are used to generate the graph [data courtesy of Richard Kay, see Kay (1984)].

1997). This relationship between SQ and diet holds for all major extant primate groups: strepsirhines, platyrrhines, cercopithecoids, and hominoids.

Shearing quotient studies have been applied to the fossil record by superimposing extinct primate values on an extant primate regression plot. In this way, results for fossils can be compared with those of a baseline series of living primates. For example, Kay and Simons (1980) examined shearing crest lengths in several Oligocene anthropoids from the Fayum Depression in Egypt. SQ values of these taxa all fell within the frugivorous extant hominoid range except for *Simonsius grangeri*, which had longer crests, suggesting folivory. Kay (1977b) also observed that SQ's of most early Miocene African catarrhines including *Proconsul*, *Dendropithecus*, and *Limnopithecus* suggest frugivory, though *Rangwapithecus* has longer crests implying folivory. In addition, Ungar and Kay (1995) applied this technique to European Miocene catarrhines, and again identified considerable variation, with SQ's suggesting a range of diets from extreme folivory in *Oreopithecus* and pliopithecids from Castell de Barbera, Spain to soft-fruit frugivory in *Dryopithecus* spp., *Anapithecus*, and *Pliopithecus platydon* to hard-object feeding in *Ouranopithecus* (Fig. 3). Further, Kay and colleagues have quantified SQ's for fossil platyrrhines (comparing values with those of extant frugivorous platyrrhines), with results again suggesting a range of diets including insects, leaves, soft fruits, and hard fruit seeds (Anthony and Kay, 1993; Fleagle *et al.*, 1996; Meldrum and Kay, 1997). Finally, using two different approaches, Williams and Covert (1994) and Strait (1993a, b, 1997) compared crests lengths of a variety of Eocene omomyids. While their results differed for individual taxa, most taxa had shearing crests of moderate length, and none had the extremely long, specialized crests of the "soft-objected" faunivore.

Shearing quotient studies have not yet been published for early hominids because their cusps are generally low and bulbous. Kay (1985) has noted (following Grine 1981, 1984) that "gracile" australopithecine cheek teeth show more occlusal relief than "robust" early hominids, but nonetheless, *Australopithecus* shows much less shearing emphasis than any living folivorous ape.

Still, it should be noted that as with incisor size comparisons, SQ values cannot be compared directly among higher-level taxa. For example, while shearing quotient accurately tracks diet in cercopithecoids or hominoids, Old World monkeys have relatively longer shearing crests than do apes when diet is controlled for (Kay and Covert, 1984). Therefore, phylogeny also plays an important role in determining crest length. So as with studies of tooth size, morphology should be interpreted within the context of phylogeny.

Kay and Ungar (1997) have recently suggested an additional twist to this phenomenon when examining the fossil record. While the molars of early Miocene apes show a substantial range of morphological variation (about equal to that of living apes), this range is "downshifted" from the modern hominoid range (Fig. 3). Much like extant hominoids have less well-developed shearing crests than cercopithecoids as a whole, these early Miocene apes have less well-developed shearing crests than extant hominoids, or middle to late

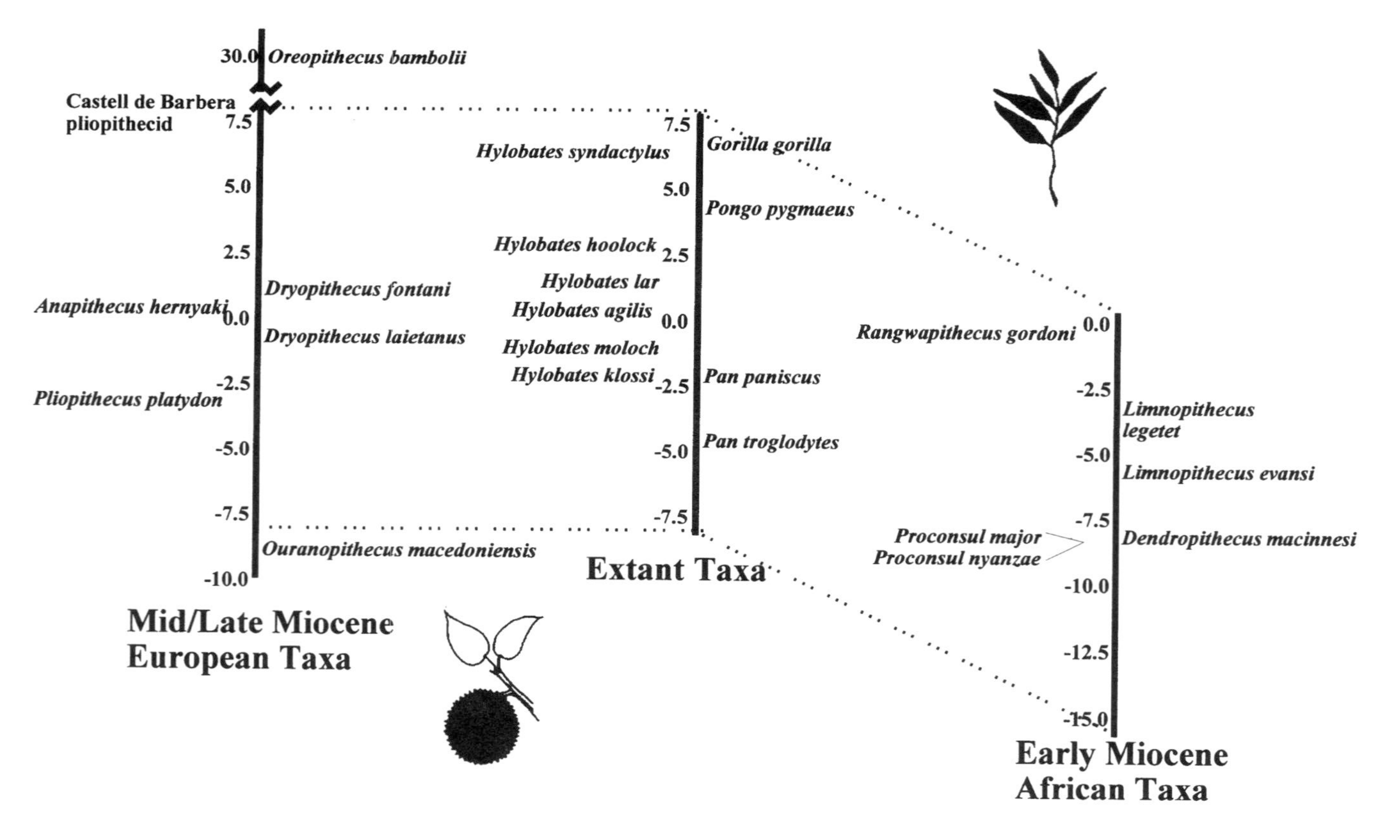

Fig. 3. Shearing quotients of extant hominoids, Early Miocene catarrhines from Africa, and Middle to Late Miocene catarrhines from Europe. Note that the SQ range for the European taxa exceeds that of living apes. Also note that while the African taxa have a comparable SQ range to the living hominoids, that range is "up-shifted" in modern apes, suggesting relatively longer shearing crests in later species [data from Kay and Ungar (1997) and Ungar and Kay (1995)].

Miocene European catarrhines for that matter. Interestingly, the shift toward longer crests seems to have occurred independently in middle to late Miocene hominoids and pliopithecids. Since this trend evidently cross-cuts phylogenetic boundaries, we may be looking at a temporal rather than a phylogenetic phenomenon, perhaps related to competition with evolving Old World monkeys or some other factor.

Dental Biomechanics

Since the 1980's, workers have begun to develop biomechanical models of primate dental morphology by considering idealized shapes for teeth designed to break down foods with given mechanical properties (Lucas, 1980; Lucas and Luke, 1984; Lucas and Teaford, 1994; Lucas *et al.*, 1991, 1994; Spears and Crompton, 1996; Strait, 1997). The basic idea is that natural selection dictates that primate tooth shape should reflect the mechanical properties of foods—occlusal morphology should be well-suited to break down the foods that primates habitually eat. As Spears and Crompton (1996) note, dental morphology affects the nature, magnitude, and distribution of stress on food particles. Thus, by understanding mechanisms of food "failure" (fracturing or crack propagation), it should be possible to predict the most efficient tooth morphology to break down a food item with given material properties.

Dental biomechanics researchers point out that the terms crush, grind, and shear all denote different types of fracture, each of which can be rather complex. Lucas and Teaford (1994), for example, prefer to think of food breakdown in mechanical terms such as strength (stress necessary to cause a structure to start to fracture, or permanently deform) and fracture toughness (the resistance to propagation of a crack). The tooth form best suited to fragment or break down a given food item depends on the strength and fracture toughness of that item.

Work to assess relationships between food properties (rather than general food categories *per se*) and actual tooth shape is now beginning. Strait (1993a, b, 1997), for example, found that mammals that eat soft moths, caterpillars, and worms have longer shearing crests than those that habitually consume hard-shelled beetles. Yamashita (1996) has extended this line of work by examining hardness and shear strength for foods eaten by five Malagasy primate taxa. While many of her individual predictions for shearing crest lengths do not hold, this approach does show great promise to allow us to test hypotheses concerning relationships between tooth shape and food properties, and ultimately, to understand dental design.

Enamel Thickness

Researchers have considered the functional implications of enamel thickness for some time. Two adaptive explanations for thick tooth enamel have usually been offered—either it evolved to prolong use-life of teeth given an abrasive (e.g., terrestrial) diet, or it evolved to minimize the risk of crown damage given high occlusal forces caused by a diet including very hard foods.

Simons and Pilbeam (1972; Simons, 1976) noted that sivapithecines and australopithecines both had thick tooth enamel. They suggested that these forms were both terrestrial, and that tough, grit-laden abrasive foods selected for thick enamel to lengthen the use-life of cheek teeth given rapid wear [much like hypsodonty accompanies the presence of tough, abrasive foods for ungulates (Janis and Fortelius, 1988)]. Workers who suggest an abrasive diet explanation frequently link thick enamel with terrestrial feeding. Smith and Pilbeam (1980) even speculated that because orangutans have thick tooth enamel, they must have been terrestrial in the past.

Kay (1981) questioned this adaptive explanation. While he acknowledged that thicker enamel should make teeth better resistant to wear, he noted that there is no tendency among hominoids or cercopithecoids for terrestrial species to have thicker enamel than arboreal ones. Dean and coauthors (1992) have further argued that thicker enamel of orangutan teeth compared with those of other great apes is not a compensatory mechanism for increased wear. If enamel thickness were related to wear intensity, one would expect *Pongo*, *Pan*, and *Gorilla* to all show comparable levels of wear. Indeed, orangutans tend to show less occlusal wear than African apes (Dean *et al.*, 1992). In sum it is unlikely that relative enamel thickness in primates clearly reflects dietary abrasiveness, and whether an animal forages primarily on the ground or in the trees.

In contrast, Kay (1981) suggested that thick enamel evolved to withstand tooth cracking and splintering during the crushing of hard foods. He noted that thick enamel does occur in catarrhines that eat hard foods regularly (e.g., *Cebus apella*, *Lophocebus albigena*, *Pongo pygmaeus*). Recent work by Dumont (1995) lends credence to this hypothesis. In her study of species pairs of closely related bats and primates (i.e., capuchins and mangabeys), hard-object feeders had thicker enamel than soft-object feeders. She noted, however, that there is no "thickness threshold" for hard-object feeders across the groups, making it difficult to interpret enamel thickness in individual fossil species, particularly those with uncertain phyletic affinities. As with dental allometry and morphology then, functional implications of enamel thickness must be interpreted within the context of phylogenetic heritage.

Given the benefits of thick enamel then, why would some primates have thin enamel? Kay (1981) speculates that thin enamel may confer an advantage to those that eat leaves and other foods that require shearing when chewed. He notes a negative correlation between enamel thickness and shearing crest development in many primates (though orangutans have thick enamel and relatively long shearing crests). Dentin islands form rapidly as thin enamel is perforated by dental wear. Because dentin is softer and wears more rapidly than enamel, the margins of these dentin islands would tend to be sharp, presumably enhancing shredding and slicing during mastication.

Teaford and Walker (1984) suggest that dental microwear could be used to test associations between thick enamel and food hardness in the fossil record because harder foods cause microwear pitting (see below). Their study indicates that *Sivapithecus* microwear resembles that of *Pan troglodytes* rather than a hard-object feeder. Indeed, shearing quotient values for *Sivapithecus* also put this species in the range of extant soft-fruit eaters (Ungar *et al.*, 1996). Teaford and Walker therefore suggest that some other adaptive explanation is necessary to explain thick enamel in *Sivapithecus*. Further, the thick-enamelled *Apidium* had microwear indicative of a soft-fruit eater with an abrasive diet (Teaford *et al.*, 1996). On the other hand, a recent microwear study of another primate with very thick enamel, *Ouranopithecus*, indicates high pit percentages and hard-object feeding (Ungar, 1996b).

Enamel thickness may be of some use for functional studies, but more work needs to be done relating diet to enamel thickness in living primates. One limiting factor at this point is sample size. Dumont's (1995) study, for example, was limited to between one and five individuals per taxon. Without larger samples, it is difficult to assess intraspecific variability in enamel thickness, and hence functional significance of this attribute.

Enamel Structure

Numerous researchers have suggested an association between enamel structure and diet in primates. This research is based on the fact that aspects of enamel structure affect tooth wear and fracture resistance (see Maas, 1991, 1993, 1994; Maas and O'Leary, 1996; Teaford *et al.*, 1996). Two aspects of enamel microstructure that have been examined are the arrangement and orientation of groups of prisms (this affects resistance of a loaded enamel surface to fracture) and the orientation of crystallites relative to the plane of wear (this affects resistance of the surface to wear). In some taxa, for example, prisms run between the enamel–dentine junction and the surface of the tooth along straight lines, while in others they decussate, or "worm about" (Aeillo and Dean, 1990). Strong decussation is said to strengthen the tooth against enamel fracture (Pfretzschner, 1986). Therefore, one might expect that primates that load their teeth with great force (e.g., hard-object feeders) should more often show decussation than those that do not.

Maas and colleagues have considered decussation patterns for a variety of fossil primates. Maas and O'Leary (1996), for example, examined microstructure in North American notharctines. They noted little variation among species in prism packing patterns or relative orientations of crystallites. On the other hand, they did find differences in decussation patterns, such that larger notharctines showed distinct decussation zones while smaller taxa had indistinct zones or no decussation. This is consistent with other reports that herbivores with larger molars usually show decussation while those with smaller cheek teeth rarely do (Koenigswald *et al.*, 1987). Maas and O'Leary relate decussation to increased chewing stresses associated with increased body size (rather than adaptations for hard-object feeding *per se*).

The same pattern holds for early anthropoids from the Fayum of Egypt. Maas and colleagues (Maas and Simons, 1995; Teaford *et al.*, 1996) note that *Aegyptopithecus*, *Propliopithecus*, and *Parapithecus* show enamel decussation while the smaller taxa *Apidium* and *Catopithecus* do not. This again suggests a size (rather than diet) related phenomenon. Teaford *et al.* (1996) in fact suggest a body size threshold (1500–2000 g) over which decussation is likely to occur and under which it is not.

Crystallite orientation differences may be more valuable for distinguishing diets among fossil taxa. For example, *Aegyptopithecus* and *Parapithecus* both have a large proportion of surface parallel crystallites at the cusp tips, but surface oblique crystallites at chewing wear facets (Teaford *et al.*, 1996). This pattern, like that seen in extant strepsirhines, optimizes resistance to abrasive wear in both areas (assuming that masticatory forces run perpendicular to cusp tips during puncture crushing, and more parallel to wear facets during chewing) (Maas, 1993). In contrast, *Apidium* crystallite orientations are more uniform, with most more-or-less oblique to the tooth surface. Because crystallites are expected surface parallel at the cusp tips when puncture-crushing is important to an animal, Teaford and coauthors suggest that *Apidium* did not specialize on hard foods—despite its comparatively thick molar enamel (see discussion above).

Nondental Adaptive Evidence for Diet in Fossil Primates

Some researchers have attempted to relate nondental adaptive evidence with diet in primates. For example, as mentioned above, body size has been associated with diet such that most insectivorous primates are less than 350 g and folivores tend to exceed 750 g in body weight (Kay, 1984). Frugivores overlap these ranges. Others (e.g., Avis, 1962; Napier, 1967) have suggested that locomotor adaptations might be useful for predicting diet such that, for example, brachiation can be related to terminal branch feeding and vertical clinging to insect foraging (but see Fleagle, 1978, 1980, 1984; Fleagle and Mittermeier, 1980; Stern and Oxnard, 1973). Yet others (Clutton-Brock and Harvey, 1980; Harvey *et al.*, 1980; Mace *et al.*, 1981; Martin, 1984; see also Aiello and Wheeler, 1995) have noted that brain size varies with diet, such that frugivores have larger brains than comparably sized folivores.

Mandibular Form

Mandibular fragments are among the most common bony remains in fossil primate assemblages. It therefore makes sense that many researchers have focused their attention on the functional anatomy of the lower jaw. The basic idea here is that the architecture of this bone has been adapted to withstand stresses and strains associated with oral food processing and, therefore, that mandibular form reflects (at least indirectly) diet. Such studies have focused on the size, shape, and degree of fusion of the mandibular symphysis and on the size and shape of the mandibular corpus.

Strepsirhines and tarsiers tend not to have fused mandibular symphyses, while anthropoids do. This fusion has been linked with the anthropoid transition, and has been suggested to imply (along with other lines of evidence, such as vertically implanted, spatulate incisors) a dietary shift toward frugivory (e.g., Hiiemae and Kay, 1972; Rosenberger, 1986). Greaves (1988, 1993), for example, has argued that symphyseal fusion stiffens the joint between the dentaries to help cope with heavy incisor loads associated with crushing small, hard objects such as fruit seeds.

Several authors (e.g., Beecher, 1977, 1979, 1983; Hylander, 1979a,b, 1981, 1984, 1985; Hylander and Johnson, 1994; Ravosa, 1996; Ravosa and Hylander, 1993, 1994; Ravosa and Simons, 1994) have countered that symphyseal fusion is not related to ingestive behaviors, but rather to stresses that occur during chewing. These workers argue, based in large part on *in vivo* studies of bone strain in various primates, that a fused mandibular symphysis serves to strengthen the joint to resist structural failure from "wishboning" and dorso-ventral shear given increased recruitment of balancing-side muscle force during mastication.

According to these authors, mandibular symphyseal fusion in early anthropoids was likely more related to increased chewing stresses than to the ingestion of small, hard objects. Indeed, recent phylogenetic work also calls into question functional links between anterior dental loading and symphyseal fusion in early anthropoids (Ravosa, 1998). It also seems that degree of fusion in early fossil primates relates in large part to body size, such that larger primates tend to show more symphyseal fusion than smaller ones. This allometric pattern holds for recent and fossil strepsirrhines (Beecher, 1977, 1979; 1983; Ravosa, 1991, 1996; Ravosa and Hylander, 1994). The larger primates would probably have recruited relatively greater amounts of balancing side jaw-muscle force than smaller taxa.

Functional implications of mandibular shape have also been considered. For example, an increase in the labio-lingual dimension of the symphysis should counter greater wishboning stresses as occur during unilateral chewing (Hylander 1984, 1985). The fact that "robust" australopithecines have unusually thick symphyses, for example, may therefore be associated with a design to resist failure due to repetitive loading of tough items, or to unusually high stresses related to crushing hard items (Hylander, 1988).

Mandibular corpus shape comparisons should also provide information concerning differences in chewing stresses among taxa. Daegling and Grine (1991), for example, have noted that australopithecines have postcanine corpora with greater transverse dimensions than found in living apes. This is consistent with an enhanced ability to resist failure due to acute twisting of the jaw. Daegling and Grine suggest that this too might be related to crushing hard foods, or to grinding tough foods.

In sum, mandibular evidence may provide important clues to jaw loading regimes. Nevertheless, it is still difficult to associate aspects of mandibular functional anatomy with specific diets. Indeed, as Brown (1997:169) recently wrote, there is not "a single useful formula whereby a given mandible can be associated with a specific diet." Nevertheless, continued analyses of mandibular biomechanics will likely help us put together, along with other lines of evidence, a more complete picture of food processing and diet in living and fossil primates.

Nonadaptive Signals for Diet in Primates

Tooth Wear Studies

Gross Tooth Wear

While many dental anthropologists have considered dietary implications of gross tooth wear in prehistoric human populations (see Rose and Ungar, 1998 for review), comparatively little work has been done to assess functional implications of gross tooth wear in nonhuman primates. That which has been done has focused on wear gradients, unusual patterns of wear, and the extent and inclination of wear facets on molar teeth. Meikle (1977), for example, noted that geladas show substantial wear gradients between anterior and posterior cheek teeth, suggesting rapid wear and a highly abrasive diet. Further, Teaford (1982) compared wear gradients between species of primates, demonstrating that macaques have steeper wear gradients than langurs. He suggested that such differences might be related to diet.

Other workers have identified unusual patterns of wear on teeth, which they have related to diet and tooth use. For example, Kilgore (1989) suggested that approximal grooves between the front teeth in chimpanzees might relate to stripping foods, such as husks of palm fronds. Indeed, approximal grooves have been reported for several hominid taxa, from *Homo habilis* (Boaz and Howell, 1977) to *Homo erectus* (Weidenreich, 1937), to Neanderals (Siffre, 1923; Martin, 1923; Brace, 1975; Frayer and Russell, 1987; Lalueza *et al.*, 1993). Such grooves have been explained by a variety of causes, from hydrolic patterns of grit-laden saliva (Wallace, 1974) to toothpick use (Lalueza *et al.*, 1993).

Yet others have examined wear facet inclination and extent. Butler (1952) noted that teeth preserve attritional facets as the result of tooth-to-tooth contact, and that these facets record a small part of jaw movement. Grine (1981) further pointed out that facet inclination indicates whether teeth contact one another steeply (as in shearing) or head-on (as in crushing). He also argued that the sizes of Phase II versus Phase I facets evince the relative importances of grinding and shearing in early hominids. Janis (1984) extended this line of reasoning, and considered relative extents of wear facets associated with crushing or shearing behaviors. She found that the seed-eater *C. satanus* has larger crushing facets than the more folivorous *C. guereza*, which has larger shearing facet areas for a given degree of wear. Studies of this sort may have potential for the inference of diet in fossil primates, but more quantitative work needs to be done on gross tooth wear in living primates to realize that potential.

Dental Microwear

Many studies have explored the functional implications of microscopic patterns of tooth wear (dental microwear). The first microwear studies set out to look for evidence of the use of teeth as guides for chewing, such that the direction of jaw movement is reflected in the orientations of microscopic, subparallel striations on molar teeth (Butler, 1952, 1973; Butler and Mills, 1959; Mills, 1955, 1963, 1967). Such studies focused more on jaw movements and the mechanics of chewing *per se* than on direct associations between foods eaten and patterns of microscopic wear on teeth.

Philip Walker (1976) can be credited with the first effort to associate microscopic wear on nonhuman primate teeth with diet *per se*. He examined the incisors of a series of living cercopithecoids using a light microscope, and found that terrestrial monkeys possess more striated dentine surfaces than do arboreal forms (because of feeding substrate, siliceous material in foods eaten, and the mechanical demands of food breakdown). Also, striation orientation differences between colobines and cercopithecines led Walker to speculate that folivores preferentially strip leaves laterally across the incisors.

The scanning electron microscope (SEM) was first used to examine dental microwear a couple of years later, when Rensberger (1978) and Walker and colleagues (1978) examined dental microwear of rodents and hyraxes. These and related pioneering studies (Walker, 1980, 1981) suggested that microwear might be of considerable significance to studies of paleobiology.

In the early 1980's, experimental work called associations between specific diets and patterns of dental microwear into question (Covert and Kay, 1981; Peters, 1982; Kay and Covert, 1983). As a result, more quantitative analyses followed, with strict control over magnifications used and surfaces examined (Gordon, 1982, 1984). Some workers began to consider the effects of enamel structure on tooth wear. Maas (1988, 1991, 1993, 1994), for example, noted that for prismatic enamel, variation in crystallite orientation can swamp striation breadth differences due to particle size. On the other hand, under compressive loads, abrasive particle size is the primary determinant of microwear feature size. Such studies further emphasized the need to understand and control for nondiet factors that might effect microwear formation.

In 1984, Teaford and Alan Walker examined molar microwear in a series of wild-caught museum specimens representing several species of extant primates with known differences in diet (Fig. 4, see Teaford and Walker, 1984). These workers quantified microscopic wear feature densities and dimensions on homologous facets on homologous teeth at a fixed magnification. They found that frugivores (*Lophocebus*, *Cebus*, *Pongo*, *Pan*) had higher ratios of microscopic pits to scratches than did folivores (*Colobus*, *Gorilla*, *Alouatta*). Further, among frugivores, hard-object specialists (*Lophocebus*, *Cebus apella*) had the highest relative frequencies of pits.

Subsequent study by Teaford (1985, 1986) identified microwear differences between the closely related taxa *Colobus guereza* and *Procolobus badius*, and even among the congeners *Cebus capucinus*, *C. olivaceus*, and *C. apella*. Differences in molar microwear pit percentages again mirrored differences in diet such that more folivorous taxa had relatively more scratches than frugivores, and among frugivores, hard-object specialists had relatively more pits. Even more encouraging, Teaford and colleagues have managed to identify *intra*specific differences in microwear, reflecting both seasonal and ecological zone differences in *Cebus olivaceus* and *Alouatta palliata* (Teaford and Robinson, 1989; Teaford and Glander, 1991)—though these microwear differences are not of a magnitude that would obscure differences between species.

Quantitative work has also progressed on examinations of microwear on incisor teeth and their relation to feeding behavior. For example, Ungar

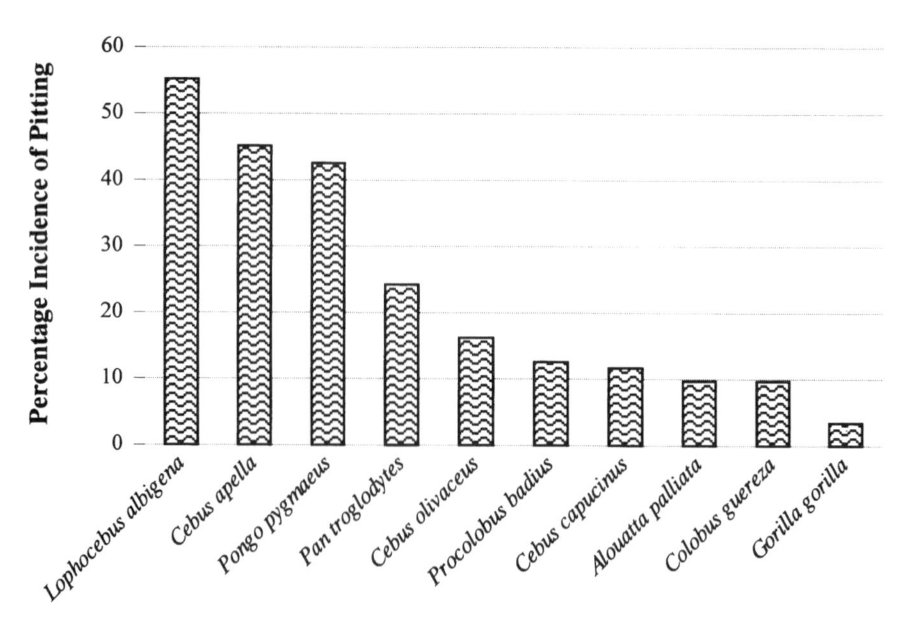

Fig. 4. Molar microwear pit percentages for various extant primates with known differences in diet (based on data reported by Teaford, 1988).

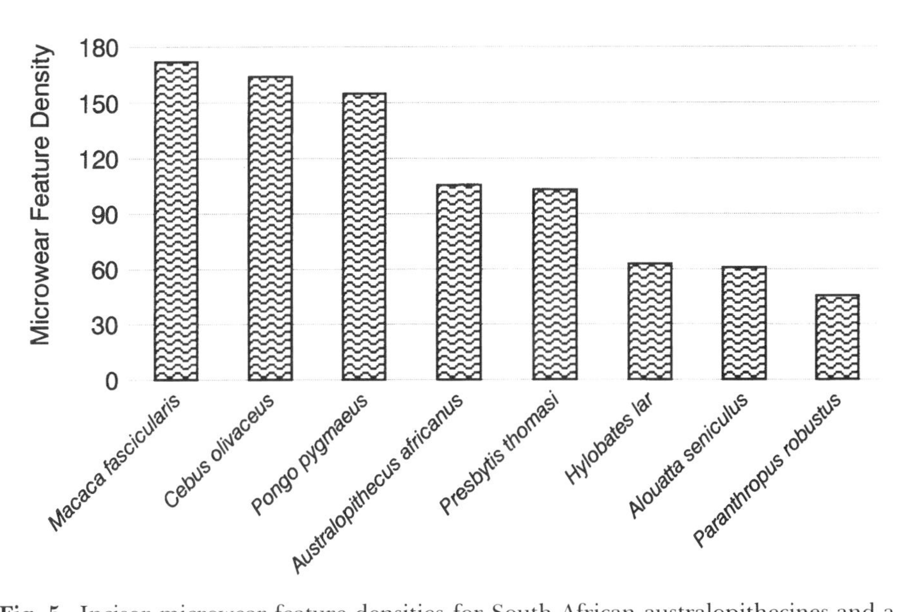

Fig. 5. Incisor microwear feature densities for South African australopithecines and a variety of living primates [data from Ungar and Grine (1991), Ungar (1992, 1994a)].

(1990, 1992) found that *Cebus olivaceus* incisors have a higher density of microwear features on their incisors than do *Alouatta seniculus* (Fig. 5). These species are found sympatrically in the Llanos of Venezuela, where capuchins use their incisors more in ingestion than do howler monkeys—even when the two species consume the same foods. Thus, microwear density can be related to degree of anterior tooth use on abrasive foods. Further work with Sumatran anthropoids (*Hylobates lar*, *Macaca fascicularis*, *Pongo pygmaeus*, and *Presbytis thomasi*) confirmed this association, and suggested that incisor microwear orientation can be related to processing techniques, and microwear feature size may be related to the sizes and/or shapes of dietary abrsives (Ungar, 1992, 1994a, 1994b, 1995a; Ungar et al., 1995).

Other work has focused on learning more about microwear formation and improving methods of data collection and analysis. For example, Teaford and Oyen (1989a, b, c) conducted an experimental study with captive vervets, *Chlorocebus aethiops*. These authors considered reliable methods of replicating occlusal surfaces, determined microwear feature longevity and rates of turnover, and assessed effects of differences in diet in a controlled environment. They also showed a direct and causal relationship between microwear and diet, such that primates fed dry, hard monkey chow had more microwear on crushing facets than those fed moistened, soft chow. These studies also demonstrated that individual microwear features can be formed and obliterated quite rapidly—in some cases within 24 hours. Even with minimum reported rates of cusp reduction (about 70 μm per year), it is clear that the average microwear striation recorded at 500× magnification (<1 μm) will be worn away less than a week after formation. This indicates both the need for large samples to get a representative cross-section of a group or species to infer dietary proclivities, and the potential of microwear to discern short-term (e.g., seasonal) variation in diet.

Further, some workers have recently begun to examine causes of wear by studying primates in their natural habitats. Such work has been conducted on cercopithecoids and hominoids in Indonesia (Ungar, 1992, 1994a, 1994b, 1995a; Ungar *et al.*, 1995), platyrrhines in Venezuela and Costa Rica (Burnell *et al.*, 1994; Pastor *et al.*, 1995; Teaford and Glander, 1991; Teaford et al., 1994; Ungar, 1990, 1992; Ungar *et al.*, 1995), and strepsirhines in Madagascar (Strait and Overdorff, 1994, 1996).

Dental microwear has now been examined for about thirty genera of extant primates, including strepsirrhines, platyrrhines, cercopithecoids, and hominoids (Ungar 1990, 1994a; Strait, 1991, 1993c; Teaford, 1993; Teaford and Runestad, 1992). Work is continuing to reveal the processes of microwear formation, and to tease apart variables that may affect microwear patterns. Work is likewise continuing to develop more repeatable, automated methods of microwear quantification (Boyde and Fortelius, 1991; Kay, 1987; Ungar, 1995b; Ungar *et al.*, 1991).

Associations between aspects of diet and microwear in living primates have also been used to infer diets of fossil primates. Early studies focused on hominids. These were largely qualitative, but still comparative in nature, either to an experimentally created microwear template (Puech, 1979, 1984, 1986; Puech and Albertini, 1984; Puech *et al.*, 1980, 1985, 1986) or to living primates (Ryan, 1980; Walker, 1981). For example, Walker (1981) suggested that East African early hominid dental microwear looked like that of living frugivores such as mandrills, chimpanzees, and orangutans. Similarly, Ryan and Johanson (1989) compared *Australopithecus afarensis* incisor microwear with that of extant primates, and proposed that these early hominids showed a mosaic of gorilla and baboon-like features indicating the use of incisors to strip gritty plant parts such as seeds, roots, and rhizomes. Further, Grine (1981) suggested that among South African early hominids, *Australopithecus africanus* deciduous molars had more striated occlusal facets, while *Paranthropus robustus* had more heavily pitted surfaces. This suggests the consumption of small, hard objects by "robust" australopithecines, and softer foods, such as fruits and immature leaves, by the "gracile" forms. He later confirmed these results in a quantitative study of microwear on permanent molars in these same taxa (Grine, 1986, 1987; Grine and Kay, 1987).

The first microwear study of a wide variety of fossil primates was a qualitative comparison of data for a variety of living primates and Miocene apes (Kelley, 1986). Kelley suggested that some fossil forms (*Proconsul nyanzae*, *Ouranopithecus macedoniensis*, and *Dryopithecus laietanus*) had little incisor wear and were probably folivorous while others (*Proconsul heseloni*, *P. major*,

Rangwapithecus gordoni, and *Sivapithecus indicus*) showed more incisor striations and were presumably frugivorous.

Quantitative microwear work has followed, and this approach has now been used to infer diets of fossil primates from all epochs from which primates are well-known. For example, Strait (1991) has examined dental microwear of several small-bodied Eocene omomyids. Her results, together with studies of molar form, suggest that these primates were mostly "hard-object" generalists. Further, Teaford and coauthors (1996) examined molar microwear in Oligocene catarrhines, including *Aegyptopithecus*, *Parapithecus*, and *Apidium*. Their evidence suggests these primates were mostly frugivorous, though there were some differences among these forms. As for Miocene primates, researchers have examined the Asian genera *Gigantopithecus* and *Sivapithecus* (Daegling and Grine, 1994; Teaford and Walker, 1984; Ungar *et al.*, 1996); the African forms *Dendropithecus*, *Micropithecus*, *Proconsul*, *Prohylobates*, and *Rangwapithecus* (Ungar *et al.*, 1996; Ungar and Teaford, 1996; Walker *et al.*, 1994); and the European taxa *Anapithecus*, *Dryopithecus*, *Oreopithecus*, *Ouranopithecus*, and *Pliopithecus* (Ungar, 1996b). Molar pit percentages suggest a broader variety of food preferences than seen in living apes, ranging from hard-object feeding (i.e., *Ouranopithecus*) to soft-fruit frugivory (i.e., *Gigantopithecus*, *Sivapithecus*, *Dendropithecus*, *Proconsul*, *Rangwapithecus*, *Anapithecus*, *Dryopithecus*, *Pliopithecus*) to diets including more leaves (i.e., *Micropithecus*, *Rangwapithecus*, *Oreopithecus*, the pliopithecid from Castell de Barbera).

Workers have also studied microwear patterns on the molars of Plio-Pleistocene cercopithecoids such as *Cercocebus*, *Cercopithecoides*, *Colobus*, *Papio*, *Paracolobus*, *Rhinopithecus*, and *Theropithecus* (Lucas and Teaford, 1995; Teaford and Leakey, 1992; Ungar and Teaford, 1996). Many of these primates show patterns comparable to closely-related modern analogs, but not all. For example, the colobine *Cercopithecoides* shows microwear on its nonocclusal surfaces consistent with terrestrial feeding, a scenario supported by its locomotor skeleton (Ungar and Teaford, 1996). Finally, workers have even begun to apply dental microwear techniques to the study of dietary diversity in subfossil primates from the late Pleistocene of Madagascar (Rafferty and Teaford, 1992).

Stable Isotope and Trace Element Analyses

A final line of evidence for dietary reconstructions involves the study of primate tissue chemistry (i.e., stable isotope and trace element analyses).

Isotope ratios in primate tissues have the potential to provide important information about both diet and the environment in which an animal lived. Stable isotopes of elements such as carbon (^{13}C and ^{12}C) and nitrogen (^{15}N and ^{14}N) typically occur in nature in predictable ratios. The basic idea here is that primate tissues should reflect isotope ratios of foods eaten. For example, because plant tissues from closed canopy, evergreen forests are more depleted in ^{13}C (i.e., have more negative $\delta^{13}C$ values) than those from drier, more open canopy forests, $^{13}C/^{12}C$ ratios in the tissues of a primate should tell us about the structure of the forest in which it fed and lived (Schoeninger *et al.*, 1997). Indeed, Ambrose (1993) has even shown a correlation between height above the forest floor and $\delta^{13}C$ values in leaves due to the combined effects of light intensity, humidity, and isotope weights at different levels of the canopy.

As another example, $^{15}N/^{14}N$ ratios in animal tissues have been said to reflect trophic level such that faunivores have higher $\delta^{15}N$ values than herbivores, and herbivores have higher $\delta^{15}N$ values than plants (Ambrose and DeNiro, 1986; DeNiro and Epstein, 1981; Schoeninger and DeNiro, 1984). Further, because legumes start out with low $\delta^{15}N$ values, those primates that eat these plants regularly should have relatively less ^{15}N than those that eat other types of plants (Ambrose and DeNiro, 1986; Schoeninger *et al.*, 1997). It therefore seems evident that stable isotope analyses have the potential to yield important insights into primate feeding behaviors.

Trace element analyses also have been used to reveal aspects of diet. Strontium to calcium ratios, for example, decrease as one moves from plant to herbivore to faunivore for given baseline levels of alkaline earths in the soil (see Sillen and Kavanagh, 1991, for review). This approach has received considerable attention by paleoanthropologists because recent studies of Sr/Ca ratios in early Pleistocene "robust" australopithecine and early *Homo* samples have suggested that the former were omnivorous while the latter were more herbivorous, perhaps specializing on underground plant storage organs (Lee-Thorp *et al.*, 1994; Sillen, 1992; Sillen *et al.*,1995).

Studies of trace elements and isotope ratios can potentially provide important information about diet independent of the other lines of evidence described in this chapter (i.e., those associated with the mechanical breakdown of food in the mouth). Still, such studies afford real challenges to the paleontologist. First, fossilization may distort trace element and isotopic signals (e.g., Nelson *et al.*, 1986; Schoeninger and DeNiro, 1982), though efforts to distinguish diagenetic from original biological tissues may help researchers to circumvent this problem in part (see Sillen and Lee-Thorp, 1994 for a review). Further, limitations to the antiquity of fossil specimens suitable for such analyses, particularly those for elements (e.g., nitrogen) found in organic phases of bone and teeth, may make them less than ideal for studies of earlier Tertiary primates.

Finally, while knowledge of environmental parameters and diet may allow one to predict isotope ratios or trace element levels in primate tissues; the reverse is not always true (Burton and Wright, 1995). For example, a folivore can have Sr/Ca levels indistinguishable from those of a carnivore (Sealy and Sillen, 1988). This problem may be compounded in primate generalists that take a wide variety of food types in a range of microhabitats at different times. Can we ascribe a diet on the basis of isotope ratios or trace element levels alone if two or more foods can yield comparable results, or if a given food type yields different results in different environments? Probably not. Still, when the results of such analyses are interpreted within the context of other lines of evidence, they may help provide us with a more complete picture of the diets of fossil primates.

Discussion

Differences among living primates in attributes including dental allometry, morphology and wear, tooth enamel thickness and structure, mandibular form and mineralized tissue chemistry have all been related to some degree to differences in diet or tooth use. Further, fossil or sub-fossil primates have been shown to vary in all of these attributes. Therefore, the evidence presented in this review should be of some value to reconstructing the feeding adaptations of past primates. Still, there are some significant limits to what this evidence can currently tell us.

Function, Phylogeny, and Clues from Nonadaptive Signals

One major "wrench in the works" for any study of adaptation is the difficulty of teasing apart the confounding effects of phylogeny from function. For example, while incisor size differences between closely related species can evidently be explained by differences in anterior tooth use, platyrrhines have relatively smaller incisors than catarrhines when diet is controlled for (Eaglen, 1984). Likewise, among closely related primates, folivores have longer molar shearing crests than frugivores, but cercopithecoids have longer shearing crests than hominoids independent of dietary differences (Kay and Covert, 1984). These results and those from other studies suggest that morphological specializations related to dietary differences can evolve in similar directions, but from different starting points—for example, the common ancestor of extant catarrhines had larger incisors than did the first platyrrhines.

So, can we relate directly differences in attributes such as enamel thickness and tooth or jaw size and shape to dietary differences in distantly related fossil taxa, or those with unknown phyletic affinities? Perhaps not. The comparative method depends on direct comparison of fossils with an appropriate neontological baseline series. Still, we can gain some control over phylogenetic effects by comparing fossil species with extant taxa from the same adaptive radiation. It is more appropriate to compare the teeth of a fossil Old World monkey to a series of extant cercopithecoids than to a series of extant platyrrhines when trying to infer diet. Still, basal taxa or those of uncertain phylogenetic affinities present a real problem.

Can we hope to know anything about the diets of basal taxa, or fossil primates with uncertain phyletic affinities? Perhaps. Kay and Ungar (1997) recently suggested one possible approach. If, as stated above, diet-related morphological differences evolve in similar ways from different starting points, the *variation* in such attributes can still be compared directly. For example, early Miocene catarrhines from East Africa show a similar range of shearing crest development to extant hominoids, though living apes generally have longer crests than do these fossil forms. It is likely that those early Miocene taxa with the longest crests (e.g., *Rangwapithecus*) were more folivorous than those with shorter crests, even though SQ values for these fossil species fall within the range of extant frugivorous hominoids (Fig. 3). Likewise, as Ungar and Kay (1995) suggest, where a group of fossils shows a greater range of shearing crest values than does an extant comparative baseline series, the fossils probably had a greater range of dietary adaptations (Fig. 3).

Unfortunately, it is difficult to test such ideas without some clue to diet that does not depend on identifying a morphological starting point. Fortunately, nonadaptive signals such as dental microwear can be of use here. Shearing quotient and microwear pit percentage values can be used to separate folivores from frugivores, and among frugivores, hard-object from soft-object feeders. For example, because *Rangwapithecus* has low microwear pit percentages, it is likely that the range of shearing quotient values for extant apes has been shifted upward as compared with the early Miocene forms.

In fact, a rank correlation test indicates that SQ and pit percentage values for fossil Miocene catarrhines covary significantly (Ungar *et al.*, 1996). Early Miocene African forms fall below a least-squares regression line, indicating that these species have relatively lower shearing crest values for a given pit percentage (Fig. 6). This is consistent with the interpretation that the early Miocene African forms as a group have shorter shearing crests than do later Eurasian apes. This may suggest an evolutionary trend (or trends) toward longer crests in later apes, independent of diet. Perhaps then, microwear and other nonadaptive signals can help tease apart function from phylogeny, even among fossil primates.

Other Limits to Dental Evidence for Diet in Primates

There are other limits to our ability to infer diet in fossil primates, and such limits suggest future avenues of research for paleobiologists. First, we must have a better handle on the mechanics of food breakdown, and how that relates to the sizes and shapes of primate teeth and jaws. Why is it that among cercopithecoids, folivores have smaller molars relative to body size than do frugivores but the reverse is true for other primates? Further, while the lengths of shearing crests are good predictors of diet when considering unworn molar teeth, how does wear figure into the equation? Wear can have a dramatic effect on tooth morphology—we need to understand how this affects tooth function and the efficiency with which different sorts of foods are eaten.

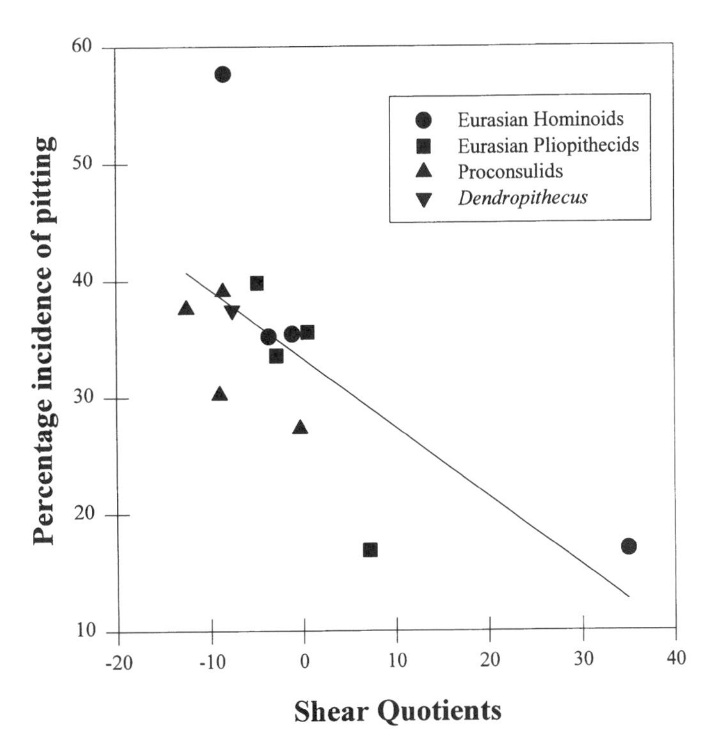

Fig. 6. Regression of shearing quotient versus dental microwear molar pit percentages for various Miocene catarrhine taxa (symbols represent taxa as indicated in the legend). Each symbol represents averaged values for one species. Data from Ungar *et al.* (1996) and work in progress by the author, Mark Teaford, and Richard Kay.

Second, we must get a better handle on primate diets and feeding behaviors. While many primate species have a strong preference for certain types of foods, most primates take a diverse array of foods (Harding, 1981; Richard *et al.*, 1981). Categorizing primates as "frugivores," "folivores," or "insectivores" may be masking considerable variation in diet and tooth use. This has important implications for the study of hard-tissue evidence for diet. For example, many workers have reasoned that fruits require more incisal preparation than leaves, so that frugivorous primates would likely have larger incisors than folivores. Naturalistic study has demonstrated that this need not be the case. While lar gibbons are fruit specialists, they tend to consume smaller fleshy fruits that require little if any incisal preparation—and they have relatively small incisors (Ungar, 1996a). It is also important to note that teeth and jaws can be adapted to foods other than those most frequently eaten if they are important dietary resources (Kinzey, 1978). Such factors need to be taken into account if the resolutions of our dietary reconstructions for fossil primates are to improve. Still, we may take some consolation in the knowledge that primates *do* show tendencies. Sussman (1978, 1987) has identified species-specific dietary adaptations, wherein primates of a given species in different forests tend to eat similar sorts of foods, even when the same plant taxa are not available in each.

Finally, we need to get a better handle on nonadaptive evidence for diet. For example, we need to work on isolating the many factors that affect wear, such as the underlying structure of tooth enamel (Maas, 1988, 1991, 1993, 1994). We must also appreciate that individual microwear features can be obliterated rather quickly, so microwear on an individual tooth gives only information on diet representing at most a few days (Teaford and Oyen, 1989a,b). While there are limits to what microwear can tell us, most of these problems can probably be obviated with good comparative baseline studies and large fossil samples. Likewise, we need a better understanding of diagenetic effects and other factors that affect isotope ratios and trace element concentrations in bone and teeth.

Summary and Conclusions

In sum then, what can the evidence tell us about the diets and feeding behaviors of fossil primates? That depends on what the evidence can tell us about the diets and feeding behaviors of living primates. If we can control for phylogeny, incisor size seems to reflect degree of front tooth use in ingestion. Incisor microwear likewise provides information about feeding behaviors, such as degree of front tooth use and directions that foods are pulled across the front of the mouth during ingestion. Such information may give us important clues to feeding adaptations, and perhaps even niche separation in primate communities if the fossil record is detailed enough to document sympatry. We can feel pretty confident in our reconstructions when two or more lines of independent evidence, such as incisor size and anterior dental microwear, lead to similar conclusions (Ungar and Grine, 1991).

As for molar morphology and microwear, these can give important clues to the material properties and, in turn, the sorts of foods habitually eaten by primates. Long molar crests and high molar microwear scratch densities reflect tough foods, such as leaves. In contrast, bunodont molars and high microwear pit frequencies are consistent with brittle, "harder" foods, such as nuts. Soft fruits tend to be of intermediate fracture toughness, and are reflected in an intermediate molar morphology and microwear pit percentage. As with ingestive behavior inferences, reconstructions of food properties are most robust when microwear and shearing quotients suggest similar diets (Ungar, 1996b).

Further, extent of bony jaw buttressing can give us some indication of an adaptation to resist structural failure of the mandible due to repetitive loading with moderate forces or occasional loading with high forces. While it may be difficult to distinguish adaptations for tough foods from those for harder foods using mandible size and shape alone, other lines of evidence, such as enamel thickness and microstructure, as well as molar morphology and microwear, can help provide an answer. Likewise, while it may be difficult to tell whether thick enamel has evolved to strengthen the tooth to allow hard-object feeding or to extend to the life of the tooth given an abrasive diet, evidence from microwear and mandibular form can help us distinguish among such competing hypotheses.

Finally, hard-tissue chemistry may give us additional insights into aspects of primate diets, such as degree of folivory or legume consumption, and perhaps degree of faunivory. While a number of factors can contribute to isotope ratios or trace element levels, robust dietary reconstructions may be best achieved when tissue chemistry is taken in the context of other lines of evidence.

There is probably a lot that we cannot hope to know about the diets of past primates unless someone figures out how to build a time machine. Still, further comparative study will likely allow us to learn more than we now know. In the end, the best approach will be to consider as many lines of evidence for diet as possible. Results from two or more methods may help us to test hypotheses that data from any one approach would not be sufficient to evaluate. In this case, the whole can be greater than the sum of its parts. This can lead to more robust reconstructions, and may allow us to increase the resolution with which we "see" the past. As Teaford, Maas, and Simons have written:

> "Each paleobiological technique, be it bone isotope analysis, standard functional morphology, dental microstructure or microwear, etc., gives us but one piece of a complicated puzzle. Our task is to put that puzzle together as accurately as possible. To do so, we need the best information we can get from all available sources" (Teaford *et al.*, 1996:541).

ACKNOWLEDGMENTS

I thank Mike Plavcan, Rich Kay, Bill Jungers, and Carel van Schaik for inviting me to participate in this volume. I would like to acknowledge the National Science Foundation, the L.S.B. Leakey Foundation, Sigma Xi, the Andrew Mellon Foundation, the University of Arkansas' Fulbright College, the State of Arkansas Information Liaison Office, and the SUNY Stony Brook Doctoral Program in Anthropological Sciences for funding my research on primate feeding behavior and the reconstruction of fossil primate diets. I am grateful to many researchers for their generous help, encouragement, and fruitful discussions on diet and feeding adaptations, especially John Fleagle, Fred Grine, Charlie Janson, Bill Jungers, Rich Kay, Peter Lucas, Jerry Rose, Mark Teaford, and Alan Walker. I also thank two anonymous reviewers, Stan Ambrose, Dave Daegling, and Matt Ravosa for their comments on parts of an earlier version of the text and Rich Kay and Mark Teaford for allowing me to use their data in figures presented here. This paper is expanded from an article published in *Evolutionary Anthropology* and is printed here with permission from Wiley Press.

References

Aiello, L., and Dean, M. C. 1990. *An Introduction to Human Evolutionary Anatomy*. Academic Press: London.

Aiello, L. C., and Wheeler, P. 1995. The expensive tissue hypothesis. *Curr. Anthropol.* **36:**199–221.

Ambrose, S. H. 1993. Isotopic analysis of paleodiets: methodological and interpretive considerations. In: M. I. Sandford (ed.), *Investigations of Ancient Human Tissue. Chemical Analyses in Anthropology*, pp. 59–130. Gordon and Breach Scientific, Langhorne, PA.

Ambrose, S. H., and DeNiro, M. J. 1986. The isotopic ecology of East African mammals. *Oecologia* **69:**395–406.

Anapol, F., and Lee, S. 1994. Morphological adaptation to diet in platyrrhine primates. *Am. J. Phys. Anthropol.* **94:**239–262.

Anthony, M. R. L., and Kay, R. F. 1993. Tooth form and diet in Ateline and Alouattine primates: Reflections on the comparative method. *Am. J. Sci.* **293-A:**356–382.

Avis, V. 1962. Brachiation: The crucial issue for man's ancestry. *Southwest J. Anthropol.* **18:**119–148.

Beecher, R. M. 1977. Function and fusion at the mandibular symphysis. *Am. J. Phys. Anthropol.* **47:**325–336.

Beecher, R. M. 1979. Functional significance of the mandibular symphysis. *J. Morphol.* **159:**117–130.

Beecher, R. M. 1983. Evolution of the mandibular symphysis in Notharctinae (Adapidae, Primates). *Int. J. Primatol.* **4:**99–112.

Boaz, N. T., and Howell, F. C. 1977. A gracile hominid cranium from upper member G of the Shungura Formation, Ethiopia. *Am. J. Phys. Anthropol.* **46:**93–108.

Boyde, A., and Fortelius, M. 1991. New confocal LM method for studying local relative microrelief with special references to wear studies. *Scanning*. **13:**429–430.

Brace, C. L. 1975. Comment on: Did La Ferrassie use his teeth as a tool? by J. A. Wallace. *Curr. Anthropol.* **16:**396–397.

Brown, B. 1997. Miocene hominoid mandibles: Functional and phylogenetic perspectives. In: D. R. Begun, C. V. Ward, and M. D. Rose (ed.), *Function, Phylogeny, and Fossils: Miocene Hominoid Evolution and Adaptations*, pp. 153–171. Plenum Press, New York.

Burnell, C. L., Teaford, M. F., and Glander, K. E. 1994. Dental microwear differs by capture site in live-caught *Alouatta* from Costa Rica. *Am. J. Phys. Anthropol.* **Suppl. 18:**62.

Burton, J. H., and Wright, L. E. 1995. Nonlinearity in the relationship between bone Sr/Ca and diet: Paleodietary implications. *Am. J. Phys. Anthropol.* **96:**273–282.

Butler, P. M. 1952. The milk molars of perissodactyla with remarks on molar occlusion. *Proc. Zool. Soc. London*. **121:**777–817.

Butler, P. M. 1973. Molar wear facets of early Tertiary North American primates. In: M. R. Zingeser (ed.), *Craniofacial Biology of Primates*, pp. 1–27. S. Karger Press, Basel.

Butler, P. M. and Mills, J. R. E. 1959. A contribution to the odontology of *Oreopithecus*. *Bull. Br. Mus. Nat. Hist. Geol.* **4:**1–26.

Clutton-Brock, T. H., and Harvey, P. H. 1980. Primates, brains and ecology. *J. Zool.* **190:**309–323.

Conroy, C. G. 1972. Problems with the interpretation of *Ramapithecus* with special reference to anterior tooth reduction. *Am. J. Phys. Anthropol.* **37:**41–48.

Corruccini, R. S., and Henderson, A. M. 1978. Multivariate dental allometry in primates. *Am. J. Phys. Anthropol.* **49:**517–532.

Covert, H. H., and Kay, R. F. 1981. Dental microwear and diet: Implications for determining the feeding behavior of extinct primates, with a comment on the dietary pattern of *Sivapithecus*. *Am. J. Phys. Anthropol.* **55:**331–336.

Crompton, A. W., and Sita-Lumsden, A. G. 1970. Functional significance of Therian molar pattern. *Nature*. **227:**197–199.

Daegling, D. J., and Grine, F. E. 1991. Compact bone distribution and biomechanics of early hominid mandibles. *Am. J. Phys. Anthropol.* **86:**321–339.

Daegling, D. J., and Grine, F. E. 1994. Bamboo feeding, dental microwear, and diet of the Pleistocene ape *Gigantopithecus blacki*. *S. Afr. J. Sci.* **90:**527–532.

Dean, M. C., Jones, M. E., and Pilley, J. R. 1992. The natural history of tooth wear, continuous eruption and periodontal disease in wild shot great apes. *J. Hum. Evol.* **22:**23–39.

Demes, B., and Creel, N. 1988. Bite force and cranial morphology of fossil hominids. *J. Hum. Evol.* **17:**657–676.

DeNiro, M. J., and Epstein, S. 1981. Influence of diet on the distribution of nitrogen isotopes in animals. *Geochim. Cosmochim. Acta* **45:**341–351.

Dumont, E. R. 1995. Enamel thickness and dietary adaptation among extant primates and chiropterans. *J. Mammal.* **76:**1127–1136.

Dunbar, R. I. M. 1976. Australopithecine diet based on a baboon analogy. *J. Hum. Evol.* **5:**161–167.

Eaglen, R. H. 1984. Incisor size and diet revisited: the view from a platyrrhine perspective. *Am. J. Phys. Anthropol.* **69:**262–275.

Eaglen, R. H. 1986. Morphometrics of the anterior dentition in strepsirhine primates. *Am. J. Phys. Anthropol.* **71:**185–202.

Fleagle, J. G. 1978. Locomotion, posture and habitat use of two sympatric leaf-monkeys in West Malaysia. In: D. J. Chivers, and J. Herbert (eds.), *Recent Advances in Primatology*, pp. 331–336. Academic Press, New York.

Fleagle, J. G. 1980. Locomotion and posture. In: D. J. Chivers (ed.), *Malayan Forest Primates: Ten Years' Study in Tropical Rainforest*, pp. 105–117. Plenum Press, New York.

Fleagle, J. G. 1984. Primate locomotion and diet. In: D. J. Chivers, B. A. Wood, and A. Bilsborough (eds.), *Food Acquisition and Processing in Primates*, pp. 105–117. Plenum Press, New York.

Fleagle, J. G. 1988. *Primate Adaptation and Evolution*. Academic Press, New York.

Fleagle, J. G., and Mittermeier, R. A. 1980. Locomotor behavior, body size and comparative ecology of seven Surinam monkeys. *Am. J. Phys. Anthropol.* **52:**301–322.

Fleagle, J. G., R. F. Kay, and M. R. L. Anthony, 1996. Fossil New World monkeys. In: R. F. Kay, R. H. Madden, R. L. Cifelli, and J. J. Flynn (eds.), *Vertebrate Paleontology in the Neotropics*, pp. 473–495. Smithsonian Institution, Washington D.C.

Frayer, D. W., and Russell, M. D. 1987. Artifical grooves on the Krapina Neandertal teeth. *Am. J. Phys. Anthropol.* **74:**393–405.

Goldstein, S., Post, D., and Melnick, D. 1978. An analysis of cercopithecoid odontometrics. *Am. J. Phys. Anthropol.* **49:**517–532.

Gordon, K. D. 1982. A study of microwear on chimpanzee molars: Implications of dental microwear analysis. *Am. J. Phys. Anthropol.* **59:** 195–215.

Gordon, K. D. 1984. Hominoid dental microwear: complications in the use of microwear analysis to detect diet. *J. Dent. Res.* **63:**1043–1046.

Gould, S. J. 1971. Geometric similarity in allometric growth: A contribution to the problem of scaling in the evolution of size. *Am. Nat.* **105:**113–136.

Greaves, W. S. 1988. A functional consequence of an ossified mandibular symphysis. *Am. J. Phys. Anthropol.* **77:**53–56.

Greaves, W. S. 1993. Reply to Drs. Ravosa and Hylander. *Am. J. Phys. Anthropol.* **90:**513–514.

Gregory, W. K. 1922. *The Origin and Evolution of Human Dentition*. Williams and Wilkins: Baltimore.

Grine, F. E. 1981. Trophic differences between 'gracile' and 'robust' australopithecines: A scanning electron microscope analysis of occlusal events. *S. Afr. J. Sci.* **77:**203–230.

Grine, F. E. 1984. Deciduous molar microwear of South African australopithecines. In: D. J. Chivers, B. A. Wood, and A. Bilsborough (eds.), *Food Acquisition and Processing in Primates*, pp. 525–534. Plenum Press, New York.

Grine, F. E. 1986. Dental evidence for dietary differences in *Australopithecus* and *Paranthropus*. *J. Hum. Evol.* **15:**783–822.

Grine, F. E. 1987. Quantitative analysis of occlusal microwear in *Australopithecus* and *Paranthropus*. *Scan. Microsc.* **1:**647–656.

Grine, F. E., and Kay, R. F. 1987. Early hominid diets from quantitative image analysis of dental microwear. *Nature*. **333:**765–768.

Groves, C. P. 1970. *Gigantopithecus* and the mountain gorilla. *Nature* **226:**973–974.

Groves, C. P., and Napier, J. R. 1968. Dental dimensions and diet in australopithecines. *Proc. VIII Int. Cong. Anthropol. Ethnogr. Sci.* **3:**273–276.

Harding, R. S. O. 1981. An order of omnivores: nonhuman primate diets in the wild. In: R. S. O. Harding, and G. Teleki (eds.), *Omnivorous Primates: Gathering and Hunting in Human Evolution*, pp. 191–214. Columbia University Press, New York.

Harrison, T. 1982. *Small-bodied Apes from the Miocene of East Africa*, Ph.D. Dissertation, University College, London.

Harrison, T. 1993. Cladistic concepts and the species problem in hominoid evolution. In: W. H. Kimbel, and L. B. Martin (eds.), *Species, Species Concepts and Primate Evolution*, pp. 345–371. Plenum Press, New York.

Harvey, P. H., Clutton-Brock, T. M., and Mace, G. M. 1980. Brain size and ecology in small mammals and primates. *Proc. Natl. Acad. Sci. USA* **77:**4387–4389.

Hiiemae, K. M. and Kay, R. F. 1972. Trends in the evolution of primate mastication. *Nature*. **240:**486–487.

Hylander, W. L. 1975. Incisor size and diet in anthropoids with special reference to Cercopithecoidea. *Science* **189:**1095–1098.

Hylander, W. L. 1979a. The functional significance of primate mandibular form. *J. Morphol.* **160:**223–240.

Hylander, W. L. 1979b. Mandibular function in *Galago crassicaudatus* and *Macaca fascicularis*: An in vivo approach to stress analysis of the mandible. *J. Morphol.* **159:**253–296.

Hylander, W. L. 1981. Patterns of stress and strain in the macaque mandible. In: D. S. Carlson (ed.), *Craniofacial Biology. Monograph No. 10, Craniofacial Growth Series*, pp. 1–37. University of Michigan Press, Ann Arbor.

Hylander, W. L. 1984. Stress and strain in the mandibular symphysis of primates: A test of competing hypotheses. *Am. J. Phys. Anthropol* **64:**1–46.

Hylander, W. L. 1985. Mandibular function and biomechanical stress and scaling. *Am. Zool.* **25:**315–330.

Hylander, W. L. 1988. Implications of *in vivo* experiments for interpreting the functional significance of "robust" australopithecine jaws. In: F. E. Grine (ed.), *Evolutionary History of the "Robust" Australopithecines*, pp. 55–58. Aldine de Gruyter, New York.

Hylander, W. L., and Johnson, K. R. 1994. Jaw muscle function and wishboning of the mandible during mastication in macaques and baboons. *Am. J. Phys. Anthropol* **94:**523–547.

Janis, C. 1984. Prediction of primate diets form molar wear patterns. In: D. J. Chivers, B. A. Wood, and A. Bilsborough (eds.), *Food Acquisition and Processing in Primates*, pp. 331–340. Plenum Press, New York.

Janis, C. M., and Fortelius, M. 1988. On the means whereby mammals achieve increased functional durability of thier dentitions, with special reference to limiting factors. *Biol. Rev.* **63:**197–230.

Jolly, C. J. 1970a. The seed-eaters: A new model of hominid differentiation based on a baboon analogy. *Man* **5:**1–26.

Jolly, C. J. 1970b. *Hadropithecus*: A lemurid small-object feeder. *Man* **5:**619–626.

Jungers, W. L. 1988. New estimates of body size in australopithecines. In: F. E. Grine (ed.), *Evolutionary History of the "Robust" Australopithecines*, pp. 115–125. Aldine de Gruyter, New York.

Kay, R. F. 1973. *Mastication, Molar Tooth Structure and Diet in Primates*, Ph.D. Dissertation, Yale University.

Kay, R. F. 1975. Allometry in early hominids. *Science* **189:**63.

Kay, R. F 1977a. The evolution of molar occlusion in the Cercopithecidae and early catarrhines. *Am. J. Phys. Anthropol.* **46:**327–352.
Kay, R. F. 1977b. Diets of early Miocene African hominoids. *Nature.* **268:**628–630.
Kay, R. F. 1978. Molar structure and diet in extant Cercopithecidae. In: P. M. Butler, and K. A. Joysey (eds.), *Development, Function, and Evolution of Teeth*, pp. 309–339. Academic Press, New York.
Kay, R. F. 1981. The nut-crackers: A new theory of the adaptations of the Ramapithecinae. *Am. J. Phys. Anthropol.* **55:**141–151.
Kay, R. F. 1984. On the use of anatomical features to infer foraging behavior in extinct primates. In: P. S. Rodman, and J. G. H. Cant (eds.), *Adaptations for Foraging in Nonhuman Primates: Contributions to an Organismal Biology of Prosimians, Monkeys and Apes*, pp. 21–53. Columbia University, New York.
Kay, R. F. 1985. Dental evidence for the diet of *Australopithecus. Annu. Rev. Anthropol.* **14:** 315–341.
Kay, R. F. 1987. Analysis of primate dental microwear using image processing techniques. *Scan. Microsc.* **1:**657–662.
Kay, R. F., and Cartmill, M. 1977. Cranial morphology and adaptations of *Palaechthon nacimienti* and other Paromomyidae (Pleisadapoidea, ?Primates), with a description of a new genus and species. *J. Hum. Evol.* **6:**19–53.
Kay, R. F., and Covert, H. H. 1983. True grit: A microwear experiment. *Am. J. Phys. Anthropol.* **91:**33–38.
Kay, R. F. and Covert, H. H. 1984. Anatomy and behavior of extinct primates. In: D. J. Chivers, B. A. Wood, and A. Bilsborough (eds.), *Food Acquistion and Processing in Primates*, pp. 467–508. Plenum Press, New York.
Kay, R. F., and Hiiemae, K. M. 1974. Jaw movement and tooth use in recent and fossil primates. *Am. J. Phys. Anthropol* **40:**227–256.
Kay, R. F., and Hylander, W. L. 1978. The dental structure of mammalian folivores with special reference to primates and Phalangeroidea (Marsupialia). In: G. G. Montgomery (ed.), *The Ecology of Arboreal Folivores*, pp. 173–191. Smithsonian Institution, Washington DC.
Kay, R. F., and Sheine, W. S. 1979. On the relationship between chitin particle size and digestibility in the primate *Galago senegalensis. Am. J. Phys. Anthropol.* **50:**301–308.
Kay, R. F., and Simons, E. L. 1980. The ecology of Oligocene African Anthropoidea. *Int. J. Primatol.* **1:**21–37.
Kay, R. F., and Ungar, P. S. 1997. Dental evidence for diet in some Miocene catarrhines with comments on the effects of phylogeny on the interpretation of adaptation. In: D. R. Begun, C. Ward, and M. Rose (eds.), *Function, Phylogeny and Fossils: Miocene Hominoids and Great Ape and Human Origins*, pp. 131–151. Plenum Press, New York.
Kelley, J. 1986. *Paleobiology of Miocene Hominoids*. Ph.D. Dissertation, Yale University.
Kilgore, L. 1989. Dental pathologies in ten free-ranging chimpanzees from Gombe National Park, Tanzania. *Am. J. Phys. Anthropol.* **80:**219–227.
Kinzey, W. G. 1974. Ceboid models for the evolution of hominoid dentition. *J. Hum. Evol.* **3:** 193–203.
Kinzey, W. G. 1978. Feeding behavior and molar features in two species of titi monkey. In: D. J. Chivers, and J. Herbert (eds.), *Recent Advances in Primatology, Volume 1, Behavior*, pp. 373–385. Academic Press: London.
Kleiber, M. 1961. *The Fire of Life*. John Wiley & Sons: New York.
Koenigswald, W. von., Rensberger, J. M., and Pfretzschner, H. U. 1987. Changes in tooth enamel of early Paleocene mammals allowing increased diet diversity. *Nature.* **328:**150–152.
Lalueza, C., P'erez-P'erez, A., and Turbon, D. 1993. Microscopic study of the Banyoles mandible (Girona, Spain): Diet, cultural activity and toothpick use. *J. Hum. Evol.* **24:**281–300.
Lee-Thorp, J. A., van der Merwe, N. J., and Brain, C. K. 1994. Diet of *Australopithecus robustus* at Swartkrans from stable carbon isotopic analysis. *J. Hum. Evol.* **27:**361–372.
Lucas, P. W. 1980. *Adaptation and Form of the Mammalian Dentition with Special Reference to the Evolution of Man*. Ph.D. Dissertation, University College, London.

Lucas, P. W., Lowrey, T. K., Pereira, B. P., Sarafis, V., and Kuhn, W. 1991. The ecology of *Mezzettia leptopoda* (Hk. f. et Thoms.) Oliv. (Annonaceae) as viewed from a mechanical perspective. *Funct. Ecol.* **5:**545–553.

Lucas, P. W., and Luke, D. A. 1984. Chewing it over: Basic principles of food breakdown. In: D. J. Chivers, B. A. Wood, and A. Bilsborough (eds.), *Food Acquisition and Processing in Primates*, pp. 283–301. Plenum Press, New York.

Lucas, P. W., and Teaford, M. F. 1994. Functional morphology of colobine teeth. In: A. G. Davies, and J. F. Oates (eds.), *Colobine Monkeys: Their Ecology, Behaviour and Evolution*, pp. 173–203. Cambridge University Press, Cambridge.

Lucas, P. W., and Teaford, M. F. 1995. Significance of silica in leaves eaten by long-tailed macaques (*Macaca fascicularis*). *Folia Primatol.* **64:**30–36.

Lucas, P. W., Peters, C. R., and Arrandale, S. R. 1994. Seed-breaking forces exerted by orangutans with their teeth in captivity and a new technique for estimating forces produced in the wild. *Am. J. Phys. Anthropol.* **94:**365–378.

Maas, M. C. 1988. *The Relationship of Enamel Microstructure and Microwear.* Ph.D. Dissertation, State University of New York at Stony Brook.

Maas, M. C. 1991. Enamel structure and microwear: An experimental study of the response of enamel to shearing forces. *Am. J. Phys. Anthropol.* **85:**31–50.

Maas, M. C. 1993. Enamel microstructure and molar wear in the greater galago, *Otolemur crassicaudatus* (Mammalia, Primates). *Am. J. Phys. Anthropol.* **92:**217–233.

Maas, M. C. 1994. A scanning electron-microscopic study of *in vitro* abrasion of mammalian tooth enamel under compressive loads. *Arch. Oral Biol.* **39:**1–11.

Maas, M. C., and O'Leary, M. 1996. Evolution of molar enamel microstructure in North American Notharctidae (primates). *J. Hum. Evol.* **31:**293–310.

Maas, M. C., and Simons, E. L. 1995. Enamel microstructure of early anthropoids from the Fayum of Africa. *Am. J. Phys. Anthropol.* **Suppl. 20:**138.

Mace, G. M., Harvey P. H., and Clutton-Brock, T. H. 1981. Brain size and ecology in small mammals. *J. Zool.* **193:**333–354.

Martin, H. 1923. *L'Homme fossile de la Quina.* Libraire Octave Doin: Paris.

Martin, R. D. 1984. Body size, brain size, and feeding strategy. In: D. J. Chivers, B. A. Wood, and A. Bilsborough (eds.), *Food Acquisition and Processing in Primates*, pp. 73–103. Plenum Press, New York.

McHenry, H. M. 1988. New estimates of body weight in early hominids and their significance to encephalization and megadontia in "robust" australopithecines. In: F. E. Grine (ed.), *Evolutionary History of the "Robust" Australopithecines*, pp. 133–148. Aldine de Gruyter, New York.

McHenry, H. M. 1992. How big were the early hominids. *Evol. Anthropol.* **1:**15–20.

Meikle, W. E. 1977. Molar wear stages in *Theropithecus gelada. Kroeber Anthropol. Soc. Papers* **50:**21–25.

Meldrum, D. J., and Kay, R. F. 1997. *Nucicruptor rubicae*, a new pitheciin seed predator from the Miocene of Colombia. *Am. J. Phys. Anthropol.* **102:**407–428.

Mills, J. R. E. 1955. Ideal dental occlusion in primates. *Dent. Practitioner* **6:**47–51.

Mills, J. R. E. 1963. Occlusion and malocclusion in the teeth of Primates. In: D. Brothwell (ed.), *Dental Anthropology*, pp. 29–53. Oxford University Press, Oxford.

Mills, J. R. E. 1967. A comparison of lateral jaw movement in some mammals from wear facets on the teeth. *Arch. Oral Biol.* **12:**645–661.

Napier, J. 1967. Evolutionary aspects of primate locomotion. *Am. J. Phys. Anthropol.* **27:**333–342.

Nelson, D. G. A., DeNiro, M. J., Schoeninger, M. J., and DePaola, D. J. 1986. Effects of diagenesis on strontium, carbon, nitrogen and oxygen concentration and isotopic composition of bone. *Geochim. Cosmochim. Acta.* **50:**1941–1949.

Pastor, R. F., Teaford, M. F., and Glander, K. E. 1995. Methods for collecting and analyzing airborne abrasive particles from neotropical forests. *Am. J. Phys. Anthropol.* **Suppl. 20:**168.

Peters, C. R. 1982. Electron-optical microscopic study of incipient dental microdamage from experimental seed and bone crushing. *Am. J. Phys. Anthropol.* **57:**283–301.

Pfretzschner, H. U. 1986. Structural reinforcement and crack propagation in enamel. In: D. E. Russell, J. P. Santoro, and D. Sigogneau-Russell (eds.), *Teeth Revisited: Proceedings of the VIIth International Symposium on Dental Morphology. Mem. Mus. Nat. Hist. Nat. (Ser C)* **53:**133–143.

Pilbeam, D., and Gould, S. J. 1974. Size and scaling in human evolution. *Science*. **186:** 892–901.

Pilbeam, D., and Gould, S. J. 1975. Allometry and early hominids. *Science* **189:**63–64.

Puech, P. F. 1979. The diet of early man: Evidence from abrasion of teeth and tools. *Curr. Anthropol.* **20:**590–592.

Puech, P. F. 1984. Acidic food choice in *Homo habilis* at Olduvai. *Curr. Anthropol.* **25:**349–350.

Puech, P. F. 1986. *Australopithecus afarensis* Garusi 1, diversitè et spècialisation des premiers Hominides d'après les caractères maxillo-dentaires. *C. R. Acad. Sci. Paris, Ser. II* **303:**1819–1824.

Puech, P. F., and Albertini, H. 1984. Dental microwear and mechanisms in early hominids from Laetoli and Hadar. *Am. J. Phys. Anthropol.* **65:**87–91.

Puech, P. F., Prone, A., Roth, H., and Cianfarani, F. 1985. Reproduction experimentale de processus d'usure des surfaces dentaires des Hominides fossils: Consèquences morphoscopiques et exoscopics avec application a l'Hominide I de Garusi. *C. R. Acad. Sci. Paris, Ser. II.* **301:**59–64.

Puech, P. F., Cianfarani, F., and Albertini, H. 1986. Dental microwear features as an indicator for plant food in early hominids: A preliminary study of enamel. *Hum. Evol.* **1:**507–515.

Rafferty, K., and Teaford, M. F. 1992. Diet and dental microwear in Malagasy subfossil lemurs. *Am. J. Phys. Anthropol.* **Suppl 14:**134.

Ravosa, M. J. 1991. Structural allometry of the mandibular corpus and symphysis in prosimian primates. *J. Hum. Evol.* **23:**197–217.

Ravosa, M. J. 1996. Mandibular form and function in North American and European Adapidae and Omomyidae. *J. Morph.* **229:**171–190.

Ravosa, M. J. 1999. Anthropoid origins and the modern symphysis. *Folia Primatol.* **70:**65–78.

Ravosa, M. J., and Hylander, W. L. 1993. Functional significance of an ossified mandibular symphysis: A reply. *Am. J. Phys. Anthropol.* **90:**509–512.

Ravosa, M. J., and Hylander, W. L. 1994. Function and fusion of the mandibular symphysis in primates: Stiffness or strength? In: J. G. Fleagle, and R. F. Kay (eds.), *Anthropoid Origins*, pp. 447–468. Plenum Press, New York.

Ravosa, M. J., and Simons, W. L. 1994. Mandibular growth and function in *Archaeolemur. Am. J. Phys. Anthropol.* **95:**63–76.

Rensberger, J. M. 1973. An occlusion model for mastication and dental wear in herbivorous mammals. *J. Paleontol.* **47:**512–528.

Rensberger, J. M. 1978. Scanning electron microscopy of wear and occlusal events in some small herbivores. In: P. M. Butler, and K. A. Joysey (eds.), *Development, Function and Evolution of Teeth*, pp. 415–438. Academic Press: New York.

Richard, A. F., Goldstein, S. H., and Dewar, R. E. 1981. Primates as weeds: The implications for macaque evolution. *Am. J. Phys. Anthropol.* **54:**267.

Robinson, J. T. 1954. Prehominid dentition and hominid evolution. *Evolutio* **8:**324–334.

Robinson, J. T. 1963. Adaptive radiation in the australopithecines and the origin of man. In: F. C. Howell, and F. Bourliere (eds.), *African Ecology and Human Evolution*, pp. 385–416. Aldine de Gruyter, Chicago.

Rose, J. C., and Ungar, P. S. 1998. Gross dental wear and dental microwear in historical perspective. In: K. W. Alt, F. W. Rosing, and M. Teschler-Nicola (eds.), *Dental Anthropology: Fundamentals, Limits, Prospects*, pp. 349–386. Stuttgart: Gustav-Fischert.

Rosenberger, A. L. 1986. Platyrrhines, catarrhines and the anthropoid transition. In: B. A. Wood, L. Martin, and P. Andrews (eds.), *Major Topics in Primate and Human Evolution*, pp. 66–88. Cambridge University Press, Cambridge.

Rosenberger, A. 1992. Evolution of feeding niches in New World monkeys. *Am. J. Phys. Anthropol.* **88:**525–562.

Rosenberger, A. J., and Kinzey, W. G. 1976. Functional patterns of molar occlusion in platyrrhine primates. *Am. J. Phys. Anthropol.* **45:**281–297.

Rosenberger, A. L., and Strasser, E. 1985. Toothcomb origins: Support for the grooming hypothesis. *Primates*. **26**:73–84.

Ryan, A. S. 1980. *Anterior Dental Microwear in Hominoid Evolution: Comparisons with Human and Nonhuman Primates*. Ph.D. Dissertation, University of Michigan.

Ryan, A. S., and Johanson, D. C. 1989. Anterior dental microwear in *Australopithecus afarensis*. *J. Hum. Evol.* **18**:235–268.

Schoeninger, M. J., and DeNiro, M. J. 1982. Carbon isotope ratios of apatite from fossil bone cannot be used to reconstruct the diets of animals. *Nature* **292**:333–335.

Schoeninger, M. J., and DeNiro, M. J. 1984. Nitrogen and carbon isotopic composition of bone collagen from marine and terrestrial animals. *Geochim. Cosmochim. Acta* **48**:625–639.

Schoeninger, M. J., Iwaniec, U. T., and Glander, K. E. 1997. Stable isotope ratios indicate diet and habitat use in New World monkeys. *Am. J. Phys. Anthropol.* **103**:69–84.

Sealy, J., and Sillen, A. 1988. Sr and Sr/Ca in marine and terrestrial foodwebs in the southwestern Cape, South Africa. *J. Archaeol. Sci.* **15**:425–438.

Seligsohn, D., and Szalay, F. S. 1978. Relationship between natural selection and dental morphology: Tooth function and diet in *Lepilemur* and *Hapalemur*. In: P. M. Butler, and K. A. Joysey (eds.), *Development, Function and Evolution of Teeth*, pp. 289–307. Academic Press, London.

Sheine, W. S. 1979. *The Effect of Variations in Molar Morphology on Masticatory Effectiveness and Digestion of Cellulose in Prosimian Primates*. Ph.D. Dissertation, Duke University.

Sheine, W. S., and Kay, R. F. 1977. An analysis of chewed food particle size and its relation to molar structure in the primates *Cheirogaleus medius* and *Galago senegalensis* and the insectivoran *Tupaia glis*. *Am. J. Phys. Anthropol.* **47**:15–20.

Siffre, A. 1923. L'alimentation des hominds mousteriens et l'usure de leurs dents. *Rev. Anthropol.* **33**:291–293.

Sillen, A. 1992. Strontium-calcium ratios (Sr/Ca) of *Australopithecus robustus* and associated fauna from Swartkrans. *J. Hum. Evol.* **23**:495–516.

Sillen, A. and Kavanagh, M. 1991. Stronium and paleodietary research: A review. *Yearb. Phys. Anthropol.* **25**:67–90.

Sillen, A., and Lee-Thorp, J. A. 1994. Trace element and isotopic aspects of predator–prey relationships in terrestrial food-webs. *Palaeogeog. Palaeoclimatol. Palaeoecol.* **107**:243–255.

Sillen, A., Hall, G., and Armstrong, R. 1995. Strontium calcium ratios (Sr/Ca) and strontium isotope ratios (^{87}Sr/^{86}Sr) of *Australopithecus robustus* and *Homo* sp. from Swartkrans. *J. Hum. Evol.* **28**:277–285.

Simons, E. L. 1976. The nature of the transition in the dental mechanism from pongids to hominids. *J. Hum. Evol.* **5**:511–528.

Simons, E. L., and Ettel, P. C. 1980. *Gigantopithecus*, the largest primate. *Dent. Abstr.* **15**:266–267.

Simons, E. L., and Pilbeam, D. 1972. Hominoid paleoprimatology. In: R. Tuttle (ed.), *The Functional and Evolutionary Biology of Primates*, pp. 36–62. Aldine-Atherion, New York.

Simpson, G. G. 1933. Paleobiology of Jurassic mammals. *Paleobiology* **5**:127–158.

Smith, R. J., and Pilbeam, D. R. 1980. Evolution of the orangutan. *Nature*. **284**:447–448.

Spears, I. R., and Crompton, R. H. 1996. The mechanical significance of the occlusal geometry of great ape molars in food breakdown. *J. Hum. Evol.* **31**:517–535.

Stern, J. T., and Oxnard, C. E. 1973. Primate locomotion: Some links with evolution and morphology. *Folia Primatol.* **4**:1–93.

Strait, S. G. 1991. *Dietary Reconstruction in Small-Bodied Fossil Primates*. Ph.D. Dissertation, State University of New York at Stony Brook.

Strait, S. G. 1993a. Differences in occlusal morphology and molar size in frugivores and faunivores. *J. Hum. Evol.* **25**:471–482.

Strait, S. G. 1993b. Molar morphology and food texture among small-bodied insectivorous mammals. *J. Mammal.* **74**:391–402.

Strait, S. G. 1993c. Molar microwear in extant small-bodied faunivorous mammals: An analysis of feature density and pit frequency. *Am. J. Phys. Anthropol.* **92**:63–79.

Strait, S. G. 1997. Tooth use and the physical properties of foods. *Evol. Anthropol.* **5**:199–211.

Strait, S. G., and Overdorff, D. J. 1994. A preliminary examination of molar microwear in strepsirrhine primates. *Am. J. Phys. Anthropol.* **Suppl. 18:**190.

Strait, S. G. and Overdorff, D. J. 1996. Physical properties of fruits eaten by Malagasy primates. *Am. J. Phys. Anthropol.* **Suppl. 22:**224.

Sussman, R. W. 1978. Foraging patterns of nonhuman primates and the nature of food preferences in man. *Fed. Proc.* **37:**55–60.

Sussman, R. W. 1987. Morpho-physiologicalanalysis of diets: species-specific dietary patterns in primates and human dietary adaptations. In: W. G. Kinzey (ed.), *The Evolution of Human Behavior: Primate Models*, pp. 157–179. SUNY Press, Albany.

Swindler, D. 1976. *Dentition of Living Primates*. Academic Press: New York.

Teaford, M. F. 1982. Differences in molar wear gradient between juvenile macaques and langurs. *Am. J. Phys. Anthropol.* **57:**323–330.

Teaford, M. F. 1985. Molar microwear and diet in the genus *Cebus. Am. J. Phys. Anthropol.* **66:**363–370.

Teaford, M. F. 1986. Dental microwear and diet in two species of *Colobus*. In: J. Else, and P. Lee (eds.), *Proceedings of the Tenth Annual International Primatological Conference. Volume II: Primate Ecology and Conservation*, pp. 63–66. Cambridge University Press: Cambridge.

Teaford, M. F. 1988. A review of dental microwear and diet in modern animals. *Scan. Microsc.* **2:**1149–1166.

Teaford, M. F. 1993. Dental microwear and diet in extant and extinct *Theropithecus*: Preliminary analyses. In: N. G. Jablonski (ed.), *Theropithecus: The Life and Death of a Primate Genus*, pp. 331–349. Cambridge University Press, Cambridge.

Teaford, M. F., and Glander, K. E. 1991. Dental microwear in live, wild-trapped *Alouatta palliata* from Costa Rica. *Am. J. Phys. Anthropol.* **85:**313–319.

Teaford, M. F., and Leakey, M. G. 1992. Dental microwear and diet in Plio-Pleistocene cercopithecoids from Kenya. *Am. J. Phys. Anthropol.* **Suppl 14:**160.

Teaford, M. F., and Oyen, O. J. 1989a. Differences in rate of molar wear between monkeys raised on different diets. *J. Dent. Res.* **68:**1513–1518.

Teaford, M. F. and Oyen, O. J. 1989b. In vivo and in vitro turnover in dental microwear. *Am. J. Phys. Anthropol.* **80:**447–460.

Teaford, M. F., and Oyen, O. J. 1989c. Live primates and dental replication: New problems and new techniques. *Am. J. Phys. Anthropol.* **80:**73–81.

Teaford, M. F., and Robinson, J. G. 1989. Seasonal or ecological zone differences in diet and molar microwear in *Cebus nigrivittatus. Am. J. Phys. Anthropol.* **80:**391–401.

Teaford, M. F., and Runestad, J. A. 1992. Dental microwear and diet in Venezuelan primates. *Am. J. Phys. Anthropol.* **88:**347–364.

Teaford, M. F., and Walker, A. C. 1984. Quantitative differences in dental microwear between primate species with different diets and a comment on the presumed diet of *Sivapithecus. Am. J. Phys. Anthropol.* **64:**191–200.

Teaford, M. F., Pastor, R. F., Glander, K. E., and Ungar, P. S. 1994. Dental microwear and diet: Costa Rican *Alouatta* revisited. *Am. J. Phys. Anthropol.* **Suppl 18:**194.

Teaford, M. F., Maas, M. C., and Simons, E. L. 1996. Dental microwear and microstructure in Early Oligocene primates from the Fayum, Egypt: Implications for diet. *Am. J. Phys. Anthropol.* **101:**527–543.

Ungar, P. S. 1990. Incisor microwear and feeding behavior in *Alouatta seniculus* and *Cebus olivaceus. Am. J. Primatol.* **20:**43–50.

Ungar, P. S. 1992. *Incisor Microwear and Feeding Behavior of Four Sumatran Anthropoids*. Ph.D. Dissertation, State University of New York at Stony Brook.

Ungar, P. S. 1994a. Incisor microwear of Sumatran anthropoid primates. *Am. J. Phys. Anthropol.* **94:**339–363.

Ungar, P. S. 1994b. Patterns of ingestive behavior and anterior tooth use differences in sympatric anthropoid primates. *Am. J. Phys. Anthropol.* **95:**197–219.

Ungar, P. S. 1995a. Fruit preferences of four sympatric primate species at Ketambe, northern Sumatra, Indonesia. *Int. J. Primatol.* **16:**221–245.

Ungar, P. S. 1995b. A semiautomated image analysis procedure for the quantification of dental microwear II. *Scanning* **17:**57–59.
Ungar, P. S. 1996a. Relationship of incisor size to diet and anterior tooth use in sympatric Sumatran Anthropoids. *Am. J. Primatol.* **38:**145–156.
Ungar, P. S. 1996b. Dental microwear of European Miocene catarrhines: Evidence for diets and tooth use. *J. Hum. Evol.* **31:**335–366.
Ungar, P. S., and Grine, F. E. 1991. Incisor size and wear in *Australopithecus africanus* and *Paranthropus robustus*. *J. Hum. Evol.* **20:**313–340.
Ungar, P. S., and Kay, R. F. 1995. The dietary adaptations of European Miocene catarrhines. *Proc. Natl. Acad. Sci. USA.* **92:**5479–5481.
Ungar, P. S., and Teaford, M. F. 1996. A preliminary examination of non-occlusal dental microwear in anthropoids: Implications for the study of fossil primates. *Am. J. Phys. Anthropol.* **100:**101–113.
Ungar, P. S., Simons, J.-C., and Cooper, J. W. 1991. A semiautomated image analysis procedure for the quantification of dental microwear. *Scanning* **13:**31–36.
Ungar, P. S., Teaford, M. F., Glander, K. E., and Pastor, R. F. 1995. Dust accumulation in the canopy: A potential cause of dental microwear in primates. *Am. J. Phys. Anthropol.* **97:**93–99.
Ungar, P. S., Kay, R. F., Teaford, M. F., and Walker, A. 1996. Dental evidence for diets of Miocene apes. *Am. J. Phys. Anthropol.* **Suppl. 22:**232–233.
Walker, A. 1980. Functional anatomy and taphonomy. In: A. K. Behrensmeyer, and A. P. Hill (eds.), *Fossils in the Making*, pp. 182–196. University of Chicago Press, Chicago.
Walker, A. 1981. Diet and teeth: Dietary hypothesis and human evolution. *Philos. Trans. R. Soc. London Ser. B* **292:**57–64.
Walker, A. C., Hoeck, H. N., and Perez, L. 1978. Microwear of mammalian teeth as an indicator of diet. *Science.* **201:**808–810.
Walker, A., Teaford, M. F., and Ungar, P. S. 1994. Enamel microwear differences between species of *Proconsul* from the early Miocene of Kenya. *Am. J. Phys. Anthropol.* **Suppl 18:**202–203.
Walker, P. L. 1973. *Great Ape Feeding Behavior and Incisor Morphology*. Ph.D. Dissertation, University of Chicago.
Walker, P. L. 1976. Wear striations on the incisors of cercopithecoid monkeys as an index of diet and habitat preference. *Am. J. Phys. Anthropol.* **45:**299–308.
Walker, P. L., and Murray, P. 1975. An assessment of masticatory efficiency in a series of anthropoid primates with special reference to Colobinae and Cercopithecinae. In: R. Tuttle (ed.), *Primate Functional Morphology and Evolution*, pp. 135–150. Mouton, The Hague.
Wallace, J. A. 1974. Approximal grooving of teeth. *Am. J. Phys. Anthropol.* **40:**285–290.
Weidenreich, F. 1937. The dentition of *Sinanthropus pekinensis*. *Paleo. Sin.* **101:**1–180.
Williams, B. A., and Covert, H. H. 1994. New early Eocene anaptomorphine primate (Omomyidae) from the Washakie Basin, Wyoming, with comments on the phylogeny and paleobiology of anaptomorphines. *Am. J. Phys. Anthropol.* **93:**323–340.
Wood, B. A. 1981. Tooth size and shape and their relevance to studies of hominid evolution. *Philos. Trans. R. Soc. London, Ser. B* **292:**65–76.
Wood, B. A., and Abbott, S. A. 1983. Analysis of the dental morphology of Plio-Pleistocene hominids. I. Mandibular molars—crown area measurements and morphological traits. *J. Anat.* **136:**197–219.
Yamashita, N. 1996. Seasonal and site specificity of mechanical dietary patterns in two Malagasy lemur families (Lemuridae and Indriidae). *Int. J. Primatol.* **17:**355–387.

Reconstructing Social Behavior from Dimorphism in the Fossil Record

8

J. MICHAEL PLAVCAN

Introduction

The social behavior of extinct species has long been a subject for speculation. Group size and composition, mating systems, territoriality, resource defense, predator defense, cooperation, and competition, among others, have all been inferred for extinct taxa (e.g., Aiello and Dunbar, 1993; Blumenberg, 1981, 1984; Fleagle *et al.*, 1980; Gingerich, 1981, 1995; Kay, 1982; Kay *et al.*, 1997; Krishtalka *et al.*, 1991; Lovejoy, 1981; McHenry 1991, 1992, 1994a,b; Moore, 1997; Plavcan and van Schaik, 1997a; Simons and Rasmussen, 1996; Simons *et al.*, 1999; Wrangham, 1987; Zihlman, 1997—see chapter by Nunn and van Schaik). Such characters are important for painting a history of the evolution and diversification of lineages, and they occasionally feature as key components of models for primate social evolution. There is an especially rich literature speculating on the evolution of human social behavior (e.g., Kinzey, 1987). The purpose of this chapter is to evaluate how social behavior can (or cannot) be inferred in extinct primates on the basis of remains preserved in the fossil record.

J. MICHAEL PLAVCAN • Department of Anatomy, New York College of Osteopathic Medicine, Old Westbury, New York 11568; Present address: Department of Anthropology, University of Arkansas, Fayetteville, Arkansas 72701.

Reconstructing Behavior in the Primate Fossil Record, edited by Plavcan *et al.* Kluwer Academic/Plenum Publishers, New York, 2002.

Most behavioral characteristics of extinct species are inferred by comparison to morphologically similar extant species. Usually, inferences are based on comparative or functional analysis of living species which identify cause/effect relations between morphology and behavior (Kay and Cartmill, 1977; Kay *et al.*, this volume, Chapter 9). These may be either adaptive or not. For example, molar tooth morphology reflects adaptations to particular diets, while dental microwear is caused by the ingestion of particular types of food (Ungar, this volume, Chapter 7). There are also cases where behavior might be inferred from nonanatomical evidence preserved in the fossil record. For example, taphonomic evidence may yield insights into the environment which a species inhabited, or its specific habits (for example, burrows indicate fossorial habits). Understanding of the habits of later hominids is greatly enhanced by the preservation of artifacts and tools, evidence of unique activities such as making fire, deliberate burial of individuals, and so on. However, for the vast majority of extinct species, behavior can only be inferred from morphology.

Inferring the social behavior of extinct species offers a particular challenge because there are few direct morphological correlates of social behavior. For this reason, theoretical modeling has proven a popular tool for reconstructing social behavior, especially in hominids (e.g., Foley and Lee, 1989; Kinzey, 1987 and references cited therein; Moore, 1997; Zihlman, 1997). For example, Foley and Lee (1989) develop a theoretical model of how social systems evolve from one type to another given a variety of ecological constraints. Then, on the basis of inferred changes in habitat and diet they reconstruct the likely changes in hominid social systems from a chimpanzee-like common ancestor to extant humans. Wrangham (1987) used cladistic methods to reconstruct likely ancestral hominid behavioral characters using an extant hominoid phylogeny. Moore (1997) uses extant savanna-dwelling chimpanzees as a model for early hominid behavior. Such models offer extremely valuable insights into the evolution of social behavior, and are indispensable for constructing reasonable hypotheses, and weeding out improbable ones. Still, the models do not provide direct evidence of social behavior from the fossils themselves, and a review of them is beyond the scope of this chapter.

Several anatomical features are indirectly related to variation in social behavior in extant primates. For example, nocturnality is associated with solitary habits (Ross, 1996). Relatively large orbits imply nocturnality, which, in turn, implies solitary habits. Unfortunately, the relation between orbit size and nocturnality only holds for small-bodied primates (Ross, 1996). At larger body size (greater than about 1300 grams), relative orbit size offers no evidence of activity cycle, and therefore no evidence of social behavior.

Relative brain size—especially when it is comparatively large—is occasionally used to infer greater social complexity (among a variety of other traits). Clutton-Brock and Harvey (1980) report loose associations between relative brain size and mating systems in primates. They suggest that monog-

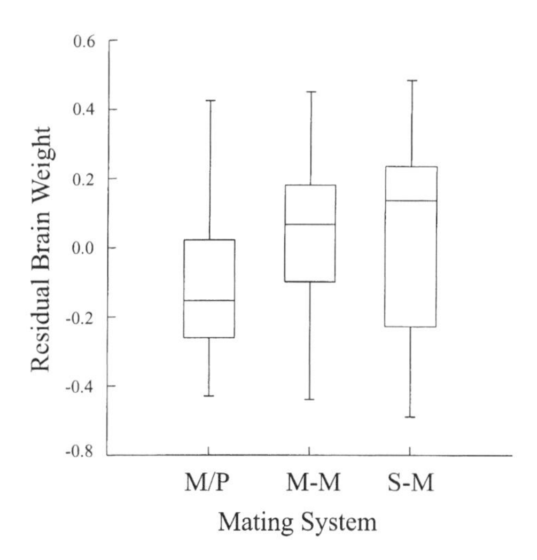

Fig. 1. Box and whiskers comparison between residual brain weight and mating system in anthropoid primates. Residual brain weight is measured as the residual from a least-squares regression of brain weight against female body mass for 74 species. Results do not substantially change if a reduced major axis line is used. Data for brain weights taken from Harvey *et al.* (1987). Abbreviations as follows: M/P = monogamous/polyandrous; M-M = multi-male, multi-female; S-M = single-male, multi-female.

amous species tend to have relatively smaller brains than polygynous species. On the basis of this finding, Blumenberg (1984) suggested that the relatively small-brained *Aegyptopithecus zeuxis* was monogamous, though sexual dimorphism suggests that this animal was polygynous (Fleagle *et al.*, 1980). A comparison between residual brain size estimates and mating system classifications demonstrates no unique relation between any given mating system (or any other measure of social behavior reviewed below), and relative brain size (Fig. 1). Thus, for most primates relative brain size alone offers no direct evidence for specific mating systems. A more restrained interpretation is offered by McHenry (1994a, p. 83), who suggests that an increase in relative brain size in hominids "...implies that a major alteration in subsistence occurred." Again, though, as noted by Smith (1996), the standard error between estimates of relative brain size and behavioral variables is so great that such inferences must be taken with a grain of salt.

Variation in diet is important to understanding variation in mating systems and social behavior in primates. For example, female dispersal patterns are important in determining male dispersal and behavior. Female distribution reflects in part the availability of resources (Wrangham, 1980). Where resources, such as fruit trees, are clumped, females can form larger groups. In contrast, where resources, such as leaves, are relatively abundant, females form smaller groups. In the former case, multi-male groups tend to form, characterized by intense male–male competition and male dominance hierarchies. In the latter case, a single male tends to be resident in each group of females. Unfortunately, these distinctions are very broad, and there is no one-to-one correspondence between broad dietary categories and particular types of mating systems. For example, *Pongo* is characterized by solitary males and females, *Pan* is characterized by solitary females and groups of males, *Hylobates concolor* is pair-living, *Macaca fascicularis* is characterized by multi-male, multi-female groups, and *Cercopithecus mitis* is characterized by single-male, multi-female groups. All of these species are primarily frugivorous. Hence, crude dietary characterizations alone cannot be used to predict social behavior in extinct species.

Among primates, the only direct anatomical correlate of social behavior is sexual dimorphism. In most anthropoid primates, males possess larger canine teeth than females, and are larger in body size and skeletal dimensions than females. Dimorphism is often used as evidence of polygyny or intense male–male competition, while its absence is used as evidence of monogamy or polyandry (e.g., Brace, 1972; Brace and Ryan, 1980; Fleagle *et al.*, 1980; Gingerich, 1981, 1995; Kappelman, 1996; Kay, 1982; Leutenegger and Shell, 1987; Lovejoy, 1981; McHenry, 1994a,b; Plavcan and van Schaik, 1997b; Simons and Rasmussen, 1996). An exception is offered by Foley and Lee (1989), who employ their model for behavior to infer male-bonded societies in *Australopithecus afarensis*. The magnitude of size dimorphism in *A. afarensis* is inconsistent with this reconstruction. They suggest that the apparent strong dimorphism actually reflects the presence of two species in the *A. afarensis* hypodigm. Hence, Foley and Lee alter the degree of dimorphism in *A. afarensis* to fit theoretical predictions, rather than alter the model to concur more closely with evidence from dimorphism.

Often dimorphism in extinct species is loosely compared to dimorphism in broadly morphologically similar extant species. For example, Simons and Rasmussen (1996) infer a *Saimiri*-like mating system for *Catopithecus* on the basis of overall morphological similarities between the two species, including canine dimorphism. Hominid dimorphism is often compared to that of chimpanzees or gorillas (e.g., Kappelman, 1996), with little consideration of variation in the relation between dimorphism and behavior across or within taxa at any level. Conversely, Plavcan and van Schaik (1997a) discuss the ambiguous behavioral implications of mass and canine dimorphism in australopithecines, with reference to a broad array of anthropoids.

Numerous recent comparative analyses evaluate the behavioral and ecological correlates of sexual dimorphism in canine tooth size and body size in primates (Cheverud *et al.*, 1985; Clutton-Brock *et al.*, 1977; Ely and Kurland, 1989; Ford, 1994; Gaulin and Sailer, 1984; Gautier-Hion, 1975; Greenfield, 1992a,b, 1996; Greenfield and Washburn, 1991; Harvey *et al.*, 1978a,b; Kappeler, 1990, 1991, 1996; Kay *et al.* 1988; Leigh, 1992, 1995a,b; Leigh and Shea, 1995; Leigh and Terranova, 1998; Leutenegger and Cheverud, 1982, 1985; Leutenegger and Kelly, 1977; Lucas *et al.*, 1986; Martin *et al.*, 1994;

Masterson and Hartwig, 1998; Mitani *et al.*, 1996; Philips-Conroy and Jolly, 1981; Plavcan, 1993, 1998b; Plavcan and van Schaik, 1992, 1994, 1997b; Plavcan *et al.*, 1995; Rowell and Chism, 1986; Smith, 1981). While these analyses have been used in varying ways to support behavioral reconstructions in the fossil record, to date there has been no systematic evaluation of how these models can actually be used to infer behavior from dimorphism in the fossil record. Hence, though we certainly know more about the correlates of dimorphism in extant species, there is no clear consensus about what dimorphism can and cannot tell us about behavior in the fossil record. The aim of this chapter is to explore this question in some depth.

Data

Data on canine tooth size dimorphism, relative male and female canine tooth size, mandibular molar tooth size, and body mass dimorphism in 100 species of primates are extracted from Plavcan (1990), Plavcan *et al.* (1995), Plavcan and van Schaik (1997b), and Smith and Jungers (1997). Sample sizes vary from character to character. All data on fossils are taken from the literature, and the sources are noted where relevant.

Six canine dimensions are used—three for the maxillary and three for the mandibular canine. For each tooth, the mesiodistal (MD) and buccolingual (BL) diameters are the longest axis of the tooth in the occlusal plane, and the widest diameter perpendicular to the length, respectively. These are referred to as "occlusal dimensions" of the canines. The crown height (Hgt) of the tooth is measured from the apex of the tooth to the cementum–enamel junction. Body mass data are extrapolated from Plavcan and van Schaik (1997b) and Smith and Jungers (1997).

For all extant species, dimorphism is measured as the ln-transformed ratio of mean male divided by female values for each measure. This is equivalent to subtracting the natural log of the female mean from that of the male, and reduces skewness and kurtosis of the data (Gaulin and Sailer, 1984).

Behavioral and ecological data are extracted from the literature. Summaries and citations are found primarily in Plavcan and van Schaik (1992, 1997b), Mitani *et al.* (1996), Kappeler (1996), and Plavcan *et al.* (1995).

Sources of Error in Estimates of Dimorphism

Before discussing the behavioral implications of dimorphism, it is important to underscore the inherent uncertainty in estimates of dimorphism for fossils (Smith, 1996). Estimates of sexual dimorphism are subject to sampling error, and geographic or temporal variation. Consequently, most estimates of dimorphism in extinct species should be approach with caution.

As with any population descriptor, ratios are estimated with error. Though standard errors of ratios can be calculated (Kendall and Stuart, 1954; Simpson *et al.*, 1960), they rarely are (Brace and Ryan, 1980; Josephson *et al.*, 1997; Phillips-Conroy and Jolly, 1981). Calculation of the standard error of a ratio requires knowledge of the within-sex variance for males and females. At the same time, sexual dimorphism is usually estimated for fossil samples without a priori knowledge of the sex of individual specimens. Where sex is unknown, dimorphism can be estimated by extrapolation from coefficients of variation (Kay, 1982; Plavcan 1994), division of the sample into "males" and "females" about the sample mean or median (Godfrey *et al.*, 1993; Plavcan, 1994; equivalent to "technique dimorphism" Lovejoy *et al.*, 1989), finite mixture analysis (Godfrey *et al.*, 1993), or the method-of-moments (Josephson *et al.*, 1997). Though division about the sample mean is most commonly used and probably most conservative method, all of these techniques introduce error into the dimorphism estimates.

Unfortunately, standard errors for ratios are quite large at the small sample sizes typically encountered in the fossil record, and this fact should be kept in mind when making assertions about the certainty of behavioral implications of dimorphism. This means that estimates of dimorphism in extinct species should be treated as biologically interesting, but very tentative. In keeping with the findings of Smith (1996), behavioral inferences based on dimorphism should not themselves be used as data for evaluating other models and hypotheses unless the samples are relatively large.

Another important consideration is whether both sexes are present in a sample—a real problem with many extinct species. With a sample of 6 one can be 95% certain that both sexes are present, following from a binomial probability distribution (and asuming no taphonomic bias). With fewer than six specimens, though, it is difficult to justify any analysis of dimorphism unless it is clear that both sexes are present.

Geographic variation is another important source of error for estimating dimorphism. Geographic variation in primates is well documented for a number of species (e.g., Albrecht, 1976, 1980; Albrecht *et al.*, 1990; Albrecht and Miller, 1993; Fooden, 1995; Shea *et al.*, 1993), as is geographic and subspecific variation in the expression of dimorphism (e.g., Hayes *et al.*, 1995; Masterson and Hartwig, 1998; Turner *et al.*, 1994; Uchida, 1998a,b). Assuming that geographic variation can serve as an analogy for temporal variation, estimates of dimorphism derived from samples with loose stratigraphic or geographic control should be approached with great caution.

Behavioral Variables

While the term "social behavior" encompasses a wide range of behavioral attributes, sexual dimorphism is commonly used as evidence of *mating system*, or *intrasexual competition*. Underlying this are the assumptions that (1) interspecific variation in the magnitude of dimorphism directly arises from variation in the strength of sexual selection, and (2) different mating systems reflect variation in the strength of sexual selection.

Modern sexual selection theory is complex (cf. Anderson, 1994), but in general there are two types of sexual selection that can occur: mate choice and mate competition. Sexual dimorphism in extant primates is commonly viewed as a consequence of male–male competition for access to mates. It is assumed that males fight to exclude other males from access to females, resulting in differential male reproductive success and strong selection for characters that help males to win fights. That male primates are usually larger and have bigger canine teeth than females, and often fight with one another seems to corroborate this. In contrast, there is little comparative evidence that female choice plays a role in the evolution of dimorphism in primates (this is not to say that female choice does not occur—only that current evidence for the phenomenon cannot be used in comparative analysis, especially for extrapolating behavior in extinct species).

A true test the sexual selection hypothesis should evaluate whether large males with large canine teeth actually tend to win fights, and whether winning fights confers a reproductive advantage to individuals. Such data are not available. Consequently, researchers must use surrogate measures of sexual selection. It is these surrogate measures of sexual selection that form the basis for inferring social behavior in the fossil record from dimorphism. In the recent primate literature, four surrogate measures of sexual selection have been used to test the sexual selection hypothesis in formal comparative analyses. These are *mating system* (e.g., Cheverud *et al.*, 1985; Harvey *et al.*, 1978b), *socionomic sex ratio* (Clutton-Brock *et al.*, 1977), *operational sex ratio* (Mitani *et al.*, 1996), and male–male *competition levels* (Kay *et al.*, 1988; Plavcan and van Schaik, 1992, 1997b).

Additionally, independent male and female *competition classifications*, based on the competition levels of Plavcan and van Schaik (1992, 1997b), have been compared to separate measures of male and female canine tooth size in primates (Plavcan *et al.*, 1995; Plavcan, 1998b). Following are brief definitions of each variable.

Mating System (MS)

Mating system is the most commonly cited behavioral correlate of sexual dimorphism. Most studies use either a simple monogamy/polygyny dichotomy, or a tripartite classification of monogamy, single-male/multi-female, and multi-male/multi-female mating systems (e.g., Clutton-Brock and Harvey, 1978). Monogamous and polyandrous species are lumped together as having little or no differential sexual selection, while polygynous species are assumed to experience stronger sexual selection in males than in females. The former are hypothesized to show no dimorphism, while the latter should be dimorphic. In the tripartite classification, single-male species were originally hypothesized to show greater dimorphism than multi-male species (Harvey *et al.*, 1978b) under the assumption that differential male reproductive success is inversely proportional to the number of males in a group.

It is well known that mate competition is highly variable among multi-male groups (Mitani *et al.*, 1996). For example, male baboons compete intensely for access to females, while chimpanzees show promiscuous mating. Furthermore, some species, such as *Presbytis* (= *Semnopithecus*) *entellus*, show either single-male or multi-male groups among populations (Struhsaker and Leland, 1987), while among guenons multi-male influxes into single male groups are common during the mating season (Cords, 1987).

Regardless, MS is still commonly used in dimorphism studies (Gittleman and Van Valkenburg, 1997; Kappeler, 1990, 1996; Leigh, 1992, 1995a,b), and inferred for extinct species (recent examples are Kappelman, 1996; Simons and Rasmussen, 1996).

Socionomic Sex Ratio (SR)

The socionomic sex ratio is simply the number of adult males typically present in a breeding group divided by the number of adult females (Clutton-Brock *et al.*, 1977). Theoretically this ratio reflects the degree to which males exclude each other from access to groups of females, and hence is proportional to the intensity of sexual selection. It is recognized that SR is a poor estimate of sexual selection, for the same reasons that mating system classifications do not adequately capture variation in mate competition (Andersson, 1994; Emlen and Oring, 1977). Sex ratio has occasionally appeared in the literature on the behavioral implications of dimorphism in the fossil record.

Operational Sex Ratio (OSR)

Emlen and Oring (1977) introduced the OSR, but only recently has it been compared to dimorphism in primates (Mitani *et al.*, 1996). Emlen and Oring (1977) pointed out that the number of receptive females relative to the number of males in a group should determine the strength of sexual selection. For example, if there are 10 females in a group, but they never become receptive at the same time, then a single male can effectively exclude other males from access to all females and monopolize matings because he only has to guard one female at a time. This should lead to large differential reproductive success among males, resulting in intense male–male competition and strong sexual selection. Conversely, if all females become receptive at once, a single male cannot simultaneously guard all females, and sexual selection is weak or absent.

Mitani *et al.* (1996) quantify the OSR with a formula requiring information on birth seasonality, length of gestation, interbirth intervals, number of estrous cycles to conception and length of estrous, in addition to socionomic sex ratios. Reliable data on all these variables for wild populations are not common. Hence, Mitani *et al.* (1996) present data for only 18 species. Nevertheless, while this is a comparatively small sample size, the measure is theoretically important.

Competition Levels (CL)

Competition levels are created from categorical estimates of the "potential frequency" and "intensity" of male–male agonistic competition (Kay *et al.*, 1988; Plavcan and van Schaik, 1992, 1994, 1997b). Competition levels assume that the strength of selection associated with male–male competition is proportional to the degree to which males actually fight. Intensity and potential frequency classifications are dichotomized into "high" and "low" classes. Species are classified as low-frequency if there is one male typically present in a breeding group, and high-frequency if there is typically more than one male present. Species are ranked as high-intensity when males form agonistic dominance hierarchies, are reported as intolerant of one another, or fight commonly. Species are ranked as low-intensity when males are reported as tolerant, dominance hierarchies are difficult to detect, and fighting is infrequent. Note that competition intensity essentially is a behavioral measure of whether males regularly escalate competition into fights, while competition frequency is essentially a demographic measure.

The four combinations of potential frequency and intensity are combined to form four competition types, or levels. Low-intensity, low-frequency species are placed into competition level 1, while high-intensity, high-frequency species are placed in competition level 4. Dimorphism is clearly predicted to be greatest in competition level 4, and least in competition level 1. Competition levels 2 and 3 are ordered on the basis of competition intensity. This ordering was initially presented in Kay *et al.* (1988), who distinguished between patterns of promiscuous mating in competition level 2, and intense male–male competition that is limited seasonally in competition level 3. In the more formal analysis, Plavcan and van Schaik (1992) empirically demonstrated a much stronger difference in dimorphism between high- and low-intensity species, than between high- and low-frequency species, justifying the initial ranking of the intermediate competition levels on the basis of intensity rather than frequency.

Competition levels are similar to mating system classifications in that most multi-male/multi-female species are ranked into competition level 4, most monogamous or polyandrous species are ranked into competition level 1, and most single-male/multi-female species are ranked into competition level 3. Important differences are that multi-male species which are characterized by "fission–fusion" social systems (e.g., *Pan*) or promiscuous mating (e.g., *Brachyteles*) are ranked into competition level 2, as are some polyandrous species (e.g., some *Saguinus*). Species with multi-male groups in which male–male competition is limited to a restricted breeding season are placed into competition level 3 (e.g., *Saimiri*), as are species characterized by single-male groups with multi-male influxes occurring during a restricted breeding season (e.g., *Cercopithecus mitis*).

Competition Classifications (CC)

Plavcan *et al.* (1995) generalize the definitions of "potential frequency" and "intensity" of agonistic competition and apply them separately to male and female primates. Canine size is predicted to be relatively larger in males or females ranked as showing high-intensity and/or high-frequency competition than in those classified as showing low-intensity or low-frequency competition.

Additionally, Plavcan *et al.* (1995) evaluated the relation between relative canine size and the "context" of competition. "Context" refers to whether competition occurs between coalitions of individuals or not. Where coalitions or alliances are regularly formed, the outcome of agonistic contests should be determined by the number of allies or partners an individual can recruit. This is common where females regularly form matrilineal dominance hierarchies. The development of large size or weaponry should be less important in determining the outcome of fights among coalitions than when individuals fight alone. Thus, canines should be smaller in "coalitionary" males or females by comparison to those classified as showing "non-coalitionary" high-intensity, high-frequency competition.

Relations between Dimorphism and Behavioral Estimates

The associations between the above measures and canine and body mass dimorphism have been reported elsewhere in the literature (e.g., Cheverud *et al.*, 1985; Clutton-Brock *et al.*, 1977; Ely and Kurland, 1989; Ford, 1994; Gaulin and Sailer, 1984; Harvey *et al.*, 1978b; Kappeler, 1990, 1991, 1996; Kay *et al.*, 1988; Leutenegger and Cheverud, 1982, 1985; Leutenegger and Kelly, 1977; Lucas *et al.*, 1986; Plavcan and van Schaik, 1992, 1997b; Plavcan *et al.*, 1995). They are briefly reviewed here.

Socionomic Sex Ratio

The socionomic sex ratio is significantly correlated with all measures of sexual dimorphism (Table I), but this is clearly driven by the lack of dimor-

Table I. Correlations and Regression Statistics for the Relationship between Mass and Canine Dimorphism Measures and Sex Ratio[a]

Dimorphism measure	n	r	Intercept	Slope	P	SEE
Mass	47	0.453	0.136	0.166	0.001	0.208
Mandibular						
MD	46	0.585	0.108	0.130	<0.001	0.112
BL	46	0.555	0.140	0.114	<0.001	0.107
Hgt	46	0.636	0.239	0.215	<0.001	0.163
Maxillary						
MD	46	0.609	0.148	0.161	<0.001	0.132
BL	46	0.426	0.120	0.096	0.003	0.127
Hgt	46	0.636	0.288	0.288	<0.001	0.218

[a] MD = mesiodistal; BL = buccolingual; Hgt = crown height; SEE = standard error of the estimate.

phism in monogamous species (Fig. 2). There is no correlation between dimorphism and SR within mating system classifications. Phylogenetic control does not improve the results.

Operational Sex Ratio

Mitani *et al.* (1996) demonstrated a significant relation between the OSR and body mass dimorphism using the phylogenetic contrast method (Felsenstein, 1985; Pagel, 1992). Using different body mass data, Plavcan and van Schaik (1997b) found that OSR correlated weakly with mass dimorphism, and is not correlated with any measure of canine dimorphism (Fig. 2). This latter result is peculiar, given that canine dimorphism is more strongly correlated with competition levels than is body mass dimorphism (see below). Most importantly for trying to infer OSR from dimorphism in fossils, there is no significant correlation between OSR and any of the variables analyzed here when simple species values are analyzed without phylogenetic control. These results may reflect the small sample sizes (between 15 and 18) in these analysis, but further work is needed to account for the discrepancy in results.

Mating System

It is well known that monogamous and polyandrous primates as a group show significantly less mass and canine dimorphism than polygynous primates, no matter how the analysis is carried out (Table II, Fig. 3). Single-male/multi-female species and multi-male/multi-female species do not significantly differ in any measure of dimorphism. While many polygynous species show strong

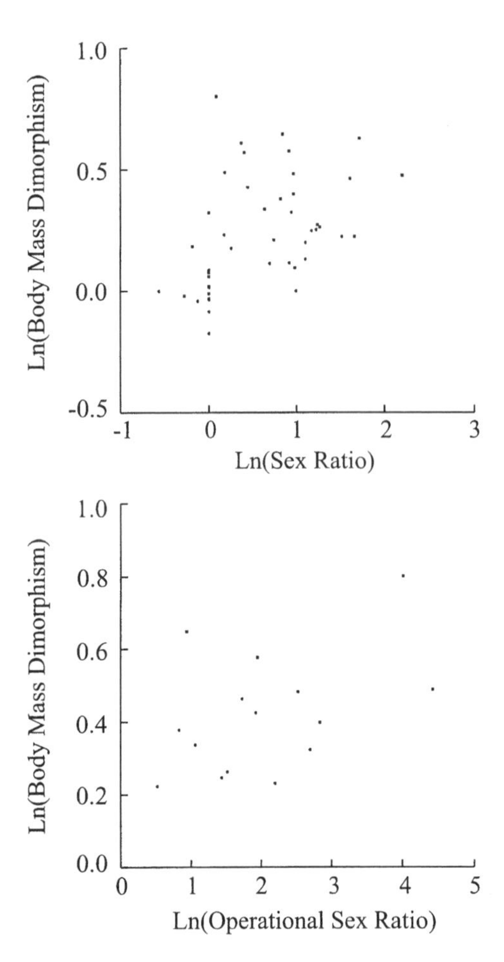

Fig. 2. Bivariate comparisons between ln-transformed body mass dimorphism and ln-transformed estimates of sex ratio (top panel) and operational sex ratio (OSR: bottom panel) for anthropoid primates. Data for sex ratio taken from sources in Smuts *et al.* (1987). Data for OSR taken from Mitani *et al.* (1996).

degrees of mass and canine dimorphism, there are a number—especially among colobines—that show relatively low degrees of mass and canine dimorphism (Leutenegger and Kelly, 1977; Plavcan and van Schaik, 1997b).

Competition Levels

Plavcan and van Schaik (1992, 1997b) demonstrate a strong relation between all measures of dimorphism and CL (Table II, Fig. 3). For mass dimorphism, the strongest distinction is between high- and low-intensity

Table II. Means, Medians, Standard Deviations, and Ranges of Ln-Transformed Dimorphism Estimates within Competition Levels and Mating System Classifications

Variable	Min	Max	Median	Mean	SD	N
Body Mass						
CL 1	−0.174	0.086	−0.020	−0.024	0.077	14
CL 2	−0.083	0.307	0.086	0.094	0.125	14
CL 3	0.000	0.802	0.273	0.305	0.213	24
CL 4	0.115	0.751	0.372	0.411	0.169	33
M/P	−0.174	0.086	−0.025	−0.024	0.068	20
S-M	0.000	0.802	0.219	0.295	0.211	24
M-M	0.077	0.751	0.329	0.372	0.179	41
C_1 MD						
CL 1	−0.010	0.077	0.025	0.026	0.028	14
CL 2	−0.030	0.191	0.099	0.087	0.083	10
CL 3	0.010	0.365	0.199	0.214	0.112	28
CL 4	0.068	0.548	0.304	0.309	0.111	34
M/P	−0.030	0.077	0.020	0.019	0.029	16
S-M	0.010	0.365	0.223	0.218	0.111	28
M-M	0.068	0.548	0.262	0.281	0.122	40
C_1 BL						
CL 1	−0.020	0.122	0.068	0.055	0.047	12
CL 2	−0.020	0.215	0.157	0.124	0.094	10
CL 3	0.039	0.582	0.243	0.251	0.119	28
CL 4	0.077	0.565	0.304	0.318	0.114	34
M/P	−0.020	0.122	0.053	0.048	0.051	16
S-M	0.039	0.582	0.239	0.250	0.119	28
M-M	0.077	0.565	0.285	0.279	0.115	40
C_1 Hgt						
CL 1	−0.020	0.351	0.086	0.103	0.100	12
CL 2	−0.020	0.455	0.270	0.218	0.175	10
CL 3	0.315	0.637	0.441	0.459	0.084	28
CL 4	0.300	1.047	0.539	0.577	0.197	34
M/P	−0.020	0.351	0.072	0.090	0.110	16
S-M	0.315	0.637	0.441	0.461	0.086	28
M-M	0.231	1.047	0.504	0.538	0.204	40
C^1 MD						
CL 1	−0.020	0.140	0.049	0.051	0.050	12
CL 2	−0.020	0.255	0.100	0.099	0.105	10
CL 3	0.030	0.457	0.296	0.295	0.115	28
CL 4	0.131	0.732	0.351	0.388	0.150	34
M/P	−0.020	0.140	0.015	0.037	0.050	16
S-M	0.030	0.457	0.296	0.297	0.114	28
M-M	0.095	0.732	0.344	0.354	0.163	40
C^1 BL						
CL 1	−0.062	0.113	0.020	0.023	0.048	12
CL 2	0.000	0.247	0.063	0.093	0.090	10

(Con't)

Table II. ***Continued***

Variable	Min	Max	Median	Mean	SD	N
CL 3	−0.094	0.399	0.182	0.180	0.126	28
CL 4	−0.010	0.565	0.278	0.302	0.130	34
M/P	−0.062	0.113	0.030	0.025	0.043	16
S-M	−0.094	0.399	0.195	0.189	0.125	28
M-M	−0.010	0.565	0.262	0.271	0.143	40
C^1 Hgt						
CL 1	−0.062	0.344	0.077	0.098	0.103	12
CL 2	−0.010	0.571	0.304	0.258	0.204	10
CL 3	0.344	0.775	0.536	0.536	0.099	28
CL 4	0.344	1.411	0.675	0.721	0.259	34
M/P	−0.062	0.344	0.058	0.085	0.098	16
S-M	0.344	0.775	0.539	0.541	0.099	28
M-M	0.285	1.411	0.634	0.669	0.269	40

male–male competition species, while the potential frequency effect is weak at best (Ford, 1994; Plavcan and van Schaik, 1997b). For canine teeth, the frequency effect is also weaker than the intensity effect, though it appears to be relatively stronger than in mass dimorphism. Among the behavioral measures evaluated, competition levels show the strongest relation to all measures of dimorphism.

Examples

Though all of the above behavioral variables are related with one or more measures of dimorphism, inferring a unique behavioral attribute from dimorphism is not straightforward.

The first obstacle is how best to actually compare the fossil and extant data. While all dimorphism estimates are continuous data, the behavioral data are either continuous (SR and OSR) or categorical (MS, CL, CC). Usually, estimates for continuous variables such as SR or OSR are extrapolated from correlated continuous variables using standard least-squares regressions, or less often major axis or reduced major axis. Smith (1994) provides strong reasons for preferring the least-squares method when predicting a value for one variable from an independent variable (see also Konigsberg *et al.*, 1998). In contrast, inferring anything about the categorical data requires comparisons to means or medians, confidence limits and/or ranges of dimorphism within the categories.

In theory, SR or OSR can be inferred by simply plugging an estimate of dimorphism into an appropriate regression equation. Unfortunately, the lack of significance in the nonphylogenetically controlled analysis of OSR renders

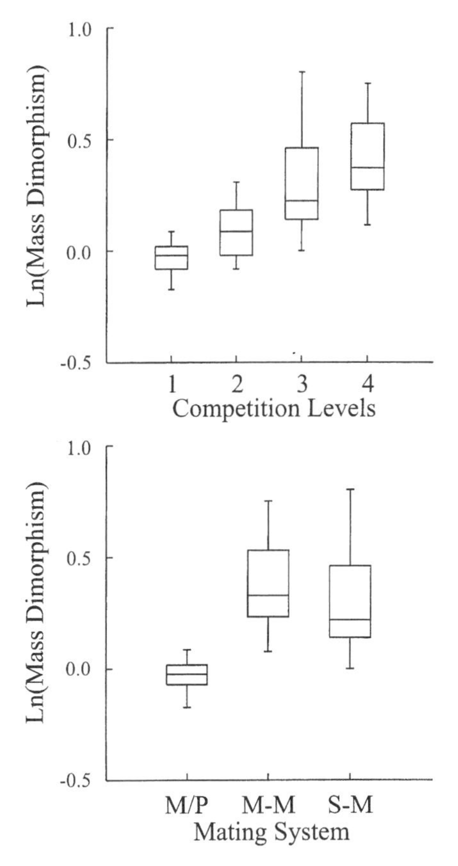

Fig. 3. Box and whiskers comparisons between ln-transformed body mass dimorphism and mating system (bottom panel) and competition levels (top panel) for anthropoid primates.

this variable of no use for evaluating fossils. For SR, confidence intervals for the estimates of the residuals about the lines are very large, rendering most of the estimates of little use (note the wide scatter of points in Fig. 2, and the large standard errors of the estimates in Table I). For example, a value of mass dimorphism of 1.5 for a fossil yields a confidence limit of 7.62 to 3.37 for SR. This covers 50% of the total range of SR. Finally, the lack of correlation between SR and dimorphism among polygynous species renders even this confidence interval of dubious biological significance. Therefore, dimorphism yields almost no information about SR. Perhaps with the advent of better estimates of OSR the situation will improve. These results underscore that a significant correlation between two variables does not necessarily mean that one variable can be meaningfully extrapolated from the other in the fossil record (cf. Smith, 1996).

Turning to the categorical data, estimates of dimorphism from fossils might be compared to means or medians within categories, confidence limits of categories, or simply the ranges of categories (Table II). Comparison to medians within categories is of some interest. For example, we can show that *Notharctus* maxillary canine MD dimorphism falls closest to that seen in competition level 3 species (Table III). However, this says nothing about whether *Notharctus* canine dimorphism is more or less than that seen in extant species of other competition levels. Comparison to confidence intervals might help in this regard. However, this is potentially confounded by the fact that variables such as MS or CL are only surrogate estimates of sexual selection. There is no reason to believe a priori that there is a normal distribution of dimorphism among species within these categories, and in fact distributions of dimorphism within categorical variables are not necessarily normal (Fig. 4). Hence confidence intervals based on means and standard deviations within such categories may be biologically naïve, and should be approached with caution.

As an alternative, fossil dimorphism can be compared conservatively to the medians and the ranges of dimorphism within categories. If dimorphism for a fossil lies outside of the range of extant species within a category, then there is no extant analogy for that type of mating system or competition level. Hence, a particular degree of dimorphism might be uniquely associated with a mating system or competition level (see Kay and Cartmill, 1977).

Extreme outliers within a category present a problem for this approach. For example, there is one CL 4 species that shows canine BL dimorphism well below the range of the rest of the species in the category (Fig. 5, bottom panel). Thus the range of CL 1 extensively overlaps that of CL 4, even though most of the overlap is due to a single value. A fossil species with dimorphism that falls comfortably within the range of CL 1, but below that of all CL 4 species except the outlier, would still fall within the range of CL 4. It is tempting to state that there is greater probability that the species is consistent with CL 1, because the outlier of CL 4 is a rare exception. This would assume that the distribution of dimorphism in extant species within the competition levels follows a predictable distribution about a mean. But if a species in CL 4 shows slight canine dimorphism, then the possibility exists that a similar degree of dimorphism could have evolved in an extinct species with a similar pattern of male–male competition. Ways around this problem are to demonstrate (1) that the condition in the extant species has evolved under special circumstances (including, for example, phylogenetic inertia), (2) that the behavioral classification is wrong or does not accurately express the relation between behavior and dimorphism, or (3) that the estimate of dimorphism in the extant species is biased.

Perhaps the greatest weakness of this approach is the uncertainty in why a particular magnitude of dimorphism should be associated with a particular mating system or competition level. As noted by Kay and Cartmill (1977), to infer a function (in this case, social behavior) from a trait in an extinct species using an extant model, there must be a causal relationship between the

Table III. Estimates of Sex Ratio (SR), Mating System (MS), and Competition Levels (CL) Based on Reported Estimates of Dimorphism in a Selected Sample of Extinct Primates[a]

Species	Dim	SR	MS	CL
Cantius torresi				
C_1 MD	1.37	2.60 ± 2.771	sm/**mm**	3/**4**
Notharctus venticolis				
C_1 MD	1.38	2.66 ± 2.77	sm/**mm**	3/**4**
C_1 BL	1.13	1.47 ± 2.85	**mp**/sm/mm	1/**2**/3/4
C^1 MD	1.47	2.56 ± 2.71	sm/**mm**	3/**4**
C^1 Bl	1.18	1.82 ± 3.12	**sm**/mm	2/**3**/4
C^1 Hgt	1.89	2.37 ± 2.24	sm/**mm**	3/**4**
Adapis magnus				
C^1 MD	1.19	1.57 ± 2.71	**sm**/mm	**2**/3/4
Mass	1.56	2.29 ± 3.20	sm/**mm**	3/**4**
Adapis paresiensis				
C^1 MD	1.13	1.40 ± 2.71	**mp**/sm/mm	1/**2**/3
Mass	1.44	2.10 ± 3.20	sm/**mm**	3/**4**
Sivapithecus indicus				
C_1 MD	1.06	1.32 ± 2.77	**mp**/sm	**1**/2/3
C_1 BL	1.16	1.58 ± 2.85	**sm**/mm	**2**/3/4
Sivapithecus sivalensis				
C_1 MD	1.14	1.60 ± 2.77	**sm**/mm	**2**/3/4
C_1 BL	1.28	2.06 ± 2.85	**sm**/mm	**3**/4
Catopithecus browni				
C_1 MD	1.22		**sm**/mm	**3**/4
C_1 BL	1.25		**sm**/mm	**3**/4
C_1 Hgt	1.33		**mp**	1/**2**
C^1 MD	1.18		**sm**/mm	**2**/3/4
C^1 BL	1.21		**sm**/mm	**2**/3/4
C^1 Hgt	1.50		**sm**/mm	**2**/3/4
Proteopithecus				
C_1 MD	1.51		**mm**	**4**
C_1 BL	1.28		**sm**/mm	**3**/4
C_1 Hgt	1.28		**mp**/mm	1/**2**/3/4

[a] Abbreviations as follows: mp = monogamous/polyandrous; mm = multi-male, multi-female; sm = single-male, multi-female. Bold mating system classifications and competition levels are those whose median for the extant species falls closest to the estimate for the fossil estimate. Data are from the following sources: *Cantius torresi* (Gingerich, 1995); *Notharctus venticolis* (Krishtalka *et al.*, 1991); *Adapis magnus* and *A. paresiensis* (Gingerich, 1981); *Sivapithecus indicus* and *S. sivalensis* (Kay, 1982); *Catopithecus browni* and *Proteopithecus* (Simons *et al.*, 1999).

trait and the function, which is present in all causal species that possess the trait. Strong degrees of dimorphism are always associated with polygyny in primates, as predicted by sexual selection theory [with the possible exception of *Cercopithecus neglectus* (Cords, 1988; Leutenegger and Lubach, 1987), though there is some evidence that this highly dimorphic species is in fact polygynous

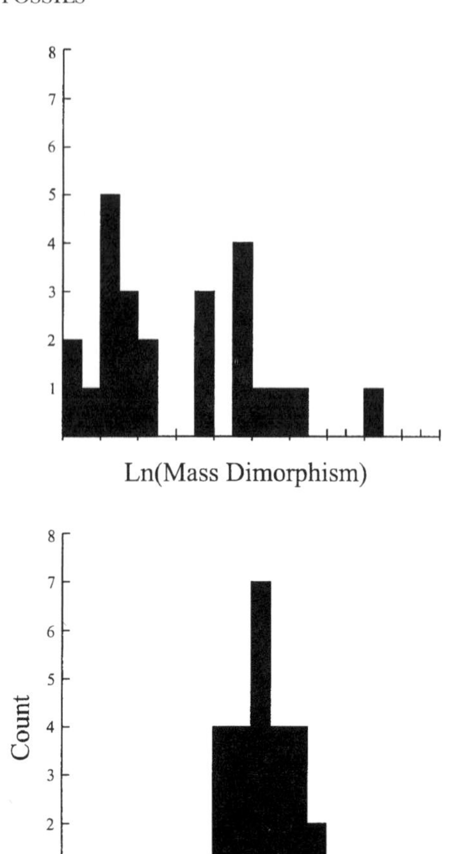

Fig. 4. Frequency histograms of body mass dimorphism (top panel) and maxillary canine crown height dimorphism (bottom panel) of anthropoid primates ranked as competition level 3. Note the normal distribution for canine dimorphism, and the non-normal distribution for body mass dimorphism. This latter result largely reflects taxonomic differences in body mass dimorphism.

(Brennan, 1985; Rowell, 1988)]. However, definitions of MS and CL used here are only imprecisely related to sexual selection. Furthermore, there is evidence that a particular degree of sexual selection does not necessarily produce a specific magnitude of dimorphism (e.g., Kappeler, 1990, 1991, 1996; Weckerly, 1998). Consequently, the most conservative approach is to present data on as many species as possible, even though this may reduce that apparent correlation between dimorphism and a behavioral variable.

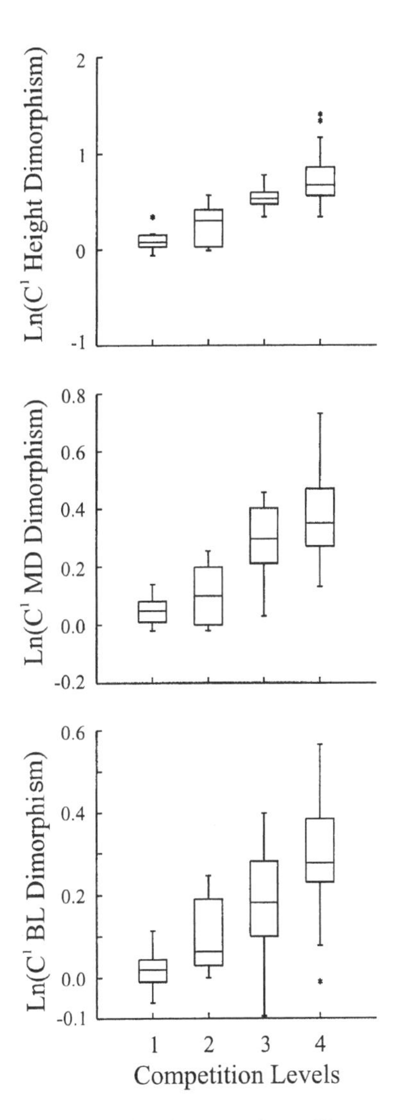

Fig. 5. Box and whisker diagrams of estimates of maxillary canine dimorphism versus competition levels. Stars represent species falling greater than 1.5 times the spread between the hinges of the box and the edge of the box (Wilkinson, 1990).

Figures 3 and 5 demonstrate that the ranges of dimorphism among either CL or MS classifications overlap extensively. For MS, very low levels of dimorphism suggest monogamy or polyandry, while high levels imply polygyny, but with no resolution between single-male and multi-male species. For CL, low degrees of dimorphism suggest low-intensity male–male competition (CL's 1 and 2), while very high degrees of dimorphism suggest intense male–male competition (CL's 3 and 4). Separation between CL or MS classifications is greatest for mass dimorphism and canine crown height dimorphism. Conversely, overlap among categories is extensive for canine occlusal dimorphism, especially for low to modest degrees of dimorphism.

Greenfield (1992a,b) and Plavcan and van Schaik (1992, 1994) both argue that crown height is functionally the most important aspect of the canines, while body mass has long been argued as important in helping males win fights (e.g., Clutton-Brock *et al.*, 1977; Kappeler, 1990, 1991). Canine occlusal dimensions are not perfectly correlated with crown height, and Plavcan and van Schaik (1997a) argue that information from canine occlusal dimensions should be ignored when information from either crown height or body mass is available.

Table IV. Estimates of Sex Ratio (SR), Mating System (MS), and Competition Levels (CL) Based on Reported Estimates of Dimorphism in a Sample of Hominids[a]

Species	Dim	SR	MS	CL
Australopithecus afarensis				
C_1 MD	1.25	2.04 ± 2.77	**sm**/mm	**3**/4
C_1 BL	1.22	1.81 ± 2.85	**sm**/mm	**2**/3/4
C^1 MD	1.11	1.34 ± 2.71	**mp**/sm/mm	1/**2**/3
C^1 BL	1.19	1.85 ± 3.12	**sm**/mm	2/**3**/4
C^1 Hgt	1.19	1.24 ± 2.24	**mp**	**1**
Mass	1.50	2.20 ± 3.20	sm/**mm**	3/**4**
A. africanus				
C_1 MD	1.09	1.42 ± 2.77	**sm**/mm	**2**/3/4
C_1 BL	1.13	1.47 ± 2.85	**mp**/sm/mm	**1**/3/4
C_1 Hgt	1.15	1.20 ± 2.64	**mp**	**1**
C^1 MD	1.15	1.46 ± 2.71	**mp**/sm/mm	1/**2**/3/4
C^1 BL	1.23	1.97 ± 3.12	**sm**/mm	2/**3**/4
C^1 Hgt	1.13	1.15 ± 2.24	**mp**	1
Mass	1.40	2.03 ± 3.20	sm/**mm**	**3/4**
A. robustus				
C_1 MD	1.16	1.68 ± 2.77	**sm**/mm	**2**/3/4
C_1 BL	1.18	1.65 ± 2.85	**sm**/mm	**2**/3/4
C^1 MD	1.13	1.40 ± 2.71	**mp**/sm/mm	1/**2**/3
C^1 BL	1.17	1.79 ± 3.12	**sm**/mm	**2**/3/4
C^1 Hgt	1.25	1.32 ± 2.24	**mp**	**1**
Mass	1.30	1.87 ± 3.20	**sm**/mm	2/**3**/4
A. boisei				
C_1 MD	1.08	1.39 ± 2.77	**mp**/sm/mm	**1**/2/3/4
C^1 MD	1.28	1.86 ± 2.71	**sm**/mm	2/**3**/4
C^1 BL	1.25	2.03 ± 3.12	**sm**/mm	2/**3**/4
Mass	1.40	2.03 ± 3.20	sm/**mm**	**3**/4

[a] Abbreviations and symbols as for Table III. Data extracted from sources listed in Plavcan and van Schaik (1997a).

Examples from the fossil record clearly illustrate these problems (Tables III and IV). For most species, dimorphism estimates of the occlusal dimensions overlaps more than one competition level or mating system classification. Considered together, dimorphism of all variables in a species usually overlaps all competition levels or mating system classifications. The only time this is not the case is when data are available for only one or two variables. For example, note that the *Cantius torresi* mandibular MD dimorphism estimate overlaps only CL's 3 and 4, implying high-intensity male–male competition. Yet scrutiny of the dimorphism estimates for *Notharctus venticolus*—which shows an almost identical level of dimorphism for the same tooth dimension—suggests that the unambiguous result for *C. torresi* reflects a lack of data.

It is difficult to follow the recommendation of Plavcan and van Schaik (1997a) to rely on canine crown height dimorphism because such data are only rarely reported (relatively unworn canines are rare in the fossil record). In the case of *Notharctus venticolus*, crown height dimorphism clearly suggests intense male–male competition, in contrast to the evidence from the occlusal dimensions (Table III). However, ignoring canine occlusal dimorphism is not necessarily a panacea. For hominids, there is a stark contrast between canine crown height dimorphism and body mass dimorphism (Table IV). This pattern has long been noted. Body mass dimorphism estimates for four species (*Australopithecus afarensis*, *A. africanus*, *A. robustus*, and *A. boisei*) overlap only the range of dimorphism of polygynous species (single-male and multi-male), and overlap only CL 3 and 4 species in three of four australopithecines. Canine crown height dimorphism overlaps only monogamous/polyandrous species, and CL 1 species. Plavcan and van Schaik (1997a) review a variety of explanations for this pattern of dimorphism, and conclude that there is no compelling comparative evidence favoring any one explanation.

The case of australopithecine dimorphism underscores the inherent problem of trying to infer any behavior from a unique morphological pattern. Unless there is a strongly supported theoretical model that can reconcile the conflicting data, a unique pattern of dimorphism can only be explained with reference to unique historical events, rendering behavioral reconstructions speculative at best.

Relations between Relative Canine Size and Competition Classifications

Plavcan *et al.* (1995) demonstrate that males or females ranked as showing "high-intensity" intrasexual competition have relatively larger canines than those ranked as showing either "low-intensity" competition or "coalitionary" competition (Fig. 6). They also demonstrate that male canine size is only

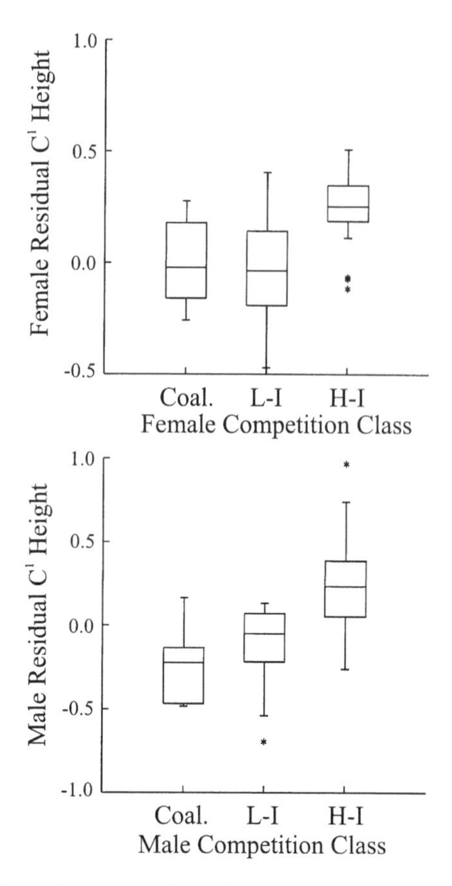

Fig. 6. Box and whiskers diagrams of residual maxillary canine crown height against competition classifications in males (bottom panel) and females (top panel). Relative canine size is measured as a least-squares residual of canine height against body mass, measured against an isometric line (see Plavcan *et al.*, 1995). Abbreviations as follows: Coal. = coalitionary competition: L-I = low-intensity competition: H-I = high-intensity competition.

associated with the frequency of competition within lower taxonomic groups, while female canine size follows the reverse pattern from that predicted.

There also is evidence that variation in female relative canine size is at least partly a result of "correlated response" to male canine size (Greenfield, 1992a, 1998; Plavcan *et al.*, 1995; Plavcan, 1998). In other words, if male and female canine size are at least partially determined by the same genes, then female canine size should covary with male canine size (Lande, 1980; Arnold, 1994). Controlling for male canine size, high-intensity competition females have relatively larger canines than low-intensity competition females (Plavcan, 1998b). In low-intensity females, male and female relative canine size are strongly correlated, suggesting either a genetic or developmental linkage of canine size between the sexes. In strepsirhines, male and female relative canine size are very strongly correlated, suggesting that the monomorphism of these species reflects an unspecified constraint on female canine reduction, rather than a lack of sexual selection.

Examples

Models evaluating relative canine size, rather than dimorphism, are difficult to apply to the fossil record because they require quantification of separate male and female components of dimorphism. Lacking comparative ontogenetic data, this cannot be done with body size dimorphism. It is possible with the canines, but only if males and females can be identified, and a reasonable estimate of body size or a surrogate of body size is available.

Where canine teeth are obviously large, application of the models is relatively trivial. The presence of very large canines in either sex implies intense intrasexual competition. For example, *Notharctus venticolus* clearly shows two canine morphologies (Krishtalka *et al.*, 1991), suggesting intense male–male competition in a polygynous mating system.

Large female canines should imply intense female–female agonistic competition. However, demonstrating that female canines are relatively large is difficult because they should be indistinguishable from large male canines. Furthermore, because of the possibility of correlated response, one should also demonstrate that female canine size is large relative to an ancestor and to male canine size. Given these problems, it is not surprising that clear cut cases among anthropoids are not common. *Dendropithecus macinnesi* seems to be characterized by large male and female canines, suggesting monogamy (Fleagle, 1998), though the canine dimorphism of this species suggests that an interpretation of an increase in female–female agonistic competition might be a more conservative interpretation.

Early hominids provide a well known example where male canines appear to be relatively small. Estimates of relative canine crown height in *A. afarensis*, *A. robustus*, and *A. africanus* fall at the low end of the male anthropoid primate range, and appear about average by comparison to most female anthropoid canines (Plavcan and van Schaik, 1997a). Importantly, the choice of the independent variable for comparison to canine size may alter results strongly, at least for hominids. Compared to body mass, the smallest relative canine crown heights among all primates are possessed by (in order from the smallest) *Gorilla*, *Homo*, *Pongo*, *P. paniscus*, and *Pan troglodytes* (Fig. 7, top panel). Hence, in this comparison, early hominids have relative canine size that is comparable to that of male *Pan* and *Pongo*, and greater than that of *Homo* and *Gorilla*. However, using mandibular molar MD length as the independent variable, hominid relative canine tooth size is somewhat greater than *Homo*, and substantially smaller than the other hominoids (Fig. 7, bottom panel). This

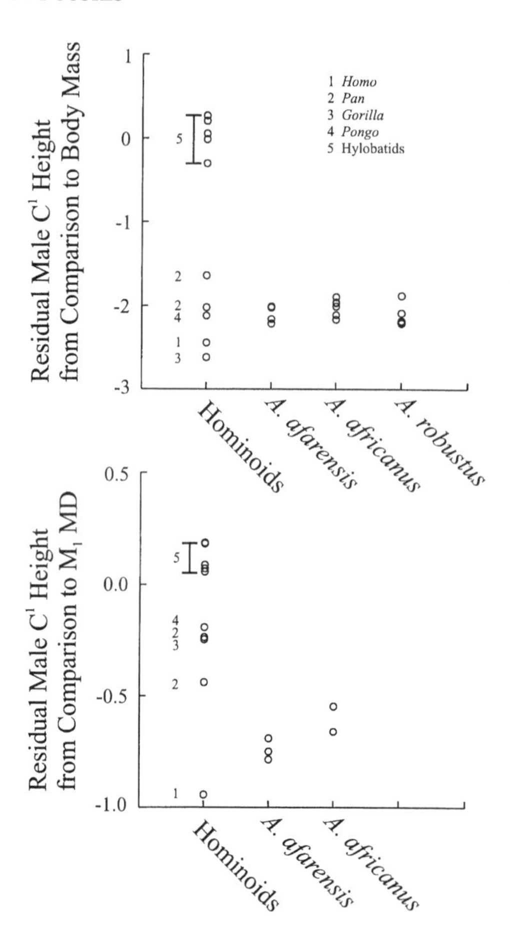

Fig. 7. Relative male canine crown height in extant hominids compared to some estimates for *A. afarensis*, *A. africanus*, and *A. boisei*. Top panel shows estimates derived from a comparison to body mass. Lower panel shows estimates derived from a comparison to mandibular molar mesiodistal length.

latter assessment seems more realistic, and strongly suggests low-intensity male–male competition.

However, apart from these examples, relative canine size offers little unambiguous information about social behavior in extinct species, primarily because of the extensive overlap in the range of relative canine sizes among competition classifications (Fig. 6).

As an aside, several extant species have large canines modified in association with dietary specialization. The best known examples are the pithecines, in which both males and females have large canines that are used to open hard fruit. *Cebuella* possess conical mandibular canines and incisors which, in concert, form a small gouge used to dig holes in tree trunks. In both cases the canines are modified from the typical anthropoid form (Greenfield, 1992a; Plavcan and Kelley, 1996). However, nothing clearly precludes large female canines from being used for a special function. Hence one can never rule out the possibility that large female canines might indicate a specialized function. Still, in the absence of obvious morphological specialization one would have little choice but to infer that large female canines imply female–female dyadic agonistic competition.

Relations between Dimorphism and Other Variables

Obviously, dimorphism alone provides only crude generalizations about social behavior from the fossil record. This begs the question; are there other factors associated with dimorphism and/or mating systems that might allow a more precise reconstruction of social behavior? Interspecific variation in dimorphism is known or hypothesized to be associated with substrate, phylogeny, body mass, diet and energetic constraints (Ford, 1994; Leutenegger and Kelly, 1977; Lucas *et al.*, 1986; Martin *et al.*, 1994; Plavcan and van Schaik, 1994). Mating systems are also known to be associated with nocturnality/diurnality (Fleagle *et al.*, 1980; Kay *et al.*, 1988; Kay *et al.*, 1997; Ross, 1996), and with phylogeny (Di Fiore and Rendall, 1994).

Substrate

Terrestrial species tend to be more dimorphic than arboreal species (Clutton-Brock *et al.*, 1977; Gautier-Hion, 1975; Leutenegger and Kelly, 1977; Plavcan and van Schaik, 1992, 1997b). Savanna-dwelling species also show stronger canine dimorphism than arboreal or terrestrial forest-dwelling species (Plavcan and van Schaik, 1992). The exact reason for this association is unclear, though the most common hypotheses are that terrestrial species are more susceptible to predators (Clutton-Brock *et al.*, 1977; Leutenegger and Kelly, 1977; Plavcan and van Schaik, 1992, 1997b), and that arboreal locomotion constrains increases in male body size (Leutenegger and Kelly, 1977; Plavcan and van Schaik, 1997b).

Knowledge of substrate use may help discriminate between MS or CL's. To begin, all monogamous/polyandrous and CL 1 species are arboreal, meaning that these classes do not occur among the terrestrial species. Conversely, there are no savanna-dwelling species in CL's 1 or 2. If there is a constraint on the evolution of monogamy among terrestrial species, then this mating system is unlikely to have characterized an extinct terrestrial species, even though dimorphism may have been low. Likewise, an extinct species from a savanna environment is more likely to have exhibited intense male–male competition. This latter observation, though, must be tempered by the fact that some populations of chimpanzees, characterized by low-intensity male–male competition in fission–fusion societies, live in relatively open habitats (Moore, 1997). For hominids, this observation could be critical.

Within both CL's 3 and 4, terrestrial species are significantly more dimorphic than arboreal species (Plavcan and van Schaik, 1992). Unfortunately, in both CL's there is a large degree of overlap in the ranges of dimorphism within the substrate classifications, rendering this observation of little use for the fossil record.

Substrate is significantly related to variation in male relative canine size (ANOVA $N = 100$, $F = 8.532$, $P < 0.001$), but, again, the overlap in the range of relative male canine size between substrate classifications and male competition classifications is so great that knowledge of substrate use offers little practical help in discriminating between male competition classifications. There is no relation between substrate classifications and female canine size.

Body Size

All measures of canine and mass dimorphism are significantly correlated with female body mass (Cheverud *et al.*, 1985; Ely and Kurland, 1989; Gaulin and Sailer, 1984; Leutenegger and Cheverud, 1982, 1985; Pickford, 1986; Plavcan and van Schaik, 1992, 1997b), though increasing size does not necessarily lead to increasing dimorphism (Godfrey *et al.*, 1993; Kappeler, 1990, 1991; Mitani *et al.*, 1996; Plavcan and van Schaik, 1992, 1997b). This raises the question of whether correcting for body size improves the resolution of any behavioral variable. To test this, residuals were calculated from least-squares regressions of each dimorphism measure against female body mass (all variables ln-transformed).

Residuals from these analyses are somewhat more poorly related to sex ratio, OSR, mating system, and competition levels than are the noncorrected dimorphism ratios (Fig. 8). Hence, allometrically adjusting for a relation between size and dimorphism offers no improvement in predicting behavior in fossils. The reason might lie in the suggestion of Mitani *et al.* (1996) that increasing body size is correlated with changes in female dispersal patterns, which in turn affects the OSR. This is supported by a significant correlation between female body mass and the OSR, and between female body mass and competition levels. If this is true, then the relation between dimorphism and body size secondarily reflects the relation between the social and demographic factors and body mass. However, Simons *et al.* (1999) demonstrate that size and canine dimorphism probably arose in very small bodied Eocene anthropoids, casting doubt on the general validity of this model. Regardless, allometric adjustment of dimorphism is not particularly helpful.

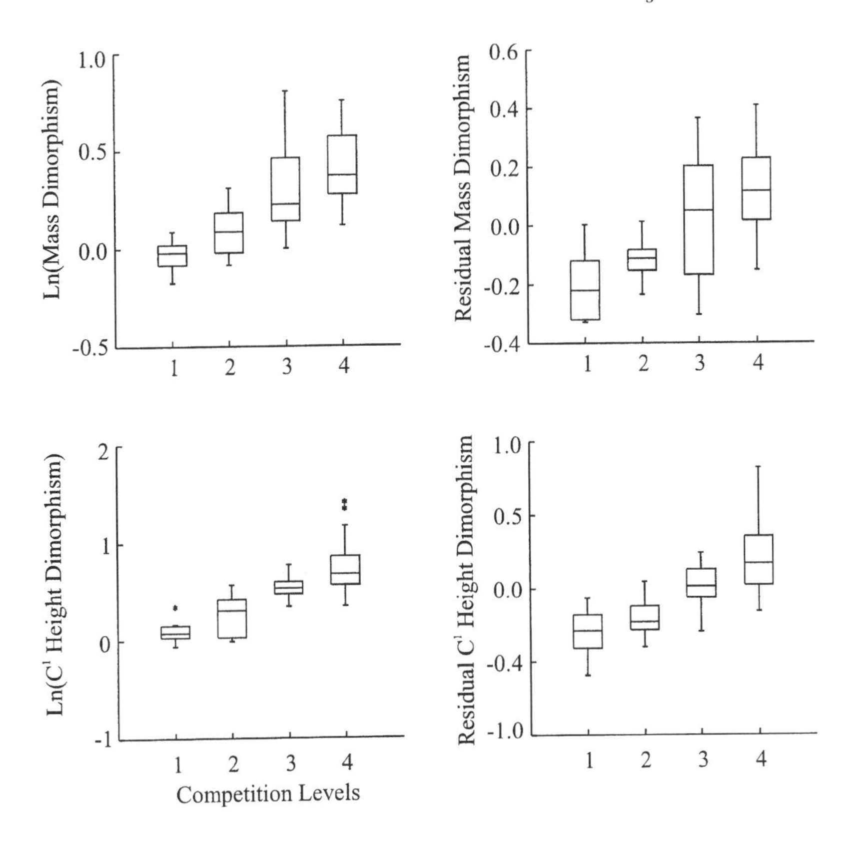

Fig. 8. Box and whiskers diagrams comparing dimorphism (the two left panels) and residual dimorphism (the two right panels) to competition levels. Residual dimorphism was calculated as the residual from a least-squares regression on ln-transformed dimorphism against ln-transformed female body mass.

Phylogeny

Canine and body mass dimorphism, as well as relative canine size in males and females, are all correlated with phylogeny (Cheverud *et al.*, 1985; Leutenegger and Kelly, 1977; Plavcan and van Schaik, 1997b). Control for phylogeny clearly can alter inferences of behavior based on comparative analysis.

Figures 9 and 10 show the range of canine and body mass dimorphism within CL's for extant cercopithecines, colobines, great apes, hylobatids, atelids, and cebids (classification of platyrrhines follows Fleagle, 1998). Within these taxonomic groups, the relation between CL's and dimorphism generally holds up, demonstrating that the significant relation between dimorphism and behavioral measures is not an artifact of the correlations between phylogeny

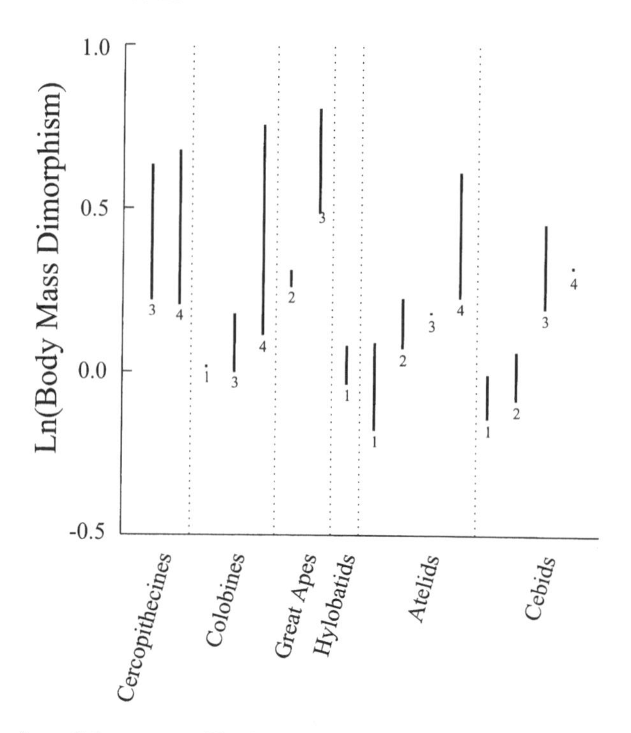

Fig. 9. Illustration of the range of body mass dimorphism within competition levels for a series of anthropoid taxonomic groups. Dotted lines separate range bars for the taxonomic groups. Competition levels are indicated under each bar. Note that all competition levels are not represented for each group.

and dimorphism or behavior. Notable exceptions are the obvious similarity in body mass dimorphism in CL 3 and 4 species in the cercopithecines and cebids (though there is only a single cebid species in CL 4, Fig. 9), and the great overlap in canine crown height dimorphism between CL 2 and 3 species in the atelids (Fig. 10).

Taxonomic differences in dimorphism are clearly responsible for some of the overlap among CL's discussed above. For example, the overlap in the range of mass dimorphism between CL's 3 and 1 is because of the low dimorphism in colobines (Plavcan and van Schaik, 1997b). The high end of the range of mass dimorphism in CL 3 species reflects the intense mass dimorphism of *Gorilla* and *Pongo*. Overlap in canine crown height dimorphism between CL's 1 and 2 is a result of the relatively strong dimorphism of *Presbytis potenziani* (compared to other CL 1 species). The range of canine crown height dimorphism of CL 2 species is similar in great apes and atelids, but is much lower in cebids, largely because of the callitrichines that are placed within CL 2.

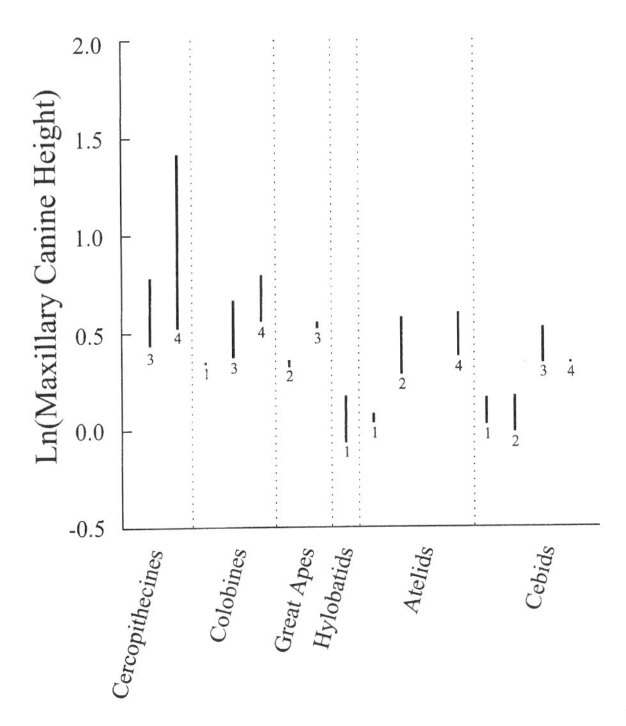

Fig. 10. Same as for Fig. 9, except using maxillary canine crown height dimorphism.

Within several groups the separation in the range of dimorphism among CL's is much greater than when all anthropoids are included together. For example, there is an absolute separation between low-intensity (levels 1 and 2) and high-intensity (levels 2 and 4) species within cebids and great apes.

Similar results are found for estimates of relative canine size. It is notable that hominoids posses relatively small canines (compared to body mass) by comparison to most other anthropoids. This is visually obvious by simply comparing a male baboon canine to that of a male gorilla.

Clearly, inferences of behavior from dimorphism should be based on comparisons to related extant taxonomic groups whenever possible—a practice that is often done but rarely justified with regard to dimorphism. The reason for restricting comparisons of, for example, dental anatomy to extant phylogenetic groups is that there are demonstrable fundamental anatomical differences among groups. Hence, because hominoids are characterized by relatively conservative molar morphology, the degree of folivory in extinct hominoids should be determined by a comparison to extant hominoids (Ungar and Kay, 1995). Regarding dimorphism, though, taxonomic differences in dimorphism—and the biological basis for such differences—are rarely evaluated.

Even though there are taxonomic differences in dimorphism among anthropoids, the practice of restricting comparisons to a limited extant group is not a panacea, and should be approached cautiously. First, by breaking a sample into lower taxonomic groups, the number of species within each CL (or any other behavioral classification) is reduced, in some cases to a single species. Hence, the range of dimorphism—and the lack of overlap between CL's—may reflect a lack of representation rather than a real biological phenomenon. This is particularly true of hominoids. Competition levels 2 and 3 contain only two hominoid species each, and competition level 4 is not represented at all. Among numerous cercopithecoids there is a wide range of dimorphism within competition levels 3 and 4, which suggests that a greater range of variation should be expected within the hominoids. This problem is illustrated in Kelley's (1993) suggestion that the upper limit of size dimorphism in extinct hominoids was greater than that seen in the most dimorphic hominoid, *Pongo*. Extreme dimorphism in *Lufengpithecus* (Kelley and Plavcan, 1998) corroborates Kelley's point, but leaves us with the question of how to interpret the biological meaning of dimorphism in *Lufengpithecus*. It could represent an extension of the range of dimorphism expected in competition level 3, but it might also imply a social system characterized by high-frequency, high-intensity male–male competition as seen in multi-male groups or multi-level societies such as *Theropithecus*. Acceptance of the latter interpretation is partly contingent on how strongly one feels that variation in hominoid social systems is constrained (e.g., Di Fiore and Rendall, 1994). Regardless, the lack of representation in social systems, competition levels, and dimorphism among extant hominoids inevitably forces one to look outside of this group in order to understand the biological significance of dimorphism in extinct hominoids. Otherwise, there is no way to recognise behavioral diversity among extinct species.

Second, restricting comparisons taxonomically assumes that the pattern seen among extant groups is manifest in the group through time. For example, cercopithecines clearly are derived in the extreme hypertrophy of their male maxillary canines. One can probably safely assume that this trait characterized the common ancestor of extant cercopithecines. The assumption obviously may not hold, however, for extinct sister taxa of cercopithecines. This might be illustrated by the presence of dimorphism in some extinct adapids (Gingerich, 1981, 1995), assuming that the adapids are the sister group to the extant strepsirrhines (Ross *et al.*, 1998), all of which are monomorphic or nearly so.

Finally, the benefit of this practice obviously breaks down as one moves deeper in a phylogeny. For example, dimorphism in *Branisella* would have to be compared to all platyrrines, since the species is not uniquely related to either the cebids or atelids (cf. Fleagle, 1998; Kay *et al.*, this volume Chapter 9). *Proteopithecus* and *Catopithecus* can only be compared to all extant anthropoids, since they lie at the base of the anthropoid radiation.

Uncertainty over anthropoid origins raises interesting questions about how to interpret dimorphism in the adapids *Notharctus* and *Cantius*. Some maintain that cercomoniine adapids gave rise to anthropoids, and interpret sexual dimorphism in adapids as a shared derived character linking adapids and anthropoids (Gingerich, 1995; Simons and Rasmussen, 1996). This would clearly justify comparing adapids to extant anthropoids. But others offer evidence that anthropoids did not arise from Eocene adapids (Beard *et al.*, 1994; Kay *et al.*, 1997; Ross *et al.*, 1998). If the latter hypothesis is correct, this would imply that dimorphism in adapids arose independently from that of anthropoids.

If adapid dimorphism is independently derived, can anthropoids be used as a model for interpreting social behavior in Eocene adapids? As discussed above, among anthropoids themselves the relation between social behavior and dimorphism can vary. Weckerly (1998) reports variation in the relation between body mass dimorphism and a variety of estimates of mating system and male–male competition among Macropodidae, Primates, Mustelidae, Pinnipedia, Elephantidae, and Ruminantia, even though variation in mass dimorphism is related to behavioral variables within groups. For example, even though Macropodids are among the most dimorphic of mammals, they are clearly not the most polygynous. This suggests that the behavioral correlates of dimorphism in taxa outside of a particular group should only be inferred in the most general terms. Thus, finding intense size dimorphism in *Cantius* or *Notharctus* may not necessarily imply as intense male–male competition as seen in extant cercopithecines.

Finally, extant strepsirhines show at best comparatively slight dimorphism in any character, even though a number are strongly polygynous (Godfrey *et al.*, 1993; Kappeler, 1990, 1991, 1996; Leigh and Terranova, 1998; Plavcan *et al.*, 1995). This lack of dimorphism may reflect different responses to selection pressures associated with sexual selection (Kappeler, 1991, 1996), similar selection pressures on males and females for the development of weaponry (Plavcan *et al.*, 1995), or genetic (Plavcan, 1998b) or developmental (Leigh and Terranova, 1998) constraints on the evolution of dimorphism. Regardless, this pattern suggests that a lack of dimorphism may not imply anything at all about the mating system in taxonomic groups with no dimorphic relatives or ancestors.

Temporal and Ontogenetic Trends in Dimorphism

Evaluation of temporal changes in dimorphism within a lineage may shed light on whether dimorphism is produced by changes in male or female values, which may in turn provide important evidence about temporal changes in behavior.

Increasing dimorphism resulting from the exaggeration of a male trait should indicate polygyny and intense male–male competition. However, the trend itself might not reflect simultaneous changes in social behavior. For example, there may be a time lag in the response of male canine size to selection pressure. It is difficult to specify the exact underpinnings of a time lag, since little is known about the genetic determinants of canine size in extant primates. One intriguing hypothesis follows from the Lande (1980; see also Arnold, 1994) model for the evolution of dimorphic traits. Lande (1980) explicitly predicts that if there is strong male and female genetic covariation in a trait, then the expression of the trait should reflect a balance between selection for the trait in one sex, and against it in the other. For example, if possession of large canines by females is too detrimental, then male canine size might be constrained by selection to maintain small female canines. Large male canine size would only evolve following a decrease in covariation in male and female canine size.

Temporal changes in dimorphism might also be linked to changes in other characters. For example, a shift to diurnal habits coupled with an increase in dimorphism should suggest a simultaneous change to a polygynous social system. Likewise, one might link an increase in dimorphism with a shift to terrestriality. Though the change in dimorphism might indeed reflect a change in mating system, the terrestrial habits also may suggest that greater predation pressure (Clutton-Brock *et al.*, 1977), or even a relaxation of locomotor constraints on male body mass (Leutenegger and Kelly, 1977; Plavcan and van Schaik, 1997b) or fighting strategies (Kappeler, 1990, 1991) could be involved.

Changes in female traits are difficult to interpret, partly because female contributions to dimorphism have received less study than those of males. Regardless, a demonstration that changes in dimorphism reflect changes in a female trait strongly suggests that a factor other than sexual selection is involved.

An increase in female canine size should reflect selection to increase female weaponry, presumably in association with female–female competition for resources (Plavcan *et al.*, 1995). However, changes in female canine size should be compared to changes in male canine size. A simultaneous increase in both male and female canine size might suggest an effect of correlated response to male canine size. Conversely, though a substantial reduction in female canine size should suggest a relaxation of selection for large female canines, again this depends on whether such changes are independent of male canine size.

Changes in female body mass are confounded by the complexity of factors that might influence body mass. Females may reduce body size because of changes in nutrition (e.g., Brace and Ryan, 1980, though they suggest that male body size should be more affected by this). Leigh and Shea (1995) suggest that, as for males, changes in either the rate or duration of development in females reflect female competition and fecundity. For example, they suggest that *Gorilla* dimorphism may be enhanced through early female maturation, while dimorphism in *Pan troglodytes* may be reduced through delayed female maturation. In folivorous primates, female fecundity can be increased by early female maturation, which in turn can lead to increased dimorphism by a reduction in female body mass. Conversely, females that compete for access to resources may gain advantages of large size, and may reduce juvenile survival risk by delaying maturation and prolonging growth. This could lead to a reduction in dimorphism through increased female mass. Leigh (personal communication) has suggested that intense body mass dimorphism in *A. afarensis* might reflect selection for early female maturation.

Though this chapter focuses on primates, several interesting hypotheses have been put forward for changes in human dimorphism. McHenry (1991) suggests that sexual dimorphism in *Homo erectus* is reduced from that of earlier hominids through an increase in female body mass. This implies first that reduced dimorphism is probably not a consequence of diminished sexual selection. It could reflect better provisioning and diet for females, resulting in greater female body mass. Conversely, it might reflect delay in female maturation to allow greater size in association with greater female competition, following the model of Leigh and Shea (1995).

Frayer (1980) suggests that a reduction in dimorphism associated with a reduction in male body mass and robusticity over a 40,000 year period in the Pleistocene reflects a transition from a hunter–gatherer subsistence strategy which requires a sharp sexual division of labor, to an agricultural society in which gender roles are not as clearly demarcated. A similar hypothesis is offered by Brace and Ryan (1980) who suggest that earlier Pleistocene hominids showed substantial mass and dental dimorphism because males required larger body mass to effectively hunt big game. With a change to subsistence on smaller game, males became smaller resulting in reduced dimorphism.

Finally, ontogenetic patterns of male and female development may offer new insights into changes in social behavior. Leigh (1995a,b) suggests that high juvenile risks associated with folivorous diet and male transfer from single-male groups may select for a high rate of growth in male anthropoids. Conversely, relatively low male juvenile risk associated with more frugivorous diets and multi-male, multi-female groups is associated with delayed male maturation and slower rates of growth. Thus, evidence about changes in the rate of male growth by comparison to an ancestral condition combined with information on diet may suggest changes in social systems and male reproductive strategies. This type of evidence is difficult to evaluate in fossils, but dental eruption patterns may offer some insight. There have been a number of studies of dental development patterns in hominids documenting general changes in growth rates. Currently, this type of evidence may as yet be too imprecise to offer a clear picture of how male and female patterns of growth differ (Smith *et al.*, 1994).

Though documentation of temporal and ontogenetic trends in dimorphism in the fossil record is rare, such trends may provide extremely valuable data for inferring behavior in the fossil record. Comparative investigations that focus on how changes in dimorphism reflect changes in social behavior may prove much more powerful in the long run than analyses that focus on a one-to-one correspondence between behavior and the magnitude of dimorphism.

Patterns of Dimorphism

Sexual dimorphism is expressed differently in different characters within and among species (e.g., Hayes *et al.*, 1995; Leiberman *et al.*, 1985; Masterson and Hartwig, 1998; Oxnard, 1984; Oxnard *et al.*, 1985). It is imperative, therefore, to consider the impact of variation in patterns of sexual dimorphism on behavioral inferences.

There are clear phylogenetic differences in canine shape and dimorphism among anthropoids (Greenfield and Washburn, 1991; Plavcan, 1993). Great apes show relatively low-crowned canines by comparison to other anthropoids, yet they also show intense canine occlusal dimorphism. In contrast, cercopithecines show extreme canine crown height dimorphism, with comparatively modest dimorphism in the occlusal dimensions (Plavcan, 1993).

Taxa differ strongly in the relation between canine dimorphism and body mass dimorphism. Figure 11 shows a plot of canine crown height dimorphism

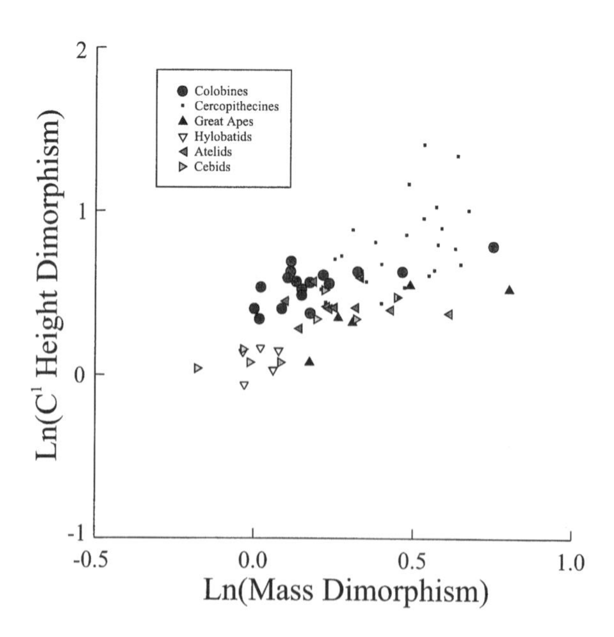

Fig. 11. Bivariate plot of maxillary canine crown height dimorphism versus body mass dimorphism in anthropoid primates. Taxonomic groups are represented by different symbols, indicated in the legend.

Table V. Pearson Product-Moment Correlations between two Species of *Cercopithecus* and two Species of *Saimiri*[a]

	C. pogonias	*C. mona*	*S. oerstedi*
C. pogonias	—		
C. mona	0.878	—	
S. oerstedi	0.509	0.563	—
S. ustus	0.560	0.589	0.869

[a] Correlations are for ln-transformed estimates of dimorphism of 39 linear craniofacial dimensions. Sample sizes are as follows (M/F): *C. pogonias grayi*, 25/17; *C. mona*, 26/13; *S. ustus*, 22/21; *S. oerstedi oerstedi* 20/13.

versus body mass dimorphism in all anthropoids. Great apes show strong degrees of mass dimorphism relative to canine dimorphism. Conversely, colobine canine dimorphism is consistent with other anthropoids of similar competition levels, while their body mass dimorphism tends to be relatively low. Thus, colobine mass dimorphism should not be compared directly with that of other anthropoids. For fossil colobines, relatively modest degrees of mass dimorphism might indicate intense male–male competition. As noted above, relative canine size also differs strongly among taxa.

Almost all analyses of size dimorphism in the fossil record refer to studies of the relation between body mass and behavioral/ecological variables in extant species. But body mass dimorphism can only be estimated indirectly from skeletal remains in the fossil record. Thus, it is crucial to understand variation in patterns of skeletal dimorphism within and among species, and to understand how skeletal dimorphism is related to body mass dimorphism. Yet explicit analyses of patterns of skeletal dimorphism across a variety of species are relatively scarce.

Table V shows correlations between two species of *Cercopithecus* and two species of *Saimiri*, using estimates of dimorphism from 39 cranial dimensions. Note that the correlations between the congeneric species are much higher than those between species of different genera. Lockwood (1999) reports a similar finding for comparisons of facial dimorphism in *Pan*, *Pongo*, *Gorilla*, and *Homo*, especially noting that while the overall pattern of dimorphism among measurements is broadly similar among species, each has a unique pattern of dimorphism. O'Higgins and Dryden (1991) and Wood *et al.* (1993) suggest that variation in patterns of dimorphism may have phylogenetic significance (at least for hominoids). Clearly, one should be careful in choosing extant reference species, with an awareness that patterns of dimorphism might differ between taxonomic groups.

Relatively little work has been done characterizing interspecific patterns of postcranial dimorphism. Several studies document variation in patterns of dimorphism, though these have been based on relatively few species, and it is not clear whether differences in the pattern of dimorphism are substantial by comparison to differences in the overall magnitude of dimorphism. Most studies seem to suggest that pattern variation in postcranial dimensions is relatively minor (though this is not to say that such differences will inevitably prove insignificant). Wood (1976) finds little variation in patterns of dimorphism in the skeleton of 5 anthropoid species, while Leutenegger and Larson (1985) report some variation in postcranial dimensions of four platyrrhines. More recently Lague and Jungers (1999) find relatively little variation in patterns of dimorphism in the distal humerus of extant hominoids. At the same time, these studies all find that dimorphism varies substantially among characters within species. Thus, it is not surprising that McHenry (1992) and McHenry and Berger (1998) report variation in the pattern of body size estimates between fore- and hindlimb elements in *Australopithecus afarensis* and *A. africanus*.

Summary and Conclusions

Though speculation on changes in social behavior are common, and often based on reasonable theoretical models for the evolution of behavior, the only trait preserved in the fossil record that is directly tied to variation in social behavior is sexual dimorphism. Over the years, models that relate dimorphism to social behavior have been refined, and there has been increasing appreciation of variation in the relation between skeletal dimorphism, mass dimorphism, and canine dimorphism.

Ironically, in spite of advances in our understanding of patterns of dimorphism and the behavioral and ecological correlates of dimorphism, dimorphism in fossils yields relatively little precise information about social behavior in extinct species. Dimorphism has long been associated with polygyny in extant primates (Crook, 1972; Darwin, 1871). Thus, a monomorphic fossil species would be interpreted as monogamous, and a dimorphic one polygynous. We now can relate both male and female canine size to estimates of male and female intrasexual competition, we can quantify covariation between male and female canine size, we can demonstrate links between dimorphism and phylogeny, substrate, and dietary specializations, we have evidence that dimorphism covaries with operational sex ratios, and we can relate variation in diet and body size to variation in patterns of female dispersal, and hence both male and female intrasexual competition. Yet, with this more sophisticated understanding of dimorphism, the best we can say is that strong dimorphism in a fossil species suggests polygyny, while low levels of dimorphism may be associated with a variety of mating systems, including monogamy, polyandry, fission–fusion societies, and polygyny. Paradoxically, increased understanding of dimorphism in extant species seems to have undermined our confidence in using dimorphism to infer behavior in the fossil record!

Does this mean that extant models for the evolution of dimorphism have failed to provide information about fossil species? The answer is emphatically no. Extant models have demonstrated that dimorphism is a complex phenomenon, and that simple equations linking dimorphism as a unitary phenomenon with simple classifications of behavior grossly oversimplify a complex biological reality. As it becomes more and more apparent that dimorphism is a complex phenomenon, studies can more precisely focus on teasing apart the factors that govern its evolution, and thereby provide more powerful tools for evaluating the biological significance of dimorphism in the fossil record.

Acknowledgments

I thank Carel van Schaik, Rich Kay, Bill Jungers, Matt Cartmill, Scott McGraw, John Hunter, Larry Witmer, Margaret Lewis, and Luke Holbrook for helpful discussions. I thank Peter Ungar, Charlie Lockwood, Carel van Schaik, and an anonymous reviewer for insightful and helpful comments on the manuscript. This paper was partly supported by NSF grant SBR 9616671.

References

Aiello, L. C., and Dunbar, R. I. M. 1993. Neocortex size, group size, and the evolution of language. *Curr. Anthropol.* **34:**184–193.

Albrecht, G. H. 1976. Methodological approaches to morphological variation in primate populations: the celebesian macaques. *Yearb. Phys. Anthropol.* **20:**290–308.

Albrecht, G. H. 1980. Latitudinal, taxonomic, sexual, and insular determinants of size variation in pigtail macaques, *Macaca nemestrina. Int. J. Primatol.* **1:**141–152.

Albrecht, G. H., and Miller, J. A. 1993 Geographic variation in primates: a review with implications for interpreting fossils. In: W. H. Kimbel and L. B. Martin (eds.), *Species, Species Concepts and Primate Evolution*, pp. 123–161. Plenum Press, New York.

Albrecht, G. H., Jenkins, P. D., and Godfrey, L. R. 1990. Ecogeographic size variation among the living and subfossil prosimians of Madagascar. *Am. J. Primatol.* **22:**1–50.

Andersson, M. 1994. *Sexual Selection*. Princeton University Press, Princeton.

Arnold, S. J. 1994. Quantitative genetic models of sexual selection. *Experientia* **41:**1296–1310.

Beard, K. C., Qi, T., Dawson, M. R., Wang, B., and Li, C. 1994. A diverse new primate fauna from middle Eocene fissure-fillings in southeastern China. *Nature* **368:**604–609.

Blumenberg, B. 1981. Observations on the paleoecology, population structure and body weight of some Tertiary hominoids. *J. Hum. Evol.* **10:**543–564.

Blumbenberg, B. 1984. Allometry and evolution of Tertiary hominoids. *J. Hum. Evol.* **13:**613–676.

Brace, C. L. 1972. Sexual dimorphism in human evolution. *Yearb. Phys. Anthropol.* **16:**31–49.

Brace, C. L., and Ryan, A. S. 1980. Sexual dimorphism and human tooth size differences. *J. Hum. Evol.* **9:**417–435.

Brennan, E. J. 1985. De Brazza's monkeys (*Cercopithecus neglectus*) in Kenya: census, distribution and conservation. *Am. J. Primatol.* **8:**269–277.

Cheverud, J. M., Dow, M. M., and Leutenegger, W. 1985. The quantitative assessment of phylogenetic constraints in comparative analysis: Sexual dimorphism in body weight among primates. *Evolution* **38:**1335–1351.

Clutton-Brock, T. M., and Harvey, P. H. 1978. Mammals, resources and reproductive strategies. *Nature* **273:**191–195.

Clutton-Brock, T. H., and Harvey, P. H. 1980. Primates, brains and ecology. *J. Zool. London* **190:**309–323.

Clutton-Brock, T. M., Harvey, P. H., and Rudder, B. 1977. Sexual dimorphism, socionomic sex ratio and body weight in primates. *Nature* **269:**797–800.

Cords, M. 1987. Forest guenons and patas monkeys: male–male competition in one-male groups. In: B. B. Smuts, D. L. Cheney, R. M. Seyfarth, R. W. Wrangham, and T. T. Struhsaker (eds.), *Primate Societies*, pp. 98–111. University of Chicago Press, Chicago.

Cords, M. 1988. Mating systems of forest guenons: a preliminary review. In: A. Gautier-Hion, F. Bourliere, J. Gautier, and J. Kingdon (eds.), *A Primate Radiation: Evolutionary Biology of the African Guenons*, pp. 323–339. Cambridge University Press, Cambridge.

Crook, J. H. 1972. Sexual selection, sexual dimorphism, and social organization in the primates. In: B. Campbell (ed.), *Sexual Selection and the Descent of Man, 1871–1971*, pp. 231–281. Aldine Publishing Company, Chicago.

Darwin, C. 1871. *The Descent of Man, and Selection in Relation to Sex*. J. Murray, London.

Di Fiore, A. and Rendall, D. 1994. Evolution of social organization: a reappraisal by using phylogenetic methods. *Proc. Natl. Acad. Sci. USA* **91:**994–995.

Ely, J., and Kurland, J. A. 1989. Spatial autocorrelation, phylogenetic constraints, and the causes of sexual dimorphism in primates. *Int. J. Primatol.* **10:**151–171.

Emlen, S., and Oring, L. 1977. Ecology, sexual selection and the evolution of mating systems. *Science* **197:**215–224.

Felsenstein, J. 1985. Phylogenies and the comparative method. *Am. Nat.* **125:**1–15.

Fleagle, J. G. 1998. *Primate Adaptation and Evolution*, 2nd Edition, Academic Press, New York.

Fleagle, J. G., Kay, R. F., and Simons. 1980. Sexual dimorphism in early anthropoids. *Nature* **287:**328–330.

Foley, R. A., and Lee, P. C. 1989. Finite social space, evolutionary pathways, and reconstructing hominid behavior. *Science* **243:**901–906.

Fooden, J. 1995. Systematic review of southeast Asian Longtail Macaques, *Macaca fascicularis* (Raffles, [1821]), *Fieldiana, Zoology New Series #81*. Field Museum of Natural History, Chicago.

Ford, S. M. 1994. Evolution of sexual dimorphism in body weight in platyrrhines. *Am. J. Primatol.* **34:**221–224.

Frayer, D. W. 1980. Sexual dimorphism and cultural evolution in the late Pleistocene and Holocene of Europe. *J. Hum. Evol.* **9:**399–415.

Gaulin, S. J. C., and Sailer, L. D. 1984. Sexual dimorphism in weight among primates: the relative impact of allometry and sexual selection. *Int. J. Primatol.* **5:**515–535.

Gautier-Hion, A. 1975. Dimorphisme sexuel et organization sociale chez les cercopithècinès forestiers africains. *Mammalia* **39:**365–374.

Gingerich, P. D. 1981. Cranial morphology and adaptations in Eocene Adapidae. I. Sexual dimorphism in *Adapis magnus* and *Adapis parisiensis*. *Am. J. Phys. Anthropol.* **56:**217–234.

Gingerich, P. D. 1995. Sexual dimorphism in earliest Eocene *Cantius torresi* (Mammalia, Primates, Adapoidea). *Cont. Mus. Paleontol. Univ. Mich.* **29:**185–199.

Gittleman, J. L., and van Valkenburg, B. 1997. Sexual dimorphism in the canines and skulls of carnivores: effects of size, phylogeny, and behavioural ecology. *J. Zool. London*. **242:**97–117.

Godfrey, L. R., Lyon, S. K, and Sutherland, M. R. 1993. Sexual dimorphism in large-bodied primates: the case of the subfossil lemurs. *Am. J. Phys. Anthropol.* **90:**315–334.

Greenfield, L. O. 1992a. Origin of the human canine: a new solution to an old enigma. *Yearb. Phys. Anthropol.* **35:**153–185.

Greenfield, L. O. 1992b. Relative canine size, behavior and diet in male ceboids. *J. Hum. Evol.* **23:**469–480.

Greenfield, L. O. 1996. Correlated response of homologous characteristics in the anthropoid anterior dentition. *J. Hum. Evol.* **31:**1–19.

Greenfield, L. O. 1998. Canine tip wear in male and female anthropoids. *Am. J. Phys. Anthropol.* **107:**87–96.

Greenfield, L. O., and Washburn, A. 1991. Polymorphic aspects of male anthropoid canines. *Am. J. Phys. Anthropol.* **84:**17–34.

Harvey, P. H., Kavanagh, M., and Clutton-Brock, T. H. 1978a. Canine tooth size in female primates. *Nature* **276:**817–818.

Harvey, P. H., Kavanagh, M., and Clutton-Brock, T. H. 1978b. Sexual dimorphism in primate teeth. *J. Zool. (London)* **186:**474–485.

Harvey, P. H., Martin, R. D., and Clutton-Brock, T. H. (1987) Life histories in comparative perspective. In: B. B. Smuts, D. L. Cheney, R. M. Seyfarth, R. W. Wrangham, and T. T. Struhsaker (eds.), *Primate Societies*, pp. 181–196. University of Chicago Press, Chicago.

Hayes, V. J., Freedman, L., and Oxnard, C. E. 1995. The differential expression of dental sexual dimorphism in subspecies of *Colobus guereza*. *Int. J. Primatol.* **16:**971–996.

Josephson, S. C., Juell, K. E., and Rogers, A. R. 1997. Estimating sexual dimorphism by method-of-moments. *Am. J. Phys. Anthropol.* **100:**191–206.

Kappeler, P. M. 1990. The evolution of sexual size dimorphism in prosimian primates. *Am. J. Primatol.* **21:**201–214.

Kappeler, P. M. 1991. Patterns of sexual dimorphism in body weight among prosimian primates. *Folia Primatol.* **57:**132–146.

Kappeler, P. M. 1996. Intrasexual selection and phylogenetic constraints in the evolution of sexual canine dimorphism in strepsirhine primates. *J. Evol. Biol.* **9:**43–65.

Kappelman, J. 1996. The evolution of body mass and relative brain size in fossil hominids. *J. Hum. Evol.* **30:**243–276.

Kay, R. F. 1982. *Sivapithecus simonsi*, a new species of Miocene hominoid, with comments on the phylogenetic status of Ramapithecinae. *Int. J. Primatol.* **3:**113–173.

Kay, R. F., and Cartmill, M. 1977. Cranial morphology and adaptations of *Palaechthon nacimienti* and other Paromomyidae (Plesiadapoidea, ? Primates), with a description of a new genus and species. *J. Hum. Evol.* **6:**19–53.

Kay, R. F., Plavcan, J. M., Glander, K. E., and Wright, P. C. 1988. Sexual selection and canine dimorphism in New World monkeys. *Am. J. Phys. Anthropol.* **77:**385–397.

Kay, R. F., Ross, C. F., and Williams, B. A. 1997. Anthropoid origins. *Science* **275:**797–804.

Kelley, J. 1993. Taxonomic implications of sexual dimorphism in *Lufengpithecus*. In: W. H. Kimbel and L. B. Martin (eds.), *Species, Species Concepts, and Primate Evolution*, pp. 429–458. Plenum Press, New York.

Kelley, J., and Plavcan, J. M. 1998. A simulation test of hominoid species number at Lufeng, China: implications for the use of the coefficient of variation in paleotaxonomy. *J. Hum. Evol.* **35:**577–596.

Kendall, M., and Stuart, A. 1954. *The Advanced Theory of Statistics*, Vol. 1. MacMillan Publishing, New York.

Kinzey, W. G. (ed.). 1987. *The Evolution of Human Behavior*. State University of New York Press, Albany.

Konigsberg, L. W., Hens, S. M., Jantz, L. M., and Jungers, W. L. 1998. Stature estimation and calibration: bayesian and maximum likelihood perspectives in physical anthropology. *Yearb. Phys. Anthropol.* **41:**65–92.

Krishtalka, L., Stucky, R. K., and Beard, K. C. 1991. The earliest fossil evidence for sexual dimorphism in primates. *Proc. Natl. Acad. Sci. USA* **87:**5223–5226.

Lague, M. R., and Jungers, W. L. 1999. Patterns of sexual dimorphism in the hominoid distal humerus. *J. Hum. Evol.* **36:**379–399.

Lande, R. 1980. Sexual dimorphism, sexual selection, and adaptation in polygenic characters. *Evolution* **34:**292–305.

Leigh, S. R. 1992. Patterns of variation in the ontogeny of primate body size dimorphism. *J. Hum. Evol.* **23:**27–50.

Leigh, S. R. 1995a. Ontogeny and the evolution of body size dimorphism in primates. *Anthropologie* **33:**17–28.

Leigh, S. R. 1995b. Socioecology and the ontogeny of sexual size dimorphism in anthropoid primates. *Am. J. Phys. Anthropol.* **97:**339–356.

Leigh, S. R., and Shea, B. T. 1995. Ontogeny and the evolution of adult body size dimorphism in apes. *Am. J. Primatol.* **36:**37–60.

Leigh, S. R., and Terranova, C. J. 1998. Comparative perspectives on bimaturism, ontogeny, and dimorphism in lemurid primates. *Int. J. Primatol.* **19:**723–749.

Leutenegger, W., and Cheverud, J. 1982. Correlates of sexual dimorphism in primates: ecological and size variables. *Int. J. Primatol.* **3:**387–402.

Leutenegger, W., and Cheverud, J. M. 1985. Sexual dimorphism in primates. The effects of size. In: W. L. Jungers (ed.), *Size and Scaling in Primate Biology*, pp. 33–50. Plenum Press, New York.

Leutenegger, W., and Kelly, J. T. 1977. Relationship of sexual dimorphism in canine size and body size to social, behavioral, and ecological correlates in anthropoid primates. *Primates* **18:**117–136.

Leutenegger, W., and Larson, S. 1985. Sexual dimorphism in the postcranial skeleton of New World Primates. *Folia Primatol.* **44:**82–95.

Leutenegger, W. and Lubach, G. 1987. Sexual dimorphism, mating system, and effect of phylogeny in De Brazza's monkey (*Cercopithecus neglectus*). *Am. J. Primatol.* **13:**171–179.

Leutenegger, W., and Shell, B. 1987. Variability and sexual dimorphism in canine size of *Australopithecus* and extant hominoids. *J. Hum. Evol.* **16:**359–367.

Lieberman, S. S., Gelvin, B. R., and Oxnard, C. E. 1985. Dental sexual dimorphisms in some extant hominoids and ramapithecines from China: A quantitative approach. *Am. J. Primatol.* **9:**305–326.

Lockwood, C. A. 1999. Sexual dimorphism in the face of *Australopithecus africanus*. *Am. J. Phys. Anthropol.* **108:**97–127.

Lovejoy, C. O. 1981. The origin of man. *Science* **211:**341–350.

Lovejoy, C. O., Kern, K. F., Simpson, S. W., and Meindl, R. S. 1989. A new method for estimation of skeletal dimorphism in fossil samples with an application to *Australopithecus afarensis*. In: G. Giacobini (ed.), *Hominidae: Proceedings of the 2nd International Congress of Human Paleontology*, pp. 103–108. Jaca Book, Milan.

Lucas, P. W., Corlett, R. T., and Luke, D. A. 1986. Sexual dimorphism in tooth size in anthropoids. *Hum. Evol.* **1:**23–39.

Martin, R. D., Wilner, L. A., and Dettler, A. 1994. The evolution of sexual size dimorphism in primates. In: R. V. Short and E. Balaban (eds.), *The Differences Between the Sexes*, pp. 159–200. Cambridge University Press, Cambridge.

Masterson, T. J., and Hartwig, W. C. 1998. Degrees of sexual dimorphism in *Cebus* and other New World monkeys. *Am. J. Phys. Anthropol.* **107:**243–256.

McHenry, H. M. 1991. Sexual dimorphism in *Australopithecus afarensis*. *J. Hum.* Evol. **20:**21–32.

McHenry, H. M. 1992. Body size and proportions in early hominids. *Am. J. Phys. Anthropol.* **87:**407–431.

McHenry, H. M. 1994a. Behavioral ecological implications of early hominid body size. *J. Hum. Evol.* **27:**77–87.

McHenry, H. M. 1994b. Sexual dimorphism in fossil hominids and its sociological implications. In: S. Shennan and J. Steele (eds.), *Power, Sex and Tradition: The Archeology of Human Ancestry*, pp. 91–109. Routledge and Kegan Paul, London.

McHenry, H. M., and Berger, L. R. 1998. Body proportions in *Australopithecus afarensis* and *A. africanus* and the origin of the genus *Homo*. *J. Hum. Evol.* **35:**1–22.

Mitani, J., Gros-Louis, J., and Richards, A. F. 1996. Sexual dimorphism, the operational sex ratio, and the intensity of male competition in polygynous primates. *Am. Nat.* **147:**966–980.

Moore, J. 1997. Savanna chimpanzees, referential models and the last common ancestor. In: W. C. McGrew, L. F. Marchant, and T. Nishida (eds.), *Great Ape Societies*, pp. 275–292. Cambridge University Press, Cambridge.

O'Higgins. P. O., and Dryden, I. L. 1993. Sexual dimorphism in hominoids: further studies of craniofacial shape differences in *Pan*, *Gorilla*, and *Pongo*. *J. Hum. Evol.* **24:**183–205.

Oxnard, C. E. 1984. *The Order of Man*. Yale University Press, New Haven.
Oxnard, C. E., Lieberman, S. S., and Gelvin, B. R. 1985. Sexual dimorphisms in dental dimensions of higher primates. *Am. J. Primatol.* **8:**127–152.
Pagel, M. 1992. A method for the analysis of comparative data. *J. Theor. Biol.* **156:**431–442.
Philips-Conroy, J. E., and Jolly, C. J. 1981. Sexual dimorphism in two subspecies of Ethiopian baboons (*Papio hamadryas*) and their hybrids. *Am. J. Phys. Anthropol.* **56:**115–129.
Pickford, M. 1986. On the origins of body size dimorphism in primates. In: M. Pickford and B. Chiarelli (eds.), *Sexual Dimorphism in Living and Fossil Primates*, pp. 77–91. Il Sedicessimo, Firenze.
Plavcan, J. M. 1990. *Sexual Dimorphism in the Dentition of Extant Anthropoid Primates*. Ph.D. Dissertation. University Microfilms, Ann Arbor.
Plavcan, J. M. 1993. Canine size and shape in male anthropoid primates. *Am. J. Phys. Anthropol.* **92:**201–216.
Plavcan, J. M. 1994. Comparison of four simple methods for estimating sexual dimorphism in fossils. *Am. J. Phys. Anthropol.* **94:**465–476.
Plavcan, J. M. 1998. Correlated response, competition, and female canine size in primates. *Am. J. Phys. Anthropol.* **107:**401–416.
Plavcan, J. M., and Kelley, J. 1996. Evaluating the "dual selection" hypothesis of canine reduction. *Am. J. Phys. Anthropol.* **99:**379–387.
Plavcan, J. M., and van Schaik, C. P. 1992. Intrasexual competition and canine dimorphism in anthropoid primates. *Am. J. Phys. Anthropol.* **87:**461–477.
Plavcan, J. M., and van Schaik, C. P. 1994. Canine dimorphism. *Evol. Anthropol.* **2:**208–214.
Plavcan, J. M., and van Schaik, C. P. 1997a. Interpreting hominid behavior on the basis of sexual dimorphism. *J. Hum. Evol.* **32:**345–374.
Plavcan, J. M., and van Schaik, C. P. 1997b. Intrasexual competition and body weight dimorphism in anthropoid primates. *Am. J. Phys. Anthropol.* **103:**37–68.
Plavcan, J. M., van Schaik, C. P., and Kappeler, P. M. 1995. Competition, coalitions and canine size in primates. *J. Hum. Evol.* **28:**245–276.
Ross, C. F. 1996. An adaptive explanation for the origin of the Anthropoidea (Primates). *Am. J. Primatol.* **40:**205–230.
Ross, C., Williams, B., and Kay, R. F. 1998. Phylogenetic analysis of anthropoid relationships. *J. Hum. Evol.* **35:**221–306.
Rowell, T. E. 1988. The social system of guenons, compared with baboons, macaques and mangabeys. In: A. Gautier-Hion, F. Bourliere, J. Gautier, and J. Kingdon (eds.), *A Primate Radiation: Evolutionary Biology of the African Guenons*, pp. 439–451. Cambridge University Press, Cambridge.
Rowell, T. E., and Chism, J. 1986. Sexual dimorphism and mating systems: jumping to conclusions. In: M. Pickford and B. Chiarelli (eds.), *Sexual Dimorphism in Living and Fossil Primates*, pp. 107–111. Il Sedicesimo, Firenze.
Shea, B. T., Leigh, S. R., and Groves, C. P. 1993. Multivariate craniometric variation in chimpanzees: implications for species identification. In: W. L. Kimbel and L. B. Martin (eds.), *Species, Species Concepts, and Primate Evolution*, pp. 265–296. Plenum Press, New York.
Simons, E. L., and Rasmussen, D. T. 1996. Skull of *Catopithecus browni*, an early Tertiary catarrhine. *Am. J. Phys. Anthropol.* **100:**261–292.
Simons, E. L., Plavcan, J. M., and Fleagle, J. G. 1999. Canine sexual dimorphism in Egyptian Eocene anthropoid primates: *Catopithecus* and *Proteopithecus*. *Proc. Natl. Acad. Sci. USA* **96:**2559–2562.
Simpson, G. G., Roe, A., and Lewontin, R. C. 1960. *Quantitative Zoology*. Harcourt Brace and Co., New York.
Smith, B. H., Crummett, T. L., and Brandt, K. L. 1994. Ages of eruption of primate teeth: a compendium for aging individuals and comparing life histories. *Yearb. Phys. Anthropol.* **37:**177–231.
Smith, R. J. 1981. Interspecific scaling of maxillary canine size and shape in female primates: relationships to social structure and diet. *J. Hum. Evol.* **10:**165–173.

Smith, R. J. 1994. Regression models for prediction equations. *J. Hum. Evol.* **26:**239–244.
Smith. R. J. 1996. Biology and body size in human evolution: statistical inference misapplied. *Curr. Anthropol.* **37:**451–481.
Smith, R. J., and Jungers, W. L. 1997. Body mass in comparative primatology. *J. Hum. Evol.* **32:**523–559.
Smuts, B. B., Cheney, D. L., Seyfarth, R. M., Wrangham, R. W., and Struhsaker, T. T. (1987) *Primate Societies.* University of Chicago Press, Chicago.
Struhsaker, T. T., and Leland, L. 1987. Colobines: infanticide by adult males. In: B. B. Smuts, D. L. Cheney, R. M. Seyfarth, R. W. Wrangham, and T. T. Struhsaker (eds.), *Primate Societies*, pp. 83–97. The University of Chicago Press, Chicago.
Turner, T. R., Anapol, F., and Jolly, C. F. 1994. Body weights of adult vervet monkeys (*Cercopithecus aethiops*) at four sites in Kenya. *Folia Primatol.* **63:**177–179.
Uchida, A. 1998a. Variation in tooth morphology of *Gorilla gorilla. J. Hum. Evol.* **34:**55–70.
Uchida, A. 1998b. Variation in tooth morphology of *Pongo pygmaeus. J. Hum. Evol.* **34:**71–79.
Ungar, P. S., and Kay, R. F. 1995. The dietary adaptations of European Miocene catarrhines. *Proc. Natl. Acad. Sci USA* **92:**5479–5481.
Weckerly, F. W. 1998. Sexual-seize dimorphism: influence of mass and mating systems in the most dimorphic mammals. *J. Mamm.* **79:**33–52.
Wilkinson, L. 1990. *SYGRAPH: The System for Graphics.* SYSTAT, Inc., Evanston.
Wood, B. 1976. The nature and basis of sexual dimorphism in the primate skeleton. *J. Zool. (London)* **180:**15–34.
Wood, B. A., Li, Y., and Willoughby C. 1991. Intraspecific variation and sexual dimorphism in cranial and dental variables among higher primates and their bearing on the hominid fossil record. *J. Anat.* **174:**185–205.
Wrangham, R. W. 1980. An ecological model of female-bonded primate groups. *Behaviour* **75:**262–300.
Wrangham, R. W. 1987. The significance of african apes for reconstructing human social evolution. In: W. G. Kinzey (ed.), *The Evolution of Human Behavior*, pp. 51–71. State University of New York Press, Albany.
Zihlman, A. 1997. Reconstructions reconsidered: chimpanzee models and human evolution. In: W. C. McGrew, L. F. Marchant, and T. Nishida (eds.), *Great Ape Societies*, pp. 293-304. Cambridge University Press, Cambridge.

The Adaptations of *Branisella boliviana*, the Earliest South American Monkey

9

RICHARD F. KAY, BLYTHE A. WILLIAMS,
and FEDERICO ANAYA

Introduction

One of the goals of paleoprimatology is to provide adaptive explanations for the origins of evolutionary novelties of the order and its major groups. For such scenarios to be more than "just-so stories," like Kipling's story of how the leopard got its spots, we need to develop and test ideas about the adaptive significance of particular morphological character states that are likely to be preserved in the fossil record. Once the adaptive context of the morphology is fully appreciated, we can go on to make inferences about the behavior of extinct primate species that possessed similar character states. But even when we know with some confidence the adaptive "meaning" of a particular morphological character state and use it to infer the behavior of an extinct species, we must be able to place that extinct species into its phylogenetic context. What

RICHARD F. KAY and BLYTHE A. WILLIAMS • Department of Biological Anthropology and Anatomy, Duke University Medical Center, Durham, North Carolina 27710. FEDERICO ANAYA • Museo Nacional de Historia Naturale, La Paz, Bolivia.

Reconstructing Behavior in the Primate Fossil Record, edited by Plavcan *et al.* Kluwer Academic/Plenum Publishers, New York, 2002.

is the distribution of the newly identified morphological peculiarity? Is it found in just one extinct species or does it characterize some larger group of species? And what does the distribution of the character state tell us about the ancestral morphological (and inferred behavioral) pattern of primate clades? Therefore, in parallel with the effort to understand adaptation of character states, there must be an effort to reconstruct the phylogenetic pattern of primates.

The issues described above are addressed by various authors in this volume in many different ways and with reference to many anatomical systems in primates. This paper deals with behavioral inferences based primarily on dental anatomy. A broad review of the dental evidence for adaptation in primates is presented by Ungar elsewhere in this volume (see also Ungar, 1998). Here, we offer a general discussion of the many meanings of the term adaptation and a review of how adaptations are recognized in the fossil record. Then we utilize the "case study" approach to focus on the strengths and limitations of the comparative method, as applied to dental anatomy. Our study focuses on one particularly important extinct taxon, the earliest known platyrrhine primate *Branisella boliviana*. Because this species is still known only from jaws and teeth and a small part of the face, and because it is so distinctive morphologically from its closest living platyrrhine relatives, such a reconstruction presents a special challenge and makes it an ideal example of the limitations and difficulties in applying our methods.

The Different Meanings of Adaptation

Analysis of adaptation in fossil animals has seen a flowering over the past 30 years. However, progress in this area has been confused somewhat by arguments about how best to define adaptation and how to recognize it in extinct species. Even before the time of Darwin it has been a common feature of biological thought that animals and plants are well suited to live in the environment where they live. In such a circumstance, an adaptation was viewed as the close fit of an organism's form and behavior to its environment and the ways in which it carries out the necessities of life (e.g., feeding, reproduction, locomotion). In pre-Darwinian thought, this good fit between an organism and its living circumstances was attributed to, and given as evidence for, a divine plan. The idea was that God had "designed" animals and plants to be adapted to their surroundings (Paley, 1854). In contrast, Darwin (1859) argued that adaptation is the consequence of natural selection: organisms having one character state are advantaged over others with another with the result that the former have more offspring than the latter. In short, adaptation is selected change of character states for better function with "nature" (=environment) as the selective agent.

In the pre-Darwinian sense, adaptation was a state of being. In the Darwinian formulation, adaptation can be understood in one of two senses. An adaptation is either (1) any character state of an organism that confers an advantage to its possessor, i.e., improves its evolutionary fitness when compared with its conspecifics that do not have that character state, or (2) a *process* by which an adaptive state is achieved, in which case a character state is only considered to be an adaptation if it evolved by natural selection for the purpose that it currently serves. These different meanings of the term led Brandon (1987) to propose that the term *adaptedness* be used for the current evolutionary "fitness" and that *adaptation* be restricted to the process.

Adaptation as a Process

Gould and Vrba (1982) define as adaptations features built by natural selection only for their current role. In this restricted definition, irrespective of its current utility, if a character state originated by processes other than natural selection—by genetic linkage, allometry, drift, etc.—it is not an adaptation. Thus, under this definition, adaptation is the historical process of change, under the direction of natural selection, for a specific set of circumstances (Gould and Vrba, 1982; Kay *et al.*, 1998a; Williams, 1966). Under this process-based definition, a character state that has current utility other than that for which it was originally selected, or that was originally fixed in a population by processes other than selection acting directly on the character state, is not an adaptation. For the category of character states that meet all the other criteria of adaptation, but originated by processes other than natural selection (e.g., by genetic drift), Gould and Vrba used the term *exaptation*.

A related view of adaptation was offered by Coddington (1988), who defined adaptation as "apomorphic function due to natural selection." This definition follows that of Gould and Vrba in the requirement that a character state is considered to be an adaptation only if it evolved by natural selection for the purpose that it currently serves. Coddington's definition casts adaptation within a phylogenetic framework and adds one further requirement: once a speciation event (cladogenesis) has occurred, the character state is then transposed from being an adaptation to being an exaptation. The daughter species possess the character state either because it continues to serve the purpose for which it was selected originally, or because the state has been co-opted for a new purpose (the exaptation), or is only phylogenetic baggage without continuing aptedness.

In Coddington's cladistic approach, for a character state to count as an adaptation, it must be an apomorphy. Symplesiomorphies are disqualified. Because synapomorphies at one level in a clade may be plesiomorphous at another level, in a sense, perhaps we should refer to adaptation at various levels (although Coddington did not do so). Under Coddington's definition, the presence of hair for thermoregulation may be taken as an adaptation of mammals, but it is not an adaptation of primates even though hair may serve the purpose of thermoregulation in primates. Such a definition seems unworkable given the uncertainty of knowing the phylogeny of the organisms under study and, even more fundamentally, of knowing the polarity of the states of a character even when phylogeny is known.

Adaptation as Fitness: A Nonhistorical Definition

Others have adopted a broader definition of what counts as an adaptation, including any feature that enhances or maintains the current fitness of an organism, regardless of its historical origin (Bock and von Wahlert, 1965; Clutton-Brock and Harvey, 1979; Simpson, 1933). For example, Reeve and Sherman (1993) define adaptation as: "... a phenotypic variant that results in the highest fitness among a specified set of variants in a given environment." This definition of adaptation would include under its umbrella a morphological or behavioral character state that was selected for one particular set of circumstances but later co-opted for a new use. It would also include character states currently under stabilizing or directional selection, but originally produced by other processes (exaptation). Under such a definition, a character state is an adaptation irrespective of the evolutionary "reason" for its origin or its phylogenetic derivation: if a state meets appropriate functional criteria, it is an adaptation whether it is an apomorphy or a plesiomorphy.

Adaptation and the Fossil Record

In the fossil record, in most cases, we can only hope to pose explanations about the behavior of a species during its lifetime. The fossil record rarely allows us to say anything about the *process* through which a character state may have been selected. The relative utility of "state" vs. "process" definitions may differ in the case of studies of adaptation of living organisms but, for the purposes of paleontological studies, the definition of adaptation as a "state" is readily investigated with empirical data, while the process by which a state achieved its (then) current adaptedness is surely not. Therefore a "state" definition of Reeve and Sherman (1993) is adopted here recognizing that it encompasses a wider range of possibilities than the definitions of adaptation offered by Brandon (1987), Gould and Vrba (1982), or Coddington (1988).

Reeve and Sherman (1993) specify three requirements that must be satisfied for a character state to be considered an adaptation.

(1) There must be a specified set of phenotypes—states of a character—the fitnesses of which must be compared (a phenotype set).
(2) There must be a measure of fitness of the various states of the character (a fitness measure).
(3) There must be an environmental context in which the phenotypes are being evaluated (an environmental context).

These three components present special problems in the context of the study of fossils. First, fitness is most directly measured by reproductive success but clearly it is impossible to measure directly reproductive fitness in an extinct organism. This problem is generally circumvented by using a state's functional design as a fitness criterion (Reeve and Sherman, 1993). For example, the possession of long, sharp shearing crests on mammalian molars is a design to increase the surface area of the foods that will be exposed to digestive processes during chewing, thus increasing the total amount of energy that can be extracted from fibrous foods and the rate at which energy can be extracted (Kay and Sheine, 1979; Sheine and Kay, 1977; Sheine and Kay, 1982). This proposition connects the possession of long, sharp shearing crests with a fitness criterion, the efficient acquisition of the energy to sustain life, and allows comparisons with alternative phenotypes such as short, blunt shearing crests, etc.

Because direct correlations between structure and function can be identified only for living organisms, what can be said about the adaptations of an extinct species must be based on field and laboratory observations of living animals with similar morphology. Kay and Cartmill (1977) formally proposed criteria for inferring adaptation of extinct species. Their criteria are similar to those offered by Reeve and Sherman, but in addition to the three lines of evidence that are essential to diagnose an adaptation in a living species (a phenotype set, a fitness measurement, and an environmental context) there are several other things required for adaptation to be inferred in an extinct species. For any extinct species possessing a particular state of a character, the inference that the state is an adaptation of an extinct species is warranted if:

(1) there are extant species that have the state;
(2) in all extant species that have the state, the state is associated with the same adaptative role; and
(3) there are no reasons for believing that the state's fixation in any lineage preceded the state's assumption of the adaptation.

These three criteria are more restrictive than those of Reeve and Sherman (1993) and certainly will lead to the exclusion of many characters that have an undoubted adaptive basis. For example, the presence of an anteroposteriorly elongate shearing crest on the anterior-most mandibular cheek tooth (plagiaulacoidy; Simpson, 1933) does not allow us to infer a significant seed-eating component in the diet of an extinct species because not all extant species with a plagiaulacoid cheek tooth are seed-eaters. Plagiaulacoidy may be in some instances an adaptation for seed-eating, but we cannot definitely infer this particular behavior in an extinct animal possessing a plagiaulacoid dentition.

Comparative Methods Designed to Eliminate the Effects of Phylogeny

Kay and Cartmill's second criterion deserves closer scrutiny. It states that in all extant species that have the (morphological) state, the state has the same role. For this requirement to be satisfied, phylogenetic history must be taken into account (Kay and Covert, 1984; Kay and Ungar, 1997; Ungar, 1998). Two related phenomena must be recognized and controlled for: the underlying *morphological substrate* of the character (bauplan), and the *phyletic independence* of the species being compared.

As an instance of what we mean by "morphological substrate," when natural selection favors the adaptive shift of a species to a new diet, or when competition forces selection for modified physiological performance, the initial structure of the dentition of the species undergoing selection influences the way in which the adaptive changes are most likely to occur in descendant species. In this way, selection for enhanced shearing on the cheek teeth would most likely act to alter a shearing device already present in an ancestral species.

The immediate practical difficulty presented by this notion is that morphometric comparisons intended to illustrate habitus-related differences (adaptations) in the anatomy of distantly related species may often pick up phylogenetic contrasts instead (Szalay, 1981). The more distant the common ancestor and the more evolutionary modifications that have occurred in the descendant lineages, the more pervasive this problem becomes. For example, among cercopithecid monkeys it is possible to relate fairly minor changes in molar proportions to dietary differences, but it is possible to gain only a broad picture of dietary specializations when comparing noncercopithecid primates as a whole (Kay and Cartmill, 1977; Kay and Covert, 1984; Kay and Ungar, 1997). Cercopithecids show variations of a single type of molar structure, bilophodonty, making heritage contrasts among the species minimal while noncercopithecid primates represent a much more diverse group with a much more distant common ancestor so that heritage contrasts complicate functional comparisons among species. Traditionally, the effects of heritage have been minimized by means of these "range restrictions" (e.g., Kay and Covert, 1984).

A second problem, "phyletic independence," becomes evident when statistical studies are used to elucidate adaptation. Statistical methods require that species be treated as statistically independent points. However, the character states of closely related species may not be phyletically (and therefore statistically) independent (Smith, 1994). For example, among platyrrhine monkeys, suppose the degree of development of shearing crests is shown to be significantly greater among folivorous than among frugivorous platyrrhines. How comfortable should we be in applying this finding to interpreting the fossil record? Perhaps we should be less confident in the generality of these results if all the extant folivores are closely related species of one genus, *Alouatta*. Intuitively, the conclusions would be more convincing if the morphological contrasts and their correlative behavioral contrasts were to be found in many phylogenetically independent pairs of frugivores and folivores. For example, if the same morphological contrast is discovered between the platyrrhines *Ateles* (frugivorous) and *Brachyteles* (folivorous) as between the cercopithecids *Cercopithecus* (frugivorous) and *Colobus* (folivorous), the comparison might be considered to have more generality.

The task of a comparative analysis is to establish an underlying pattern of correlation between a morphological character and a behavioral attribute with the confounding effects of phylogenetic history removed. For this purpose, we have utilized an analytical program called "Comparative Analysis by Independent Contrasts" (CAIC) (Purvis and Rambaut, 1995) to elucidate the adaptive properties of two quantifiable morphological characters of anthropoid molars—shearing crest development and crown height—in relation to two behavioral attributes—diet and substrate preference. The findings of the comparison of the living platyrrhine species will serve as a basis for explaining two morphological peculiarities of the molar structure of *Branisella*; the weak development of its shearing crests and its very high crowns. Further details and an example of this procedure are found in the appendix.

The statistical approach mentioned above can provide solid evidence that a specific character state (or states) is an adaptation. However, even if adaptation has been inferred in living species by statistical means (correlation), having made such an inference is only the first step toward making an inference about behavior in an extinct species. In fact, Kay and Cartmill (1977) have argued that the association between having the character state and having the behavior must be absolute before an inference about fossils is warranted.

Branisella boliviana, ***the Earliest-Known Platyrrhine Primate***

Since its initial description in 1969, *Branisella boliviana* has played a pivotal role in interpretations of platyrrhine evolution. In his original description, Hoffstetter (1969, 1980) regarded *Branisella* as a primitive monkey, based on the type and then only known specimen, a maxilla with roots, partial alveoli, or dental crowns of C-M3. He was impressed by certain primitive (omomyid-like) structures of its dentition and also by its supposed great age. He suggested that the small hypocone of *Branisella* and extant Callitrichinae (tamarins and marmosets*) could mean that they form a clade. However, he rejected this hypothesis partly on the basis of a much younger split time of callitrichines from other cebids, then proposed to have occurred at about 15 Ma (Sarich and Cronin, 1976). Hoffstetter concluded (following W. K. Gregory, 1922) that the

* Rosenberger was the first to advocate a cladistic classification of Platyrrhini (Rosenberger, 1979, 1981b; Szalay and Delson, 1979). For clarity in what follows, our classification of extant platyrrhines follows Schneider *et al.* (1993) and is a modification of Rosenberger's scheme. They divide Platyrrhini into three monophyletic families Atelidae, Pithecciidae, and Cebidae. Schneider *et al.*'s Cebidae includes the subfamilies Cebinae (*Cebus* and *Saimiri*), Aotinae (Aotus), and Callitrichinae (*Callimico*, *Callithrix*, *Cebuella*, *Saguinus*, and *Leontopithecus*). (See also Figure 6, p. 357.)

teeth of callitrichids have reverted to a more primitive form as a result of ongoing trends toward body size reduction. Hoffstetter preferred the interpretation that *Branisella* represents an ancestral group out of which evolved the living platyrrhine groups.

Hershkovitz (1977) took a very different view of these remains. He placed *Branisella* in its own family but considered it to represent a very primitive primate and declined to place it in Platyrrhini, nor indeed even into Anthropoidea or Haplorhini. Hershkovitz's views were conditioned in part by the supposition that haplorhine primates became segregated on the South American continent at an early time as a result of continental drift and separately achieved a simian grade of organisation. If so, *Branisella* need not be specially related to any extant platyrrhine primate but could equally represent a 'proimian' grade.

In the 1980's, four other fragments of *Branisella* were discovered or newly recognized (Rosenberger, 1981a; Rosenberger *et al.*, 1990; Wolff, 1984). Consideration of these led Rosenberger and colleagues to identify a second taxon, *Szalatavus attricuspis* (Rosenberger *et al.*, 1991; Rosenberger *et al.*, 1990). Since 1990, a number of other specimens have been collected during expeditions to Salla by M. Takai, R. Kay, F. Anaya, and their colleagues. Additional material of *Branisella* includes maxillary and mandibular fragments and includes some new details of the anatomy of the premolars not hitherto known. This material does not support recognition of "*Szalatavus*" as distinct at either the genus or species level from *B. boliviana* (Kay and Williams, 1995; Takai and Anaya, 1996). Whatever the ultimate outcome of the debate about the generic or specific distinctness of *Szalatavus* from *Branisella*, it is agreed that the two are closely related.

Rosenberger and others (Rosenberger *et al.*, 1990) place *Branisella* in a distinct subfamily of Cebidae. Takai and Anaya (1996) have advanced the position that *Branisella* is an early callitrichine (see the previous footnote) based on their belief that the small size of its hypocone is a shared-derived feature with callitrichines. In this study, we follow Hoffstetter's position that *Branisella* is a platyrrhine but is a sister taxon to all living taxa. Hoffsetter's view receives new support from DNA studies that indicate an antiquity of callithrichines no greater than about 18.5 Ma (Barroso *et al.*, 1997; Schneider *et al.*, 1993), and that the Cebinae differentiated no earlier than 21.3 Ma. New radiometric data indicates a probable age of the *Branisella* level between 25.8 and 27.0 Ma, in other words at least five million years earlier than the branch time for Cebinae and seven million years older than the branch time for Callitrichinae (Kay *et al.*, 1998a).

If, as we believe, *Branisella* is phylogenetically a sister-taxon to extant platyrrhines, are its adaptations representative of the ancestral platyrrhine stock? Most workers have assumed that the greater species richness of neotropical primates in humid tropical forests indicated that they had their origin in that environment (Ford, 1988). MacFadden (1990) has pointed out that the sedimentological evidence and the dental structure of other mammals at Salla suggest a more arid environment for *Branisella*. Does this mean that the ancestral stock of platyrrhines inhabited a more arid environment than Ford's hypothesis suggests? To explore this question more thoroughly, we review available evidence about the paleoenvironment in which *Branisella* dwelt and present a detailed reconstruction of its adaptations. Our data support MacFadden in suggesting that this primate inhabited a more open, less humid environment with some trees—certainly not a rainforest—and, unlike any extant platyrrhine, may have been at least partly terrestrial. Neither of these findings is consistent with what would have been predicted for the last common ancestor of crown platyrrhines. Therefore, we propose that *Branisella* is an aberrantly specialized species not representative of the adaptations of earliest platyrrhines. This conclusion is supported by the DNA evidence and dates for Salla.

The Salla Paleoenvironment

Any interpretation of the morphology of *Branisella* must be consistent with the environment in which it dwelt, as elucidated from physical and biotic evidence. It has been suggested that *Branisella* lived in a dry environment (MacFadden, 1990). MacFadden suggests two lines of evidence in support of this interpretation. First, the mammalian fauna at Salla contains a number of species of notoungulates that have high-crowned cheek teeth (MacFadden, 1990; Shockey, 1997). The marsupial *Proargyrolagus* is also very high-crowned (Sánchez-Villagra and Kay, 1997). The presence of many high-crowned species is indicative of species feeding on plants very close to the ground in "open" habitats where substantial amounts of dust and grit are encountered (Janis, 1988; Shockey, 1997).

The very low level of primate species richness at Salla also is suggestive of a dry environment. In spite of the recovery of many specimens of fossil mammals of similar size at Salla (>2000 specimens from this level), *Branisella* is the only primate represented. The state of preservation of the fossils suggests that *Branisella* was buried where it died rather than having been transported from elsewhere, because none of the specimens shows evidence of having been rolled in streams before deposition and several parts of the same animal (e.g., maxillae with mandibles) have been found together. The enamel surfaces of the available specimens are often pitted as though they had been exposed to soil acids at the time of deposition. Therefore, the fact that *Branisella* is the only known primate from Salla is probably not an artifact of sampling. Primate species richness today is strongly influenced by a number of factors, including altitude, latitude, seasonality, and rainfall, all of which have plant productivity as an underlying cause (Kay *et al.*, 1997).

The number of primate species decreases with increasing altitude (Eisenberg *et al.*, 1979), but does not do so significantly until elevations reach about 1000 m above sea level, above which it drops off dramatically (Kay *et al.*, 1997). Today the Salla fauna is situated in a mountain valley at between 3700 and

3900 m above sea level. If Salla was at high altitude in the late Oligocene, this would explain its impoverished primate richness. However, geologic evidence of the timing and rate of mountain building near La Paz, Bolivia, 90 km to the north-east, suggests that most of the uplift of the Bolivian Andes occurred in the middle to late Miocene—much more recently than the Salla fauna existed. Some volcanic activity was going on in the Salla vicinity in late Oligocene times, but the Salla beds were deposited at altitudes of less than 1000 m above sea level (Benjamin *et al.*, 1987; Kay *et al.*, 1998a; MacFadden, 1990).

Although latitude is a major factor in primate species richness (Eisenberg *et al.*, 1979), the paleolatitude of Salla, by itself, does not explain why there should be so few primate species. Data from Kay *et al.* (1997) show that nearly 40% of the variance in primate species richness is explained by latitude. If Salla had a paleolatitude of approximately 20 degrees south (MacFadden, 1990; Tarling, 1980), and if climatic zones were the same in late Oligocene as they now are, we would expect to find three to four primate species at Salla (Fig. 1).

The likeliest explanation of the low primate species richness at Salla is reduced plant productivity brought about by low annual rainfall. Kay *et al.* (1997) show that local species richness of neotropical primates displays a unimodal relationship with rainfall, rising to a peak at annual rainfall levels of about 2500 mm and then declining. A similar unimodal pattern is seen for Madagascar primate faunas (Ganzhorn *et al.*, 1997) and revealed by a reanalysis of Reed and Fleagle's (1995) south Asian primate richness data using a nonlinear model (Kay *et al.*, 1997). This pattern fits very well with what we know about plant productivity. Kay *et al.* (1997) note that both plant productivity and neotropical primate species richness increase with rainfall up to a

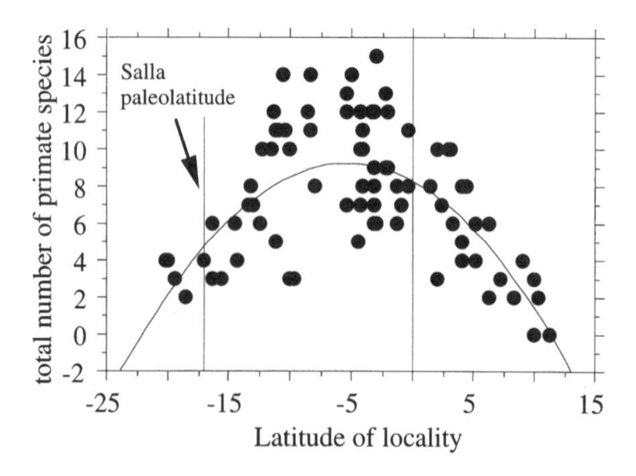

Fig. 1. Relationship between extant neotropical primate species richness and latitude. Localities vary in rainfall but are all below 1000 m in elevation. 39.7% of the variance in primate richness is explained by latitude (data from Kay *et al.*, 1997).

maximum richness at approximately 2500 mm per year and then decline at higher rainfall levels. Although primate richness is in decline above about 2500 mm of rainfall, no localities were reported where it declined below 5 species, while single species of primates were reported at two tropical localities in coastal Brazil with highly seasonal rainfall in the range of 1000–1500 mm/yr. Thus, the extremely low primate richness at Salla might be explained by very low levels of rainfall.

The presence of pedogenic carbonate, or calcrete, nodules interbedded in the Salla beds, including at the principal fossil horizon with *Branisella*, has been advanced to support the possibility of Salla having had seasonally low levels of rainfall. This was suggested to indicate a "semi-arid or arid" environment of deposition (MacFadden, 1990). However, precipitation of carbonate nodules in modern soil horizons requires only that the water table drops seasonally. When the water table rises again, carbonate is unlikely to be redissolved. Over a number of seasonal fluctuations in the water table, nodules form and persist (Kraus and Aslan, 1993). As a modern example, carbonate nodules are common in the modern soils in the Mississippi River flood plain of Louisiana. Nodules in and of themselves merely indicate seasonal fluctuations in the water table. Whether Salla had a semi-arid or arid environment *sensu* MacFadden (1990) is not addressed by this evidence (Aslan, 1994; Farrell, 1987).

Hitz (1997) reported finding very few carbonate nodules at Salla. The bulk of the carbonate was in the form of veins, which raises the question of whether the carbonate was precipitated diagenetically rather than in an active soil profile. At one level just below the *Branisella* fossil level there occurs a couplet of carbonate layers called the El Planiemento horizon. On thin sections of the El Planimiento parent rock, Hitz found features that could be poorly preserved remnants of ostracods. The presence of these fresh water invertebrates would imply persistent standing water at Salla. However, Hitz could not dismiss the possibility that the El Planiemento horizon could also be a calcrete. The latter forms today only under arid conditions.

On balance then, the low species richness of primates, the common occurrence of high-crowned species, and the presence of pedogenic carbonate nodules suggest that the paleoenvironment at Salla may have been an open woodland experiencing seasonal rainfall.

Reconstructing Branisella*'s Niche from Morphological Evidence*

Paleoenvironmental indicators indicate that *Branisella* lived in an open-woodland environment with seasonal rainfall. Using morphological evidence, we will next present a reconstruction of the niche this species may have occupied within that environment—how large was *Branisella* and what were its activity pattern, diet, and substrate preferences?

Body Size

Determination of body weight for *Branisella* is based on its lower first molar (designated m1) occlusal area, 8.56 mm (n = 4; standard error = 0.28 mm). From mean body weights of females of 15 platyrrhine species drawn from the literature and molar areas of different specimens of the same species, Kay *et al.* (1998b) derive the following least-squares regression with an r^2 of 0.935:

$$\ln \text{female body weight} = \ln \text{m1 area} (1.565) + 3.272$$

From this equation, based on our sample of four first molars, the mean body weight of *Branisella* was 759 g with a range of 721 to 759 g. This makes *Branisella* about the same size as the living tamarin *Leontopithecus* or squirrel monkey *Saimiri*.

Orbit Size

Orbit size permits an inference about activity pattern (Kay and Cartmill, 1977; Martin, 1990). Relatively large orbits are associated with a nocturnal activity pattern while relatively smaller orbits are associated with a diurnal pattern. The orbits of *Branisella* are not sufficiently preserved in any specimen to estimate an actual diameter, but two lines of evidence suggest that its orbits were relatively small and diurnally adapted. First, a part of the ventral orbital margin, preserved in several specimens, is rounded—that is, the bony plane of the face ventral to the orbit is not delineated by a sharp projecting lip from the bony orbit as it is in primate species with large orbits. Second, considerable spongy alveolar bone is interposed between the ventral surface of the orbit and the oral surface of the hard palate such that the roots of the cheek teeth do not invade the orbit, again similar to the morphology of primates with small orbits. Overall, the closest approximation we have seen to the form of the lower face and ventral orbital margin is found in the extant callitrichine *Leontopithecus*, a species with small orbits and a diurnal activity pattern (Rowe, 1996). The latter condition is what we would have expected for a platyrrhine because all living species except *Aotus* are diurnal. *Aotus* displays a number of morphological features consistent with its having had a diurnal ancestor (see Martin, 1990). It also has the genetic pathways for producing color photopigments used in diurnal vision yet has shut off these pathways (Jacobs *et al.*, 1993).

Lower Incisors

In anthropoids, incisors are used for ingestion of foods (Hylander, 1975; Kay and Hylander, 1978). All platyrrhines have two upper and lower incisors. The lower incisors have a spatulate shape and occlude against, or lingual to,

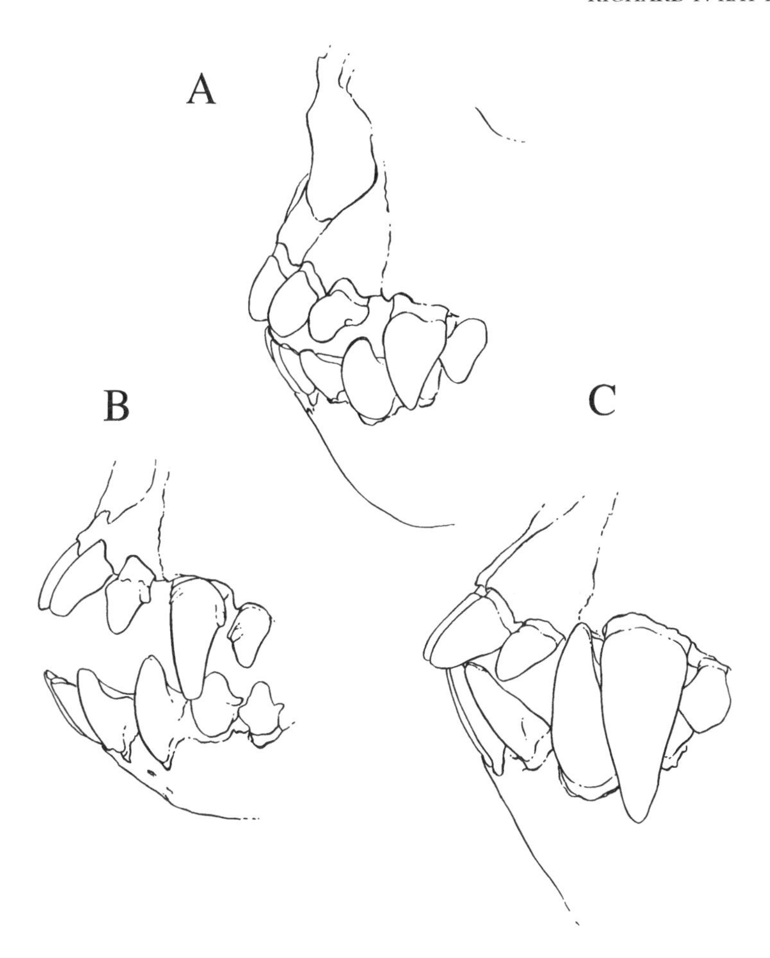

Fig. 2. Fronto-lateral views of the anterior dentitions of extant platyrrhines: A, *Saimiri sciureus*, B, *Callithrix jacchus*, C, *Pithecia pithecia*. Irregularly dashed line indicates the premaxilla–maxilla suture. The large tooth behind the suture is the upper canine. The lower occludes in the embrasure front of the upper canine and behind the upper lateral incisor. Drawings not to same scale.

the edges of the upper incisors (this occlusal pattern is exemplified by *Saimiri*, Fig. 2A). Some variation is encountered in the size of these teeth in relation to diet, with the more frugivorous species tending to have relatively larger incisors (relative to body weight) than their more folivorous close relatives (Anthony and Kay, 1993; Eaglen, 1984). These differences in incisor size, however, probably relate more to degree of tooth use in ingestion than to diet *per se* (Ungar, 1994). One should approach with caution reconstructions of diet based principally on incisor size, particularly when referring to specific, possibly distantly related fossil taxa. As Eaglen (1984) has demonstrated in comparisons of platyrrhines and catarrhines, incisor size differences among distantly related anthropoids in part reflect phylogenetic history.

Sometimes, platyrrhine incisors are specialized in shape for ingestion of particular sorts of foods and can provide a further guide to these special dietary preferences. The incisors of platyrrhine gum-eaters like the marmosets *Cebuella* and *Callithrix* (Fig. 2B) have a convex lateral profile, are mesio-distally compressed, and have a thick layer of enamel on their buccal surfaces but lack enamel on the lingual surfaces (Kay, 1990; Rosenberger, 1978). The lower canines are modified to closely resemble the structure of the incisors. Together, these lower teeth interdigitate with widely-spaced upper incisors to serve the function of gouging bark (Coimbra-Filho and Mittermeier, 1976).

Yet another specialization of the incisors and canines of platyrrhines is for husking tough-skinned nuts or fruits. This is found in *Pithecia*, *Chiropotes*, and *Cacajao* (Fig. 2C). The lower incisors of these extant pitheciines are long and styliform. The teeth are procumbent and the crowns are pressed closely together. The upper incisors also are procumbent. A wide diastema separates the lower lateral incisors from the lower canines. Likewise, a large diastema occurs between the procumbent upper incisors and laterally splayed upper canines. When incisors occlude edge-against-edge, both lower incisors occlude with the upper central incisor. The lower canines of pitheciins bear a strong crest running lingually from the tip. The crest occludes against the mesial surface of the upper canine with the effect of a chisel. These modifications are an adaptation for splitting open, or gouging holes in, tough or hard-shelled fruits which form an important part of the pitheciin diet (Kinzey and Norconk, 1990; Van Roosmalen *et al.*, 1981).

The incisors of *Branisella* are known only from one specimen in which the crowns are broken off at their bases so that some of the cemento-enamel junction is still visible buccally (Fig. 3). The teeth are somewhat compressed mesiodistally, but not to the degree seen in gum-eating *Callithrix* or *Cebuella*. Compared with the size of the molars, the mesiodistal root diameters of i1 and i2 were smaller than those of platyrrhines such as *Aotus* or *Saguinus* and much smaller than those of *Callithrix* or *Cebuella* (Fig. 4). The incisor-to-molar size proportions of *Branisella* do not suggest that the front teeth of *Branisella* were adapted for harvesting gum by means of gouging bark.

There is practically no space between the incisors and canine roots of *Branisella* (Fig. 3). The canine of *Branisella* is not unusually large for a species of this body size and the canines would not have been especially useful for husking fruits.

Molars

Molar shearing and diet. The functional design of the cheek teeth of platyrrhines and other primates, especially of the molars, is selectively modified to best deal with the physical properties of the foods the species eats

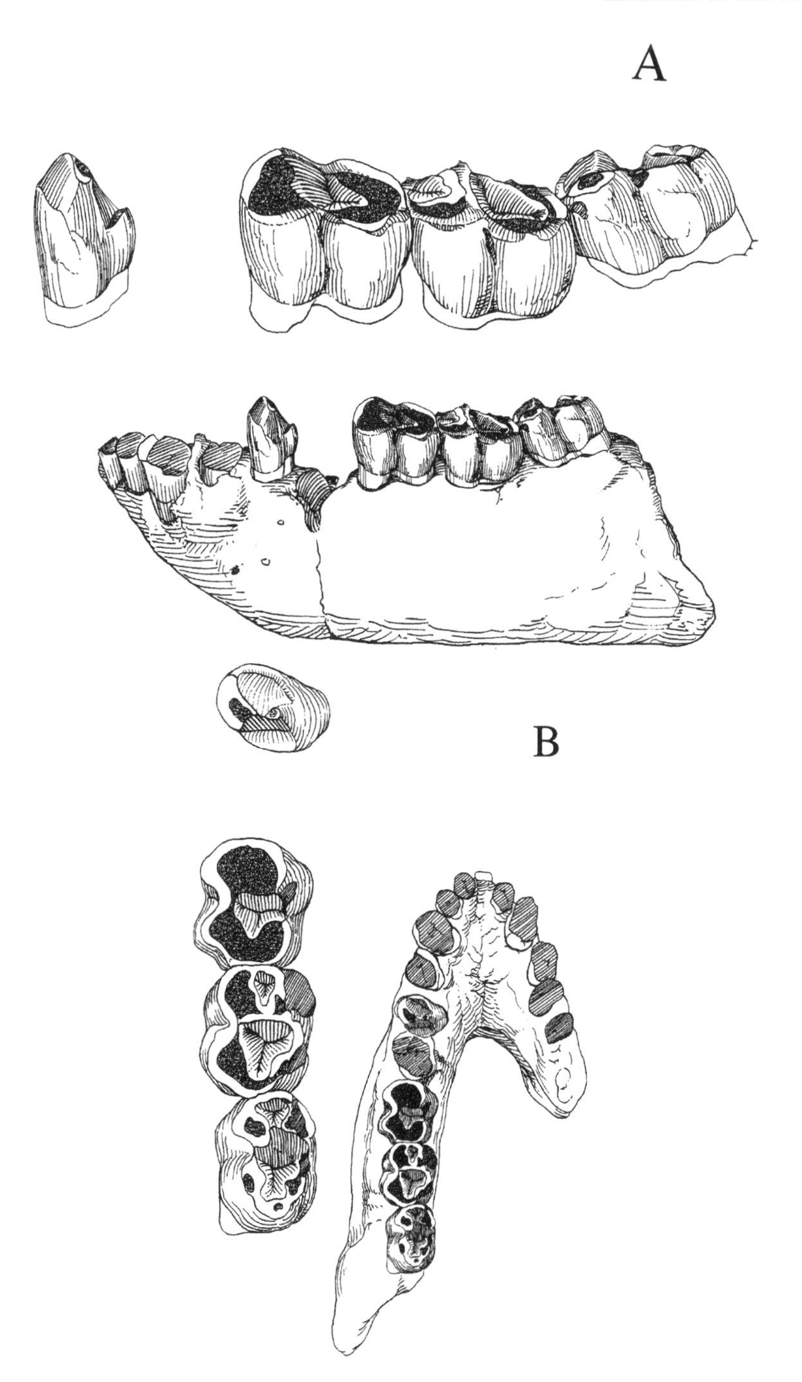

Fig. 3. Views of the mandible and dentition (p3 and m1-3) of *Branisella boliviana*, based on MNHN BOL-Y-4006, field number 96-101, from the *Branisella* "zone" at Salla: A, lateral view; B, occlusal view.

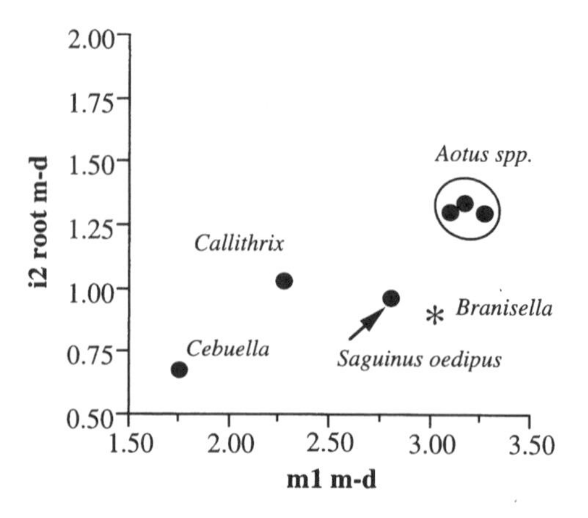

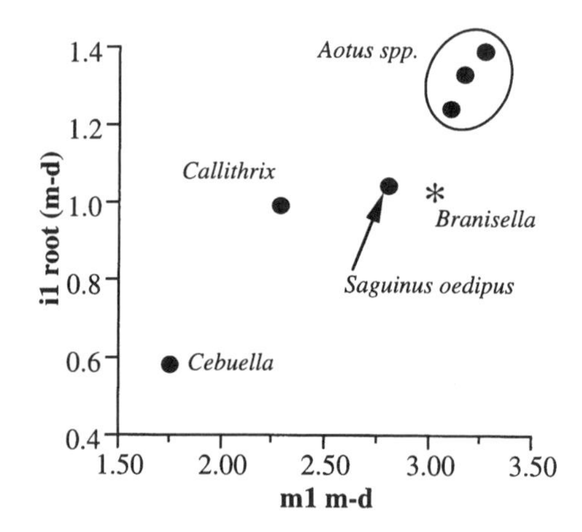

Fig. 4. Incisor size versus molar size in *Branisella* and selected extant small-bodied platyrrhines. Measurements are mesiodistal dimensions in millimeters.

(Anthony and Kay, 1993; Fleagle *et al.*, 1997; Kay and Covert, 1984; Meldrum and Kay, 1997; Rosenberger and Kinzey, 1976; Strait, 1991). These differences have been quantified by examining the relative development of the shearing crests on the molars after molar size is taken into account. Table 1 quantifies the development of molar shearing in living platyrrhines as a "shearing quotient." As calculated, the better developed the shearing crests, the larger the shearing quotient (see note in Table 1). In Fig. 5, the shearing quotient for living species is broken down by dietary preference. Species that eat considerable amounts of fibrous foods such as leaves high in cellulose (*Alouatta*, *Brachyteles*), or chitinous insects (*Saimiri*, *Callimico*), have well-developed shearing crests on the molars. In contrast, species that feed on less fibrous, soft fruits (*Ateles*, *Callicebus*), or tree gum (*Callithrix*, *Cebuella*) have relatively flatter teeth

Table I. Relative Development of Shearing Crests on m1 in Platyrrhines

Taxon	n	ml length	ml shear	Expected shear	Shearing quotient	Major dietary feature
Alouatta caraya	6	6.72	13.09	12.20	7.33	Leaves
Alouatta fusca	6	6.70	12.94	12.16	6.42	Leaves
Alouatta palliata	10	6.92	13.91	12.56	10.76	Leaves
Aotus trivirgatus	10	3.06	6.16	5.55	10.92	Fruit/Leaves
Ateles geoffroyi	10	5.26	9.31	9.55	−2.47	Fruit
Brachyteles arachnoides	9	7.22	15.19	13.10	15.93	Leaves
Branisella boliviana	2	3.07	5.72	5.57	2.66	—
Cacajao melanocephalus	2	3.97	5.90	7.20	−17.70	Seeds/Fruit
Callicebus moloch	10	3.18	5.50	5.77	−4.70	Fruit
Callimico goeldii	3	2.60	5.48	4.72	16.14	Insects
Callithrix argentata	4	2.22	4.08	4.03	1.27	Fruit/Gum
Cebuella pygmaea	4	1.78	3.26	3.23	0.92	Gum/Fruit
Cebus apella	5	4.79	7.71	8.69	−11.31	Fruit/Seeds
Chiropotes satanas	5	3.64	5.50	6.61	−15.53	Seeds/Fruit
Lagothrix lagotricha	8	5.47	10.12	9.93	1.94	Fruit/Leaves
Leontopithecus rosalia	5	3.09	5.62	5.61	0.22	Fruit/Insects
Pithecia monachus	4	4.00	6.78	7.26	−6.60	Fruit/Seeds
Saguinas mystax	5	2.52	4.03	4.57	−11.88	Fruit/Insects
Saimiri sciureus	5	2.87	5.54	5.21	6.36	Insects/Fruit

[a]The estimate of shearing development is based on measurements of six lower molar crests on m1 (see Kay, 1975, for anatomical details). A line with slope 1.0 was assigned to a bivariate cluster of the natural log of M_1 length ($\ln M_2L$) versus the natural log of the sum of the measured shearing crests (ln SH), and passing through the mean lnm2L and mean ln SH for extant taxa. The equation expressing this line is:

$$\ln SH = 1.0(\ln M_1L) + 0.596$$

For each taxon, the expected ln SH was calculated from this equation. The observed (measured) ln SH for each species was compared with the expected and expressed as a residual (shearing quotient, or SQ):

$$SQ = 100 \times (\text{observed} - \text{expected})/(\text{expected})$$

Data for extant species abstracted from Fleagle *et al.* (1997), Kay (1975), Kay (1977), and Meldrum and Kay (1997). Data for *Branisella* is based on MNHN BOL-3465 and MNHN BOL-3468.

with shorter, more rounded shearing crests. The teeth of species that specialize on eating hard seeds or in splitting open tough, hard fruits (pitheciins and *Cebus*) tend to resemble those of the soft-fruit eaters but may have thicker enamel (Kay, 1981) or specialized enamel microstructure (Martin *et al.*, 1994).

In the above assessment of shearing crest development, the potentially confounding effects of phylogeny have not been accounted for. The molar structure of each of the platyrrhine species examined above is not purely a response to the present selective demands placed upon it; it also owes its basic design to its own phylogenetic history. While the phylogenetic effect has been noted by many authors (Anthony and Kay, 1993; Kay and Ungar, 1997; Rosenberger and Kinzey, 1976; Smith, 1994; Strait, 1991; Szalay, 1981), there has been no systematic effort to separate out its effects. In an attempt to do so,

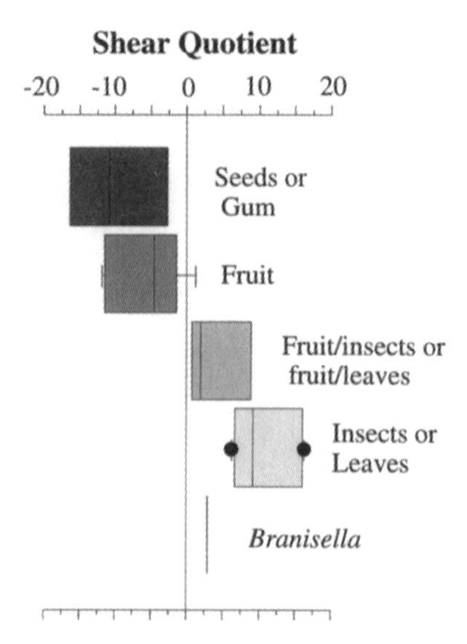

Fig. 5. The shearing quotient for living platyrrhine species and *Branisella* broken down by dietary preference (Box and whisker plot). Table I summarizes the data.

we used the CAIC program (Appendix). For the analysis, we utilize a phylogeny of platyrrhines illustrated in Fig. 6 (see caption for references). The behavioral data (diet) were organized into an ordered four-state array as follows: State 0 = gums or seeds, each of which is soft and/or very low in fiber; state 1 = fruits low in fiber; state 2 = fruit with insects or fruit with leaves (a diet with more fiber—chitin or cellulose); and state 3 = predominantly leaves or insects, high in structural carbohydrate (fiber).

The null hypothesis predicts that whenever there is a dietary shift along a lineage segment leading toward more dietary fiber, e.g., from state 1 to 2 or from 2 to 3, etc., there should be a corresponding increase in the size of the shearing quotient. Likewise, a shift in any lineage segment toward less fiber should give a corresponding decrease in the shearing quotient. Treating the four-state variable as though it were continuous, we find that the correlation between shearing change and diet shift is 0.70, so that 49% of the variance in shearing quotient is explained by a change in diet. More conservatively, because there is no reason why the difference between one pair of morphological and behavioral states should be of comparable magnitude to the difference between any other pair, (e.g., there is no reason to believe that a shift from state 1 to state 2 is equivalent in the magnitude of its effect to a shift from state 2 to state 3), two other contrast analyses were run—the Kendall Rank-Correlation test and the Spearman Rank-Correlation test. Each gives highly significant p-values of 0.004 and 0.006 further indicating that a change in diet along the lines described is strongly correlated with a change in shearing quotient. These two analyses provide powerful confirming evidence that the

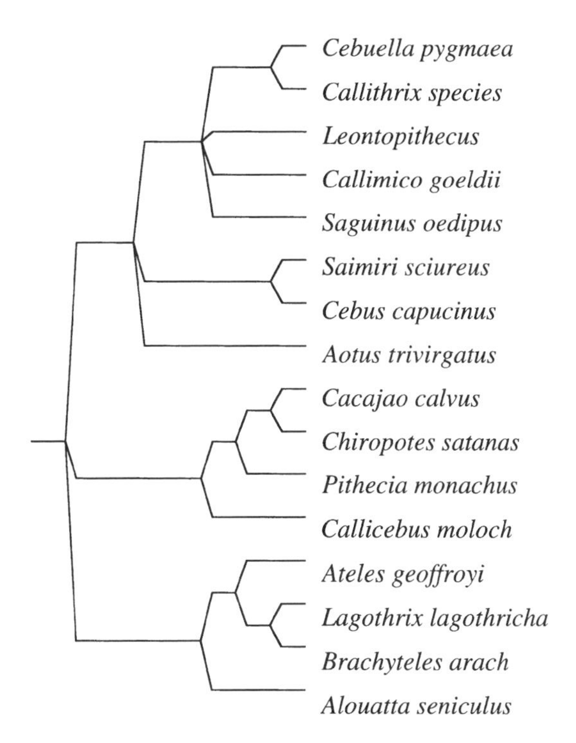

Fig. 6. A molecular phylogeny for extant Platyrrhini. After Barroso *et al.*, (1997) and Schneider *et al.*, (1993).

change in molar shearing quotient is driven by natural selection for dietary adaptation.

Two specimens of *Branisella* have relatively unworn lower first molars allowing unbiased estimates of shearing-crest length (Table 1). Unworn molars of *Branisella* have poorly developed shearing crests. As already noted (Fig. 5), quantitative estimates of shearing-crest development, expressed as "shearing quotients" or SQ, allow some dietary groups of living platyrrhines to be distinguished from one another. The SQ of *Branisella* is within the range of extant platyrrhine species that eat primarily fruit. Given a body size of about 700 to 800 grams, *Branisella* could well have supplemented its diet with substantial amounts of either insects or leaves as do the extant platyrrhines *Callicebus* and *Aotus*.

Molar Crown Height and Substrate Preference. A striking feature of *Branisella* is the exceptionally high molar crowns of its cheek teeth compared to any extant platyrrhines (Fig. 7). Having high molar crowns is associated with terrestriality, grass-eating, or both (Janis, 1988; Janis and Fortelius, 1988; Kay and Covert, 1984), but the effects of either diet or substrate preference are also confounded by phylogeny. For example, horses tend to have higher tooth crowns than do either grazing or browsing bovids (Janis, 1988; Janis and

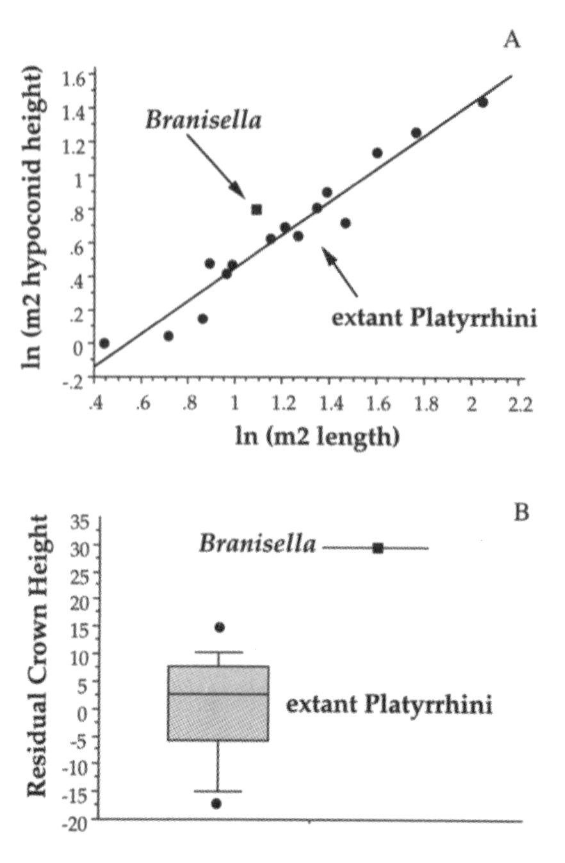

Fig. 7. Crown height of *Branisella boliviana* and 16 species of extant Platyrrhini (one species of each genus): A, log–log bivariate plot of crown height (hypoconid height) versus m1 length; B, box and whisker plot of residual crown height. For data in A, line is fit to the log hypoconid height versus log m1 length. The equation expressing this line is: ln m1 hypoconid height = 0.986 (ln m1 length) – 0.536. For each taxon, the expected ln hypoconid height was calculated from this equation. The observed (measured) ln hypoconid height for each species was compared with the expected and expressed as a residual crown height (RCH):RCH = 100 × (observed-expected)/(expected).

Fortelius, 1988). As with the diet analysis, we attempted to tease apart adaptive from phylogenetic effects using the CAIC program. Because no extant platyrrhine is terrestrial, we selected Old World monkeys (Cercopithecidae) as an adaptive array containing both arboreal and terrestrial foragers against which to test the hypothesis that crown height and terrestriality are correlated, with the effects of phylogeny removed. For the analysis, we utilize a phylogeny of Cercopithecidae illustrated in Fig. 8. The behavioral data on foraging patterns, enumerated in Table II, are organized into two states: arboreal or terrestrial.

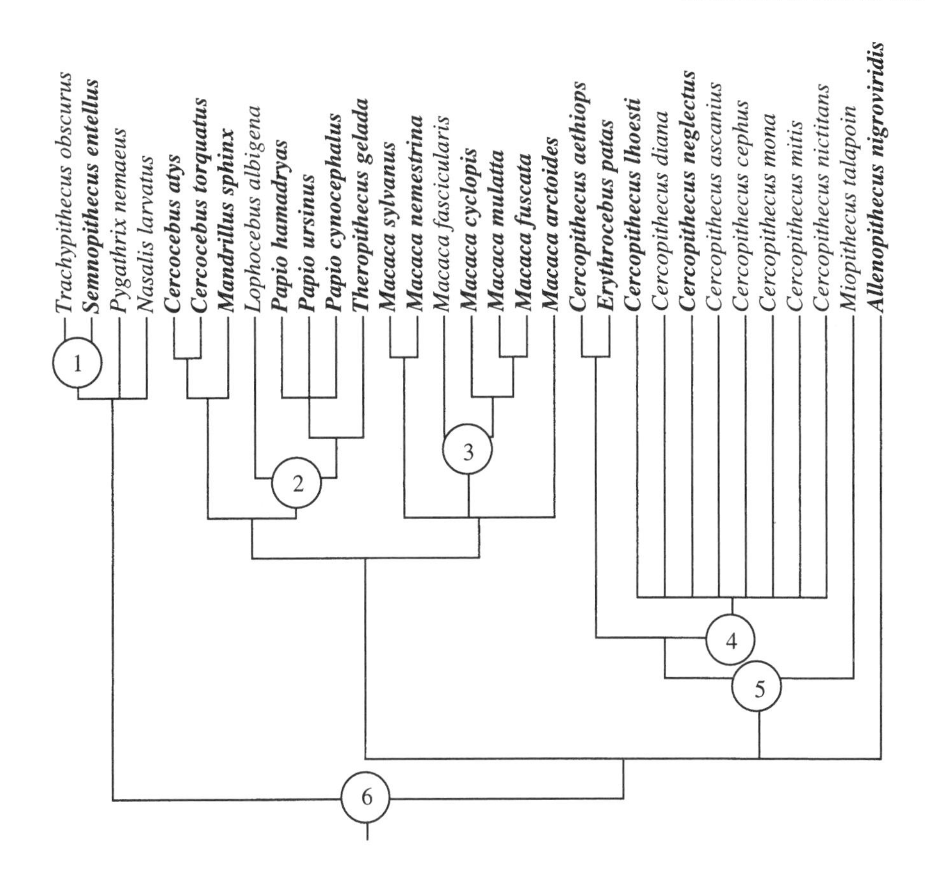

Fig. 8. A composite phylogeny for extant Cercopithecidae, after (Delson, 1980; Delson, 1988; Delson, 1994; Disotell, 1996). Taxa in bold lettering are considered to be terrestrial for the purposes of the analysis. Nodes identified by number are instances where a behavioral shift is presumed to have occurred.

If better qualitative data were available, we would have preferred a third state, semi-terrestrial, or we would have treated the behavioral variable as a continuous variable—percentage of time spent foraging on the ground versus in the trees. Relative crown height is calculated in a manner similar to that described for diet (see Table II for further details).

The null hypothesis predicts that, given a shift from arboreality to terrestriality, there should be a corresponding increase in relative crown height. Likewise, a shift in any lineage segment toward arboreality should give a corresponding decrease in relative crown height. In our data set, there are six transitions from terrestrial to arboreal (state 0 to state 1) (Fig. 8). The mean drop in residual height is −3.75%. When the effects of phylogeny are removed, arboreal species have significantly lower molar crowns than do terrestrial species (Student-t test statistic = −3.64 with 5 degrees of freedom,

Table II. Crown Height and Activity Patterns of Selected Cercopithecidae[a]

Taxon (sample size)	Mean m2 length	Mean hypoconid ht	Feeding level	log m2 length	log hypoconid height	Expected hyd ht	Crown height residual
Macaca sylvanus (16)	9.4	6.78	terrestrial	0.973	0.831	5.902	14.877
Macaca nemestrina (22)	8.5	6.5	terrestrial	0.929	0.813	5.415	20.041
Macaca fascicularis (47)	7.35	5.4	terrestrial	0.866	0.732	4.781	12.941
Macaca mulatta (21)	7.76	5.62	terrestrial	0.89	0.75	5.009	12.206
Macaca cyclopis (10)	7.63	5.49	terrestrial	0.883	0.74	4.937	11.207
Macaca esctoides (13)	9.54	7.71	terrestrial	0.98	0.887	5.977	28.991
Macaca fuscata (5)	8.94	6.69	terrestrial	0.951	0.825	5.654	18.326
Cercocebus albigena (35)	7.18	4.81	arboreal	0.856	0.682	4.686	2.637
Cercocebus torquatus (19)	8.22	6.69	terrestrial	0.915	0.825	5.262	27.144
Cercocebus atys (8)	8.88	7.01	terrestrial	0.948	0.846	5.621	24.703
Papio cynocephalus (37)	12.1	9.08	terrestrial	1.083	0.958	7.326	23.943
Papio ursinus (10)	12.78	8.64	terrestrial	1.107	0.937	7.677	12.544
Papio hamadryas (5)	12.53	9.58	terrestrial	1.098	0.981	7.548	26.917
Mandrillus sphinx (9)	12.42	9.1	terrestrial	1.094	0.959	7.491	21.471
Theropithecus gelada (8)	12.93	9.91	terrestrial	1.112	0.996	7.754	27.804
Cercopithecus aethiops (10)	6.65	4.64	terrestrial	0.823	0.667	4.389	5.726
Cercopithecus cephus (18)	5.76	4.04	arboreal	0.76	0.606	3.881	4.102
Cercopithecus diana (10)	6.41	4.14	arboreal	0.807	0.617	4.253	−2.651
Cercopithecus lhoesti (14)	6.58	4.55	terrestrial	0.818	0.658	4.349	4.619
Cercopithecus mitis (32)	6.47	4.15	arboreal	0.811	0.618	4.287	−3.191
Cercopithecus mona (30)	5.86	4.11	arboreal	0.768	0.614	3.938	4.357
Cercopithecus neglectus (26)	6.21	4.34	arboreal	0.793	0.637	4.139	4.858
Cercopithecus nictitans (34)	6.55	4.15	arboreal	0.816	0.618	4.332	−4.204
Cercopithecus ascanius (57)	5.59	3.59	arboreal	0.747	0.555	3.783	−5.09
Miopithecus talapoin (19)	4.04	2.86	arboreal	0.606	0.456	2.865	−0.16
Allenopithecus nigroviridis (16)	6.27	5.28	terrestrial	0.797	0.723	4.173	26.524
Erythrocebus patas (16)	7.74	5.42	terrestrial	0.889	0.734	4.998	8.452

[a] Measurements were taken only on unworn specimens; substrate preference from Rowe (1996).

p = 0.007). This is strong evidence that a change in relative crown height is an adaptation for substrate preference. It should be noted that we have not analyzed Cercopithecidae to determine whether the increased crown height is associated with a diet shift or terrestriality *per se*. Perhaps all terrestrial primates have higher crowns because they incorporate gritty items in their diets, or alternatively, perhaps terrestrial species eat more abrasive foods, such as high-silica grasses, leading to selection for higher crowns than their close relatives that eat less abrasive herbs or fruit. Such possibilities need to be tested with larger data sets, for example among rodents and artiodactyls.

Returning to *Branisella*, the crowns of the molars are very high, higher in fact than any platyrrhine, and in the range of Old World monkeys. A comparison of crown heights in Old World monkeys suggests an association with terrestriality, either because terrestrial species ingest more dietary silica or more exogenous grit, so *Branisella* may have been partly terrestrial in its feeding niche.

Molar Wear. The conclusion that *Branisella* may have fed on the ground, at least in part, gains further support from the exceptionally heavy rate of tooth wear that *Branisella* experienced in life. By the time the third molar was erupted, the first was nearly worn down to the gumline. Comparable gradients of tooth wear are common in terrestrial species (as observed in terrestrial Old World monkeys, for example), but not in arboreal platyrrhines. Table III presents a comparison of tooth wear in a sample of *Callicebus* and *Aotus*, two arboreal platyrrhines, and *Erythrocebus patas*, a terrestrial Old World monkey. As shown in Table IV, there are no significant differences in m3 wear among these samples. The arboreal *Callicebus* and *Aotus* do not differ significantly in m1 wear in these samples but each has significantly less m1 wear than terrestrial *Erythrocebus*. The rate of cheek-tooth wear in *Branisella* more closely resembles that of our terrestrial species in having an exceptionally high wear gradient. This again suggests that *Branisella* may have been terrestrial.

Mandibular Morphology. The mandible of *Branisella* is quite shallow and broad and the rami are divergent from a narrow symphysis (v-shaped) (Fig. 3). The mandibular symphysis of *Branisella* is stoutly fused and antero-posteriorly elongate in cross-sectional outline. Its cross-sectional profile features a long

Table III. Lower First Molar Wear in Samples of Anthropoid Taxa Matched for m3 Wear[a]

Taxon	m1 wear ratio	m3 wear ratio	Foraging pattern
Branisella boliviana (n = 2)	34.72 (9.72)	2.11 (2.12)	?
Callicebus sp. (n = 3)	1.33 (0.08)	1.43 (0.47)	Arboreal
Aotus trivirgatus (n = 5)	1.80 (0.69)	1.00 (0.59)	Arboreal
Erythrocebus patas (n = 6)	12.29 (2.20)	0.74 (0.13)	Terrestrial

[a]Wear is expressed as a percentage of exposed dentine to the outline area of the tooth. Value in parentheses is the standard error.

Table IV. Wear Data for *Branisella* Molars and Selected Platyrrhines and Catarrhines[a]

	Aotus sp	*Callicebus* sp	*Branisella*	*Erythrocebus patas*
Aotus sp	—	0.63	**0.001**	**0.002**
Callicebus sp	0.64	—	**0.02**	**0.01**
Branisella	0.49	0.71	—	**0.01**
Erythrocebus patas	0.65	0.10	0.22	—

[a]Student's t tests compare wear on samples of m1 wear and m3 wear. Values below the diagonal are the significance levels of the differences between means of m3 wear. Values above the diagonal are significance levels of the difference between the means of m1 wear.

alveolar plane and a deep genioglossal pit with stout superior and inferior transverse tori. This arrangement would have allowed the efficient transfer of muscle forces from the balancing to the working side of the mandible. The cross-sectional shape is reminiscent of that of the callitrichine *Callithrix*. However, the symphyseal pattern of the two would appear to result from very different sets of functional circumstances—powerful biting in the former and large incisors in the latter. *Branisella* has small incisor roots and more bone in its symphysis, perhaps in response to selection for efficiently transferring muscular forces from balancing to working side for powerful mastication between the rapidly wearing cheek teeth. In contrast, in *Callithrix*, symphyseal shape and size seem to be conditioned by the large size of the procumbent incisor roots which fill the symphyseal space. The cheek teeth of *Callithrix* are small, with the third molar lost, implying that large bite forces between the molars are not commonly engendered.

Summary and Conclusions

In this chapter, we have presented a brief review of the meanings of the term "adaptation" and of the ways in which we can recognize adaptations in the fossil record. We explore the adaptive context of two dental features in living anthropoids—molar crest development and crown height. With the effects of phylogeny removed, we show that the extent of development of molar shearing is naturally selected for by diet and that high-crowned molars are associated with terrestriality.

Branisella boliviana, the oldest known platyrrhine primate, now known from considerable facial, dental, and mandibular material (Kay and Williams, 1995; Takai and Anaya, 1996) is used as an example to demonstrate the reconstruction of a behavioral profile of an extinct animal. Paleoenvironmental indicators (low primate species richness, the prevalence of mammals with high-crowned cheek teeth, paleosols with carbonate nodules) suggest that

Branisella lived in a seasonally dry open woodland. It was quite a small monkey, about the size of the living tamarin *Leontopithecus*. It appears to have had small orbits, implying a diurnal activity preference.

The poorly developed molar shearing on the molars of *Branisella* is readily explained by analogy with living platyrrine species and invariably associated with diets such as fruit or gum that are low in structural fiber. However, *Branisella* departs from all other known platyrrines living or fossil, in having a combination of high molar crowns, and extremely rapid tooth wear, coupled with its weakly-developed molar-crown crests. High-crowned molars that were rapidly worn down in life is generally seen in living species having an abrasive diet, especially one that includes tree or grass leaves. Opel phytoliths (abrasive silica) are more abundant in leaves, particularly in grass leaves, but less prevalent in fleshy fruits, for example.

Branisella's combination of molar features is found only in one group of primates, the Old World monkeys, and it is here that we seek an adaptive explanation. Terrestrially-feeding Old World monkeys have a considerable range of molar shearing development with the more frugivorous species having less molar shearing than the more folivorous species. But terrestrial OW monkey species have higher molar crowns than their arboreal relatives, and this is the case irrespective of dietary preference: terrestrial folivorous species such as *Semnopithecus entellus* have high crowns as do terrestrial frugivorous species such as *Lophocebus albigena*. Therefore, we propose to resolve this apparent conflict—crown height and wear pointing to an abrasive, fibrous diet but shearing suggesting a low-fiber diet—by proposing that *Branisella* was feeding on or close to the ground where the incorporation of exogenous grit would be likely, even if it were eating fruit.

Beyond the general agreement that the last common ancestor of platyrrhines was diurnal (e.g., Clark, 1971; Martin, 1990), only Ford (1986, 1988) and Rosenberger (1992) have attempted to reconstruct the pattern of evolution of platyrrhine behavior. How does our reconstruction of the behavior of *Branisella* stand in relation to those efforts? Ford analyzed the morphology of living platyrrhines and reconstructed the locomotor behavior of their last common ancestor. She argued that the last common ancestor of extant platyrrhines was about the size of an owl monkey (*Aotus*). It was an arboreal monkey that progressed quadrupedally, emphasizing branch-running and walking with infrequent leaping or climbing. Rosenberger reconstructs the ancestor of extant platyrrhines to have been frugivorous. The important point to be made about these reconstructions is that it is the last common ancestor of the *crown* group that is being reconstructed and not necessarily the morphology and behavior of the *stem* New World monkey. A crown group contains all the living platyrrhines, their last common ancestor (LCA), and any extinct species the lineage of which can be traced back to that LCA (Williams and Kay, 1995). *Branisella* falls *outside* the crown group of platyrrhines, but is still more closely related to extant platyrrhines than to the Catarrhini. This interpretation is consistent with its age, between 25.8 and 27.0 Ma, placing it several million years earlier than the presumed age of the platyrrhine crown LCA. Therefore, *Branisella* need not be representative of the basal adaptations of crown platyrrhines. Instead, we envision the possibility that New World monkeys may have undergone a substantial and undocumented (except for *Branisella*) adaptive radiation in South America before the close of the Oligocene, to be replaced by the present adaptive radiation beginning about 22–25 million years ago. Available evidence does not permit us to speculate about the niche breadth or temporal depth of this radiation.

ACKNOWLEDGMENTS

This work was supported by an NSF grant to R.F. Kay, R. H. Madden, and B. J. MacFadden. Ms. Marsha Dohrman drew Figure 3. We thank Eric Delson, Callum Ross, Susan Williams, and an anonymous reviewer for helpful comments.

Appendix: Testing Adaptive Hypotheses Using CAIC

This computer program distributed by Purves and Rambaut (1995) is based on the "independent contrasts" approach of Felsenstein (1985). Our description of the results of analyses using CAIC follows and paraphrases that of Purves and Rambaut. The program can be downloaded from the website http://evolve.zps.ox.ac.uk/. It requires three inputs: (1) a set of taxa for which the phylogenetic relationships have been worked out; (2) for each taxon, a continuous or dichotomous variable expressing a behavioral attribute; and (3) a morphological character expressed as a continuous variable. The first of these requirements may be relaxed to the extent that the program accommodates partially resolved (polytomous) nodes. The data expressing the behavioral attribute can have more than two states provided that the relationship among the three (or more) states is ordered. An example of a three-state dietary variable could be 0 = mostly frugivorous, 1 = mixed frugivorous/folivorous, 2 = mostly folivorous.

The output of the program is a reconstruction of the cladogenetic nodes at which a behavioral change may have occurred and the direction and amount of change at that node. In a hypothetical example (Fig. 9A), we wish to investigate whether there is a relationship between molar crown height and substrate preference. In our example, nine taxa (A to I), of which five are categorized as arboreal and four as terrestrial, are arranged in a fully resolved phylogeny. For each taxon we assign a hypothetical molar crown height by a numerical value ranging from 0 to 12. The example is arranged so that the crown heights of the terrestrial species are consistently higher than those of their closest relatives.

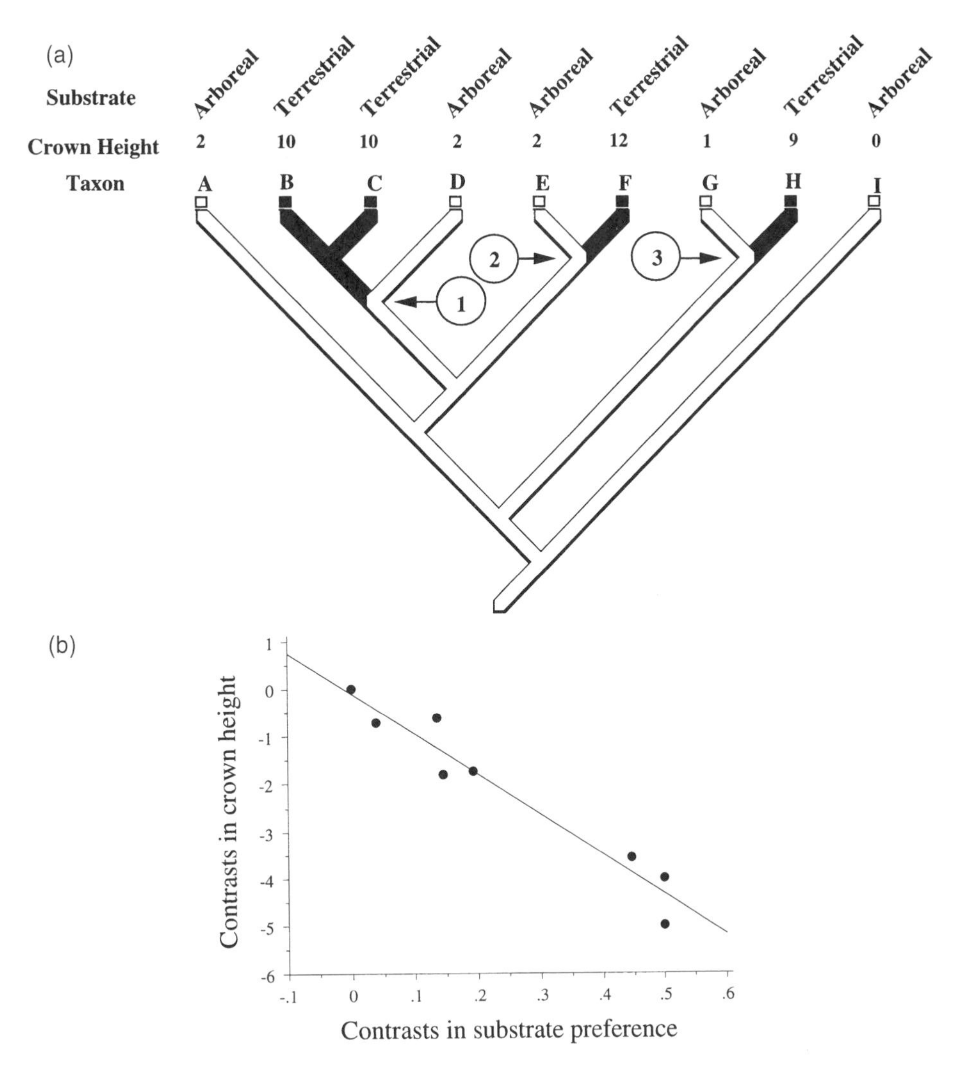

Fig. 9. (A) Nine hypothetical taxa (A to I), of which five are arboreal and four terrestrial, are arranged into the fully resolved phylogeny. Numbers above the taxon names signify crown height. Arrows with numbers indicate three places where a behavioral shift has occurred. (B) Bivariate plot of reconstructed contrasts at various nodes of the phylogenetic tree in (A) versus the reconstructed crown height at each of these eight nodes in the phylogeny.

We analyze this hypothetical data set in two ways—with the behavioral variable treated as either a continuous or discrete variable. When CAIC is used with continuous variables, the output consists of a set of "contrasts." Contrasts in the predictor variable (crown height) are always equal to or greater than zero. If the dependent "contrast" variable (substrate preference) is also positive, this means that the variations in the two are going in the same direction. In the example, whenever no change occurs in substrate preference, there is also no change in crown height, but when there is a change at the node from terrestrial to arboreal, there is a corresponding decrease in crown height. The strength of the relationship between the variables is assessed by means of linear regression through the origin. Figure 9B is a bivariate plot of the substrate contrasts vs. the crown height contrasts with the regression passing through zero. The value of the slope (-8.49) equals the true relation between the variables in the absence of phylogenetic effects (Pagel, 1993). Thus, independent of phylogenetic effects, the crown height of terrestrial species should be 8.49 and that of arboreal species should be 0. In the hypothetical example, 94% of the variance in crown height is "explained" by the arboreal/terrestrial contrast.

If, alternatively, we treat the substrate variable as a two-state character, at three nodes (indicated by arrows in Fig. 9A), a positive contrast means that the substrate shift from arboreal to terrestrial is varying in the same direction as the predictor variable (crown height). Under the null hypothesis, that evolution in the crown height is *not* linked in any way to the evolution of the arboreal to terrestrial shift, we should expect half the contrasts in crown height to be positive and half negative, and the mean value of the contrasts to be zero. We can test the null hypothesis using a t-test on the mean of the contrasts or with a randomization test. In this case, the mean value for the contrast (-4.19) yields a t-value of -9.94 with $p<0.0005$, highly significant.

In one of our analyses in the text, the behavioral variable takes four ordered states, corresponding to proposed dietary changes that reflect an increase in dietary fiber. Because there is no reason why the difference between one pair of states should be of comparable magnitude to the difference between any other pair, the discrete contrast analysis utilizes a sign-rank test rather than a t-test.

References

Anthony, M. R. L., and Kay, R. F. 1993. Tooth form and diet in ateline and alouattine primates: reflections on the comparative method. *Am. J. Sci.* **283A:**356–382.

Aslan, A. 1994. *Holocene Sedimentation, Soil Formation, and Floodplain Evolution of the Mississippi River Floodplain, Ferriday, Louisiana.* Ph.D. thesis. University of Colorado, Boulder, Colorado.

Barroso, C. M. L., Schneider, H., Schneider, M. P. C., Sampaio, I., Harada, H. L., Czelusniak, J., and Goodman, M. 1997. Update on the phylogenetic systematics of New World monkeys: Further DNA evidence for placing pygmy marmoset (*Cebuella*) within the genus *Callithrix*. *Int. J. Primatol.* **18:**651–674.

Benjamin, M. T., Johnson, N. M., and Naeser, C. W. 1987. Recent rapid uplift in the Bolivian Andes: evidence from fission-track dating. *Geology* **15:**680–683.

Bock, W. J., and von Wahlert, G. 1965. Adaptation and the form-function complex. *Evolution* **19:**269–299.

Brandon, R. 1987. *Adaptation*. Princeton University Press, Princeton, NJ.

Clark, W. E. Le Gros 1971. *The Antecedents of Man*, Third Edition. Edinburgh University Press, Edinburgh.

Clutton-Brock, T. H., and Harvey, P. H. 1979. Comparison and adaptation. *Proc. R. Soc. London, Ser. B* **205:**547–565.

Coddington, J. A. 1988. Cladistic tests of adaptational hypotheses. *Cladistics* **4:**3–22.

Coimbra-Filho, A. F., and Mittermeier, R. A. 1976. Exudate eating and tree gouging in marmosets. *Nature* **262:**630.

Darwin, C. 1859. *The Origin of Species*, Sixth Edition (1872). Reprinted by D. Appleton and Company, London.

Delson, E. 1980. Fossil macaques, phyletic relationships and a scenario of deployment. In: D. G. Lindburg (eds.), *The Macaques: Studies in Ecology*, pp. 10–30. Van Nostrand, New York.

Delson, E. 1988. Cercopithecidae. In: I. Tattersall, E. Delson, and J. Van Couvering (eds.), *Encyclopedia of Human Evolution and Prehistory*. pp. 119–121. Garland, New York.

Delson, E. 1994. Evolutionary history of the colobine monkeys in paleoenvironmental perspective. In: A. G. Davies and J. F. Oates (eds.), *Colobine Monkeys*, pp. 11–43. Cambridge University Press, New York.

Disotell, T. R. 1996. The phylogeny of Old World monkeys. *Evol. Anthropol.* **5:**18–24.

Eaglen, R. H. 1984. Incisor size and diet revisited: the view from a platyrrhine perspective. *Am. J. Phys. Anthropol.* **64:**263–275.

Eisenberg, J. F., O'Connell, M. A., and August, P. V. 1979. Density, productivity, and distribution of mammals in two Venezuelan habitats. In: J. F. Eisenberg (ed.), *Vertebrate Ecology in the Northern Neotropics*. pp. 187–207. Smithsonian Institution Press, Washington, D.C.

Farrell, K. M. 1987. Sedimentology and facies architecture of overbank deposits of the Mississippi River, False River region, Louisiana. In: F. G. Ethridge, R. M. Flores, and M. D. Harvey (eds.), *Fluvial Sedimentology*, pp. 1111–129. Soc. Econ. Paleontol. Mineral.

Felsenstein, J. 1985. Phylogenies and the comparative method. *Am. Nat.* **125:**1–15.

Fleagle, J. G., Kay, R. F., and Anthony, M. R. L. 1997. Fossil New World monkeys. In: R. F. Kay, R. H. Madden, R. L. Cifelli, and J. J. Flynn (eds.), *Vertebrate Paleontology in the Neotropics*. pp. 473–495. Smithsonian Institution Press, Washington, D.C.

Ford, S. M. 1986 Systematics of the New World monkeys. In: D. R. Swindler and J. Erwin (eds.), *Comparative Primate Biology, I, Systematics, Evolution, and Anatomy*, pp. 73–135. Alan R. Liss, New York.

Ford, S. M. 1988. Postcranial adaptations of the earliest platyrrhine. *J. Hum. Evol.* **17:**155–192.

Ganzhorn, J., Malcomber, S., Andrianantoanina, O., and Goodman, S. M. 1997. Habitat characteristics and lemur species richness in Madagascar. *Biotropica* **29:**331–343.

Gould, S. J., and Vrba, E. S. 1982. Exaptation—a missing term in the science of form. *Paleobiology* **8:**4–15.

Gregory, W. K. 1922. *The Origin and Evolution of the Human Dentition*. Williams and Wilkins, New York.

Hershkovitz, P. 1977. *New World Monkeys (Platyrrhini)*, Vol. I. University of Chicago Press, Chicago.

Hitz, R. B. 1997. *Contributions to South American Mammalian Paleontology: New Interatheres (Notoungulata) from Chile and Bolivia, Typothere (Notoungulata) Phylogeny, and Paleosols from the Late Oligocene Salla Beds*. Ph.D. dissertation. University of California, Santa Barbara.

Hoffstetter, R. 1969. Un primate de l'Oligocène inferieur Sud-Amèricain: *Branisella boliviana* gen. et sp. nov. *C. R. Acad. Sci., Paris* **269:**434–437.

Hoffstetter, R. 1980. *Los monos platirrinos (Primates): origen, extension, filogenia, taxonomia*. Actas II Congreso Argentino de Paleontologïa y Bioestratigrafïa y I Congreso Latinoamericano de Paleontologïa, Buenos Aires, 1978.

Hylander, W. L. 1975. Incisor size and diet in anthropoids with special reference to Cercopithecidae. *Science* **189:**1095–1098.

Jacobs, G. H., Deegan, J. F. I., Neitz, J., Crognale, M. A., and Neitz, M. 1993. Photopigments and color vision in the nocturnal monkey, *Aotus*. *Vision Res.* **33:**1773–1783.

Janis, C. M. 1988. An estimation of tooth volume and hypsodonty indices in ungulate mammals, and the correlation of these factors with dietary preference. *Mém. Mus. Nat. Hist. Nat. Paris (Sér. C)* **53:**367–387.

Janis, C. M., and Fortelius, M. 1988. On the means whereby mammals achieve increased functional durability of their dentitions, with special reference to limiting factors. *Biol. Rev.* **63:**197–230.

Kay, R. F. 1975. The functional adaptations of primate molar teeth. *Am. J. Phys. Anthropol.* **43:**195–216.

Kay, R. F. 1977. The evolution of molar occlusion in Cercopithecidae and early catarrhines. *Am. J. Phys. Anthropol.* **46:**327–352.

Kay, R. F. 1981. The Nut-crackers: a new theory of the adaptations of the Ramapithecines. *Am. J. Phys. Anthropol.* **55:**141–151.

Kay, R. F. 1990. The phyletic relationships of extant and fossil Pitheciinae (Platyrrhini, Anthropoidea). *J. Hum. Evol.* **19:**175–208.

Kay, R. F., and Cartmill, M. 1977. Cranial morphology and adaptations of *Palaechthon nacimienti* and other Paromomyidae (Plesiadapoidea, ?Primates), with a description of a new genus and species. *J. Hum. Evol.* **6:**19–53.

Kay, R. F., and Covert, H. 1984. Anatomy and the behavior of extinct primates. In: D. Chivers, B. A. Wood, and A. Bilsborough (eds.), *Food Acquisition and Processing in Primates*. pp. 467–508. Plenum Press, New York.

Kay, R. F., and Hylander, W. L. 1978. The dental structure of mammalian folivores with special reference to primates and Phalangeroidea (Marsupialia). In: G. G. Montgomery (ed.), *The Ecology of Arboreal Folivores*. pp. 173–191. Smithsonian Institution Press, Washington, D.C.

Kay, R. F., and Sheine, W. S. 1979. On the relationship between chitin particle size and digestibility in the primate *Galago senegalensis*. *Am. J. Phys. Anthropol.* **50:**301–308.

Kay, R. F., and Ungar, P. 1997. Dental evidence for diet in some Miocene catarrhines with comments on the effects of phylogeny on the interpretation of adaptation. In: D. R. Begun, C. Ward, and M. Rose (eds.), *Function, Phylogeny and Fossils: Miocene Hominoids and Great Ape and Human Origins*, pp. 131–151. Plenum Press, New York.

Kay, R. F., and Williams, B. A. 1995. Recent finds of monkeys of monkeys from the Oligocene/Miocene of Salla, Bolivia. *Am. J. Phys. Anthropol.* **Suppl. 20:**124.

Kay, R. F., Madden, R. H., Van Schaik, C., and Higdon, D. 1997. Primate species richness is determined by plant productivity: implications for conservation. *Proc. Natl. Acad. Sci. USA.* **94:**13023–13027.

Kay, R. F., MacFadden, B. J.,Madden, R. H., Sandeman, H., and Anaya, F. 1998a. Revised age of the Salla beds, Bolivia, and its bearing on the age of the Deseadan South American Land Mammal 'Age'. *J. Vert. Paleontol.* **18:**189–199.

Kay, R. F., Johnson, D. J., and Meldrum, D. J. 1998b. A new pitheciin primate from the middle Miocene of Argentina. *Am. J. Primatol.* **45:**317–336.

Kinzey, W. G., and Norconk, M. A. 1990. Hardness as a basis of fruit choice in two sympatric primates. *Am. J. Phys. Anthropol.* **81:**5–15.

Kraus, M., and Aslan, A. 1993. Eocene hydromorphic paleosols: significance for interpreting ancient floodplain processes. *J. Sediment. Petrol.* **63:**453–463.

MacFadden, B. J. 1990. Chronology of Cenozoic primate localities in South America. *J. Hum. Evol.* **19:**7–22.

Martin, L. C., Kinzey, W. G., and Maas, M. C. 1994. Enamel thickness in pitheciine primates. *Am. J. Phys. Anthropol.* **Suppl. 18:**138.

Martin, R. D. 1990. *Primate Origins and Evolution: A Phylogenetic Reconstruction.* Chapman and Hall, London.

Meldrum, D. J., and Kay, R. F. 1997. *Nuciruptor rubricae*, a new pitheciin seed predator from the Miocene of Colombia. *Am. J. Phys. Anthropol.* **102:**407–427.

Pagel, M. 1993. Seeking the evolutionary regression coefficient: an analysis of what comparative methods measure. *J. Theor. Biol.* **164:**191–205.

Paley, W. 1854. *Natural Theology*. American Tract Society, New York.

Purvis, A., and Rambaut, A. 1995. Comparative analysis by independent contrasts (CAIC): an Apple Macintosh application for analysing comparative data. *Comput. Appl. Biosci.* **11:**247–251.

Reed, K., and Fleagle, J. G. 1995. Geographic and climatic control of primate diversity? *Proc. Natl. Acad. USA* **92:**7874–7876.

Reeve, H. K., and Sherman, P. W. 1993. Adaptation and the goals of evolutionary research. *Quart. Rev. Biol.* **68:**1–32.

Rosenberger, A. L. 1978. Loss of incisor enamel in marmosets. *J. Mammal.* **59:**207–208.

Rosenberger, A. L. 1979. *Phylogeny, Evolution and Classification of New World Monkeys (Platyrrhini, Primates).* Ph.D. Dissertation. City University of New York.

Rosenberger, A. L. 1981a. A mandible of *Branisella boliviana* (Platyrrhini, Primates) from the Oligocene of South America. *Int. J. Primatol.* **2:**1–17.

Rosenberger, A. L. 1981b. Systematics: The higher taxa. In: A. F. Coimbra-Filhoo and R. A. Mittermeier (eds.), *Ecology and Behavior of Neotropical Primates, Vol. 1*, pp. 9–27. Academia Brasileira de Ciências, Rio de Janeiro.

Rosenberger, A. L. 1992. The evolution of feeding niches in New World monkeys. *Am. J. Phys. Anthropol.* **88:**525–562.

Rosenberger, A. L., and Kinzey, W. G. 1976. Functional patterns of molar occlusion in platyrrhine primates. *Am. J. Phys. Anthropol.* **45:**261–298.

Rosenberger, A. L., Setoguchi, T., and Shigehara, N. 1990. The fossil record of callitrichine primates. *J. Hum. Evol.* **19:**209–236.

Rosenberger, A. L., Hartwig, W. C., and Wolff, R. C. 1991. *Szalatavus attricuspis*, an early platyrrhine primate. *Folia Primatol.* **56:**225–233.

Rowe, N. 1996. *The Pictorial Guide to the Living Primates*. Pogonias Press, East Hampton, NY.

Sánchez-Villagra, M., and Kay, R. F. 1997. A skull of *Proargyrolagus*, the oldest argyrolagid (late Oligocene Salla Beds, Bolivia), with brief comments concerning its paleobiology. *J. Vert. Paleontol.* **17:**717–724.

Sarich, V. M., and Cronin, J. E. 1976. Molecular systematics of the primates. In: M. Goodman and R. E. Tashian (eds.), *Molecular Anthropology*, pp. 139–157. Plenum Press, New York.

Schneider, H., Schneidder, M. P. C., Sampiao, I., Harada, M. L. Stanhope, M., Czelusniak, J., and Goodman, M. 1993. Molecular phylogeny of the New World monkeys (Platyrrhini, Primates). *Mol. Phylog. Evol.* **2:**225–242.

Sheine, W. S., and Kay, R. F. 1977. An analysis of chewed food particle size and its relationship to molar structure in the primates *Cheirogaleus medius* and *Galago senegalensis*, and the insectivoran *Tupaia glis*. *Am. J. Phys. Anthropol.* **47:**15–20.

Sheine, W. S., and Kay, R. F. 1982. A model for comparison of masticatory effectiveness in primates. *J. Morphol.* **172:**139–149.

Shockey, B. J. 1997. Two new notoungulates (Notohippidae) from the Salla beds of Bolivia (Deseadan, late Oligocene): systematics and functional morphology. *J. Vert. Paleontol.* **17:**584–599.

Simpson, G. G. 1933. The "Plagiiaulacoid" type of mammalian dentition. *J. Mammal.* **14:** 97–107.

Smith, R. J. 1994. Degrees of freedom in Interspecific allometry: an adjustment for the effects of phylogenetic constraint. *Am. J. Phys. Anthropol.* **93:**95–107.

Strait, S. G. 1991. *Dietary Reconstruction in Small-Bodied Fossil Primates*. Ph. D. dissertation. State University of New York, Stony Brook.

Szalay, F. S. 1981. Phylogeny and the problem of adaptive significance: the case of the earliest primates. *Folia Primatol.* **36:**157–182.

Szalay, F. S., and Delson, E. 1979. *Evolutionary History of the Primates*. Academic Press, London.

Takai, M., and Anaya, F. 1996. New specimens of the oldest fossil platyrrhine, *Branisella boliviana* from Salla, Bolivia. *Am. J. Phys. Anthropol.* **99:**301–318.

Tarling, D. H. 1980. The geologic evolution of South America with special reference to the last 200 million years. In: R. L. Ciochon and B. Chiarelli (eds.), *Evolutionary Biology of the New World Monkeys and Continental Drift*, pp. 1–42. Plenum Press, New York.

Ungar, P. 1994. Patterns of ingestive behavior and anterior tooth use differences in sympatric anthropoid primates. *Am. J. Phys. Anthropol.* **94:**339–363.

Ungar, P. 1998. Dental allometry, morphology, and wear as evidence for diet in fossil primates. *Evol. Anthropol.* **6:**205–217.

Van Roosmalen, G. M., Mittermeier, R. A., and Milton, K. 1981. The bearded sakis, genus *Chiropotes*. In: A. F. Ciombra-Filho and R. A. Mittermeier (eds.), *Ecology and Behavior of Neotropical Primates, Vol. 1,* pp. 419–442. Academia Brasileira de Ciências, Rio de Janeiro.

Williams, B. A., and Kay, R. F. 1995. The taxon Anthropoidea and the crown clade concept. *Evol. Anthropol.* **3:**188–190.

Williams, G. C. 1966. *Adaptation and Natural Selection*. Princeton University Press, Princeton, NJ.

Wolff, F. G. 1984. New specimens of the primate *Branisella boliviana* from the early Oligocene of Salla, Bolivia. *J. Vert. Paleontol.* **4:**570–574.

Ecomorphology and Behavior of Giant Extinct Lemurs from Madagascar

10

WILLIAM L. JUNGERS, LAURIE R. GODFREY, ELWYN L. SIMONS, ROSHNA E. WUNDERLICH, BRIAN G. RICHMOND, and PRITHIJIT S. CHATRATH

Introduction

Inferring the behavior of extinct organisms is a formidable task, even under the best of circumstances (Rudwick, 1964; Stern and Susman, 1983; Kay, 1984; Thomason, 1995). Nevertheless, and in spite of inevitable complications and limitations, such inferences remain the ultimate goal of paleobiologists if we are to understand fossils as integrated organisms rather than isolated bones and atomized character states. In this chapter we attempt to breathe life back into the osteological remains of recently extinct (or "subfossil") prosimian

---

WILLIAM L. JUNGERS • Department of Anatomical Sciences, Health Sciences Center, State University of New York at Stony Brook, New York 11794–8081. LAURIE R. GODFREY • Department of Anthropology, University of Massachusetts, Amherst, Massachusetts 01003–9278. ELWYN L. SIMONS • Duke University Primate Center, Durham, North Carolina 27705. ROSHNA E. WUNDERLICH • Department of Biology, James Madison University, Harrisburg, Virginia 22807. BRIAN G. RICHMOND • Department of Anthropology, George Washington University, Washington DC 20052; Present address: Department of Anthropology, University of Illinois, Urbana, Illinois 61801. PRITHIJIT S. CHATRATH • Duke University Primate Center, Durham, North Carolina 27705.

Reconstructing Behavior in the Primate Fossil Record, edited by Plavcan *et al.* Kluwer Academic/Plenum Publishers, New York, 2002.

primates from the Quaternary of Madagascar. Subfossil lemurs provide many special opportunities to the optimistic functional morphologist, but they also present their own unusual set of complications and potential frustrations. Approximately one-third of Madagascar's known primate species were driven to extinction in the late Holocene by the lethal interaction of aridification and human colonization (Burney, 1997; Dewar, 1997; Simons, 1997), including all taxa of large body size (>9 kg). Two new extinct species from northern Madagascar (*Babakotia radofilai* and *Mesopropithecus dolichobrachion*) have been discovered and described in the last decade (Godfrey *et al.*, 1990; Simons *et al.*, 1995), and a third new species from the northwest will be diagnosed soon (Jungers *et al.*, in prep.). Sixteen currently recognized subfossil species of Malagasy primates are represented in museum collections, most by numerous individuals, including a growing tally of specimens with associated craniodental and postcranial elements (e.g., MacPhee *et al.*, 1984; Simons *et al.*, 1992, 1995; Wunderlich *et al.*, 1996). Table I summarizes the current taxonomy of the extinct lemurs. Aspects of morphology suggest that cheirogaleids are more closely related to galagos and lorises than to other Malagasy primates (Szalay and Katz, 1973; Cartmill, 1975; Schwartz and Tattersall, 1985; Yoder, 1992). Molecular results, as well as "total evidence" analyses that combine morphological and molecular data, argue instead that the Malagasy primates are probably monophyletic (Yoder, 1994, 1996). Regardless of the placement of the cheirogaleids within strepsirrhines, the precise relationships among the various ancient clades of Malagasy primates remain somewhat fuzzy, even from a biomolecular perspective (Yoder, 1997; Yoder *et al.*, 1999).

Before moving on to a detailed consideration of our reconstructions, we want to make it clear that we are trying to recover information on probable *behaviors* as revealed by morphology within an ecological context (viz., "ecological morphology" or *ecomorphology*; Wainright and Reilly, 1994). Our efforts lead us to consider both primitive and derived features of subfossil lemurs. We regard all aspects of an organism as potential sources of valuable information about probable behaviors and associated adaptations. "Adaptation" means very different things to different workers (e.g., Bock and von Wahlert, 1965; Stern, 1970; Lewontin, 1978; Rose and Lauder, 1996; Ross *et al.*, this volume, Chapter 1). The "current utility" perspective or definition of adaptation (Reeve and Sherman, 1993) comes closest to our occasional use of the term here. Although we acknowledge that anatomical specializations are especially useful in any attempt to reconstruct the behavior of an extinct species, there is more to an animal's full adaptive repertoire than behavioral apomorphies.

One of the most obvious differences between living and extinct Malagasy lemurs is body size (Godfrey *et al.*, 1993, 1997a,b; Smith and Jungers, 1997). The largest extant species (*Indri indri* and *Propithecus diadema*) both average less than 7 kg; all subfossil species are larger, with many estimated to have been as large as the most massive living anthropoids (Table II). We emphasize that this

Table I. Taxonomy of Malagasy Leumrs[a]

Family Cheirogaleidae	Family Megaladapidae
Microcebus murinus	†*Megaladapis (Megaladapis) grandidieri*
Microcebus rufus	†*Megaladapis (Megaladapis) madagascariensis*
Microcebus myoxinus	†*Megaladapis (Peloriadapis) edwardsi*
Microcebus ravelobensis	Family Archaeolemuridae
Mirza coquereli	†*Archaeolemur edwardsi*
Allocebus trichotis	†*Archaeolemur majori*
Cheirogaleus major	†*Hadropithecus stenognathus*
Cheirogaleus medius	Family Indridae
Phaner furcifer (four subspecies)	*Avahi laniger*
Family Lemuridae	*Avahi occidentalis*
†*Pachylemur insignis*	*Propithecus verreauxi* (four subspecies)
†*Pachylemur jullyi*	*Propithecus diadema* (four subspecies)
Hapalemur griseus (three subspecies)	*Propithecus tattersalli*
Hapalemur aureus	*Indri indri*
Hapalemur simus	Family Palaeopropithecidae
Lemur catta	†*Mesopropithecus globiceps*
Eulemur fulvus (six subspecies)	†*Mesopropithecus pithecoides*
Eulemur macaco (two subspecies)	†*Mesopropithecus dolichobrachion*
Eulemur coronatus	†*Babakotia radofilai*
Eulemur rubriventer	†*Palaeopropithecus ingens*
Eulemur mongoz	†*Paleopropithecus maximus*
Varecia variegata (two subspecies)	†*Palaeopropithecus* sp. (region near Mahajanga)
Family Lepilemuridae	†*Archaeoindris fontoynontii*
Lepilemur mustelinus	Family Daubentoniidae
Lepilemur microdon	*Daubentonia madagascariensis*
Lepilemur leucopus	†*Daubentonia robusta*
Lepilemur ruficaudatus	
Lepilemur edwardsi	
Lepilemur dorsalis	
Lepilemur septentrionalis	

[a] Families after Jenkins (1987); extinct species indicated by dagger (†).

pervasive difference in overall size between extant and all subfossil lemurs is a major complication in making commensurate comparisons and robust behavioral inferences. We are required frequently to extrapolate into body size ranges simply unknown to extant strepsirrhines and are left, therefore, with "size-controlled" comparisons only to anthropoid primates or other large-bodied mammals (e.g., Godfrey *et al.* 1995; Jungers, 1999). Although we briefly discuss some of the many implications of large body size for reconstructing subfossil behaviors (e.g., sexual dimorphism and social behavior, activity cycle, substrate preferences), we also have reasons to suspect that body size is overrated as a general predictor variable of behavior in extinct species, especially in larger forms (also see Godfrey *et al.*, this volume, Chapter 4). In this chapter we discuss our behavioral inferences and adaptive reconstructions in the following sequence: sexual dimorphism and its correlates, likely activity

Table II. Body Size of Subfossil Lemurs[a]

Species	Body mass (kg)	Anthropoid size analogy
Mesopropithecus globiceps	9.4	*Ateles chamek* Male *Cercocebus torquatus*
Mesopropithecus pithecoides	9.7	Male *Colobus angolensis* Male *Brachyteles arachnoides*
Mesopropithecus dolichobrachion	10.6	Male *Colobus satanus* Female *Hylobates syndactylus*
Pachylemur insignis	10.0	Female *Papio hamadryas* Female *Rhinopithecus bieti*
Pachylemur jullyi	12.8	Male *Erythrocebus patas* Female *Mandrillus sphinx*
Daubentonia robusta	13.5	Female *Papio anubis*
Babakotia radofilai	16.2	Male *Papio hamadryas*
Archaeolemur majori	13.9	Male *Colobus guereza*
Archaeolemur edwardsi	24.5	Male *Papio anubis*
Hadropithecus stenognathus	27.1	Male *Papio ursinus*
Palaeopropithecus sp. (Anjohibe and Amparihingidro)	25.8	Male *Papio anubis*
Palaeopropithecus ingens	45.4	Male *Pan paniscus*
Palaeopropithecus maximus	52.3	Male *Pan troglodytes*
Megaladapis madagascariensis	38.0	Female *Pongo pygmaeus* Female *Pan troglodytes*
Megaladapis grandidieri	63.0	Male *Homo sapiens*
Megaladapis edwardsi	75.4	Female *Gorilla gorilla* Male *Pongo pygmaeus*
Archaeoindris fontoynontii	197.5	Male *Gorilla gorilla*

[a] Subfossil data from Godrey *et al.* (1995, 1997b); new calculations for *Palaeopropithecus* sp. and *Palaeopropithecus ingens*. See Smith and Jungers (1997) for body masses of the living anthropoid analogies.

cycles (diurnal, nocturnal, or cathemeral), oral behaviors (from grooming to diet), and finally positional behaviors (posture, locomotion, and habitat preferences). We will attempt to summarize prior work on each of these problems, but we have chosen to emphasize our new data and analyses whenever possible. We also acknowledge at the outset that there is much we do not yet fully understand.

Body Size and Sexual Dimorphism

Using geometrical data from the long bones (humeral and femoral midshaft circumferences) and body mass in living primates and other mammals, it is possible to calculate regressions that estimate body mass for the subfossil lemurs. The data in Table II are a summary of predictions made this way by Godfrey *et al.* (1995, 1997b); each value is the mean of four predictions for that species (two from a reference sample of primates, two from a broader mammalian reference sample). As a generalization, the mammalian regressions tended to predict slightly larger masses than did the primate-specific equations. Although we present these data as point estimates, each value obviously carries with it a *substantial* confidence interval (Smith, 1996). In order to provide a familiar, qualitative sense of how large the subfossils might have been, we have added a final column to Table II that lists anthropoid primates of comparable body size. We estimate that subfossil lemurs ranged in body size from just under 10 kg up to almost 200 kg. Body mass estimates based on tooth and cranial size are even higher for many of the largest species (Fleagle, 1988, 1999; Jungers, 1990; Martin, 1990). Virtually the full range of primate body masses known to exist today would have been represented on the island of Madagascar in the early Holocene.

Increasing body size among species of living anthropoid primates is known to be correlated with increasing degrees of sexual size dimorphism (Leutenegger and Cheverud, 1982; Cheverud *et al.*, 1985; Smith, 1999), although the relationship is complex and outliers abound (Gaulin and Sailer, 1984; Ely and Kurland, 1989; Godfrey *et al.*, 1993). Sexual size dimorphism is highly but not perfectly correlated with canine tooth size dimorphism in anthropoids, and both have been linked to degrees of intrasexual competition among males (e.g., marked sexual dimorphism is correlated positively with intense and frequent competition; see Plavcan and van Schaik, 1992, 1997). Anti-predator defense has also been invoked to account for sexual dimorphism in some terrestrial anthropoids (Clutton-Brock *et al.*, 1977). Despite these apparent generalizations and the very large body sizes of some of the subfossil lemurs (plus a clear signal of terrestriality in a few species), there is evidence for little, if any, sexual size dimorphism in any species. More specifically, using skull length as a proxy for body size and an array of exploratory statistical methods, Godfrey *et al.* (1993) discovered that a lack of sexual dimorphism is a general phenomenon for Malagasy primates (for extant species, also see Jenkins and Albrecht, 1991; Kappeler, 1991), regardless of overall size or substrate preference. Figure 1 is taken from Godfrey *et al.* (1993) and clearly illustrates in subfossils the lack of the bimodality that is so typical of sexually dimorphic anthropoids (especially catarrhines). Similarly, sexual dimorphism in the canines (or caniniform lower premolars, or any other skeletal element) has never been demonstrated in a compelling fashion for any subfossil taxon. Canine monomorphism probably implies little differentiation in sex-specific agonistic profiles (see Plavcan *et al.*, 1995; Godfrey *et al.*, this volume, Chapter 4), but species-specific differences in relative canine height do suggest that the intensity of agonistic behavior varied among the subfossils (e.g., higher levels of agonism is inferred for *Archaeolemur* when compared to *Hadropithecus* because relative canine height is much greater in the former). Even in the most terrestrial of the subfossil lemurs, the archaeolemurids (see below), there is no evidence for significant sexual size dimorphism, implying that their anti-

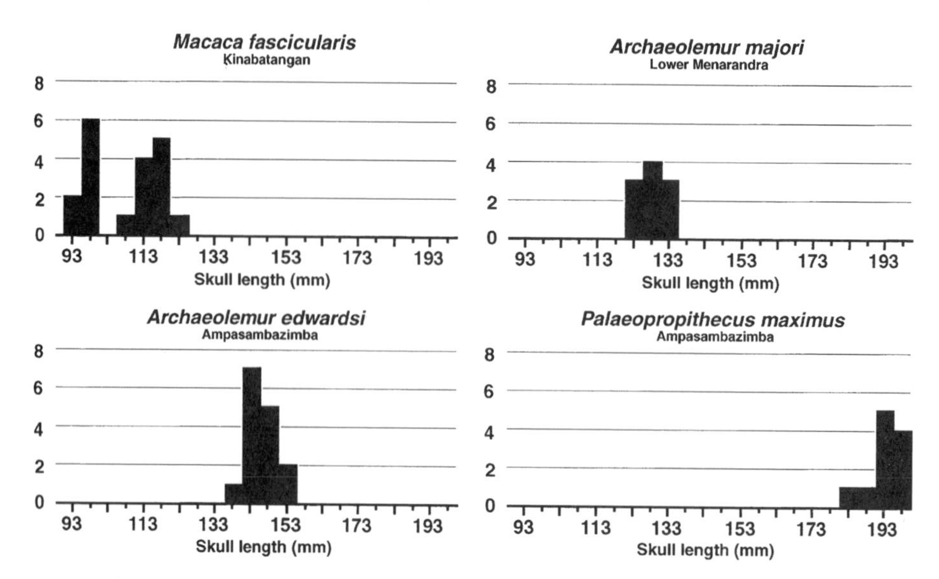

Fig. 1. Frequency distributions of skull length in samples of restricted geographic origins. Three subfossil lemur species are compared to macaques. Note the obvious bimodality in the macaques that reflects sexual size dimorphism (males>females). Despite large body sizes, the subfossil lemurs are best described as monomorphic (after Godfrey *et al.*, 1993).

predator strategies were probably unlike those linked to dimorphism in terrestrial anthropoids. If we acknowledge the high likelihood of phylogenetic inertia (and the subfossils are strepsirrhines after all), it seems probable to us that the metabolic, ontogenetic, and ecological contraints which operate on living Malagasy primates also impacted on subfossil lemur life histories to preclude the achievement of anthropoid-like degrees of sexual size dimorphism (Richard, 1992; Godfrey *et al.*, 1993, this volume, Chapter 4; Kappeler, 1996; Leigh and Terranova, 1998; also see Daniels, 1984; Richard and Nicoll, 1987; Schmid and Ganzhorn, 1996).

Activity Cycles

Walker (1967) concluded that all of the extinct lemurs of Madagascar were diurnal. This conclusion followed logically from a bivariate plot of orbit diameter (the Y variable) versus maximum skull length (the X variable) in living and subfossil lemurs; nocturnal species were above an extrapolated dashed line, and diurnal species were below—including all nine of the subfossil lemur species examined. Nocturnal species appeared to possess relatively large orbits (and eyeballs, presumably), while diurnal ones were characterized by relatively small orbits. Subfossil lemurs deviated in the diurnal direction very strongly, especially *Megaladapis*. Subsequently, Gingerich and Martin (1981) extrapolated a slightly negatively allometric regression of orbital diameter (using the average of superoinferior and mediolateral diameters) on condylobasal skull length for eight species of nocturnal lemurs, and corroborated Walker's findings; all nine of the species of subfossils they examined fell far below and to the right of the extrapolation. Martin (1990) considered fewer subfossil species, but his methods and findings were similar: subfossil lemurs possess relatively small orbits, even in comparison to large-bodied, diurnal anthropoids. However, both Martin (1990) and Gingerich and Martin (1981) acknowledged that extrapolations of this nature can be misleading, especially in view of the finding by Kay and Cartmill (1977) that relative orbit size is a decreasingly reliable indicator of activity cycle for species with skull lengths greater than 75 mm. All subfossil lemurs have skulls longer than this threshhold. At approximately 2.5 kg and a skull length of roughly 85 mm (Smith and Jungers, 1997; Albrecht *et al.*, 1990) the living aye-aye is the largest strictly nocturnal primate; this implies that extrapolation to the largest of the giant subfossil lemurs from a nocturnal (or any other extant lemur) baseline is risky indeed.

Independent of these orbital considerations, Godfrey *et al.* (1997a) argued that there are good ecological reasons to conclude that subfossil lemurs were probably diurnal. If activity patterns are constrained largely by predators (i.e., nocturnality helps small-bodied arboreal mammals avoid predation by diurnal, visually directed raptors), it is conceivable that the large subfossil lemurs exceeded the prey-size of the largest avian raptors (and perhaps all other diurnal predators) of Madagascar (cf. Goodman *et al.*, 1993). On the other hand, Rasmussen (personal communication) recently discovered an extinct eagle in the fossil collections from SW Madagascar (Ankilitelo) that appears to share adaptations similar to those of monkey-eating eagles; a large lemur-eating eagle may have preyed on some of the largest living lemurs and the smallest individuals of some subfossil species. Diurnality might still have been a viable option for very large lemurs, but immature animals might have remained vulnerable. Accordingly, Godfrey *et al.* (1997a) tentatively reconstructed all subfossil species as diurnal except the giant aye-aye (*Daubentonia robusta*), which was considered to fill the nocturnal niche of its very similar but much smaller congener (*D. madagascariensis*) based on a suite of uniquely shared–derived dental and postcranial adaptations (Simons, 1994).

Several factors complicate all of these arguments and inferences, including the need for extreme extrapolation, choice of a body size surrogate, and the realization that activity cycles in living lemurs are much more complicated than the simple diurnal–nocturnal dichotomy might suggest. As field information has accumulated over the last two decades, it has become apparent that a variety of living lemurs are really neither strictly diurnal nor completely nocturnal; they are also not merely crepuscular. Rather, numerous species

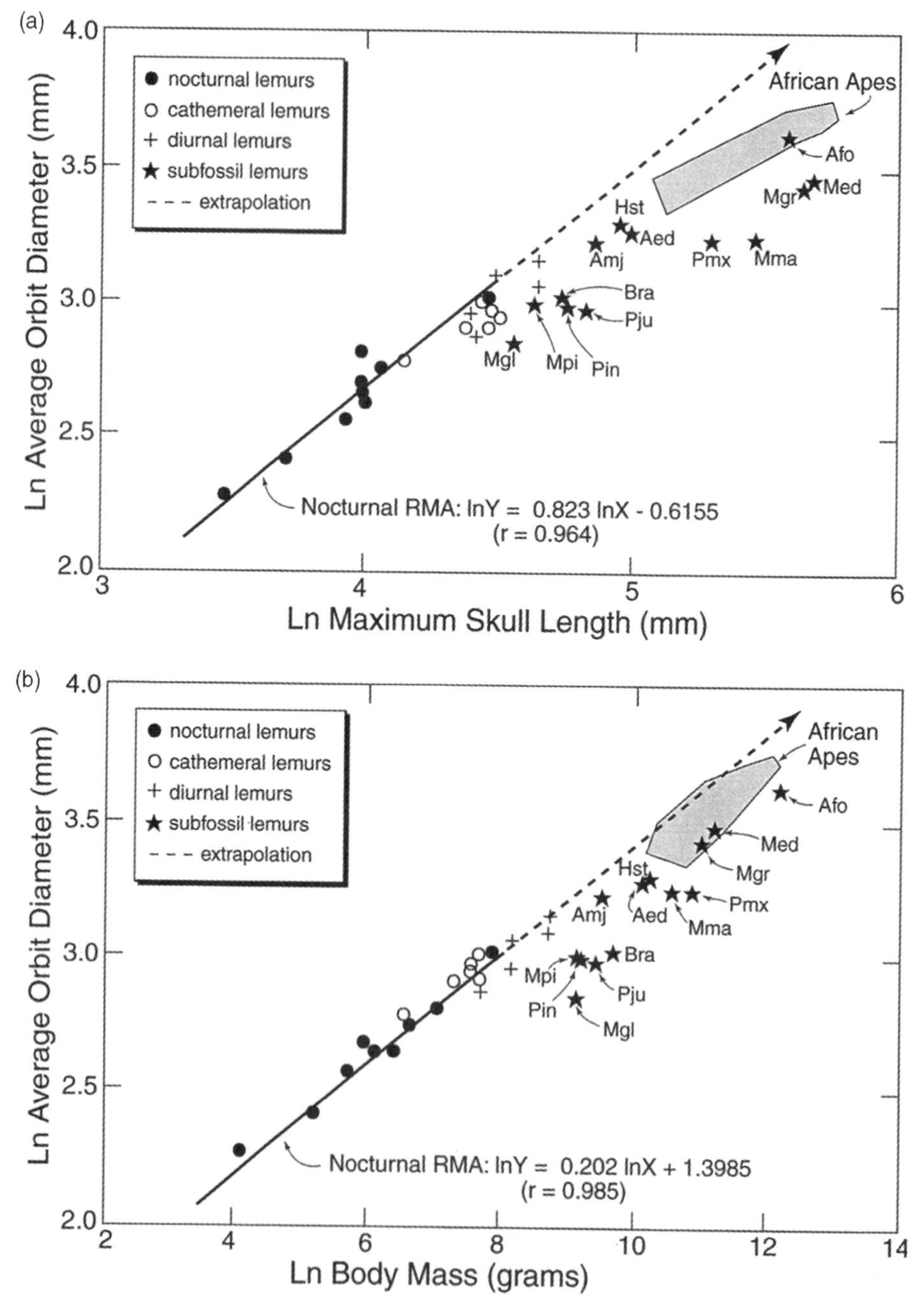

Fig. 2 (a) Orbit diameter is plotted against skull length, and the nocturnal regression is extrapolated. Cathemeral lemurs, most diurnal lemurs, African apes, and all subfossil lemurs have smaller orbits than would be predicted for their skull length using the nocturnal relationship as a guide. (b) Orbit diameter is now plotted against body mass,

distribute their travel, feeding, and other behaviors throughout the full 24-hr cycle, if sometimes only seasonally (Engqvist and Richard, 1991; van Shaik and Kappeler, 1993; Curtis *et al.*, 1999; Wright, 1999). Tattersall (1988) has coined the term "cathemerality" for this condition, to distinguish it from nocturnality and diurnality. It now appears that only the cheirogaleids, *Lepilemur*, *Avahi*, and *Daubentonia*, are truly nocturnal. Only *Indri*, *Propithecus*, and *Lemur* are best described as strictly diurnal. All species of *Eulemur* are probably cathemeral, and *Varecia* (*V. variegata variegata*) might be as well (Wright, 1999). *Hapalemur simus* and *H. griseus alaotrensis* appear to be cathemeral, but *H. griseus griseus* is better characterized as diurnal (Mutschler, 1999; Wright, 1999). Based on the presence of the reflecting tapetum, poorly defined or lack of retinal fovea, and lenticular enzyme activity levels characteristic of nocturnal mammals (Rathburn *et al.*, 1994; Godfrey *et al.*, 1997a), it is almost certainly the case that the ancestral condition for all lemurs was nocturnality. If we use Yoder's (1997) phylogeny for strepsirrhines as a guide to character evolution, this implies in turn that cathemerality evolved either once in the ancestor of all lemurids or once in the ancestor of the *Hapalemur–Lemur–Eulemur* clade if *Varecia* is really diurnal as previously thought. It also suggests that diurnality probably evolved independently in both indrids and lemurids (unless it can be demonstrated that they are really sister clades), and that strict nocturnality in *Avahi* is probably a reversal (Yoder, 1997). Because palaeopropithecids are the sister clade of living indrids (Jungers *et al.*, 1991; Yoder *et al.*, 1999), it is perhaps most parsimonious to infer that they were also diurnal. If *Pachylemur* is indeed the sister of *Varecia* (Crovella *et al.*, 1994), diurnality might be another independent acquisition in *Pachylemur* unless it was cathemeral too. If *Lepilemur* is the sister of *Megaladapis*, diurnality in the latter would represent still another parallelism. The precise degree of homoplasy is hard to assess, but its existence must be acknowledged in virtually any evolutionary scenario of activity cycles.

We have re-examined the relationship between overall body size and orbit size in the following manner. Using average orbit diameter (the mediolateral diameter was usually smaller than the superoinferior) and maximum skull length as a proxy for size, we calculated a reduced major axis regression for nine strictly nocturnal lemur species (Fig. 2a; *Microcebus murinus*, *Mirza coquereli*, *Cheirogaleus medius*, *C. major*, *Phaner furcifer*, *Avahi laniger*, *Lepilemur ruficaudatus*, *L. dorsalis*, and *Daubentonia madagascariensis*). Negative allometry is

---

and the nocturnal regression is again extrapolated. Note that all but one of the cathemeral lemurs fall above the extrapolation, which also intersects the diurnal apes. Abbreviations: Pin, *Pachylemur ingens*; Pju, *Pachylemur jullyi*; Mpi, *Mesopropithecus pithecoides*; Mgl, *Mesopropithecus globiceps*; Bra, *Babakotia radofilai*; Amj, *Archaeolemur majori*; Aed, *Archaeolemur edwardsi*; Hst, *Hadropithecus stenognathus*; Mma, *Megaladapis madagascariensis*; Mgr, *Megaladapis grandidieri*; Med, *Megaladapis edwardsi*; Pmx, *Palaeopropithecus maximus*; Afo, *Archaeoindris fontoynontii*.

implied by the slope of 0.823, but the 95% confidence interval does include the isometric value of 1.0. This line is extrapolated (as did Gingerich and Martin, 1981), and six cathemeral lemurs [*Hapalemur griseus* (in part), *H. simus*, *Eulemur rubriventer*, *E. fulvus*, *E. mongoz*, and *E. macaco*] and five diurnal lemurs [*Indri indri*, *Propithecus diadema*, *P. verreauxi*, *Varecia variegata* (perhaps), and *Lemur catta*] are then added to the plot. All six cathemeral species and all but one (*P. diadema*) of the diurnal lemurs fall below and to the right of the line, indicating that they possess relatively smaller orbits than would be predicted for their skulls from the nocturnal relationship. Thirteen subfossil lemurs are then added (the stars in Fig. 2a,b), as well as a sample of African apes (a convex polygon of bonobos, chimpanzees, and gorillas). As predicted from the findings of prior investigators, the subfossil lemurs and the large-bodied, diurnal apes all possess smaller orbits than would be predicted from the nocturnal extrapolation. We ignore the fact that the eyeball itself fills less of the orbit in large-bodied primates (Martin, 1990). These findings might suggest that the addition of even a modest amount of daytime activity such as that seen in the cathemeral species is sufficient to reduce relative orbit size in the diurnal direction (although *P. diadema* is something of an anomaly), and all of the subfossils with measurable orbits are best regarded as diurnal. The giant extinct aye-aye remains an unknown because its skull has never been recovered. However, using body mass itself rather than a cranial surrogate as the X variable and repeating the same exercise complicates this picture somewhat (Fig. 2b). Negative allometry of orbit size is again observed for nocturnal lemurs (and the 95% confidence interval to the slope of 0.202 now does not include the isometric value of 0.333), and *P. diadema* is still the only diurnal species to fall above the extrapolation. But five of the six cathemeral lemurs now fall above the line into presumptive nocturnal space. The new extrapolation intersects the convex polygon for African apes, but we can remain confident that none of them is really nocturnal. All of the subfossils continue to fall to the right of the extended reduced major axis, but the lesson from the cathemeral lemurs and African apes warrants caution in making precise inferences about activity cycles from the allometry of orbit size alone, especially when extreme extrapolation is required. These findings also suggest that skull length is a rather biased substitute for overall body size in all prior interspecific comparisons. We know, for example, that *P. diadema* and *Indri indri* are more or less the same body mass (Smith and Jungers, 1997), but the skull of *I. indri* is significantly longer (Albrecht *et al.*, 1990); *Varecia* and *Indri* have skulls of similar length but the body mass of the former is just over half of the latter. The situation is even worse with respect to some subfossil lemurs; e.g., the skull length of *Megaladapis edwardsi* greatly overestimates probable body mass if Martin's (1990) prediction of 390 kg is any indication! This probably tells us more about relative skull size rather than relative orbit size in *Megaladapis*.

More recently, Gonzalez *et al.* (1998) and Kay and Kirk (2000) have demonstrated that if information on the size of the optic canal is also added to data on orbit size and skull size, important new insights into primate activity cycles and visual acuity emerge. These authors note that given the low levels of light intensity typical of nighttime, most nocturnal species increase sensitivity at the expense of visual acuity (see also Rohen and Castenholz, 1967; Wolin and Massopust, 1970). The tapetum no doubt also helps nocturnal lemurs in this regard, but a very important part of the tradeoff is in degree of retinal summation. Nocturnal species have retinas with populations of photoreceptor cells that project to relatively few ganglion cells, thereby summating the limited information and enhancing sensitivity. Diurnal anthropoids have more than adequate amounts of light with which to work, so they possess low-summation optic systems that place a premium instead on high visual acuity (i.e., lower ratios of photoreceptor to ganglion cells). High summation like that seen in nocturnal species is known to be correlated with relatively smaller optic nerves (Stephan *et al.*, 1984), and this difference can be detected by the relative size of the optic canal or foramen (Gonzalez *et al.*, 1998; Kay and Kirk, 2000). This novel synthesis permits valuable new inferences about vision and activity cycles in fossils if orbit size and optic canal size are considered together. Subfossil lemurs will eventually be scrutinized within this promising framework, and casts of optic canals have now been collected for virtually all available subfossil skulls along with associated orbital dimensions. Much work remains to be done, including the addition of large-bodied, nonprimate mammals, but preliminary results are intriguing. Qualitative assessment of the size of optic canals (and by inference, optic nerves) relative to orbital diameters (roughly proportional to eyeball and retinal size) suggests that extremely high levels of summation characterized some subfossil taxa (e.g., *Megaladapis* and *Palaeopropithecus*) because their canals are remarkably small, especially in comparison to anthropoids of comparable body size (e.g., African apes). Optic canals are relatively larger and retinal summation may have been less pronounced in the archaeolemurids. Whether or not extreme summation necessarily implies nocturnality (or perhaps cathemerality) in some of the subfossils remains to be evaluated in a rigorous manner, but if they all were diurnal, it seems highly plausible that visual acuity was greatly reduced in some of these giants (especially if the subfossil species retained the plesiomorphic tapetum lucidum, which itself can compromise acuity). Perhaps lack of an anthropoid degree of visual acuity mattered little to slow-moving, arboreal folivores—at least until the time people colonized the island.

Oral Behaviors

Parafunction

Various aspects of oral behavior are not related directly to food acquisition or processing. Such behaviors are sometimes lumped under the category of "parafunction" (Teaford, 1994). The most obvious of these behaviors in living strepsirrhine primates is dental grooming of the fur (Roberts, 1941; Buettner-Janusch and Andrew, 1962). One of the morphological synapomorphies of all extant strepsirrhines except the aye-aye is the "tooth comb," a procumbent modification of the lower anterior dentition used specifically for auto- and allogrooming (Martin, 1990; Fleagle, 1999). The tooth comb consists of either six teeth (two incisiform canines and four incisors) or four teeth (incisors only); the indrids alone among extant lemurs exhibit the latter condition. Our goal here is not to debate the evolutionary origins of the tooth comb; it matters little for our purposes whether its design emerged as an exaptation (*sensu* Gould and Vrba, 1982) co-opted from a prior dental cropping/gouging/scraping adaptation (Avis, 1961; Gingerich, 1975; Maier, 1980; Martin, 1990) or directly for its grooming function (Szalay and Seligsohn, 1977; Rosenberger and Strasser, 1985). We also acknowledge that direct microwear evidence for grooming that involves the lower anterior dentition can exist in primates even when there is no real tooth comb, as in tarsiers (Musser and Dagosto, 1987) and some Eocene omomyids (Schmid, 1983). However, whenever the characteristic tooth-comb morphology is found, whether in living lemurs and lorises (Rose *et al.*, 1981), Miocene fossil lorises such as *Nycticeboides* (Jacobs, 1991), or in the Eocene arctocyonid *Thryptacodon* (Gingerich and Rose, 1979; Rose *et al.*, 1981), there is always a dental microwear signature of fur-grooming—hair grooves or striae resulting from the hairs contacting the mesial and distal margins of the teeth repeatedly. These grooves or "whip marks" can be very extensive in some individuals and appear on the lingual surface of the teeth in addition to interproximal locations (see Fig. 3 of *Varecia variegata*).

As noted above, *Daubentonia* has sacrificed the tooth comb and replaced it with rodent-like, gnawing incisors (Tattersall, 1982); the giant extinct aye-aye is similar in this respect (MacPhee and Raholimavo, 1988). Other subfossil lemurs have modified the lower anterior dentition away from the typical tooth-comb morphology, although most have preserved some degree of procumbency. As in indrids, *Archaeolemur* has lost the lower canines, and the incisors are stout, cylindrical and sometimes wear flat; *Hadropithecus* is similar to *Archaeolemur* in many respects, although the lower anterior dentition is considerably smaller (Tattersall, 1973; Jungers, 1978b). The lower anterior teeth of Palaeopropithecus are "short and stubby" and also no longer part of a tooth comb (Tattersall, 1982). The same is true of *Archaeoindris* (Standing, 1909), which is very similar to *Palaeopropithecus* in many other craniodental respects (Tattersall, 1973). However, despite their large body sizes, *Pachylemur* (6 teeth), *Mesopropithecus* (4 teeth), *Babakotia* (4 teeth), and *Megaladapis* (6 teeth) all preserve the typical strepsirrhine tooth-comb morphology (see Fig. 4). Electron microscopic inspection of high-resolution casts of *Babakotia* and *Megaladapis* teeth (Fig. 5) reveals the telltale grooves and confirms the use of these teeth in fur-combing behaviors. In other words, the tooth-comb morphology of these subfossils definitely implies grooming parafunction, but we cannot say whether the behavior was in an auto- or allogrooming context.

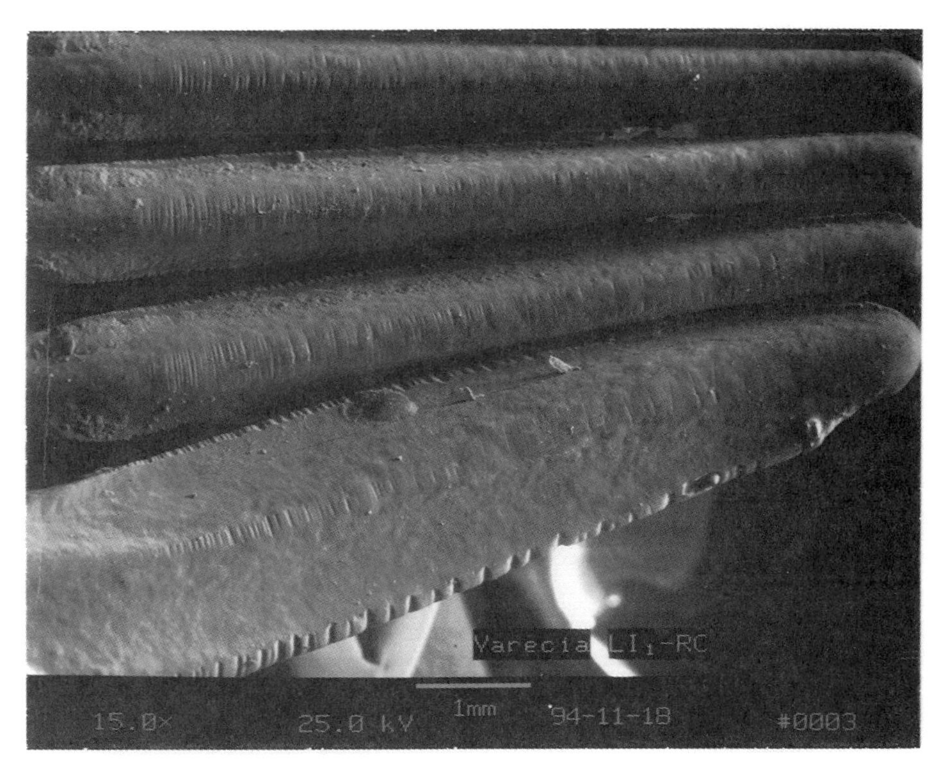

Fig. 3. An electron microphotograph of the tooth comb of *Varecia vareigata*. Four of the six teeth are shown, starting with the central left incisor at top and ending with the right canine at bottom. Note the striations and whip marks on mesial, distal, and lingual surfaces. These are the signatures of using the tooth comb to groom fur.

We do not wish to imply that the tooth comb of subfossil lemurs was used exclusively in parafunctional behaviors, but such activities do complicate efforts to extract unambiguous dietary information and signs of feeding-ingestive behavior from the lower anterior dentition (Teaford, 1994). We have also examined the lower anterior dentitions of *Archaeolemur*, *Hadropithecus*, and *Palaeopropithecus* under the electron microscope to see whether these teeth might still have been used in grooming behaviors despite their obvious modifications (Gilmartin and Jungers, 1995). Focusing on interproximal and labial surfaces, we found no evidence for grooves or striae remotely similar to those seen routinely in taxa with *bona fide* tooth combs. We conclude that these species no longer engaged in significant amounts of dental grooming and used their lower anterior dentitions instead in food procurement and processing (see Fig. 6 for a comparison of *Babakotia* and *Archaeolemur*). Although we are confident from associated foot bones that *Babakotia* also retained the typical strepsirrhine grooming claw of the second pedal digit, we strongly suspect that it was absent in *Archaeolemur* based on the shape of a large number of distal

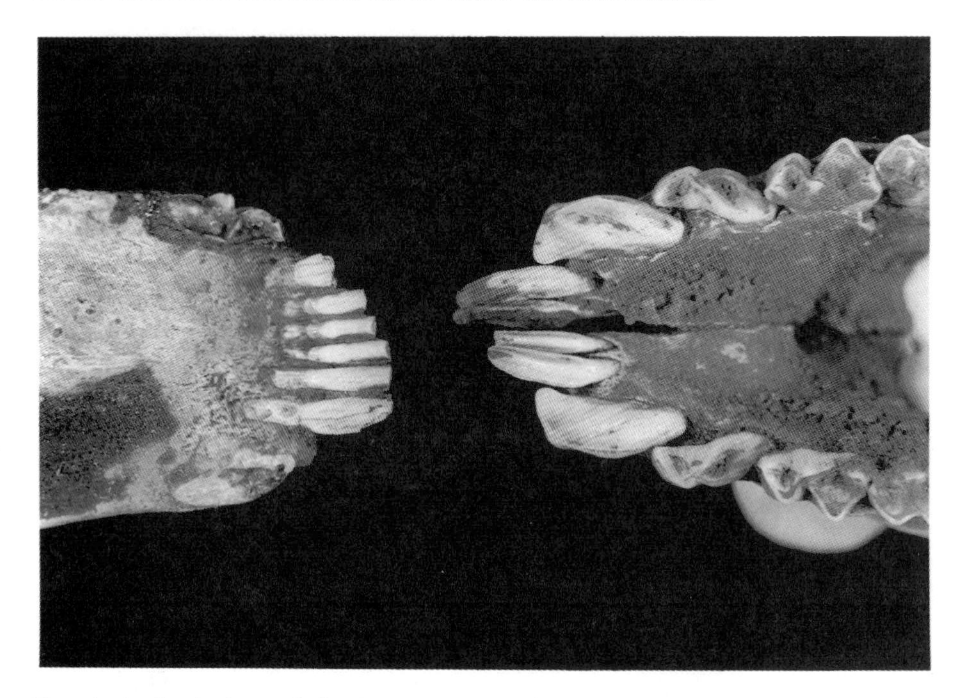

Fig. 4. Occlusal views of the lower anterior dentitions of *Babakotia* (left) and the Anjohibe *Megaladapis* (right). Despite large body size, both preserve the typical strepsirrhine tooth comb.

phalanges, also from associated pedal remains (Jungers *et al.* 1998; see Fig. 9 farther on). Perhaps these dental and digital differences among subfossil species also reflect differences in the nature of grooming behaviors (or lack thereof). For example, the unusually broad apical tufts on all the digits of *Archaeolemur* might reflect enhanced digital grooming capabilites in a species that had lost its toothcomb. Future quantification of anterior dental microwear in both lower and upper anterior teeth of subfossil species lacking the tooth comb should shed new light on aspects of diet, ingestive behaviors, and perhaps even feeding heights (see Ryan, 1981; Walker, 1976; Kelley, 1990; Ungar, 1990, 1994).

Feeding Behavior

Oral behaviors related to food acquisition and ingestion (i.e., feeding or dietary behavior) can be reconstructed in a variety of ways, some indirect (analogy, theoretical biomechanics) and some more direct (e.g., dental microwear). For example, despite major differences in body size, the uniquely shared–derived dental (continuously growing, chisel-like incisors) and postcranial (the elongate, filiform third digit of the hand) characters seen in the living and extinct aye-ayes make a strong case for a similar feeding niche in the subfossil form (Grandidier, 1929; Lamberton, 1934; MacPhee and Raholimavo, 1988; Simons, 1994). Although much has been made of the percussive foraging for insect larvae by *Daubentonia* (e.g., Erickson, 1991), it is now clear from field work that the dietary specializations of aye-ayes have more in common with squirrels than woodpeckers. The living aye-aye's diet consists primarily of ramy nuts (*Canarium*), and includes nectar from the Traveller's Tree (*Ravenala*) and a relatively small amount of fungus and insect grubs (Iwano and Iwakawa, 1988; Sterling, 1994; Iwano and Rakotoarisoa, 1998; Garbutt, 1999). The unusual, filiform third digit is useful for extractive foraging in many feeding contexts. *D. robusta* probably included insects on its menu too, perhaps including termites and giant burrowing crickets in addition to grubs (Simons, 1994), but we suspect that nuts were its primary dietary staple, especially in view of its much larger body size.

Similar reasoning connects the feeding behavior of *Lepilemur* and *Megaladapis*; the former is a small-bodied folivore (Charles-Dominique and Hladik,

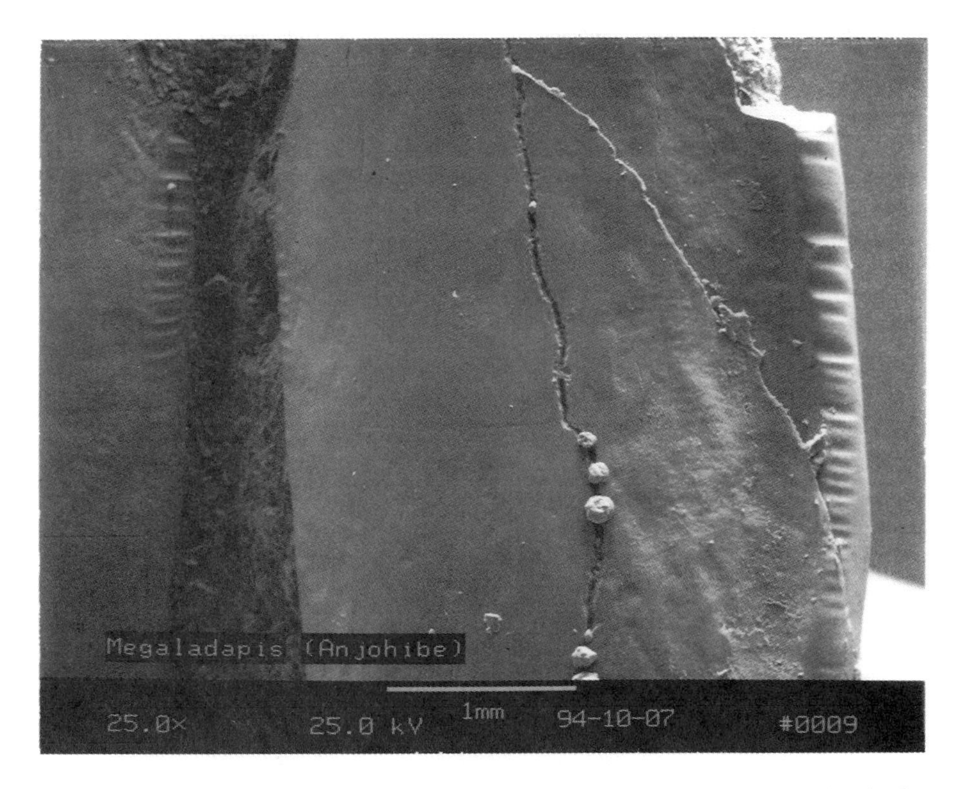

Fig. 5. An electron microphotograph of the right lateral incisor and right canine in the Anjohibe *Megaladapis*. Note the whip marks indicative of fur combing.

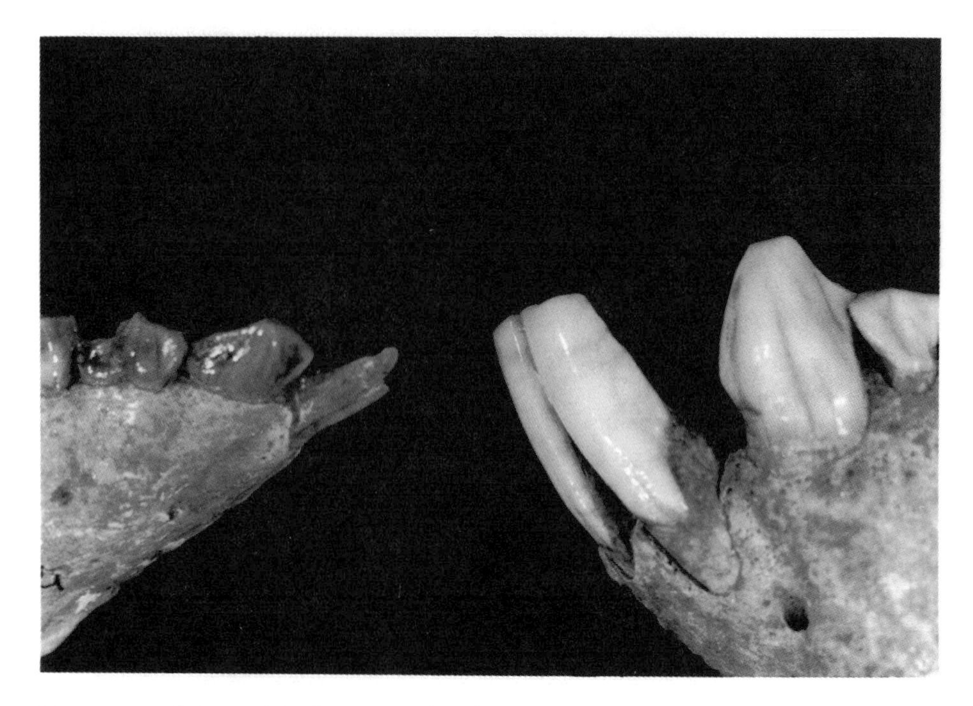

Fig. 6. Lateral views of the lower anterior dentitions of *Babakotia* (left) and *Archaeolemur* (right). While the former preserves the typical tooth comb morphology, the latter has modified it into a stout, more vertically implanted battery of teeth that are no longer used in grooming.

1971; Ganzhorn, 1993) . Despite enormous differences in body size between them (almost two orders of magnitude), they share several craniodental apomorphies linked presumably to a unique style of folivory (e.g., loss of permanent upper incisors in lieu of a plucking pad, similar molar morphology and occlusal wear, and the posterior extension of the mandibular condyle's articular surface). These similarities are so marked that some workers have placed these genera within the same taxonomic family (e.g., Petter *et al.*, 1977; Jungers, 1980; Schwartz and Tattersall, 1985; Mittermeier *et al.*, 1994; Garbutt, 1999). Additional functional analogies can be drawn between crania of *Megaladapis* and the nocturnal, folivorous koala, *Phascolarctos* of Australia (Tattersall, 1972)—including airorhynchy of the facial skeleton, maxillary and mandibular diastemata, and a vertical nuchal plane. The elongate, downward flexed nasal bones that extend beyond prosthion are unique to *Megaladapis* among primates; this anatomy is more pig-like than tapir-like (Witmer, personal communication), and the degree of proboscis prehensility/mobility it might imply is very difficult to gauge. Unpublished microwear analysis (Rafferty *et al.*, cited in Godfrey *et al.*, 1997a) is consistent with a primarily folivorous diet; e.g., narrow scratches and fewer pits are found on the molars of *Megaladapis edwardsi* than on any other primate folivore yet examined. The two smaller and more closely related species of *Megaladapis*, *M. grandidieri*, and *M. madagascariensis* (Vuillaume-Randriamanantena *et al.*, 1992) exhibit more pitting on their molars, and perhaps ingested foliage with more grit or experienced more tooth to tooth contact during chewing. Regardless, arboreal browsing and leaf-eating are quite reasonable inferences for all three species of *Megaladapis* (also see the discussion of positional behavior below).

We reconstruct the diet of *Pachylemur* to have been largely frugivorous (Godfrey *et al.*, 1997a). The dentition of *Pachylemur* is very similar to the predominantly fruit-, flower-, and nectar-consuming *Varecia* (Rigamonti, 1993; Garbutt, 1999), and both of the extinct species sport a high frequency of caries. While Ravosa (1992) believes that the size and design of the mandible indicate a more folivorous diet than observed today in *Varecia*, Seligsohn and Szalay (1974) interpret *Pachylemur*'s molar design as reflective of greater emphasis on harder fruits and more stems. Primary frugivory, no doubt seasonally supplemented to varying degrees by leaves and other foliage, is the dietary signal we perceive most strongly from the dentition.

The archaeolemurids are very derived in cranial and dental form (Tattersall, 1973). Frugivory is most often reported as the likely feeding adaptation of *Archaeolemur* (e.g., Jolly, 1970; Tattersall, 1982). This inference derives from a variety of convergences on frugivorous cercopithecines, including very broad and spatulate upper incisors (Hylander, 1975) and relatively small, low-crowned but bilophodont molars (Kay, 1975, 1977). The premolars of *Archaeolemur* form a long shearing blade that could handle fruits with tough pericarps, and we have speculated earlier that seed predation may have been relatively common in the archaeolemurids (Godfrey *et al.*, 1997a). *Archaeolemur* may have also contributed to the dispersal of some seeds, including the Malagasy baobabs (Baum, 1995; Godfrey *et al.*, 1995). Molar microwear in *Archaeolemur* is extensive, but bears no special resemblance to baboons (Teaford, personal communicaton; Rafferty *et al.*, cited in Godfrey *et al.*, 1997a); abundant scratches could imply more "foliage" in the diet of *Archaeolemur*, but might be related instead to frequent feeding on the ground. New and somewhat surprising information about diet in *Archaeolemur* comes from fecal pellets recovered in the caves of NW Madagascar by Burney *et al.* (1997).Their taphonomic context is persuasive, and if they are attributed correctly to *Archaeolemur*, their contents disclose a much more eclectic diet than the dentition alone might imply: pollen of savanna plants, plant fibers, small seeds, but also the bones of bats and other small vertebrates. "Omnivory" might then be a better description. Perhaps it is obvious, but we suspect that morphological evidence alone is probably conservative and often underestimates the full range of feeding behaviors practiced by extinct organisms. The dentition of *Hadropithecus* recalls that of *Archaeolemur* but is very derived. Stout and more vertically implanted incisors and the aforementioned premolar shearing blade are preserved, but the cheek teeth are enlarged and covered by a complex occlusal surface seen in no other prosimian. Dietary analogies have been made between *Hadropithecus* and *Theropithecus*, a known graminivorous and granivorous papionin, based on numerous craniodental convergences. The available evidence points to a capability for powerful crushing and grinding by the postcanine dental battery of *Hadropithecus* (Jolly, 1970; Tattersall, 1973). The limited molar microwear information (Rafferty *et al.*, cited in Godfrey *et al.*, 1997a) is consistent with terrestrial hard-object feeding and seed predation, but larger scratches and a higher incidence of pitting could be interpreted as evidence of much more hard-object feeding in *Hadropithecus* than in the gelada (and/or a diet with more grit, perhaps including tubers and other subterranean resources).

The palaeopropithecids are the sister clade to living indrids (Jungers *et al.*, 1991; Yoder *et al.*, 1999). Extant indrids are all usually classified as "leaf-eaters," but field data suggest that this is an oversimplification. Despite its small body size, *Avahi* does feed primarily on leaves (Albignac, 1981; Ganzhorn, 1985; Harcourt, 1991), supplemented by small amounts of fruit and flowers. Fruits are a nontrivial component of sifaka diets in some seasons (e.g., *Propithecus verreauxi*; Richard, 1978), and seeds are a preferred food resource for *Indri indri*, *P. tattersalli*, and *P. diadema* (Pollock, 1977; Meyers, 1993; Hemingway, 1996; Powzyk, 1997). Both *P. verreauxi* and *P. diadema* prefer young leaves when they are available, but *P. verreauxi* shifts to more mature leaves and *P. diadema* to seeds when young leaves are not on the menu (Yamashita, 1996). Based on many detailed similarities in their dentitions (and probably dental developmental schedules), we suspect that the palaeopropithecids were also folivorous seed-predators that supplemented their diet with a variety of fruits (Godfrey *et al.*, 1997a). Unpublished molar microwear analyses (Rafferty *et al.*, cited in Godfrey *et al.*, 1997a) do not contradict this inference for either *Palaeopropithecus* or *Babakotia*, although differences in amount of microwear (especially more pits in the latter) may reflect subtle species-specific differences in food properties like those suggested above for species of *Megaladapis*. The modification of the tooth comb into shorter, stouter teeth in *Palaeopropithecus* and *Archaeoindris* probably implies something novel in their feeding strategies, but what that might have been remains obscure.

One of the most useful metrics ever developed for the analysis of dietary adaptations in living and fossil primates is Kay's "shearing quotient" or SQ (Kay, 1975, 1984; Covert, 1986). Its application to strepsirrhines (Kay, 1984; Covert, 1986; also see Seligsohn, 1977) involves computing the sum of the lengths of crests 1–6 on the second lower molars (Kay and Hiiemae, 1974), and computing the percentage difference between the observed sum and that predicted by molar length in prosimians. Higher values indicate more total length of shearing blades for a given molar length. The shearing quotient (SQ) effectively sorts more faunivorous African lorisoids (higher SQ's) from more frugivorous ones (lower SQ's). Inspection of Table 4 in Covert (1986) discloses that frugivorous prosimians have lower SQ's on average than do folivores and faunivores, presumably because the structural carbohydrates in leaves and chitin require more cutting and overall reduction before swallowing (see Kay and Scheine, 1979; Lucas, 1979; Strait, 1997). There are a few anomalies like *Lepilemur*, a folivore with an unusually low SQ, but the general approach is robust and has proven useful in application to fossils (e.g., adapids; Covert, 1986). Using an analysis-of-covariance design, Yamashita (1998) recently demonstrated that the living indrids have more total shearing crest length for a given molar area than does a combined sample of *Lemur*, *Eulemur*, and *Varecia*; the most folivorous lemurid, *Hapalemur*, was not included in her analysis, but we predict that all three bamboo lemur species would fall much closer to indrids than to other lemurids. Strait (1993a,b) has adapted this concept into a series of shearing ratios (SR) that use the same numerator as the SQ but use a variety of denominators (molar length, molar area, and body mass). We have explored the utility of these measures of shearing capacity in application to five species of subfossil lemurs suspected of folivory as discussed above (*Megaladapis edwardsi*, *M. grandidieri*, *Palaeopropithecus maximus*, *P. ingens*, and *Babakotia radofilai*). Sixteen high-resolution casts were prepared and then measured under a Reflex Microscope (MacLarnon, 1989). The prosimian SQ and SR were calculated relative to second molar length, and the results are summarized in Table III.

All five subfossil species are characterized by relatively very high SQ's and SR's; mean SQ's range from 18% in *P. maximus* to an extreme of 50% in *Babakotia*, and the SR's parallel this spread. Individual differences within a species are to be expected, and the ranges seen in the subfossils do not seem to be exceptional in comparison to living taxa. All of the subfossils have much more shearing capability than do the largest living Malagasy folivore-seed predators, *Indri* and *Propithecus*. Only the *Hapalemur simus* SQ and SR overlap the range of average subfossil values, and this species is known to be a specialized bamboo feeder that has other dental adaptations linked to gaining access to and processing the inner pith (especially that of the giant bamboo; Glander *et al.*, 1989; Elisabete and Palka, 1994). These data therefore corroborate other lines of evidence for significant folivory in the feeding behavior and dietary adaptations of these five subfossil species. The largest lemurid, *Varecia*, and the largest cheirogaleid, *C. major*, both have much less shearing capacity. *Megaladapis* has much higher values than does the small folivore *Lepilemur*, the genus to which *Megaladapis* is often compared. The highly modified crown morphology of the archaeolemurids makes it difficult to place them within this shearing framework, but we plan to add *Pachylemur*, *Mesopropithecus*, and *Archaeoindris* to our data base in the near future. We also note that the correlation between SQ and SR across the dozen species in our Table III is virtually perfect at $r = 1.0$; this is to be expected when a linear regression like that used to calculate SQ's is an isometric relationship (Jungers *et al.*, 1995).

Table III. Shearing Capacity of Lower Second Molars in Subfossil Lemurs and Selected Living Malagasy Strepsirrhines[a]

Species		Shearing quotient	Shearing ratio
Megaladapis edwardsi	Mean	+20%	2.31
(N = 4)	Range	+11 to +26	2.12 to 2.42
Megaladapis grandidieri	Mean	+31%	2.50
(N = 2)	Range	+28 to +33	2.46 to 2.54
Palaeopropithecus maximus	Mean	+18%	2.27
(N = 2)	Range	+8 to +28	2.08 to 2.45
Palaeopropithecus ingens	Mean	+29%	2.48
(N = 4)	Range	+22 to +36	2.35 to 2.61
Babakotia radofilai	Mean	+50%	2.91
(N = 4)	Range	+38 to +60	2.66 to 3.09
Indri indri	Mean	−0.7%	1.92
(N = 7)	Range	−6 to +6.8	
Propithecus diadema	Mean	+3.9%	2.02
(N = 4)	Range	+2.2 to +5.4	
Propithecus verreauxi	Mean	+1.4%	1.97
(N = 5)	Range	−1.8 to +8.4	
Lepilemur mustelinus	Mean	−14.3%	1.67
(N = 4)	Range	−17.2 to −11.5	
Varecia variegata	Mean	−9.4%	1.76
(N = 5)	Range	−21.3 to +3.0	
Hapalemur simus	Mean	+28.8%	2.50
(N = 2)	Range	+27.7 to +29.0	
Cheirogaleus major	Mean	−22.1%	1.52
(N = 4)	Range	−32.1 to −15.1	

[a]The *shearing quotient* is calculated as 100 × (observed shear − predicted shear)/(predicted shear) after Kay *et al.* (1978), Kay (1984), and Covert (1986). Extant data are from Table 4 in Covert (1986). The prediction equation that relates second molar length to total shearing blade length (sum of 6 crests) is $\ln(\text{shear}) = 0.989 \times \ln(\text{molar length}) + 0.684$, courtesy of Dr. Covert. *Shearing ratios* are simply summed shearing crest lengths divided by molar length; these ratios for the living species were calculated from the equation above and the data on molar lengths in Covert (1986).

Positional Behavior

We suspect that total positional repertoires (posture, locomotion, and habitat preferences) are also usually, and perhaps necessarily, underestimated in fossil reconstructions. We are confident in our inferences about the *dominant* element(s) of posture and locomotion (cf. Godfrey *et al.*, 1997a) in subfossil lemurs, and we think it is possible to conclude that some types of locomotion can be excluded or minimized from the bony clues left behind (e.g., flight, swimming and burrowing). While saltatory locomotion is a hallmark of many living lemurs (Walker, 1974; Gebo, 1987; Oxnard *et al.*, 1990), both large body size and limb proportions indicate that leaping was rarely, if ever, practiced by most of the subfossils (Jungers, 1980; Godfrey *et al.*, 1995). We agree with others (Jouffroy, 1963; Jolly, 1970, Walker, 1974; Godfrey, 1988) that terrestrial quadrupedalism was more common in archaeolemurids than in any extant lemur, but we see no evidence for cursoriality in any subfossil (Godfrey *et al.*, 1997b). Although large body size may dictate some degree of terrestriality in the largest subfossils (e.g., *Megaladapis edwardsi*, *Palaeopropithecus maximus*, and *Archaeoindris fontoynontii*), if only for travel between trees, none of these taxa exhibit skeletal signatures that serve to place them exclusively or even comfortably on the ground. We believe that they would have been relatively awkward and slow-moving when they were terrestrial. Overall, we conclude that arboreal behaviors remained very important in most subfossil lemurs despite their large body masses; however, this is not to say that their skeletal designs are all that similar to arboreal anthropoids of comparable body size. After a thorough assessment of limb dominance and and humerofemoral articular indices in mammals, Godfrey *et al.* (1995: 31) noted that most "giant lemurs of Madagascar deviate from other primates in a manner that allies them with nonprimate slow arboreal climbers and hangers" (e.g., sloths, phalangerids, and porcupines). After we revisit limb proportions in subfossils, we will conclude our consideration of positional behavior with a focus on the hands and feet, those musculoskeletal elements at the biomechanical interface between the environment and the rest of the organism. Despite their great value to functional morphologists, manual and pedal remains of subfossils were very rare until quite recently.

Daubentonia robusta is true to its specific nomen; it is a much larger, more robust version of the living aye-aye (Lamberton, 1934; Simons, 1994; Godfrey *et al.*, 1995). *Daubentonia madagascariensis* is a versatile, somewhat deliberate arboreal quadruped that can also leap and climb with ease (Garbutt, 1999). The extinct aye-aye may have come to the ground more often (Godfrey *et al.*, 1997a), and it was probably an even more deliberate quadruped when foraging and traveling in the forest canopy. Its upper limb is longer relative to its lower limb than in its living congener (Jungers, 1980), but both limbs are relatively shorter and when body mass differences are taken into account (see Table IV below). Leaping was probably practiced less frequently by *D. robusta*, but we see nothing that would suggest an increase in suspensory behaviors. However, Godfrey *et al.* (1995) did discover proportional similarities to the striped possum and some mammalian diggers.

Pachylemur was to *Varecia* what *D. robusta* was to *D. madagascariensis*, in the sense that the former is for the most part a much larger, more robust version of the latter in each case. Limb proportions differ in the same way too (Jungers, 1980; Table IV below), but details of appendicular anatomy are remarkably similar in *Pachylemur* and *Varecia*. Ravololonarivo (1990) did find noteworthy differences, however, in the axial skeleton; *Pachylemur* has relatively shorter vertebral bodies and reduced anticliny of its spinous processes. We interpret these axial differences, in conjunction with larger body size, as evidence for less bounding and more emphasis on slow quadrupedalism and

Table IV. Relative Humeral and Femoral Lengths in Subfossil Lemurs and Selected Anthropoids of Known Mass[a]

Species	Humerofemoral[b] index	Mass-adjusted[c] humerus length	Mass-adjusted[d] femur length
Megaladapis edwardsi	110	57.5	52.5
M. grandidieri	113	53.7	46.8
M. madagascariensis	112	55.0	49.0
Pachylemur insignis[e]	86	57.5	67.6
Pachylemur. jullyi	81	52.5	64.6
Daubentonia robusta[e]	80	51.3	63.1
Archaeolemur edwardsi	82	52.5	63.1
Archaeolemur majori[f]	85	57.5	67.6
Hadropithecus stenognathus	103	63.1	61.7
Mesopropithecus pithecoides	91	61.7	67.6
Mesopropithecus globiceps	90	57.5	63.1
Mesopropithecus dolichobrachion	104	66.1	63.1
Babakotia radofilai	115	70.8	61.7
Palaeopropithecus maximus	150	79.7	53.0
Palaeopropithecus ingens	149	79.8	53.7
Palaeoprithecus sp.nov.	155	92.7	59.9
Chimpanzees[g]	102	79.4	78.0
Bonobos[g]	98	85.5	87.1
Gorillas[g]	118	83.4	70.8
Orang-utans[g]	130	93.5	72.1
Siamang[g]	129	118.0	91.8
Spider Monkeys[h]	101	103.3	102.6
Baboons[h]	88	78.7	89.7
Proboscis Monkeys (m)	88	78.5	89.5

[a] Hominoid data are from Jungers (1987).
[b] (Mean maximum humerus length/mean maximum femur length) × 100.
[c] Mean maximum humerus length in mm)/(cube root of body mass in kg).
[d] Mean maximum femur length in mm)/(cube root of body mass in kg).
[e] Specimens limited to Tsirave only.
[f] Specimens limited to Anavoha only.
[g] Males and females pooled.
[h] Males only.

climbing. *Varecia* engages in a considerable amount of suspensory behavior, including hind-limb hanging (Dagosto, 1994; Meldrum *et al.*, 1997), and by analogy and in view of its relatively large femoral head (Godfrey *et al.*, 1995) *Pachylemur* probably did too. Arboreal quadrupedalism would have been the dominant feature of the positional repertoire of *Pachylemur* (both species).

Based on limb proportions and appendicular anatomy, the posture and locomotion of *Megaladapis* has been likened to *Phascolarctos*, the koala of Australia (Walker, 1974; Jungers, 1977, 1978a). The modest leaping abilities of the koala would have no doubt been sacrificed to much larger body size in *Megaladapis*, and terrestrial travel was probably more common in the subfossil species (especially in the largest, *M. edwardsi*) than in the koala. Although there are two distinct morphs in *Megaladapis* that have been recognized at the subgeneric level by Vuillaume-Randriamanantena *et al.* (1992), all three species share the same total biomechanical pattern indicative of arboreal climbing, vertical clinging, cautious quadrupedalism, and probably some beneath-branch suspension. The very reduced spinous processes seen throughout the thoracolumbar vertebral column (Lamberton, 1934; Shapiro *et al.*, 1994) are also consistent with some nontrivial frequency of antiprongrade behaviors in this genus. We defer the discussion of the hands and feet of *Megaladapis* to below, but note here that their relatively huge size and overall grasping design corroborate these conclusions.

The palaeopropithecids exhibit a morphocline of skeletal characters from *Mesopropithecus* through *Babakotia* to *Palaeopropithecus* that reflects increasing amounts of suspensory positional behaviors. *Mesopropithecus* was an arboreal quadruped with better developed antipronogrady capabilities than those seen in *Pachylemur* (Walker, 1974; Godfrey, 1988; Simons *et al.*, 1995), and *Babakotia* was even more suspensory, but hang-feeding specializations culminate in the highly derived skeleton of *Palaeopropithecus* (Lamberton, 1947; Walker, 1974; Jungers, 1980; Jungers *et al.*, 1991, 1997). As a group, the palaeopropithecids have been dubbed the "sloth lemurs" (Simons *et al.*, 1992; also see Carleton, 1936) in view of their constellation of suspensory adaptations and remarkable convergences with arboreal sloths of South America. Limb proportions, limb joint mobility, progressive reduction of vertebral spinous processes, pelvic anatomy, and many details of the hands and feet support this characterization (also see Tardieu and Jouffroy, 1979). If *Palaeopropithecus* was the most sloth-like of the sloth lemurs, then *Archaeoindris* was perhaps the closest thing to a "ground sloth" to evolve in Madagascar (Lamberton, 1946; Jungers, 1980). Nevertheless, by analogy to male gorilla body size and positional behavior (Remis, 1995; Doran, 1997) and in view of its unusual femoral anatomy (e.g., very high collodiaphyseal angle and huge, globular femoral head), we suspect that *Archaeoindris* still frequented the trees to feed and perhaps to nest.

As was noted above, the archaeolemurids exhibit a variety of features in their postcrania that are best interpreted as evidence for a substantial amount of terrestrial quadrupedalism. For example, the elbow displays a retroflexed olecranon process and dorsally wrapped medial epicondyle; these and other aspects of the forelimb are convergences on terrestrial cercopithecines (Jolly, 1970; Walker, 1974; Godfrey, 1988). By analogy to other mammalian quadrupeds, a more extended limb posture is implied by this articular configuration (Biewener, 1990). Other parts of the skeleton (e.g., scapula and pelvis, carpus), however, argue for the retention of scansorial/arboreal abilities (Godfrey, 1988; Hamrick, 1995; Hamrick *et al.*, 2000), and it is now clear that the familiar baboon analogy for *Archaeolemur* is overstated and potentially misleading (Godfrey *et al.*, 1997a). A recent reassessment of materials attributed to *Hadropithecus* disclosed prior errors of attribution that had influenced earlier reconstructions, especially those based on fore- and hindlimb proportions

(Godfrey *et al.*, 1997b). Although the forelimb to hindlimb ratio is higher in *Hadropithecus* than in *Archaeolemur*, the two are very similar overall in their stocky and robust postcrania, and neither exhibits the hallmarks of high-speed, overground locomotion. New manual and pedal specimens of *Archaeolemur* point to a higher incidence of terrestrial locomotion than observed in any living lemur, but they also fail to reveal any special resemblance to either baboons or geladas in degree of prehensility (Jungers *et al.*, 1998).

Table IV summarizes much of what we now know about interlimb proportions in subfossil lemurs. Forelimb to hindlimb ratios (humerofemoral and intermembral indices) are typically higher in all subfossils than in any living lemurid or indrid. Because leaping prowess is associated with lower ratios in all primates (Jungers, 1985), it is reasonable to conclude that saltatory locomotion was a relatively rare feature of any subfossil's positional repertoire. Although the humerofemoral index of *Archaeolemur* appears similar (>80) to large-bodied cercopithecoids, mass-adjusted measures of long bone length (Jungers, 1985, 1987; Vogel, 1988) indicate that both elements are relatively much shorter in the subfossils (including *Hadropithecus*). The same is true for *Pachylemur* and the giant extinct aye-aye; the humerofemoral index of each is higher than that seen in either *Varecia* (70.5) or *D. madagascariensis* (69.5), but both limb segments are actually shorter relative to body mass. Although sloth lemurs are closely related to living indrids, their humerofemoral proportions are decidedly different (Fig. 7) and more closely resemble living hominoids and atelids (as well as living sloths). The same is true of *Megaladapis* with humerofemoral indices well over 100 in all three species. However, the high humerofemoral indices seen in both palaeopropithecids and *Megaladapis* have a fundamentally different allometric basis than those of anthropoids: with the exception of the undescribed new species of *Palaeopropithecus*, the subfossils have relatively shorter humeri (mass-adjusted) than most living apes in conjunction with relatively *very* short femora. In fact, all subfossil lemurs have relatively short femora in comparison to hominoids, many cercopithecoids, atelids, and the largest living lemurs. This implies that subfossil lemurs distributed their body mass quite differently than anthropoids and probably had relatively and absolutely "massive axial skeletons" (Godfrey *et al.*, 1995).

Relatively few hand and foot bones were known for the subfossils until recently, and even fewer were associated (Lamberton, 1939; Decker and Szalay, 1974; Szalay and Delson, 1979). Recovery of associated and relatively complete specimens over the last two decades from cave deposits in the north (Simons *et al.*, 1990) and northwest (MacPhee *et al.*, 1984; Burney *et al.*, 1997) have remedied this gap to a large extent. We now have digital elements from all genera except *Pachylemur*, and it is possible to assemble more or less complete hands and feet for *Babakotia*, *Megaladapis* (Fig. 8), *Archaeolemur* (Fig. 9), and *Palaeopropithecus*. The foot of *Megaladapis* has been described and analyzed (Wunderlich *et al.*, 1996), and the carpal bones of several species have recently been scrutinized (Hamrick *et al.*, 2000), but much work remains to be done. One of the most obvious differences between living and most subfossil lemurs

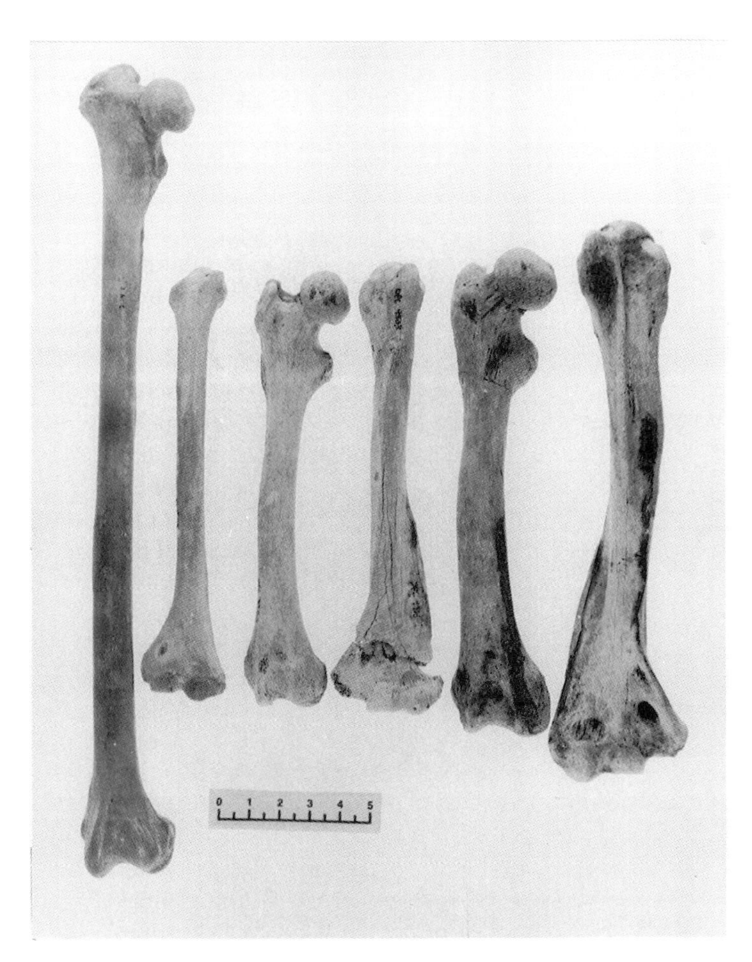

Fig. 7. Humeri and femora of (left to right) *Indri indri*, *Mesopropithecus dolichobrachion*, and *Babakotia radofilai*. The hind-limb dominance of extant indrids is lost in the more suspensory sloth lemurs.

is the *relative* size of the hands and feet. As a percentage of the rest of the fore and hind limbs, the hands and feet of *Babakotia*, *Megaladapis*, and *Palaeopropithecus* are very long (Fig. 10). Among other primates, only the orang-utan approximates the condition seen in these subfossil genera. The foot of *Megaladapis* is enormous compared to its femur and tibia, and the very divergent hallux is huge and robust (Wunderlich *et al.*, 1996); the impression is that of a powerful pincer or clamp-like grasping organ. The hand of *Megaladapis* is similar in this design.

The hands and feet of the sloth lemurs are also very elongate, but the hallux and pollex are reduced in contrast to the long metapodials and phalanges; this hook-like configuration implies hanging rather than vise-like

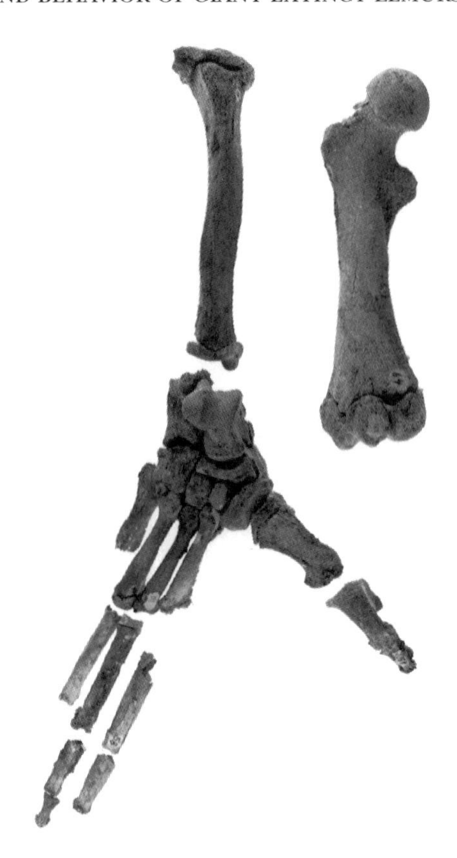

Fig. 8. The associated foot, tibia, and femur of a subadult *Megaladapis madagascariensis*. Note the enormous size of the foot and the strongly divergent hallux.

gripping. The hindfoot of *Megaladapis* remains large and is loris-like in some respects (e.g., curvature of the calcaneal tuberosity; inverted set or orientation), while the tarsus of sloth lemurs is relatively reduced and highly derived, especially in *Palaeopropithecus* (Wunderlich *et al.*, 1994). In other words, when it comes to hands and feet, there was more than one way to be a successful, large-bodied arboreal folivore in Madagascar.

The hands and feet of *Archaeolemur* are relatively very short. Relative foot length is most similar to that seen in the living indrids (Fig. 10), but this similarity is more apparent than real. Indrids have large grasping feet (Gebo and Dagosto, 1988), but they are short relative to their very long, leaping hind limbs; relative to body size, *Archaeolemur* has relatively short feet *and* short hindlimbs (Table IV). All digits are also short in *Archaeolemur*, including the hallux and pollex, and overall intrinsic proportions are more mandrill-like than baboon-like (Jungers et al., 1998). The very broad apical tufts of the distal phalanges (Fig. 9) distinguish *Archaeolemur* from all other prosimians and most

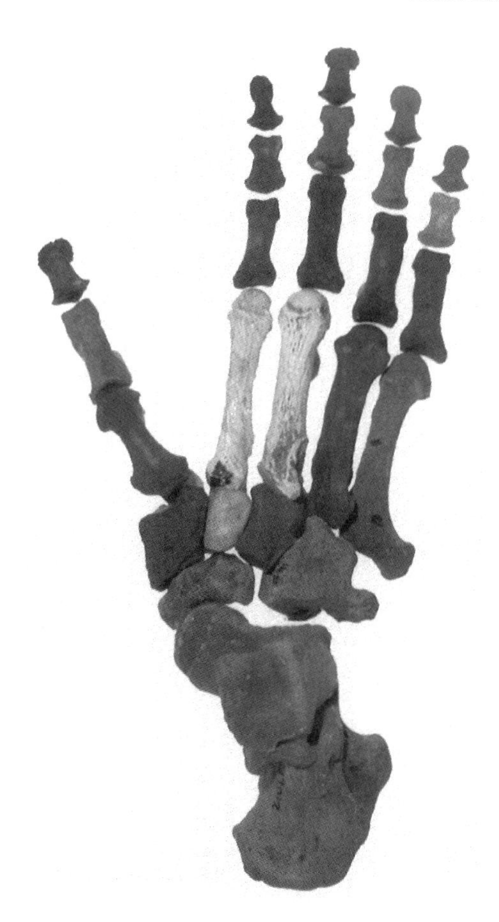

Fig. 9. The associated pedal remains of *Archaeolemur* from Anjohikely. The second and third metatarsals as well as the mesocuneiform are from another individual. The hallucal metatarsal and all phalanges are short and robust; note the expanded apical tufts of the distal phalanges. The lateral margin of the foot sports pronounced tuberosities on the calcaneus, cuboid, and fifth metatarsal.

anthropoids, but we are not sure what their functional significance might have been. They occur on the digits of both hand and foot, and any attempt to associate them directly with sensitive fingertips for refined manipulation is offset by relatively low indices of manual prehensility. The dorsal extension of the articular surfaces of the metapodial heads implies that both hands and feet frequently adopted digitigrade postures. The ankle and subtalar joints point to a more everted set to the foot than seen in living lemurs (Decker and Szalay, 1974), a geometry compatible with more frequent use of terrestrial substrates.

We conclude our discussion of positional behavior and habitat use by consideration of proximal phalangeal curvature in subfossil lemurs. Phalangeal

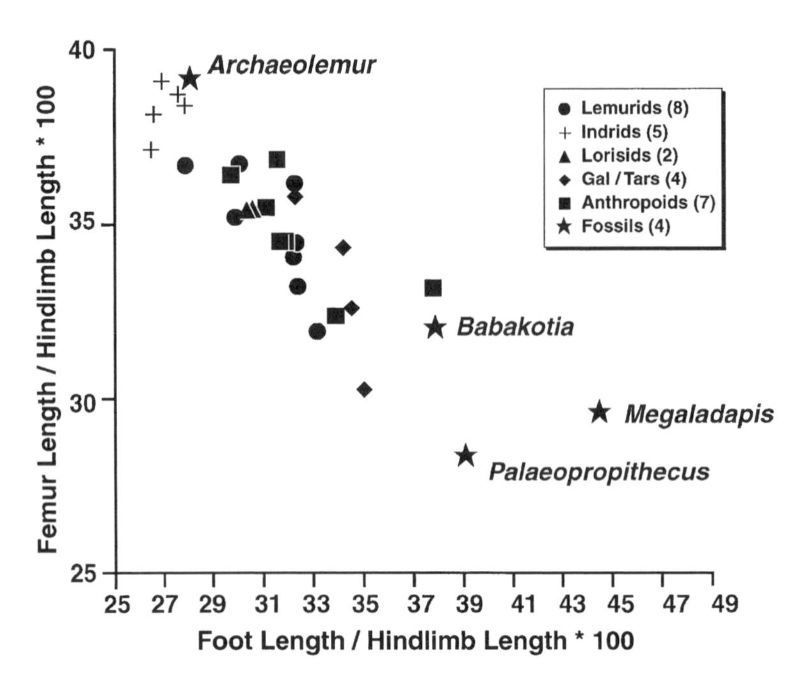

Fig. 10. A bivariate plot of relative foot length (X axis) versus relative femur length (Y axis). *Megaladapis* and the sloth lemurs have relatively very long feet, while that of *Archaeolemur* is quite short. Comparative data derive for the most part from Jouffroy and Lessertiseur (1978).

curvature has proven to be an especially valuable tool in the analysis of fossil form and function (see Stern *et al.*, 1995 for a recent review). It is relatively easy to quantify as the "included angle" (theta) with either coordinate calipers or from video images (Fig. 11). Analogy alone has been useful in connecting degree of curvature to substrate use; straight phalanges occur in more terrestrial species and curved phalanges in more arboreal species, especially in suspensory ones (e.g., Susman *et al.*, 1984; Hamrick *et al.*, 1995; Jungers *et al.*, 1997). Recent work on the ontogeny and biomechanics of phalangeal curvature provides causal links between its expression and distribution among primates (Richmond, 1998, in press). For example, finite element analysis (FEA) has been used to test the longstanding hypothesis that a curved phalanx experiences lower bending stresses than a straight phalanx (Preuschoft, 1970, 1973). FEA is a powerful numerical tool in which a structure is divided into numerous simple elements with defined mechanical properties, loads are applied at specific points (nodes), and stress and strain are calculated for each element to obtain an approximate solution of the structure as a whole (Cook *et al.*, 1989; Beaupre and Carter, 1992). Figure 12 (top) presents an FEA of a siamang third proximal phalanx during unimanual suspensory loading; a second, straight version (bottom) was created by mathematically translating

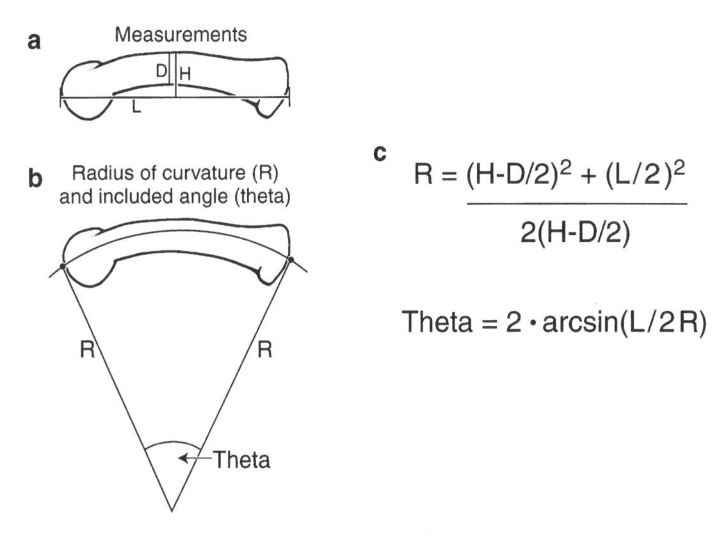

Fig. 11. Measurements used to calculate phalangeal curvature as the included angle and incidental ingestion of grit (theta).

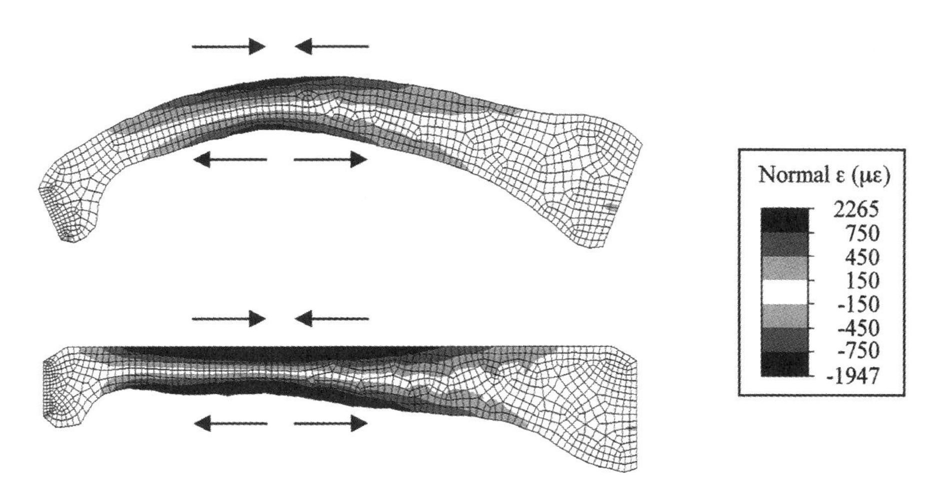

Fig. 12. Under the loading conditions of unimanual suspension, finite element models indicate that the siamang proximal phalanx experiences dorsal concave bending such that the dorsal surface is compressed (inward arrows) and the palmar surface tensed (outward arrows). The normally curved phalanx (at top) experiences substantially lower strains than a mathematically straightened version (at bottom) of the same length, depth, and area. Normal strain along the long axis of the bone is depicted in units of microstrain.

and rotating the digitized points of a CAT scan to retain the same length, depth, and area (Richmond, in press). ALGOR software was employed in the meshing and analytical steps, and mechanical properties of cortical bone were used (Young's modulus = 18 GPa; Poisson's ratio = 0.3). Loads were estimated from a prior biomechanical model that incorporates lesser ape kinematics, electromyography, and manual pressures along with siamang body weight and anatomy. When computed under the same loading conditions, both models experience dorsally concave bending, with dorsal compression and palmar tension (Fig. 12). However, peak strain magnitudes (in Mpa) in the straight model are twice those seen in the curved, realistic model. The difference in curvature alone accounts for this difference, essentially verifying Preuschoft's theoretical hypothesis. The high forces and resultant stresses generated during digital flexion and gripping in many different arboreal contexts are best resisted by some degree of phalangeal curvature.

The ontogeny of phalangeal curvature also confirms that the final form of an adult phalanx is the result of bony modeling and remodeling that takes place throughout life as positional behavior changes. Figure 13 plots both third phalangeal curvature (solid circles) and percentage of arboreal support use (open squares) at successive dental stages from birth to adulthood in gorillas. The gorilla locomotor data are taken from Doran (1997); the curvature data are from Richmond (1998). After an initial increase in both following birth, there is a steady decline to adult levels of modest degree of curvature in

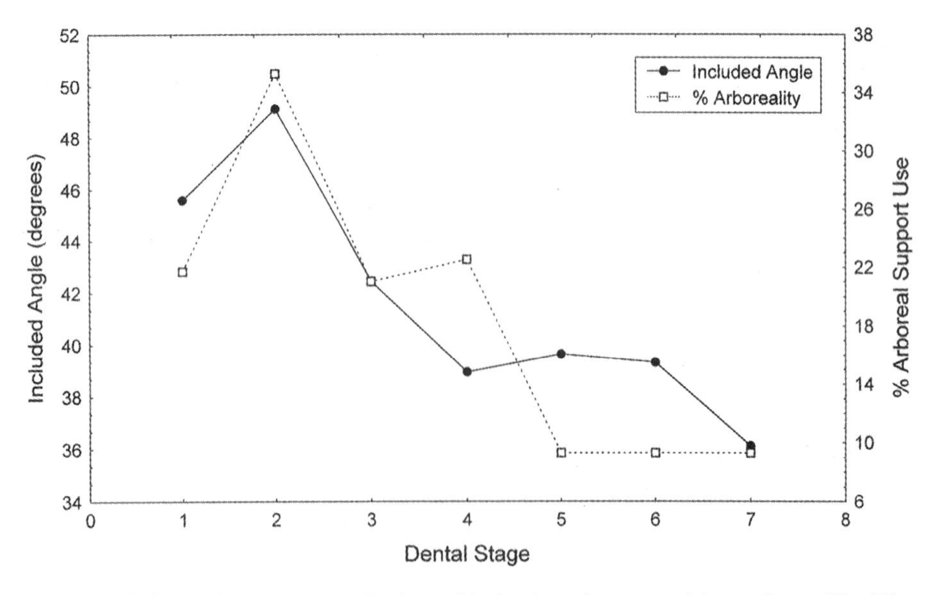

Fig. 13. Phalangeal curvature and arboreal behaviors decrease with age in gorillas.The ontogenetic plasticity of phalangeal curvature indicates that form is intimately linked to mechanical function and substrate use.

conjunction with occasional arboreality. Paciulli (1995) found similar trends in chimpanzee phalanges and positional behavior, including an increase in curvature from fetal stages to early infancy. Clearly, degree of phalangeal curvature is sensitive to frequency and context of mechanical stresses, and as such provides a valuable glimpse into hand and foot use in extinct species.

Table V summarizes the available information on phalangeal curvature in subfossil lemurs and an assortment of living primates. Manual and pedal phalanges are pooled, unless noted otherwise. The southwestern sample of *Megaladapis* phalanges was subdivided into a large morph (cf. *M. edwardsi*) and a small morph (cf. *M. madagascariensis*), but the differences are not significant. The sloth lemurs have the highest degree of phalangeal curvature among the subfossils, roughly twice that seen in their closest living relatives, the indrids. Only highly suspensory anthropoids approach or equal the sloth lemurs in this respect (e.g., orang-utans, siamang, atelines). The three *Megaladapis* species all exhibit a moderate degree of curvature, not unlike that seen in chimpanzees and bonobos. *Archaeolemur* has the straightest phalanges of any of the subfossil species, but they are not as straight as the pedal phalanges of baboons; they do overlap with *Nasalis* to some extent. These findings are consistent with and serve to corroborate many of the previously mentioned inferences about the positional behaviors of extinct lemur species. Hang-feeding and other anti-pronograde behaviors were most common in sloth lemurs, powerful arboreal grasping was commonplace in *Megaladapis*, and *Archaeolemur* engaged in a significant amount of terrestrial quadrupedalism. In other words, there was much greater diversity in lemur positional behavior in the recent past, so much so that we must reach far beyond the order Primates in our search for suitable analogues.

Summary

All of the recently extinct (subfossil) lemurs were substantially larger than the largest living Malagasy species, and some of them were truly giant in comparison. Despite such large body size, behavior analogies to comparably sized anthropoids are of limited utility due to phylogenetic inertia and alternative adaptive solutions. Not unlike their living counterparts on Madagascar but very different from large-bodied anthropoids, the subfossils exhibit no sexual dimorphism in either adult body size or canine size. We suspect that this monomorphism is related to similar levels of agonistic behavior in both sexes within a species, while species-specific differences in relative canine height suggest more overall aggression and competition in some (e.g., *Archaeolemur*) and less in others (e.g., *Hadropithecus*).

Activity cycles are complex and sometimes seasonal among living lemurs, and cathemerality is common. The largest living nocturnal primate is the aye-aye, so any extrapolation to much larger body size is problematic. Orbits

Table V. Curvature of Proximal Phalanges in Subfossil Lemurs and Selected Living Primates[a]

Species	Included angle (degrees)	Coefficient of variation
Megaladapis (SW, cf. *edwardsi*) N = 18	49.2	15.9
Megadadapis (SW, cf. *madagascariensis*) N = 23	46.1	17.6
Megaladapis grandidieri N = 3	42.4	—
Megaladapis sp. (N & NW) N = 5	43.9	15.3
Archaeolemur sp. (N & NW) N = 28	27.9	22.6
Babakotia radofilai N = 38	59.1	10.7
Mesopropithecus dolichobrachion N = 5	65.8	10.2
Palaeopropithecus maximus N = 13	57.2	17.1
Palaeopropithecus ingens N = 45	60.3	9.9
Palaeopropithecus sp. (Anjohibe) N = 11	73.3	10.0
Indri indri N = 130	33.1	18.7
Propithecus diadema N = 51	31.8	16.7
Varecia variegata N = 40	51.6	9.5
Ateles species N = 78	54.5	11.2
Lagothrix lagotricha N = 16	59.1	11.2
Nasalis larvatus N = 26	36.5	13.4
Papio hamadryas N = 19 (foot only)	11.1	60.4
Hylobates syndactylus N = 20 (hand only)	53.1	10.5
Pan paniscus N = 92	41.5	14.5
Pn troglodytes N = 100	42.0	13.3
Gorilla gorilla N = 119	35.1	12.1
Pongo pygmaeus N = 112	75.0	13.3

[a]Values are for pooled manual and pedal proximal phalanges, except where noted. Coefficient of variation = 100 × (standard deviation/mean).

of subfossils are relatively small compared to extant baselines, regardless of the size variable selected, but optic canals are also relatively very small. The former observation is consistent with the reconstruction of a diurnal activity pattern for most of the subfossils, but the latter points to reduced visual acuity in some taxa (e.g., *Megaladapis* and *Palaeopropithecus*) and might imply high degrees of retinal summation suitable for nocturnal behaviors.

While some subfossils have modified their anterior dentitions for food procurement and ingestion, and thereby sacrificed the typical strepsirrhine tooth comb (archaeolemurids, *Palaeopropithecus* and *Archaeoindris*), microwear analysis confirms fur-grooming behavior in those that preserved the tooth comb (*Megaladapis*, *Babakotia*, and probably *Mesopropithecus* and *Pachylemur*). The giant extinct aye-aye possessed ever-growing, chisel-like incisors and a filiform third manual digit just like those in its living congener; a diet of nuts and fruits supplemented by insects is implied thereby. Primary frugivory is reconstructed for *Pachylemur*, hard-object feeding for *Hadropithecus*, and omnivory for *Archaeolemur* from a variety of sources (e.g., dental morphology, incidence of caries, molar microwear, and fecal pellets). The exceptionally well-developed shearing crests on the molars of *Megaladapis*, *Palaeopropithecus*, and *Babakotia* confirm a high degree of folivory consistent with other anatomical details and molar microwear, although we suspect that the sloth lemurs were probably seed-predators too (not unlike living indrids).

The reconstructed positional repertoires of subfossil lemurs greatly expand the observed extant envelope of such behaviors. Although *Archaeolemur* and *Hadropithecus* exhibit features indicative of a substantial amount of terrestrialism (articular shape and digital design in particular), more so than in any living prosimian, neither should be described as cursorial or baboon-like. Unlike the archaeolemurids, most other subfossil lemurs are reconstructed here as primarily arboreal. In our assessment, some were deliberate quadrupeds (*Pachylemur* and *Daubentonia robusta*), while others were more suspensory to varying degrees (including all sloth lemurs), culminating in the extreme hook-like hanging adaptations of *Palaeopropithecus*. Sloth lemurs tended to de-emphasize pollical and hallucal grasping, but the species of *Megaladapis* relied instead on enormous, pincer-like hands and feet to negotiate arboreal pathways (and the hands and feet were relatively long in both groups). The leaping locomotion so characteristic of many living lemurs appears to have been sacrificed in virtually all of the subfossils; the hindlimbs of all of the extinct lemurs are relatively very short and robust. We need to reach far beyond anthropoid analogies to conjure up the probable behaviors and adaptations of the giant extinct lemurs of Madagascar.

ACKNOWLEDGMENTS

The research reported here was supported in part by funds from the National Science Foundation (Grant SBR-9630350 to ELS and Grant SBR-

9624726 to WLJ and BGR), the Boise Fund, the Margot Marsh Foundation and the School of Medicine at Stony Brook (to WLJ), and the University of Massachusetts at Amherst (to LRG). Luci Betti-Nash prepared Figures 1, 2, 10, and 11. Fred Grine provided expert assistance in obtaining the electron micrographs. Thanks are due to David Burney and Helen James for access to remarkably complete *Archaeolemur* hands and feet. We are deeply indebted to our Malagasy colleagues and the many curators of museum collections for access to valuable specimens in their care. This is Duke University Primate Center Publication number 701.

References

Albignac, R. 1981. Variabilite dans l'organisation territoriale et l'ecologie de *Avahi laniger*. C. R. Acad. Sci. Paris **292:**331–334.

Albrecht, G. H., Jenkins, P. D., and Godfrey, L.R. 1990. Ecogeographic size variation among the living and subfossil prosimians of Madagascar. *Am. J. Primatol.* **22:**1–50.

Avis, V. 1961. The significance of the angle of the mandible: an experimental and comparative study. *Am. J. Phys. Anthropol.* **19:**55–61.

Baum, D. A. 1995. A systematic revision of *Adansonia* (Bombacaceae). *Ann. Missouri Bot. Gard.* **82:**440–470.

Beaupre, G. S., and Carter, D. S. 1992. Finite element analysis in biomechanics. In: A. A. Biewener (ed.), *Biomechanics—Structures and Systems: A Practical Approach,* pp. 149–174. IRL Press, Oxford.

Biewener, A. A. 1990. Biomechanics of mammalian terrestrial locomotion. *Science* **250:**1097–1103.

Bock, W. J., and von Wahlert, G. 1965. Adaptation and the form-function complex. *Evolution* **19:**269–299.

Buettner-Janusch, J., and Andrew, R. J. 1962. The use of the incisors by primates in grooming. *Am. J. Phys. Anthropol.* **20:**127–129.

Burney, D. A. 1997. Theories and facts regarding Holocene environmental change before and after human colonization. In: S. M. Goodman, and B. D. Patterson (eds.), *Natural Change and Human Impact in Madagascar*. pp. 75–89. Smithsonian Press, Washington, DC.

Burney, D. A., James, H. F., Grady, F. V., Rafamantanantsoa, J.-G., Ramilisonina, Wright, H. T., and Cowart, J. B. 1997. Environmental change, extinction, and human activity: Evidence from caves in NW Madagascar. *J. Biogeogr.* **24:**755–767

Carleton, A. 1936. The limb-bones and vertebrae of the extinct lemurs of Madagascar. *Proc. Zool. Soc. London* **106:**281–307.

Cartmill, M. 1975. Strepsirhine basicranial structures and the affinities of the Cheirogaleidae. In: W. P. Luckett and F. S. Szalay (eds.), *Phylogeny of the Primates: A Multidisciplinary Approach*. pp. 313–354. Plenum Press, New York.

Charles-Dominique, P., and Hladik, C. M. 1971. Le *Lepilemur* du sud de Madagascar: Ecologie, alimentation et vie sociale. *La terre et la vie* **25:**3–66.

Cheverud, J. M., Dow, M. M., and Leutenegger, W. 1985. The quantitative asessment of phylogenetic constraints in comparative analyses: sexual dimorphism in body weight among primates. *Evolution* **39:**1335–1351.

Clutton-Brock, T. H., Harvey, P. H., and Rudder, B. 1977. Sexual dimorphism, socionomic sex ratio, and body weight in primates. *Nature* **269:**191–195.

Cook, R. D., Malkus, D. S., and Plesha, M. E. 1989. *Concepts and Applications of Finite Element Analysis*. John Wiley and Sons, New York.

Covert, H. H. 1986. Biology of early Cenozoic primates. In: D. R. Swindler, and J. Erwin (eds.), *Comparative Primate Biology*. Vol.1. *Systematics, Evolution, and Anatomy*, pp. 335–359. Alan R. Liss, New York.

Crovella, S., Montagnon, D., Rakotosamimanana, B., and Rumpler, Y. 1994. Molecular biology and systematics of an extinct lemur: *Pachylemur insignis*. *Primates* **35:**519–522.

Curtis, D. J., Zaramody, A., and Martin, R. D. 1999. Cathemerality in the mongoose lemur, *Eulemur mongoz*. *Am. J. Primatol.* **47:**279–298.

Dagosto, M. 1994. Testing positional behavior of Malagasy lemurs: a randomization approach. *Am. J. Phys. Anthropol.* **94:**189–202.

Daniels, H. L. 1984. Oxygen consumption in *Lemur fulvus*: deviation from the ideal model. *J. Mammal* **65:**584–592.

Decker R. L, and Szalay F. S. 1974. Origins and function of the pes in the Eocene Adapidae (Lemuriformes, Primates). In: F. A. Jenkins, Jr. (ed.), *Primate Locomotion*, pp. 261–291. Academic Press, New York.

Dewar, R. E. 1997. Were people responsible for the extinction of Madagascar's subfossils, and how will we ever know? In: S. M. Goodman, and B. D. Patterson (eds.) *Natural Change and Human Impact in Madagascar*. pp. 364–377. Smithsonian Press, Washington, DC.

Doran, D. M. 1997. Ontogeny of locomotion in mountain gorillas and chimpanzees. *J. Hum. Evol.* **32:**323–344.

Elisabete, M., and Palka, S. 1994. Feeding behavior and activity patterns of two Malagasy bamboo lemurs, *Hapalaemur simus* and *Hapalemur griseus*, in captivity. *Folia Primatol.* **63:**44–49.

Ely, J., and Kurland, J. A. 1989. Spatial autocorrelation, phylogenetic constraints, and the causes of sexual dimorphsim in primates. *Int. J. Primatol.* **10:**151–171.

Engqvist, A., and Richard, A. 1991. Diet as a possible determinant of cathemeral activity patterns in primates. *Folia Primatol.* **57:**169–172.

Erickson, C. J. 1991. Percussive foraging in the aye-aye (*Daubentonia madagascariensis*). *Anim. Behav.* **41:**793–801.

Fleagle, J. G. 1988. *Primate Adaptation and Evolution*, First edition. Academic Press, New York.

Fleagle, J. G. 1999. *Primate Adaptation and Evolution*, Second edition. Academic Press, New York.

Ganzhorn, J. U. 1985. Some aspects of the natural history and food selection of *Avahi laniger*. *Primates* **26:**452–463.

Ganzhorn, J. U. 1993. Flexibility and constraints of *Lepilemur* ecology. In: P. M. Kappeler, and J. U. Ganzhorn (eds.) *Lemur Social Systems and their Ecological Basis*. pp. 153–165. Plenum Press, New York.

Ganzhorn, J. U., Abraham, J. P., and Razanahoera-Rakotomalala, M. 1985. Some aspects of the natural history and food selection of *Avahi laniger*. *Primates* **26:**452–463.

Garbutt, N. 1999. *Mammals of Madagascar*. Yale University Press, New Haven.

Gaulin, S. J. C., and Sailer, L. D. 1984. Sexual dimorphism in weight among the primates: the relative impact of allometry and sexual selection. *Int. J. Primatol.* **5:**515–535.

Gebo, D. L. 1987. Locomotor diversity in prosimian primates. *Am. J. Primatol.* **13:**271–281.

Gebo, D. L., and Dagosto, M. 1988. Foot anatomy, climbing, and the origin of the Indriidae. *J. Hum. Evol.* **17:**135–154.

Gilmartin, A. M., and Jungers, W. L. 1995. Form and function of the "toothcomb" of subfossil lemurs: a dental microwear study. *Am. J. Phys. Anthropol.* **Suppl. 20:**99.

Gingerich, P. D. 1975. Dentition of *Adapis parisiensis* and the evolution of lemuriform primates. In: I. Tattersall, and R. W. Sussman (eds.), *Lemur Biology*, pp. 65–80. Plenum Press, New York.

Gingerich, P. D., and Martin, R. D. 1981. Cranial morphology and adaptations in Eocene Adapidae.II. The Cambridge skull of *Adapis parisiensis*. *Am. J. Phys. Anthropol.* **56:**235–257.

Gingerich P. D., and Rose, K. D. 1979. Anterior dentition of the Eocene condylarth *Thryptacodon*: convergence with the tooth comb of lemurs. *J. Mammal.* **60:**16–22.

Glander, K. E., Wright, P. C., Seigler, D. S., and Randrianasolo, B. 1989. Consumption of cyanogenic bamboo by a newly discovered species of bamboo lemur. *Am. J. Primatol.* **19:**119–124.

Godfrey, L. R. 1988. Adaptive diversification of Malagasy strepsirrhines. *J. Hum. Evol.* **17:**93–134.

Godfrey, L. R., Simons, E. L., Chatrath, P. S., and Rakotosamimanana, B. 1990. A new fossil lemur (*Babakotia*, Primates) from northern Madagascar. *C. R. Acad. Sci. Paris* **310** (Sèrie II):81–87.

Godfrey, L. R., Lyon, S. K., and Sutherland, M. R. 1993. Sexual dimorphism in large-bodied primates: the case of the subfossil lemurs. *Am. J. Phys. Anthropol.* **90:**315–334.

Godfrey, L. R., Sutherland, M. R., Paine, R. R., Williams, F. L., Boy, D. S., and Vuillaume-Randriamanantena, M. 1995. Limb joint surface areas and their ratios in Malagasy lemurs and other mammals. *Am. J. Phys. Anthropol.* **97:**11–36.

Godfrey, L. R., Jungers, W. L., Reed, K. E., Simons, E. L., and Chatrath, P. S. 1997a. Subfossil lemurs: Inferences about past and present primate communities In: S. M. Goodman, and B. D. Patterson (eds.), *Natural Change and Human Impact in Madagascar*, pp. 218–256. Smithsonian Institution, Washington, DC.

Godfrey, L. R., Jungers, W. L., Wunderlich, R. E., and Richmond, B. G. 1997b. Reappraisal of the postcranium of *Hadropithecus* (Primates, Indroidea). *Am. J. Phys. Anthropol.* **103:**529–556.

Gonzalez, W. G., Kay, R. F., and Kirk, E. C. 1998. Optic canal and orbit size—implications for the origins of diurnality and visual acuity in primates. *Am. J. Phys. Anthropol.* **Suppl. 26:**87.

Goodman, S. M., O'Connor, S., and Langrand, O. 1993. A review of predation on lemurs: implications for the Evolution of social behavior in small, nocturnal primates. In: P. M. Kappeler, J. U. Ganzhorn (eds.) *Lemur Social Systems and Their Ecological Basis,* pp. 51–66. Plenum Press, New York.

Gould, S. J., and Vrba, E. S. 1982. Exaptation—a missing term in the science of form. *Paleobiology* **8:**4–15.

Grandidier, G. 1929. Une variete du *Cheiromys madagascariensis* actuel et un nouveau *Cheiromys* subfossile. *Bull. Acad. Malgache* **11:**101–107.

Hamrick, M. W. 1995. *Carpal Joint Morphology and Function in the Strepsirhine Primates*. Doctoral dissertation, Northwestern University, Evanston, IL.

Hamrick, M. W., Meldrum, D. J., and Simons E. L. 1995. Anthropoid phalanges from the Oligocene of Egypt. *J. Hum. Evol.* **28:**121–145.

Hamrick, M. W., Simons, E. L., and Jungers, W. L. (2000). New wrist bones of Malagasy giant subfossil lemurs. *J. Hum. Evol.* **38:**635–650.

Harcourt, C. 1991. Diet and behavior of a nocturnal lemur, *Avahi laniger*, in the wild. *J. Zool. London* **223:**667–674.

Hemingway, C. A. 1996. Morphology and phenology of seeds and whole fruit eaten by Milne-Edwards' sifakas, *Propithecus diadema edwardsi*, in Ranomafana National Park, Madagascar. *Int. J. Primatol.* **17:**637–659.

Hylander, W. 1975. Incisor size and diet in anthropoids with special reference to Cercopithecidae. *Science* **189:**1095–1098.

Iwano, T., and Iwakawa, C. 1988. Feeding behavior of the aye-aye (*Daubentonia madagascariensis*) on nuts of ramy (*Canarium madagascariensis*). *Folia Primatol.* **50:**136–142.

Iwano, T., and Rakotoarisoa, G. 1998. On the niche of the aye-aye (*Daubentonia madagascariensis*). Abstracts XVIIth Cong. Int. Primatol. Soc., no. 295.

Jacobs, L. L. 1981. Miocene lorisid primates from the Pakistan Siwaliks. *Nature* **289:**585–587.

Jenkins, P. D. 1987. *Catalogue of the Primates in the British Museum (Natural History) and elsewhere in the British Isles. Part IV. Suborder Strepsirrhini, including the Subfossil Madagascan Lemurs and Family Tarsiidae*. British Museum (Natural History), London.

Jenkins, P. D., and Albrecht, G. H. 1991. Sexual dimorphism and sex ratios in Madagascan prosimians. *Am. J. Primatol.* **24:**1–14.

Jolly, C. J. 1970. *Hadropithecus*: a lemuroid small-object feeder. *Man* **5:**619–626.

Jouffroy, F.-K. 1963. Contribution à la connaissance du genre *Archaeolemur* Filhol, 1895. *Ann. Paléontol.* **49:**129–155.

Jouffroy, F.-K., and Lessertiseur, J. 1978. Etude ecomorphologiqie des proportions des membres des Primates et specialement des Prosimiens. *Ann. Sci. Nat. Zool. Biol. Animale.* **20:**99–128.

Jungers, W. L. 1977. Hindlimb and pelvic adaptations to vertical climbing and clinging in *Megaladapis*, a giant subfossil prosimian from Madagascar. *Yearb. Phys. Anthropol.* **20** (for 1976):508–524.

Jungers, W. L. 1978a. The functional significance of skeletal allometry in *Megaladapis* in comparison to living prosimians. *Am. J. Phys. Anthropol.* **49:**303–314.

Jungers, W. L. 1978b. On canine reduction in early hominids. *Curr. Anthropol.* **19:**155–156.

Jungers, W. L. 1980. Adaptive diversity in subfossil Malagasy prosimians. *Z. Morph. Anthropol.* **71:**177–186.

Jungers, W. L. 1985. Body size and scaling of limb proportions in primates. In: W. L. Jungers (ed.), *Size and Scaling in Primate Biology*, pp. 345–381. Plenum Press, New York.

Jungers, W. L. 1987. Body size and morphometric affinities of the appendicular skeleton in *Oreopithecus bambolii* (IGF 11778). *J. Hum. Evol.* **16:**445–456.

Jungers, W. L. 1990. Problems and methods in reconstructing body size in fossil primates. In: J. Damuth, and B. J. MacFadden (eds.), *Body Size in Mammalian Paleobiology*, pp. 103–118. Cambridge University Press, Cambridge.

Jungers, W. L. 1999. Brain size and body size in subfossil Malagasy lemurs. *Am. J. Phys. Anthropol.* **Suppl.28:**163.

Jungers, W. L., Godfrey, L. R., Simons, E. L., Chatrath, P. S., and Rakotosamimanana, B. 1991. Phylogenetic and functional affinities of *Babakotia radofilai*, a new fossil lemur from Madagascar. *Proc. Natl. Acad. Sci. USA* **88:**9082–9086.

Jungers, W. L., Falsetti, A. B., and Wall, C. E. 1995. Shape, relative size and size-adjustments in morphometrics. *Yearb. Phys. Anthropol.* **38:**137–161.

Jungers, W. L. Godfrey, L. R., Simons, E. L., and Chatrath, P. S. 1997. Phalangeal curvature and positional behavior in extinct sloth lemurs (Primates, Palaeopropithecidae). *Proc. Natl. Acad. Sci. USA* **94:**11998–12001.

Jungers, W. L., Wunderlich, R. E., Lemelin, P., Godfrey, L. R., Burney, D. A., Simons, E. L., Chatrath P. S., and James, H. F. 1998. New hands and feet for an old lemur (*Archaeolemur*). *Am. J. Phys. Anthropol.* **Suppl. 26:**131–132.

Kappeler, P. M. 1991. Patterns of sexual dimorphism in body weight among prosimian primates. *Folia Primatol.* **57:**132–146.

Kappeler, P. M. 1996. Intrasexual selection and phylogenetic constraints in the *Evolution* of sexual canine dimorphism in strepsirhine primates. *J. Evol. Biol.* **9:**43–65.

Kappeler, P. M., and Ganzhorn, J. U. 1993. *Lemur Social Systems and Their Ecological Basis*. Plenum Press, New York.

Kay, R. F. 1975. The functional adaptations of primate molar teeth. *Am. J. Phys. Anthropol.* **43:**195–216.

Kay, R. F. 1977. The *Evolution* of molar occlusion in the Cercopithecidae and early catarrhines. *Am. J. Phys. Anthropol.* **46:**327–352.

Kay, R. F. 1984. On the use of anatomical features to infer foraging behavior in extinct primates. In: P. S. Rodman, and J. G. H. Cant (eds.), *Adaptations for Foraging in Nonhuman Primates*, pp. 21–53. Columbia University Press, New York.

Kay, R. F, and Cartmill, M. 1977. Cranial morphology and adaptations of *Palaechthon nacimienti* and other Paromomyidae (Plesiadapoidea, ? Primates), with a description of a new genus and species. *J. Hum. Evol.* **6:**19–35.

Kay, R. F., and Hiiemae, K. M. 1974. Mastication in *Galago crassicaudatus*, a cineflourographic and occlusal study. In: R. D. Martin, G. A. Doyle, and A. C. Walker (eds.), Prosimain Biology, pp. 501–530. Duckworth, London.

Kay, R. F., and Kirk, E. C. (2000). Osteological evidence for the evolution of activity pattern and visual acuity in primates. *Am. J. Phys. Anthropol.* **113:**235–262.

Kay, R. F., and Scheine,W. S. 1979. On the relationship between chitin particle size and digestibility in the primate *Galago senegalensis*. *Am. J. Phys. Anthropol.* **50:**301–308.

Kay, R. F., Sussman, R. W., and Tattersall, I. 1978. Dietary and dental variations in the genus *Lemur*, with comments concerning dietary-dental correlations among Malagasy primates. *Am. J. Phys. Anthropol.* **49:**119–128.

Kelley, J. 1990. Incisor microwear and diet in three species of *Colobus*. *Folia Primatol.* **55**:73–84.

Lamberton, C. 1934. Contribution à la connaissance de la faune subfossile de Madagascar. L'emuriens et Ratites. *Mém. Acad. Malgache* **17**:1–168.

Lamberton, C. 1939. Contribution à la connaissance de la faune subfossile de Madagascar: L'emuriens et cryptoproctes. *Mém. Acad. Malgache* **27**:1–193.

Lamberton, C. 1946. Contribution à la connaissance de la faune subfossile de Madagascar. XX. Membre posterieur des Neopropitheques et des Mesopropitheques. *Bull. Acad. Malgache* **27**:24–28.

Lamberton, C. 1947. Contribution à la connaissance de la faune subfossile de Madagascar. Note XVI. *Bradytherium* ou Palaeopropithèque? *Bull. Acad. Malgache (nouv. série)* **26**:89–140.

Leigh, S. R., and Terranova,C. J. 1998. Comparative perspectives on bimaturism, ontogeny, and dimorphism in lemurid primates. *Int. J. Primatol.* **19**:723–749.

Leutenegger, W., and Cheverud, J. 1982. Correlates of sexual dimorphism in primates: ecological and size variables. *Int. J. Primatol.* **3**:387–402.

Lewontin, R. C. 1978. Adaptation. *Sci. Am.* **239**:156–169.

Lucas, P. W. 1979. The dental-dietary adaptations of mammals. *Neues Jahrb. Geol. Palaeontol.* **8**:486–512.

MacLarnon, A. M. 1989. Applications of the reflex instruments in quantitative morphology. *Folia Primatol.* **53**:33–49.

MacPhee, R. D. E., and Raholimavo, E. M. 1988. Modified subfossil aye-aye incisors from southwestern Madagascar: Species allocation and paleoecological significance. *Folia Primatol.* **51**:126–142.

MacPhee, R. D. E., Simons, E. L., Wells, N., and Vuillaume-Randriamanantena, M. 1984. Team finds giant lemur skeleton. *Geotimes* **29**:10–11.

Maier, W. 1980. Konstruktions-morphologische Untersuchungen am Gebiss der rezente Prosimiae (Primates). *Abh. Senckenb. Naturforsch. Ges.* **538**:1–158.

Martin, R. D. 1990. *Primate Origins and Evolution*. Princeton University Press, Princeton.

Meldrum, D. J., Dagosto, M., and White, J. 1997. Hindlimb suspension and hind foot reversal in *Varecia variegata* and other arboreal mammals. *Am. J. Phys. Anthropol.* **103**:85–102.

Meyers, D. M. 1993. The Effects of Resource Seasonality on Behavior and Reproduction in the Golden-Crowned Sifaka (*Propithecus tattersalli*, Simons 1988) in Three Malagasy Forests. Ph.D. Dissertation, Duke University.

Mittermeier, R. A., Tattersall, I., Konstant, W. R., Meyers, D. M., and Mast, R. B. 1994. *Lemurs of Madagascar*. Conservation International, Washington DC.

Musser, G. G., and Dagosto, M. 1987. The identity of *Tarsius pumilus*, a pygmy species endemic to the montane mossy forest of Central Sulawesi. *Am. Mus. Novit.* **2867**:1–53.

Mutschler, T. 1999. *The Alaotran Gentle Lemur* (*Hapalemur griseus alaotrensis*): *A Study in Behavioral Ecology*. Doctoral dissertation. University of Zurich, Zurich, Switzerland.

Oxnard, C., Crompton, R., and Lieberman, S. 1990. *Animal Lifestyles and Anatomies: The Case of the Prosimian Primates*. University of Washington Press, Seattle.

Paciulli, L. 1995. Ontogeny of phalangeal curvature and positional behavior in chimpanzees. *Am. J. Phys. Anthropol.* **Suppl. 20**:165.

Petter, J. J., Albignac, R., and Rumpler, Y. 1977. *Faune de Madagascar*, Vol. 44: *Mammifères lemuriens* (Primates, Prosimiens). ORSTOM, CNRS, Paris.

Plavcan, J. M., and van Schaik, C. P. 1992. Intrasexual competition and canine dimorphism in anthropoid primates. *Am. J. Phys. Anthropol.* **87**:461–477.

Plavcan, J. M., and van Schaik, C. P. 1997. Interpreting hominid behavior on the basis of sexual dimorphism. *J. Hum. Evol.* **32**:345– 374.

Plavcan, J. M., van Schaik, C. P., and Kappeler, P. M. 1995. Competiiton, coalitions and canine size in primates. *J. Hum. Evol.* **28**:245–276.

Pollock J. I. 1977. The ecology and sociology of feeding in *Indri indri*. In: T. Clutton-Brock, (ed.), *Primate Ecology: Studies of Feeding and Ranging Behavior in Lemurs, Monkeys and Apes*, pp 37–69. Academic Press, London.

Powzyk, J. A. 1997. *The Socio-Ecology of Two Sympatric Indriids: Propithecus diadema and Indri indri. A Comparison of Feeding Strategies and their Possible Repercussions on Species-Specific Behaviors.* Doctoral Dissertation, Duke University, Durham, NC.

Preuschoft, H. 1970. Functional anatomy of the lower extremity. In: J. Bourne (ed.), *The Chimpanzee*. Vol. 3, pp. 221–294. Karger, Basel.

Preuschoft, H. 1973. Functional anatomy of the upper extremity. In: J. Bourne (ed.), *The Chimpanzee*. Vol. 6, pp. 34–120. Karger, Basel.

Rafferty, K., Teaford, M. F. and Jungers, W. L. In prep. Diet and dental microwear in Malagasy subfossil lemurs.

Rathburn, W. B., Holleschau, A. M., and Alterman, L. 1994. Glutathione metabolism in primate lenses: A phylogenetic study of glutathione synthesis and glutathione redox cycle enzyme activities. *Am. J. Primatol.* **33:**101–120.

Ravololonarivo, G. 1990. Contribution à l'etude de la colonne vertebrale du genre *Pachylemur* (Lamberton 1946): Anatomie et analyse cladistique. Doctorat 3e Cycle, Universite d'Antananarivo, Madagascar.

Ravosa, M. 1992. Allometry and heterochrony in extant and extinct Malagasy primates. *J. Hum. Evol.* **23:**197–217.

Reeve, H. K., and Sherman, P. W. 1993. Adaptation and the goals of evolutionary research. *Q. Rev. Biol.* **68:**1–32.

Remis, M. J. 1995. Effects of body size and social context on the arboreal activities of Lowland Gorillas in the Central African Republic. *Am. J. Phys. Anthropol.* **97:**413–434.

Richard, A. F. 1978. *Behavioral Variation: Case Study of a Malagasy Lemur.* Buckwell University Press, Lewisberg, PA.

Richard, A. F. 1992. Aggressive competition between males, female-controlled polygyny and sexual monomorphism in a Malagasy primate, *Propithecus verreauxi*. *J. Hum. Evol.* **22:**395–406.

Richard, A. F., and Nicoll, M. E. 1987. Female social dominance and basal metabolism in a Malagasy primate, *Propithecus verreauxi*. *Am. J. Primatol.* **12:**309–314.

Richmond, B. G. 1998. *Ontogeny and Biomechanics of Phalangeal Form in Primates.* Doctoral Dissertation, SUNY at Stony Brook, NY.

Richmond, B. G. (in press). Finite element methods in paleoanthropology: the case of phalangeal curvature. *J. Hum. Evol.*

Rigamonti, M. M. 1993. Home range and diet in red ruffed lemurs (*Varecia variegata rubra*) on the Masoala Pensinsula, Madagascar. In: P. M. Kappeler, and J. U. Ganzhorn (eds.), *Lemur Social Systems and Their Ecological Basis.* pp. 25–39. Plenum Press, New York.

Roberts, D. 1941. The dental comb of lemurs. *J. Anat.* **75:**236–238.

Rohen, J. W., and Castenholz, A. 1967. Uber die Zentralisation der Retina bei Primaten. *Folia Primatol.* **5:**92–147.

Rose, K. D., Walker, A. C., and Jacobs, L. L. 1981. Function of the mandibular tooth comb in living and extinct mammals. *Nature* **289:** 583–585.

Rose, M. R., and Lauder, G. V. 1996. *Adaptation.* Academic Press, New York.

Rosenberger, A. L., and Strasser, E. 1985. Toothcomb origins: support for the grooming hypothesis. *Primates* **26:**73–84.

Rudwick, M. J. S. 1964. The inference of function from structure in fossils. *Br. J. Philos. Sci.* **15:**27–40.

Ryan, A. S. 1981. Anterior dental microwear and its relationship to diet and feeding behavior in three African primates (*Pan troglodytes*, *Gorilla gorilla*, and *Papio hamadryas*). *Primates.* **22:**533–550.

Schmid, J., and Ganzhorn, J. U. 1996. Resting metabolic rates of *Lepilemur ruficaudatus*. *Am. J. Primatol.* **38:**169–174.

Schmid, P. 1983. Front dentition of the Omomyiformes (Primates). *Folia Primatol.* **40:**1–10.

Schwartz, J. H., and Tattersall, I. 1985. Evolutionary relationships of living lemurs and lorises (Mammalia, Primates) and their potential affinities with European Adapidae. *Anthropol. Pap. Am. Mus. Nat. Hist.* **60:**1–100.

Seligsohn, D. 1977. Analysis of species-specific molar adaptations in strepsirhine primates. *Contrib. Primatol* **11:**1–116.

Seligsohn, D., and Szalay, F. S. 1974. Dental occlusion and the masticatory apparatus in *Lemur* and *Varecia*: their bearing on the systematics of living and fossil primates. In: R. D. Martin, G. A. Doyle, and A. C. Walker (eds.), *Prosimian Biology*. pp. 543–561. Duckworth, London.

Shapiro, L., Jungers, W. L., Godfrey, L. R., and Simons, E. L. 1994. Vertebral morphology of extinct lemurs. *Am. J. Phys. Anthropol.* **Suppl.18:**179–180.

Simons, E. L. 1994. The giant aye-aye *Daubentonia robusta*. *Folia Primatol.* **62:**14–21.

Simons, E. L., Godfrey, L. R., Vuillaume-Randriamanantena, M., Chatrath, P. S., and Gagnon, M. 1990. Discovery of new giant subfossil lemurs in the Ankarana Mountains of Northern Madagascar. *J. Hum. Evol.* **19:**311–319.

Simons, E. L., Godfrey, L. R., Jungers, W. L., Chatrath, P. S., and Rakotosamimanana, B. 1992. A new giant subfossil lemur, *Babakotia*, and the evolution of the sloth lemurs. *Folia Primatol.* **58:**197–203.

Simons, E. L., Godfrey, L. R., Jungers, W. L., Chatrath, P. S., and Ravaoarisoa, J. 1995. A new species of *Mesopropithecus* (Primates, Palaeopropithecidae) from Northern Madagascar. *Int. J. Primatol.* **16:**653–682.

Simons, E. L. 1997. Lemurs: Old and new. In: S. M. Goodman, and B. Patterson (eds.), *Natural Change and Human Impact in Madagascar*. pp. 142–166. Smithsonian Institution Press, Washington, DC.

Smith, R. J. 1996. Biology and body size in human evolution. Statistical inference misapplied. *Curr. Anthropol.* **37:**451–481.

Smith, R. J. 1999. Statistics of sexual size dimorphism. *J. Hum. Evol.* **36:**423–458.

Smith, R. J., and Jungers, W. L. 1997. Body mass in comparative primatology. *J. Hum. Evol.* **32:**523–559.

Standing, H-F. 1909. Subfossiles provenant des fouilles d'Ampasambazimba. *Bull. Acad. Malgache* **6:**9–11.

Stephan, H., Frahm, H. D., and Baron, G. 1984. Comparison of brain volumes in primates. VI. Noncortical visual structures. *J. Hirnforsch.* **25:**385–403.

Sterling, E. J. 1994. Aye-ayes: specialists on structurally defended resources. *Folia Primatol.* **62:**142–154.

Stern, J. T. Jr. 1970. The meaning of "adaptation" and its relation to the phenomenon of natural selection. *Evol. Biol.* **4:**39–66.

Stern, J. T. Jr, and Susman, R. L. 1983. Locomotor anatomy of *Australopithecus afarensis*. *Am. J. Phys. Anthropol.* **60:**279–317.

Stern J. T. Jr, Jungers W. L., and Susman R. L. 1995. Quantifying phalangeal curvature: an empirical comparison of alternative methods. *Am. J. Phys. Anthropol.* **97:**1–10.

Strait, S. G. 1993a. Molar morphology and food texture among small-bodied insectivorous mammals. *J. Mammal.* **74:**391–402.

Strait, S. G. 1993b. Differences in occlusal morphology and molar size in frugivores and faunivores. *J. Hum. Evol.* **25:**471–484.

Strait, S. G. 1997. Tooth use and the physical properties of food. *Evol Anthropol.* **5:**199–211.

Susman, R. L., Stern, J. T. Jr, and Jungers, W. L. 1984. Arboreality and bipedality in the Hadar hominids. *Folia Primatol.* **43:**113–156.

Szalay, F. S., and Delson, E. 1979. *Evolutionary History of the Primates*. Academic Press, New York.

Szalay, F. S., and Katz, C. C. 1973. Phylogeny of lemurs, galagos and lorises. *Folia Primatol.* **19:**88–103.

Szalay, F. S., and Seligsohn, D. 1977. Why did the strepsirhine tooth comb evolve? *Folia Primatol.* **27:**75–82.

Tardieu, C. and Jouffroy, F. K. 1979; Les surfaces articulaires femorales du genou chez les Primates. Etude preliminaire. *Ann. Sci. Nat. Zool. Paris* **1:**23–38

Tattersall, I. 1972. The functional significance of airorhynchy in Megaladapis. *Folia Primatol.* **18:**20–26.

Tattersall, I. 1973. Cranial anatomy of the Archaeolemurinae (Lemuroidea, Primates). *Anthropol. Pap. Am. Mus. Nat. Hist.* **52:**1–110.

Tattersall, I. 1975. Notes on the cranial anatomy of the subfossil Malagasy lemurs. In: I. Tattersall, and R. W. Sussman (eds.), *Lemur Biology*, pp. 111–124. Plenum Press, New York.
Tattersall, I. 1982. *The Primates of Madagascar*. Columbia University Press, New York.
Tattersall, I. 1988. Cathemeral activity in primates: a definition. *Folia Primatol.* **49:**200–202.
Teaford, M. 1994. Dental microwear and dental function. *Evol. Anthropol.* **3:**17–30.
Thomason, J. J. 1995. *Functional Morphology in Vertebrate Paleontology*. Cambridge University Press, Cambridge.
Ungar, P. S. 1990. Incisor microwear and feeding behavior in *Alouatta seniculus* and *Cebus olivaceus*. *Am. J. Primatol.* **20:**43–50.
Ungar, P. S. 1994. Incisor microwear of Sumatran anthropoid primates. *Am. J. Phys. Anthropol.* **94:**339–363.
Van Schaik, C. P., and Kappeler, P. M. 1993. Life history, activity period and lemur social systems. In: P. M. Kapperler, and J. U. Ganzhorn (eds.), *Lemur Social Systems and Their Ecological Basis*. pp. 241–260. Plenum Press, New York.
Vogel, S. 1988. *Life's Devices: The Physical World of Animals and Plants*. Princeton University Press, Princeton.
Vuillaume-Randriamanantena, M., Godfrey, L. R., Jungers, W. L., and Simons, E. L. 1992. Morphology, taxonomy and distribution of *Megaladapis*—giant subfossil lemur from Madagascar. *C. R. Acad. Sci. Paris* **315** (Sèrie II):1835–1842.
Wainright, P. C., and Reilly, S. M. 1994. *Ecological Morphology*. University of Chicago Press, Chicago.
Walker, A. C. 1967. Patterns of extinction among the subfossil Madagascar lemuroids. In:. R. S. Martin, and H. E. Wright (eds.), *Pleistocene Extinctions: The Search for a Cause*. pp. 407–424. Yale University Press, New Haven.
Walker, A. C. 1974. Locomotor adaptations in past and present prosimian primates. In: F. A. Jenkins, Jr. (ed.), *Primate Locomotion*, pp. 349–382. Academic Press, New York.
Walker, P. L. 1976. Wear striations on the incisors of cercopithecoid monkeys as an index of diet and habitiat preference. *Am. J. Phys. Anthropol.* **45:**299–308.
Wolin, L. R., and Massopust, L. C. 1970. Morphology of the primate retina. In: C. R. Noback, and W. Montagna, (eds.), *The Primate Brain*, pp. 1–27. Appleton-Century-Crofts, New York.
Wright, P. C. 1999. Lemur traits and Madagascar ecology: coping with an island environment. *Yearb. Phys. Anthropol.* **42:**31–72.
Wunderlich, R. E., Jungers, W. L., Godfrey, L. R., and Simons, E. L. 1994. Pedal form and function in subfossil indroids. *Am. J. Phys. Anthropol.* **Suppl. 18:**211–212.
Wunderlich, R. E., Simons, E. L., and Jungers, W. L. 1996. New pedal remains of *Megaladapis* and their functional significance. *Am. J. Phys. Anthropol.* **100:**115–139.
Yamashita, N. 1996. Seasonality and site-specificity of mechanical dietary patterns in two Malagasy lemur families (Lemuridae and Indriidae). *Int. J. Primatol.* **17:**355–387.
Yamashita, N. 1998. Molar morphology and variation in two Malagasy lemur families (Lemuridae and Indriidae). *J. Hum. Evol.* **35:**137–162.
Yoder, A. D. 1992. The application and limitations of ontogenetic comparisons for phylogeny reconstruction: the case of the strepsirhine internal carotid artery. *J. Hum. Evol.* **23:**183–195.
Yoder, A. D. 1994. Relative position of the Cheirogaleidae in strepsirhine phylogeny: A comparison of morphological and molecular methods and results. *Am. J. Phys. Anthropol.* **94:**25–46.
Yoder, A. D. 1996. The use of phylogeny for reconstructing lemuriform biogeography. *Biogeographie de Madagascar* **1996:**245–258.
Yoder, A. D. 1997. Back to the future: a synthesis of strepsirrhine systematics. *Evol. Anthropol.* **6:**11–22.
Yoder, A. D., Rakotosamimanana, B., and Parsons, T. 1999. Ancient DNA in subfossil lemurs: methodological challenges and their solutions. In: B. Rakotosamimanana, H. Rasaminanana, J. Ganzhorn, and S. M. Goodman (eds.), *New Directions in Lemur Studies*. pp. 1–17. Plenum Press, New York.

Conclusions: Reconstructing Behavior in the Fossil Record

11

J. MICHAEL PLAVCAN, RICHARD F. KAY, WILLIAM L. JUNGERS, and CAREL P. VAN SCHAIK

Introduction

Paleontology is often seen as a descriptive science that documents the morphological diversity of life in the past. But description is just the first step. An animal's morphology can offer information about locomotion, habitats, development, diet, social behavior, and other life-history parameters of species, allowing paleontologists to understand the diversity of adaptations and behaviors of the past. This historical perspective can give a greater understanding of the origins of adaptations and diversity of extant species. For example, Richmond and Strait (2000) recently argued that features of the distal radius of *Australopithecus afarensis*—an obligate biped—derived from a knuckle-

---

J. MICHAEL PLAVCAN • Department of Anatomy, New York College of Osteopathic Medicine, Old Westbury, New York 11568; Present address: Department of Anthropology, University of Arkansas, Fayetteville, Arkansas 72701. RICHARD F. KAY • Department of Biological Anthropology and Anatomy, Duke University Medical Center, Durham, North Carolina 27710. WILLIAM L. JUNGERS • Department of Anatomical Sciences, Health Sciences Center, State University of New York at Stony Brook, Stony Brook, New York 11794-8081. CAREL P. VAN SCHAIK • Department of Biological Anthropology and Anatomy, Duke University, Box 90383, Durham, North Carolina 27708-0383.

Reconstructing Behavior in the Primate Fossil Record, edited by Plavcan *et al.* Kluwer Academic/Plenum Publishers, New York, 2002.

walking ancestor. Understanding the primitive locomotor behavior of the ancestor of hominids helps us to formulate models of the adaptive significance of bipedality in early hominids. Likewise, such paleontological data can shed light on whether or not traits shared in common by an extant taxonomic group are merely shared through common heritage, or are maintained through stabilizing selection. For example, Ross *et al.* (this volume, Chapter 1) point out that the altered tooth combs of several extinct lemurs indicate that the tooth comb of living strepsirrhines is maintained by stabilizing selection, rather than being a phylogenetic constraint.

The Comparative Approach

As shown by the papers presented here, there are a variety of approaches used to infer behavior in extinct species. These range from simple informed speculation to extrapolation from well-established form–function relationships. Reconstructing behavior in any fossil species must strike a balance between the enthusiastic inference of how animals lived in the past, and the skepticism necessary to understand when we cross the boundary between supported hypothesis and unsupported speculation. It seems that the skeptical part of the equation is sometimes missing—while there are many hypotheses put forward about the potential adaptations and life histories of extinct species, there are few studies that critique just how far we can go in making such inferences on the basis of comparative data. But skepticism is healthy in that it points to the weaknesses of our arguments and generates new research and new ways of looking at old data. For example, Smith (1996) demonstrated clearly that extrapolation of behaviors from empirical statistical relationships can be deceptive if confidence intervals are ignored. Plavcan (2000, this volume) notes that empirical studies of the relationship between sexual dimorphism and mating systems often do not support commonly held reconstructions of social behavior in extinct species. Hylander and Johnson (this volume) demonstrate convincingly that experimental data can undermine commonly held functional interpretations of skeletal morphology. Even so, reconstructions of behavior in the fossil record have become much more rigorous in recent years as investigators probe the weaknesses of particular models, and seek ways to strengthen comparative models used to infer behavior in extinct species. Ungar (this volume) demonstrates the great variety of information that can be brought to bear on dietary inferences, largely as a result of years of careful investigation. In this volume, Godfrey *et al.*, Nunn and van Schaik, Rovosa and Vinyard, and Reed all discuss innovative and relatively unexplored approaches to reconstructing behavior using comparative models.

By necessity, reconstructions of the behavior of extinct species are limited to the evidence available in the fossil record. For the most part, this constitutes osteological remains, but other types of evidence are sometimes available. For example, the environmental context in which the fossil was found may offer insight into what the species was, or was not, doing (Reed, this volume; Kay *et al.*, this volume). Thus, a species that is consistently found in savanna environments probably was not arboreal. Kay *et al.* (this volume) use this type of information to infer that *Branisella boliviensis* was probably at least partly terrestrial—a surprising reconstruction given that all living platyrrhines are almost entirely arboreal. Likewise, Reed (this volume) points out that early hominids seemed to have lived not in savannas—as long assumed—but in open woodland habitats. This has profound implications for models of early human evolution, forcing scholars to rethink the diet, resource exploitation, and locomotion of early hominids.

The strongest inference of behavior is derived from the fossils themselves, and is usually based on an adaptive or form–function analogy to extant species. In this volume, Ungar, Ross *et al.*, and Kay *et al.* all point to one of the most widely-cited and best-known examples—the fact that well-developed shearing crests on primate molar teeth indicate a folivorous diet, because extant folivorous primates show enhanced shearing on the molars, while nonfolivorous primates do not (Kay, 1975, 1978; Ungar, 1998). Over the last several decades, there has been an explosion in understanding the relation between form and function among extant primates in all anatomical regions. This knowledge offers an enormous data base for reconstructing behavior in the fossil record, and potentially allows the synthetic reconstruction of behaviors that extend beyond one-to-one correspondences between a particular form and a particular function (e.g., Godfrey *et al.*, Jungers *et al.*, and Kay *et al.*, this volume).

It is still common practice to use individual extant species as analogues for the behavior and ecology of similar extinct species (e.g., Simons and Rasmussen, 1996; Susman, 1987; Tanner, 1987). This practice will always play a vital role in helping to visualize and understand the biology of extinct species. For example, the image of *Catopithecus* as resembling *Saimiri* or *Miopithecus* in social behavior is powerful (Simons and Rasmussen, 1996). But it is clear from all the papers in this volume that such analogies unto themselves do not constitute strong evidence for the behavior of extinct species—there is simply too much variation in the relation between behavior and morphology among extant species.

The behavior of extinct species is usually inferred on the basis of comparative analyses that seek to test hypotheses of adaptation and function in extant forms. The most widely cited method for using comparative analysis to infer adaptation and function in extinct forms is that of Kay and Cartmill (1977). To briefly summarize their approach, if among extant species a trait, T, can be shown to always be related to a function, F, and the lack of trait T is always associated with the lack of function F, then if a fossil taxon has trait T it is most parsimonious to assume that function F was also present. This view was put forward in response to the view that adaptations of any species must be understood in the context of the adaptations of its ancestors, and that adaptation is best understood as a function of newly acquired characters which indicate changes in form associated with changes in function. For example,

Szalay and Delson (1979) argued that even though *Roonyia* possessed relatively large orbits by comparison to extant diurnal species, it was, nevertheless, probably diurnal because it inherited large orbits from a common ancestor. Kay and Cartmill (1977) pointed out that this type of argument creates an infinite regress. If one must understand the adaptations of a species' ancestor in order to understand its adaptations, then how is one to reconstruct the adaptations of the ancestor? Presumably by referring to its ancestor, and so on and so on in an infinite regress.

While the argument of Kay and Cartmill is relatively straightforward, recent advances in the definition and recognition of adaptations, and in comparative methods employed to test adaptive hypotheses, demand a re-evaluation of the methods used to reconstruct behavior in the fossil record (Ross *et al.*, this volume). For example, Gould and Vrba (1982) pointed out that "adaptation" as a term is over-broad. They suggested that adaptation instead be restricted to traits that develop in specific response to selective pressures. The term "exaptation" was put forward for characters that are associated with some current function, but either developed in response to some other selective pressure not associated with the current function, or developed as a consequence of some factor other than natural selection. This suggests that the presence of a character in an extinct species may have been associated with a different function than that seen in its descendent species. Furthermore, the point made by Gould and Lewontin (1979) that all structures do not necessarily develop in response to natural selection sparked intense debates about the meaning of form/function relations, adaptation, and the evolutionary mechanisms producing interspecific morphological diversity. At the very least the result has been the identification of a variety of mechanisms for altering form without reference to natural selection—the most familiar of these being allometric effects. The implication is that one-to-one form–function relations are not necessarily adequate for inferring behavior in the fossil record, and an understanding of evolutionary mechanisms leading to particular forms becomes more important for inferring behavior from form. For example, Lande's (1980) model for correlated response between male and female characters has been used to explain the development of elaborate facial projections and tusks in the females of several extinct suids, even though such structures probably served little or no function in the females (Wright 1993). In this sense, the classic form–function relation traditionally used to study adaptation becomes a subset of possible causes of the generation of morphological form.

On the other hand, Ross *et al.* and Kay *et al.* (this volume) point out that, for example, the existence of different definitions of adaptation does not necessarily undermine the use of the Kay and Cartmill formula for inferring behavior. Current utility, as long as it is universally present, can suggest that stabilizing selection maintains features, validating the inference of form/function relationships in fossils. Likewise, Ross *et al.* point out that the "allometry as a criterion of subtraction" formulation commonly used in comparative studies misrepresents Gould's (1966) original formulation, and ignores the fact that allometric relationships may reflect adaptive changes in structures that are correlated with size. The end result seems to be that careful scrutiny of what is meant by adaptation, and the biological basis of form/function relationships, leads to more robust behavioral inferences.

Phylogeny

Most of the contributions to this volume discuss the importance of phylogeny to reconstructions of behavior in the fossil record. It is now well appreciated that primate morphology represents a mosaic between adaptation and phylogenetic history. Different species do not respond identically to similar selective pressures, and each species faces a unique set of adaptive pressure and historical constraints. For example, the relationship between molar shearing crest development and folivory varies among hominoids, platyrrhines, and cercopithecoids (Kay and Ungar, 1996). The latter have higher shearing quotients than the former two groups. Kay and Ungar (1996) also demonstrated a shift in the relationship between folivory and shearing crest development between Miocene and extant hominoids, illustrating that form–function relationships can vary both among taxa and through time.

Recent years have seen dramatic advances in the formulation of comparative analyses, especially with regards to phylogenetic control, and many of these are freely available on the web. For example, Kay *et al.* (this volume) use the CAIC (comparative analysis using independent contrasts) program of Pagel (1992), available on the Oxford University web site, and based on Felsenstein's (1985) method of phylogenetic contrasts. These methods provide powerful tests of adaptive hypotheses controlling for the confounding effects of phylogeny. Methods based on "known" phylogenies and divergence times have been critiqued on the basis that the phylogeny and divergence times of many groups is uncertain. However, simulation studies have demonstrated that even incorrect phylogenies provide better comparative results than simple analysis of species values (Nunn, 1995; Purvis and Webster, 1999). Hence, it is now widely considered essential to employ some sort of phylogenetic control in comparative analysis, unless there is an explicitly justified reason for ignoring phylogeny.

Ross *et al.* (this volume) point out that recent years have witnessed some scrutiny about the assumptions of phylogenetic analysis which is relevant to inferring behavior in fossils. The demonstration of phylogenetic correlations does not necessarily prove phylogenetic constraints, just as any correlation does not prove causation. It has long been acknowledged that the maintenance of characters by stabilizing selection obviates the need for phylogenetic control (Felsenstein, 1985; Pagel and Harvey, 1988), though the use of phylogenetic analysis will not produce a spurious result in such a case (Pagel, 1992). Nevertheless, as far as reconstructing behavior in fossils goes, actually identifying and quantifying phylogenetic correlations may be more important than simply controlling for potential phylogenetic bias in a comparative analysis.

Unfortunately, at this point there is no standard method to account for phylogenetic effects on behavioral reconstructions in a manner analogous to allometric correction for the effects of size. The primary concern of these methods is proper statistical control for nonindependence of data points in comparative analyses and elimination of the role of confounding factors in correlated evolution (Felsenstein, 1985; Kay *et al.*, this volume; Nunn, 1995; Pagel and Harvey, 1988; Purvis and Webster, 1999; Ross *et al.*, this volume; Smith, 1996). The methods are not easily amenable to predicting unknown values in a trait in a manner analogous to estimating an unknown value from an empirically derived regression equation.

A commonly used technique for phylogenetic control often applied to the fossil record is to generate comparative relationships within restricted taxonomic groups. One widely known example of this is the extrapolation of body size from tooth size relationships in different taxonomic groups of primates (Conroy, 1987). But this only goes so far, in that it cannot account for morphological shifts over time (as demonstrated for molar shearing quotients by Kay and Ungar, 1996), and cannot accommodate species that are not clearly or uniquely related to a particular taxonomic group (for example, if tooth size–body size relationships differ between colobines and cercopithecines, to which group does one compare the common ancestor of the group?).

Kay and Ungar's (1996) work suggests that evolutionary trends in the development of structures through time are potentially as meaningful as form/function relationships in living forms. Hence, extinct apes may have relatively little shearing by comparison to modern apes, but the development of shearing that is relatively greater than that seen in either contemporaries or older taxa may be strongly indicative of folivory. Application of such a model may be difficult, but potentially fruitful. Essentially, one would compare relative values of a trait among contemporary species, and assume that the total variation in behavior seen among extant taxa was similar in the extinct taxa. Such an assumption would itself rest on the assumption that changes in the character state over time reflect quantitative changes in the morphological character, and not qualitative changes in the basic relationship between form and function. Speculative as this may be, Kay and Ungar's example demonstrates that such an approach can offer powerful insight into the potential adaptive meaning of traits.

Standard Error from Comparative Analyses

The previous discussion centered on the fact that most behavioral inferences in extinct species are based on extrapolation from a comparative analysis. Smith (1996) pointed out that standard errors in comparative relationships are often insufficiently appreciated by those inferring life-history parameters from morphology, leading to statistically meaningless inferences. Smith's logic is not limited to statistical extrapolations, though. Unless morphological variation exactly corresponds to any aspect of behavior, there is always error associated with a reconstruction (see also Nunn and van Schaik, this volume; Plavcan, this volume).

There are two types of error that are associated with comparative analyses—statistical error associated with measuring and estimating form and function, and biological error reflecting the lack of a precise relation between form and function. The former is familiar to most biologists. Any measurement of morphology or function is only an estimate of the true population parameter, and estimates of the relation between some form and function among species will consequently vary. However, there is also biological error in such relationships. As noted by Jungers *et al.* (this volume), estimates of behavior based on morphology are likely to be highly conservative, because morphology does not capture all of the functions and behaviors that are possible with a given form. Animals can easily modify their behavior in order to adapt to changes in an environment. Ross *et al.* (this volume) point to Kinzey's (1978) hypothesis that some features are adapted not to common functions, but rather for the exploitation of "keystone" resources. Likewise, as noted by Ross *et al.*, and Hylander and Johnson, bony morphology can represent a compromise between multiple functions. As discussed above, phylogeny can have a significant impact on the exact design of a feature. Consequently, there is normally a large variation in the relationship between morphology and behavior across species.

Smith's message, while pointing to logical problems with extrapolating behavior from morphology, does not undermine behavioral inferences. Rather, understanding the underlying biological basis of error in the relationships should lead to more precise behavioral inferences for extinct species. Knowing that there are taxonomic differences in form–function relationships of the dentition should improve dietary reconstructions, by limiting the comparative data to those animals that serve as the most appropriate comparative model (Kay *et al.*, this volume). Knowing that some relationships are based on operational behavioral categorizations (see below) that are more or less biologically realistic can allow one to place greater weight on certain types of evidence, and can help resolve potential conflicts in behavioral reconstructions.

Body Mass and Allometry

In one form or another, the importance of body mass for reconstructing behavior in the fossil record appears in all chapters of this volume. Body mass clearly plays a role in reconstructing almost any kind of behavior through the ubiquitous influence of allometry on most aspects of an animal's morphology, behavior, and life history. Many comparative analyses control for the allometric effects of body mass on morphology. Some studies control for allometric effects on behavior too (e.g., Harvey *et al.*, 1987). Most allometric control is straightforward—least-squares regression is carried out on a comparative sample, using the regression line as a "criterion of subtraction." Residuals are then used as the actual data. There is a large literature on problems associated with predicting variables from least-squares regressions (see Jungers *et al.*, 1995 for a discussion of the effects of different equations on size and shape).

Body mass is clearly related to behavioral/ecological variables in living species, and can play an important role in predicting the behavior of extinct species, especially when used in conjunction with other evidence. For example, "Kay's threshold" is a body size criterion for distinguishing between insectivorous and folivorous diets of animals with well-developed molar shearing (Kay, 1975; Ungar, 1998, this volume). But, as pointed out by Nunn and van Schaik (this volume) and by others, relationships between body mass and behavioral/ecological variables, while statistically significant, often are not strong enough to use body mass alone to predict behavior in the fossil record. In this volume, Godfrey *et al.* and Jungers *et al.* both give explicit examples indicating how body mass can be a deceptive variable for reconstructing behavior.

Smith (1996) has strongly critiqued the use of body mass in predicting behavioral/ecological traits in fossils on the basis of allometric equations. Apart from the large standard errors typical of relationships between behavioral data and body mass data, there is a problem in that comparative analyses of extant species evaluate the relationship between body mass and behavioral/ecological variables, while analyses of fossils must extrapolate body mass from skeletal remains. The extra step involved in extrapolating body mass from the skeletal remains introduces substantial error into the process (see also Fig. 1 in Nunn and van Schaik, this volume). Smith recommends that more studies evaluate the direct relationship between behavioral/ecological variables and the remains that are actually preserved in the fossil record. Unfortunately, this implies that those wishing to infer anything about behavior from body mass in the fossil record will need to empirically examine relations between skeletal variables and behaviors that covary with body mass.

Importantly, Ross *et al.* (this volume) note that the use of "allometry as a criterion of subtraction" ignores the fact that allometric relationships themselves can provide meaningful information about size-related adaptation. An understanding of the underlying causes of an allometric relationship can at best greatly strengthen confidence in behavioral or functional inferences, and at worst inspire caution in such inferences.

Plavcan (this volume) presents an example of how allometric correction of data can actually remove biological information critical to behavioral reconstruction. Occasionally, studies allometrically correct estimates of sexual dimorphism by regressing them against body mass (e.g., Ford, 1994; Mitani *et al.*, 1996), because it is well known that sexual dimorphism tends to increase with increasing body size in primates. Importantly, several studies suggest that body mass has a causal influence on male mate competition, such that sexual selection tends to be greater in larger species. This means that the correlation between dimorphism and body size might reflect only covariation between body size and behavior. If true, size adjustment of dimorphism cannot be justified for predicting behavior in the fossil record, because it inadvertently removes the variation in dimorphism that is caused by behavior.

Incomplete Extant Models

Incompleteness of extant models is a difficult topic to address. Species representation alone may preclude the development of an adequate comparative model (Foley, 1999). For example, comparative analyses of sexual dimorphism in primates suggest that there is phylogenetic inertia in the pattern of dimorphism, that dimorphism is positively correlated with body mass, and the intensity of male–male competition, or the operational sex ratio, are positively correlated with body mass. Yet *Catopithecus* and *Apidium*, both small, early anthropoid primates, are sexually dimorphic (Fleagle *et al.*, 1980; Simons and Rasmussen, 1996; Simons *et al.*, 1999). These species demonstrate that dimorphism arose early in anthropoid history, that genetic de-coupling of male and female characters occurred early in anthropoid history, and that neither dimorphism nor male mate competition are necessarily tied to evolutionary increases in body mass (Fleagle *et al.*, 1980; Simons *et al.*, 1999). Hence, it appears that the extant data alone can give a somewhat misleading picture of the relation between certain variables.

This same phenomenon can be noted for phylogenetic reconstructions of behavior or morphology based on extant species (see Nunn and van Schaik, this volume). For example, Wrangham (1987) reconstructed several likely behavioral characteristics of early hominids using a straightforward assumption of parsimony. However, Foley (1999) notes that phylogenetic analysis predicts certain morphological traits in early hominids that are demonstrably not true. By extrapolation, the same *must* be true for behavioral traits that can be reconstructed by parsimony analysis. This should also serve as a caution to those using phylogenetic constraints as the underpinnings of behavioral reconstructions. For example, Maryanski (1996) invokes phylogenetic constraint in female dispersal patterns of hominoids as the basis for a model of the evolution of human mating systems. Just as allometry should not be automatically used as "a criterion of subtraction" without a biological model for understanding the basis of the allometric relation, the causal mechanisms underlying phylogenetic correlations of behavior need to be understood if phylogeny is to be used to reconstruct behavior.

Foley (1999) points out that the extrapolation of characters in extinct species from extant taxonomic patterns, while a powerful tool for generating hypotheses, can be strongly influenced by the differential extinction of species. Foley argues that the fossils themselves can be used to assess weaknesses in extant comparative data. As noted above, the evidence that Kay *et al.* (this volume) muster for terrestriality in *Branisella boliviensis*, and the evidence of sexual dimorphism in *Apidium* and *Catopithecus*, indicate that fossils can play an important role in assessing the validity of extant behavioral models.

Problems with Defining Behavior

All of the chapters in this book point out that the loose observed relationship between morphology and behavior among species is a major obstacle to reconstructing behavior in an extinct species.

Variation in morphology is not necessarily directly associated with variation in common behaviors. Jungers *et al.* emphasize that skeletal morphology underestimates the total behavioral repertoire of an animal, and hence should be highly conservative for reconstructing the variety of behaviors engaged by an animal. Most features of an animal can and do serve more than one function; the identification of adaptations based on particular behaviors is a challenge for comparative analysis (Ross *et al.*, this volume). Hylander and Johnson pointedly note that skeletal design is not necessarily associated with variation in common skeletal functions.

It is important to remember how evolution acts on morphology. Selection should favor the change in a morphological structure that brings the greatest net fitness benefit. In many cases, this need not be a change that improves the performance of the structure during routine situations, but rather one that improves performance during emergencies or times of environmental stress. Thus, selection may favor a slight modification of musculature that would allow faster escape from a predator, even though this improvement would only be used once or twice a year, and actually might reduce locomotor efficiency during routine activity. Dental structure might reflect adaptations for feeding on critical, keystone food items that must be exploited at certain times of the year (Kinzey, 1978; Ross *et al.*, this volume). Generally, this means that the most common activity performed using the structure may not be the one that has provided the strongest selective pressure on its final form. This widely recognized principle has implications for how we think about the dentition and foraging, forcing us to consider the characteristics of keystone resources (e.g., highly fibrous bark or piths) rather than those of foods eaten during times of abundance. More attention to these kinds of relationships may help to selectively increase the resolution of some reconstructed traits, while also helping us to better recognize the limitations for other traits (Ungar, this volume).

A related problem is that animals can adapt by simply modifying their behavior. A classic example was the observation by Kay (1978) that female cercopithecoids had relatively large teeth for their body size. Kay speculated that this reflected the need for females to process greater amounts of food in response to greater metabolic demands imposed by pregnancy and lactation. Subsequent studies showed that the observation did not hold for other primates, and it was soon acknowledged that females can meet increased metabolic demands simply by chewing faster, eating more, or exploiting more nutritious resources (Cochard, 1987; Kay, 1984).

Behavior itself tends to be more ontogenetically flexible [i.e., it has a broader norm of reaction, as demonstrated by generally lower heritability scores (cf. Mousseau and Roff, 1987)] than morphology. This is especially true for behaviors not directly linked to morphology, such as range use or group size. For example, group composition in *Semnopithecus entellus* varies across India, with some groups showing multi-male, multi-female groups, and others showing single-male harem defense polygyny (Struhsaker and Leland, 1987). Such behavioral plasticity within species without any morphological correlates helps explain the poor success of reconstructing such behaviors.

Even behaviors directly linked to morphology can show considerable flexibility. For example, it is widely accepted that the high degrees of molar shearing in colobine primates reflects an adaptation to a folivorous diet. While all colobine primates rely on leaves in their diets, most also include substantial proportions of fruit (Struhsaker and Leland, 1987). Dietary components of many primate species can differ at different times of year, depending on abundance of preferred food items. This dietary variation can be picked up by direct measures of diet such as microwear analysis (Teaford and Glander, 1991; Teaford and Robinson, 1989; Ungar, 1998, this volume). Hence, while molar morphology and general microwear patterns may re-enforce each other in reconstructing the general dietary adaptations of a species, supplemental information such as microwear analysis might help point to dietary variation that could not be detected by study of morphology alone.

A point that seems to be lost in behavioral reconstructions is the reliance on categorical definitions of behavior in many comparative studies (Nunn and van Schaik, this volume; Plavcan, this volume). Comparative analyses commonly classify species as monogamous or polygynous, folivorous or frugivorous, arboreal quadrupeds or leapers. Such categories reflect underlying biological realities, but they are also operational approximations of biological phenomena. A look at historical changes in the way these categories are presented underscores the fact that they are human constructs. For example, early quantitative comparative studies of sexual dimorphism recognized either a dichotomous (monogamous vs. polygynous; Leutenegger and Cheverud, 1982) or trichotomous (monogamy vs. single-male vs. multi-male; Clutton-Brock and Harvey, 1978) mating system classification. Later, multi-male communities were recognized that differed dramatically in male–male relationships, male and female intrasexual competition, and the structure of male–female relationships (Kay *et al.*, 1988; Plavcan and van Schaik, 1992). It was also recognized that the structure of male and female relationships can vary within species. While none of this completely undermines the utility of the use of categories, it does underscore the behavioral complexity and variation that belies any such categorizations. Thus, behavioral categories should not necessarily be treated as biological entities unto themselves, but only as approximations of biological phenomena.

The need to categorize behavior can lead to confusion about how to apply comparative models to extinct data. Some systems of categorization reflect constructs of observed patterns of behavior. Hence, the "vertical clinger and leaper" locomotor classification was constructed in recognition of a suit of locomotor behaviors shared by a number of primates (Napier and Walker, 1967). Behavioral categorizations are also constructed specifically to test hypotheses about the evolution of morphology. Hence, mating system or competition levels are only surrogate measures of differential male reproductive success arising from male mate competition (Plavcan, 2000, this volume). Some categories originally erected on the basis of observation are retained for comparative analyses, even though more recent observations indicate that they subsume significant biological behavioral variation. Dietary categories such as "folivore" refer only to major components of diet, and do not quantify precisely variation in either the properties of food, or the particular types of food that may be most critical to survival of a species (Ungar, this volume). Demonstrating a relationship between morphology and these categories can offer powerful evidence about the adaptive significance of morphology. But turning these relationships around to predict the behavioral categories from the morphology should be done with caution appropriate to the original design of the behavioral categories.

Multiple Lines of Evidence

Several papers in this volume (Godfrey *et al.*; Jungers *et al.*; Kay *et al.*; Ungar) emphasize the use of multiple lines of evidence for reconstructing behavior. This approach is obviously far more powerful than simple extrapolation of behaviors from a single variable or type of evidence. Not only can multiple lines of evidence re-enforce behavioral reconstructions, but they can also point to conflicts. A relatively simple and well-known example of how multiple lines of evidence can re-enforce each other is the relation between molar shearing crest development, body mass, and either folivory or insectivory (Kay, 1975; Ungar, this volume). Highly developed shearing crests are associated with either a folivorous or frugivorous diet. Folivores tend to be large while insectivores tend to be small. Thus, relatively high shearing in small primates is usually interpreted as evidence for insectivory, while high shearing in larger primates is interpreted as evidence of folivory.

Unique traits or combinations of traits offer a particular challenge to reconstructing any type of behavior. The best that can be done with unique traits is to offer a plausible hypothesis based on known biomechanical or behavioral-ecological principals, perhaps supplemented with evidence against certain functions. Hominids offer good examples of the problems with inferring behavior from unique traits. Apart from humans, the only bipedal vertebrates are dinosaurs and birds, neither of which offer a clearly appropriate model for human bipedality. Other primates are capable of bipedal locomotion, but only engage in the behavior facultatively. Any primate-based model of bipedality faces the critical problem of why the comparative species did not themselves evolve bipedality. Consequently, the best that can be done is to identify the advantages that bipedality confers (freeing of the hands, possible increased reach when foraging from the ground, thermoregulatory advantages, etc.), and exclude models on the basis of conflicting data [for example, Lovejoy's (1981) provisioning model relies on the assumption of monogamy associated with a lack of dimorphism—a contention not supported by the fossils (Plavcan and van Schaik, 1997)].

Unique *combinations* of traits that offer conflicting behavioral reconstructions potentially offer an opportunity to scrutinize the validity of the comparative models. In the first place, it may be possible to dismiss some reconstructions as more or less likely depending on the strength of an extant model. For example, Blumenberg (1984) suggested that *Aegyptopithecus* was characterized by a relatively small brain, folivory, and strong canine dimorphism. Small brain size and folivory were purported to be associated with monogamous mating systems, while canine dimorphism was thought to be indicative of polygyny. Blumenberg favored the former interpretation. Current comparative models strongly suggest that the associations between mating system and either brain size or folivory are so weak that neither diet nor relative brain size should be used as evidence for mating systems in the fossil record.

In the second place, conflicts might indicate that the extant models are either flawed (for example, in assuming cause–effect relations) or incomplete. On the basis of paleoenvironmental and dental data, Kay *et al.* (this volume) suggest that *Branisella boliviensis* may have been terrestrial, unlike any living platyrrhine. Nunn and van Schaik (this volume) challenge this assertion on the basis of a model for suits of co-occuring behavioral traits. Regardless of which model is ultimately correct, the conflict clearly underscores the need for closer scrutiny of the data and models used in each reconstruction.

Conclusions

Behavioral inferences of extinct species are always tentative. It is easy to loose sight of this in our excitement to make inferences about the adaptations and life history of extinct species, and there are numerous examples in the anthropological literature where inferences about the behavior or life history of extinct humans or primates exceed the evidence in support of such inferences. But over the years, there have been great strides in the methods of comparative analyses, in the number and variety of comparative analyses that have been carried out, in the formulation of models for understanding the functional and adaptive significance of numerous traits, and in the understanding of the behavioral ecology of living primates. As shown in this volume, evidence from behavioral models, paloecommunity, paleoecolgical and taphonomic evidence, data from ontogeny and life history traits, and non-adaptive morphological data such as dental microwear can all be used in addition to standard comparative form/function data. Application of such advances to the question of reconstructing behavior in extinct species has underscored both what can and cannot be said about behavior on the basis of morphology alone. A greater understanding of the limitations of comparative models for inferring behavior is the natural indication of the progress being made toward making more rigorously supported inferences about behavior. Conflicts between fossil evidence and behavioral models clarify important areas for further research. Reconstructing behavior in extinct species has never been easy, but the blossoming of comparative analyses of primate behavior and morphology offers enormous promise for a deeper understanding of the evolution of primates and humans.

References

Blumenberg, B. 1984. Allometry and evolution of Tertiary hominoids. *J. Hum. Evol.* **13:**613–676.

Clutton-Brock, T. M., and Harvey, P. H. 1978. Mammals, resources and reproductive strategies. *Nature* **273:**191–195.

Cochard, L. R. 1987. Postcanine tooth size in female primates. *Am. J. Phys. Anthropol.* **74:**47–54.

Conroy, G. C. 1987. Problems of body-weight estimation in fossil primates. *Int. J. Primatol.* **8:**115–137.

Felsenstein, J. 1985. Phylogenies and the comparative method. *Am. Nat.* **125:**1–15.

Fleagle, J. G., Kay, R. F., and Simons, E. L. 1980. Sexual dimorphism in early anthropoids. *Nature* **287:**328–330.

Foley, R. A. 1999. Homninid behavioral evolution: missing links in comparative primate socioecology. In: P. C. Lee (ed.), *Comparative Primate Socioecology*, pp. 363–386. Cambridge University Press, Cambridge.

Ford, S. M. 1994. Evolution of sexual dimorphism in body weight in platyrrhines. *Am. J. Primatol.* **34:**221–224.

Gould, S. J. 1966. Allometry and size in ontogeny and phylogeny. *Biol. Rev.* **41:**587–640.

Gould, S. J., and Lewontin, R. C. 1979. The Spandrels of San Marco and the Panglossian Paradigm: a critique of the adaptationist programme. *Proc. R. Soc. London, Ser. B* **205:**582–598.

Gould, S. J., and Vrba, E. 1982. Exaptation: a missing term in the science of form. *Paleobiology*. **8:**4–15.

Harvey, P. H., Martin, R. D., and Clutton-Brock, T. H. 1987. Life histories in comparative perspective. In: B. B. Smuts, D. L. Cheney, R. M. Seyfarth, R. W. Wrangham, and T. T. Struhsaker (eds.), *Primate Societies*, pp. 181–196. University of Chicago Press, Chicago

Jungers, W. L., Falsetti, A. B., and Wall, C. E. 1995. Shape, relative size, and size-adjustments in morphometrics. *Yearb. Phys. Anthropol.* **38:**137–161.

Kay, R. F. 1975. The functional adaptations of primate molar teeth. *Am. J. Phys. Anthropol.* **43:**195–216.

Kay, R. F. 1978. Molar structure and diet in extant Cercopithecidae. In: P. M. Butler and K. A. Joysey (eds.), *Development, Function and Evolution of Teeth*, pp. 309–339. Academic Press, New York.

Kay, R. F. 1984. On the use of anatomical features to infer foraging behavior in extinct primates. In: P. S. Rodman and J. G. H. Cant (eds.), *Adaptations for Foraging in Nonhuman Primates: Contributions to an Organismal Biology of Prosimians, Monkeys, and Apes*, pp. 21–53. Columbia University Press, New York.

Kay, R. F., and Cartmill, M. 1977. Cranial morphology and adaptations of *Palaechthon nacimienti* and other Paromomyidae (Plesiadapoidea, ?Primates), with a description of a new genus and species. *J. Hum. Evol.* **6:**19–53.

Kay, R. F., and Ungar, P. S. 1996. Dental evidence for diet in some Miocene catarrhines with comments on the effects of phylogeny on the interpretation of adaptation. In: D. R. Begun, C. V. Ward, and M. D. Rose (eds.), *Function, Phylogeny and Fossils: Miocene Hominoid Evolution and Adaptations*, pp. 131–151. Plenum Press, New York.

Kay, R. F., Plavcan, J. M., Glander, K. E., and Wright, P. C. 1988. Sexual selection and canine dimorphism in New World monkeys. *Am. J. Phys. Anthropol.* **77:**385–397.

Kinzey, W. G. 1978. Feeding behavior and molar features in two species of titi monkey. In: D. J. Chivers and J. Herbert (eds.), *Recent Advances in Primatology, Vol. 1: Behavior*, pp. 373–385. Academic Press, London.

Lande, R. 1980. Sexual dimorphism, sexual selection, and adaptation in polygenic characters. *Evolution* **34:**292–305.

Leutenegger, W., and Cheverud, J. 1982. Correlates of sexual dimorphism in primates: ecological and size variables. *Int. J. Primatol.* **3:**387–402.

Lovejoy, C. O. 1981. The origin of man. *Science* **211:**341–350.

Maryanski, A. 1996. African ape social networks: a blueprint for reconstructing early hominid social structure. In: J. Steele and S. Shennan (eds.), *The Archeology of Human Ancestory: Power Sex and Tradition*, pp. 67–90. Routledge, London.

Mitani, J., Gros-Louis, J., and Richards, A. F. 1996. Sexual dimorphism, the operational sex ratio, and the intensity of male competition in polygynous primates. *Am. Nat.* **147:**966–980.

Mousseau, T. A., and Roff, D. A. 1987. Natural selection and the heritability of fitness components. *Heredity* **59:**181–197.

Napier, J. R., and Walker, A. C. 1967. Vertical clinging and leaping—a newly recognised category of locomotor behaviour of primates. *Folia primatol.* **6:**204–219.

Nunn, C. L. 1995. A simulation test of Smith's "degrees of freedom" correction for comparative studies. *Am. J. Phys. Anthropol.* **98:**355–367.

Pagel, M. 1992. A method for the analysis of comparative data. *J. Theor. Biol.* **156:**431–442.

Pagel, M. D., and Harvey, P. H. 1988. Recent developments in the analysis of comparative data. *Q. Rev. Biol.* **63:**413–440.

Plavcan, J. M. 2000. Inferring social behavior from sexual dimorphism in the fossil record. *J. Hum. Evol.* **39:**

Plavcan, J. M., and van Schaik, C. P. (1992) Intrasexual competition and canine dimorphism in anthropoid primates. *Am. J. Phys. Anthropol.* **87:**461–477.

Plavcan, J. M., and van Schaik, C. P. (1997) Interpreting hominid behavior on the basis of sexual dimorphism. *J. Hum. Evol.* **32:**345–374.

Purvis, A. and Webster, A. J. 1999. Phylogenetically independent comparisons and primate phylogeny. In: P. C. Lee (ed.), *Comparative Primate Socioecology*, pp. 44–70. Cambridge University Press, Cambridge.

Richmond, B. G., and Strait D. S. 2000. Evidence that humans evolved from a knuckle-walking ancestor. *Nature* **404:**382–385.

Simons, E. L., and Rasmussen, D. T. 1996. Skull of *Catopithecus browni*, an early Tertiary catarrhine. *Am. J. Phys. Anthropol.* **100:**261–292.

Simons, E. L., Plavcan, J. M., and Fleagle, J. G. 1999. Canine sexual dimorphism in Egyptian Eocene anthropoid primates: *Catopithecus* and *Proteopithecus*. *Proc. Natl. Acad. Sci. USA* **96:**2559–2562.

Smith, R. J. 1996. Biology and body size in human evolution: statistical inference misapplied. *Curr. Anthropol.* **37:**451–481.

Struhsaker, T. T., and Leland, L. 1987. Colobines: infanticide by adult males. In: B. B. Smuts, D. L. Cheney, R. M. Seyfarth, R. W. Wrangham, and T. T. Struhsaker (eds.), *Primate Societies*, pp. 83–97. University of Chicago Press, Chicago.

Susman, R. L. 1987. Pygmy chimpanzees and common chimpanzees: models for the behavioral ecology of the earliest hominids. In: W. G. Kinzey (ed.), *The Evolution of Human Behavior: Primate Models*, pp. 72–86. State University of New York Press, Albany.

Szalay, F. S., and Delson, E. 1979. *Evolutionary History of the Primates*. Academic Press, New York.

Tanner, N. M. 1987. The chimpanzee model revisited and the gathering hypothesis. In: W. G. Kinzey (ed.), *The Evolution of Human Behavior: Primate Models*, pp. 3–27. State University of New York Press, Albany.

Teaford, M. F., and Glander, K. E. 1991, Dental microwear in live, wild-trapped *Alouatta palliata* from Costa Rica. *Am. J. Phys. Anthropol.* **85:**313–319.

Teaford, M. F., and Robinson, J. G. 1989. Seasonal or ecological zone differences in diet and molar microwear in *Cebus nigrivittatus*. *Am. J. Phys. Anthropol.* **80:**391–401.

Ungar, P. S. 1998. Dental allometry, morphology, and wear as evidence for diet in fossil primates. *Evol. Anthropol.* **6:**205–217.

Wrangham, R. W. 1987. The significance of African apes for reconstructing human social evolution. In: W. G. Kinzey (ed.), *The Evolution of Human Behavior: Primate Models*, pp. 51–71. State University of New York Press, Albany.

Wright, D. B. 1993. Evolution of sexually dimorphic characters in peccaries (Mammalia, Tyassuidae). *Paleobiology*. **19:**52–70.

Index

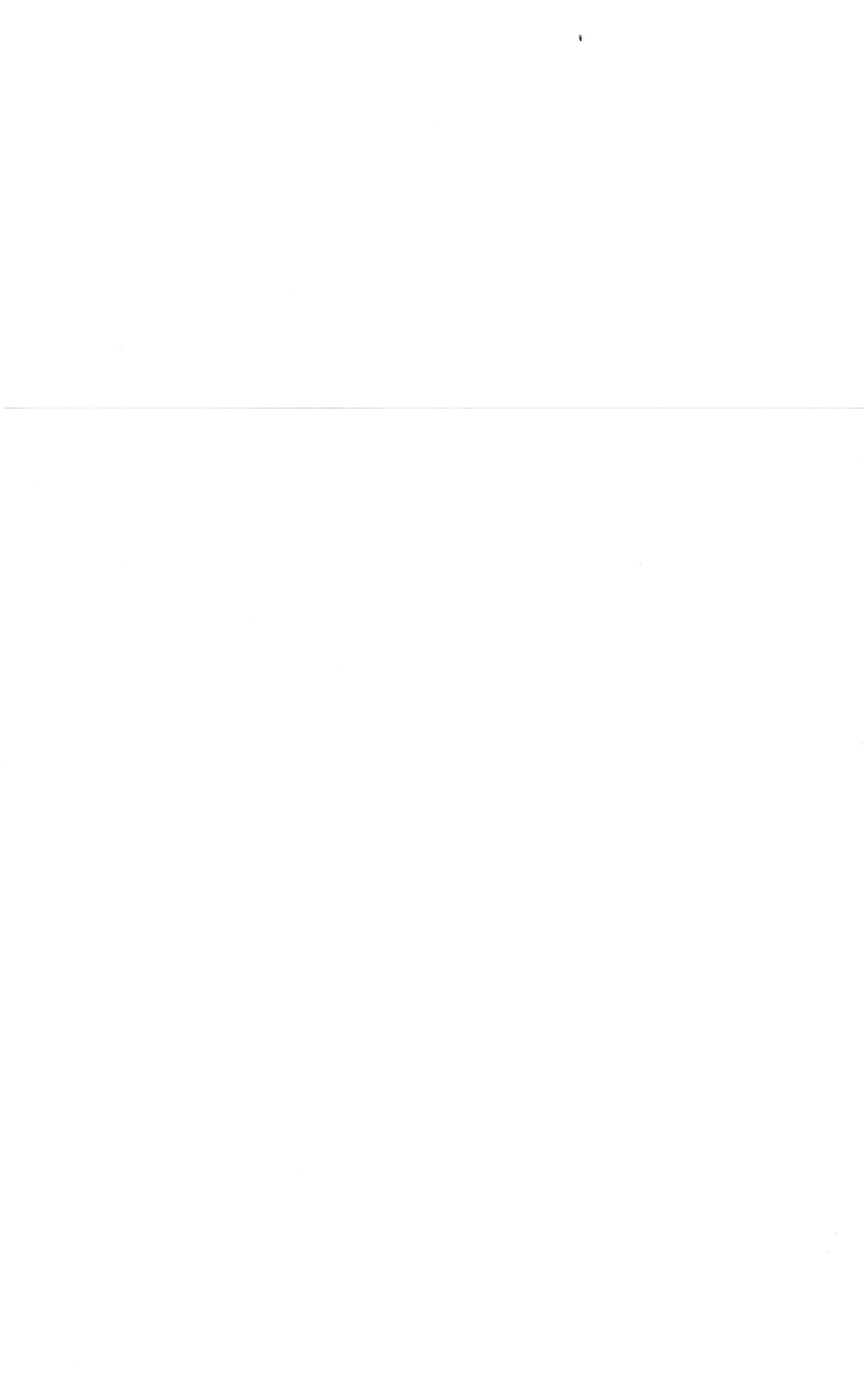